THEORY OF NONLINEAR STRUCTURAL ANALYSIS

THEORY OF NONLINEAR STRUCTURAL ANALYSIS

THE FORCE ANALOGY METHOD FOR EARTHQUAKE ENGINEERING

Gang Li
Dalian University of Technology, China

Kevin K.F. Wong
Ph.D., University of California Los Angeles, USA

WILEY

This edition first published 2014

Registered Office
John Wiley & Sons, Singapore Pte. Ltd., 1 Fusionopolis Walk, #07-01 Solaris South Tower, Singapore 138628.

For details of our global editorial offices, for customer services and for information about how to apply for permission to reuse the copyright material in this book please see our website at www.wiley.com.

Library of Congress Cataloging-in-Publication Data applied for.

ISBN: 978-1-118-71806-3

A catalogue record for this book is available from the British Library.
Printed and bound in Singapore by Markono Print Media Pte Ltd.

1 2014

Contents

Preface

Although the seismic design for buildings is currently based on elastic analysis, nonlinear structural analysis has become increasingly important in the investigation of structural response to environmental loads, especially during earthquakes. Nonlinear structural analysis in civil engineering is not a new topic, but the existing method used for calculating the nonlinear behavior of civil engineering structures is often by changing the structural member stiffness. With respect to the dynamic analysis algorithms, the typical method is the time integral of variable stiffness matrices, such as the Wilson-θ and Newmark-β methods. For these conventional methods, the primary problem is that significant iterative computations have to be carried out to ensure numerical convergence once the structure experiences yielding and nonlinear deformation. As a result, the iterative operation is time consuming and the entire dynamic analysis process becomes practically uneconomical. The Force Analogy Method as a relatively new algorithm was first developed in 1999 for solving nonlinear dynamic analysis problems, in which the state transition matrix needs to be computed only once due to the consistent use of initial stiffness, and this greatly simplifies the overall computation and makes the nonlinear analysis readily available for solving various practical problems.

This book focused on the Force Analogy Method, a novel method for nonlinear dynamic analysis and simulation. A review of the current nonlinear analysis method for earthquake engineering is summarized and its importance explained. Additionally, how the force analogy method can be used in nonlinear static analysis will be discussed through several nonlinear static examples. The emphasis of this book is to extend and develop the force analogy method to performing dynamic analysis on structures under earthquake excitations, where the force analogy method is incorporated in the flexural element, axial element, shearing element and so on will be exhibited. Moreover, the geometric nonlinearity into nonlinear dynamic analysis algorithm based on the force analogy method is included in this book. The application of the force analogy method in seismic design for buildings and structural control area will be discussed combined with practical engineering. This book will be a milestone of nonlinear dynamic analysis and may bring about significant advancement in earthquake engineering.

The authors would like to express their appreciation to Professor Hong-Nan Li at the Dalian University of Technology (DUT), for his advice and support, Professor Larry A. Fahnestock at University of Illinois at Urbana & Champaign, Mr. Yu Zhang, a Ph. D candidate at DUT, and Mr. Feng Zhang, a graduate student at DUT, for their collaborative works, and Mr. Ying Li, for his early contributions. In addition, many thanks go to Zhi-Qian Dong, Yong-Qiang Jin, Jia-Long Li and Li-Hua Zhu, present students at DUT, for their work on the figures.

Gang Li

About the Authors

Gang Li is an associate professor of the Faculty of Infrastructure Engineering, Dalian University of Technology, China. He received a B.S. degree in civil engineering from the Hebei University of Engineering, China and Ph.D. degree in civil engineering from the Dalian University of Technology, China. His research interest is in the earthquake engineering, structural control, structural dynamic analysis, steel structures, large-scale structural experiments, etc. Some of his techniques have been applied to many engineering structures.

Kevin K.F. Wong is a research structural engineer at the National Institute of Standards and Technology (NIST) in the United States. He received a B.S. degree in civil engineering from the University of Hawaii and M.S. and Ph.D. degrees in civil engineering from the University of California, Los Angeles (UCLA). His research areas include structural dynamics, nonlinear structural analysis, earthquake engineering, and structural control.

Dr. Wong was first exposed to the Force Analogy Method as a graduate research assistant under the supervision of Dr. T. H. Lin at UCLA in 1992 while conducting research on the development of plastic strains in high-cycle fatigue of materials. He successfully captured the essence of the method and applied it to the nonlinear analysis of civil engineering structures, where he summarized the work in his Ph.D. thesis titled *Optimal Linear Control of Inelastic Building Response During Earthquakes* in 1996. Since then, he has published over 30 journal articles and 30 conference papers related to the method.

Prior to joining NIST, Dr. Wong was an assistant professor at the University of Utah, USA, for four years, and the Nanyang Technological University, Singapore, for six years. He coauthored the textbook *Structural Dynamics for Structural Engineers* for John Wiley & Sons in 2000 with his Ph.D. advisor, Dr. Gary Hart. Currently at NIST, he is a member of the National Earthquake Hazards Reduction Program research group, conducting research on the development of Codes and Standards on the use of nonlinear analysis for performance-based seismic engineering.

Disclaimers:

The opinions expressed in this book are those of the authors and do not necessarily reflect the opinions of the National Institute of Standards and Technology or the United States Government. While the information in this book is believed to be correct, the National Institute of Standards and Technology and the authors assume no liability for, nor express or imply any warranty with regard to, the information contained herein. Users of the information contained in this book assume all liability arising from such use.

1

Introduction

1.1 History of the Force Analogy Method

The force analogy method (FAM) is an analytical tool for solving structural analysis problems with material nonlinearity. It uses the concept of "inelastic displacement", or more commonly known as the "residual displacement" in the formulation, where the nonlinear stiffness force due to material nonlinearity is represented by a change in displacement instead of a change in stiffness. The original concept of FAM was first introduced by Lin (1968), where the proposed method was actually applied to stress and strain in continuum mechanics with the inelastic behavior defined by plastic strain. Unfortunately, this concept only found limited acceptance because it was developed at approximately the same time as researchers were focusing their attention on studying the deformation of solids using numerical simulation methods, such as the finite element method with the inelastic behavior defined by changing stiffness. Although the finite element method is a powerful tool and widely used, the procedure of the step-by-step numerical integration is unavoidable, time consuming, cumbersome, and costly for practical design in 1980s and even today.

Recognizing that nonlinear finite element method of analysis is a time-consuming process, many structural engineers are constantly seeking a simplified dynamic analysis approach for analyzing nonlinear multi-degree-of-freedom (MDOF) systems to carry out their structural designs. One simplified approach is to represent the nonlinear MDOF system as an elastic system, in which structural response can be estimated by response spectra analysis of using the convenient and efficient modal superposition method. Newmark (1970) proposed a well-known method of extending the elastic response spectra analysis to engineering design of nonlinear systems through the use of inelastic response spectra. However, the method is strictly valid for single-degree-of-freedom (SDOF) systems and thus is inadequate for the analysis

Theory of Nonlinear Structural Analysis: The Force Analogy Method for Earthquake Engineering, First Edition. Gang Li and Kevin K.F. Wong.

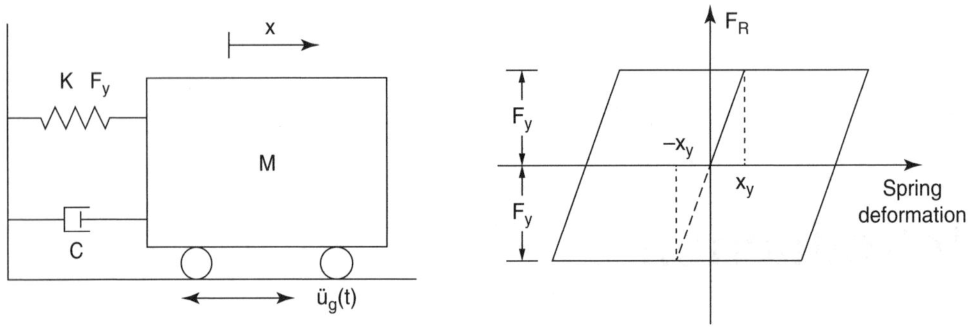

Figure 1.1 SDOF system and restoring force model.

of nonlinear MDOF systems due to the changing stiffness matrix. The changing stiffness matrix in the equations of motion for the nonlinear MDOF system is the drawback of this method, since the nonlinearity effect is coupled in each and every mode. Thus, significant effort was spent towards extending the modal superposition method in elastic analysis to inelastic analysis. One effort similar to the FAM, where the restoring force term of nonlinear MDOF systems was expressed by the sum of the elastic restoring force and additional external force, was presented by Villaverde (1988, 1996). After moving the additional external force term to the right-hand side of the equation of motion, the left-hand side of the equation is interpreted as an equivalent linear system. A different approximate modal decomposition method for the equation of motion was subsequently presented by Georgoussis (2008). While the above works emphasized the development of simplified analysis methods, only simple system models and load–deformation relationships, such as those shown in Figure 1.1, were selected to explore the physical significance of the external force term. The relationship between the external force term and inelastic behavior of structural members were ignored at that time.

The same problem was encountered by Wong (1996) during his study on the structural control of nonlinear structures. Since the theory of state space dynamic analysis, as a computing platform for performing structural control calculation, was only applicable to elastic systems, it was an obvious barrier when apply the structural control technique in nonlinear structures. Thus, a method of analyzing the inelastic response of the building by recovering the forces from the states of the building was introduced. Subsequently, Wong and Yang (1999) formally published the first application of the FAM for civil structures where the method was formulated in force–deformation space for inelastic dynamic analysis. The fundamental concept of the FAM is that each inelastic deformation in the structure is formulated as a degree of freedom such that the initial stiffness matrix is computed only once at the beginning and can be used throughout the inelastic analysis. Coupling the FAM with the state space formulation for dynamic analysis provides an accurate, efficient, and stable solution algorithm such that it can be used to analyze structures with various material properties, not only for elastic–plastic property but also for both hardening and softening properties. In addition, the external force term was interpreted as the force analogy, which causes inelastic deformation of structural members at certain locations in the structure. The inelastic deformation includes nonlinear extension of the braces in a braced frame, plastic rotation of the beams and columns in a moment resisting frame, or yielding of the base isolators in a base isolation system. Since then, Zhao and Wong (2006) further

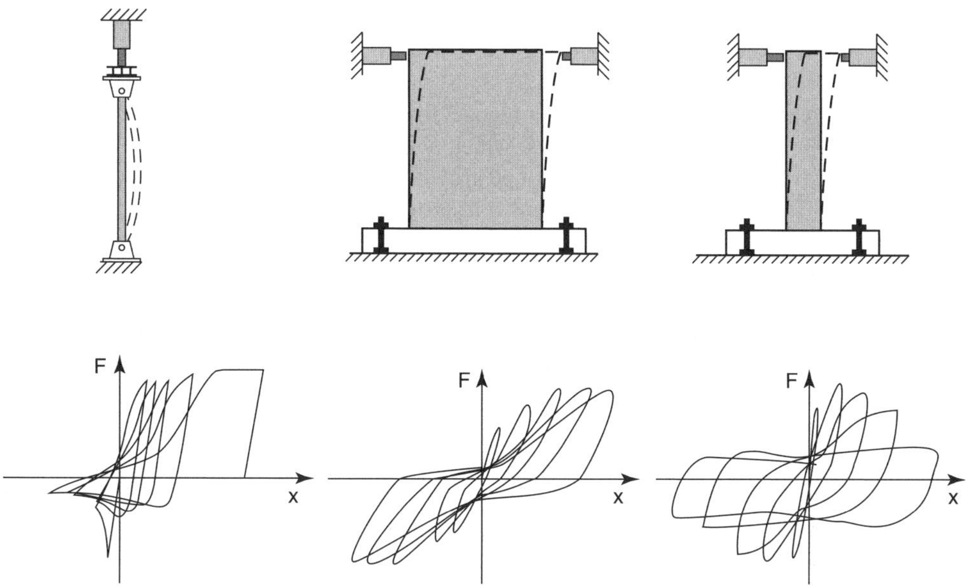

Figure 1.2 Complex cyclic behaviors of structural members.

developed the FAM by incorporating the geometric nonlinear effect and presented a comprehensive nonlinear approach for inelastic framed structures, including geometric nonlinearity and material nonlinearity. The approach uses finite element formulation to derive the elemental stiffness matrices, particularly to derive the geometric stiffness matrix in a general form. However, the geometric stiffness matrix used in the nonlinear formulation was not exact, and further improvement by Wong (2012, 2013) was recently conducted and will be presented in Chapter 7.

Although Wong and Yang (1999) pointed out that all material properties can be used in the FAM and they have no influence on the algorithm stability, only nonlinear response of a steel moment-resistant frame with simple bilinear moment versus plastic rotation relationship was mentioned in the study. In fact, structural members often exhibit complex cyclic inelastic behavior (i.e. buckling of braces, strength degradation of reinforced concrete members, as shown in Figure 1.2) when they undergo excessive dynamic loadings, and some well-known models have been proposed and developed. It is clear that updating element local stiffness matrices, re-assembling them and performing static condensation to derive system global tangent stiffness matrix is not necessary in the FAM. However, existing material models cannot be applied in the FAM directly because they often reveal a highly nonlinear relation of the external force and total deformation rather than plastic deformation. Since the global behavior of nonlinear structures is closely associated with the relationship between the internal force and plastic deformation, some investigations were carried out to extend the application of the FAM for structural members with different material behaviors.

Chao and Loh (2007) used three different plastic mechanisms to simulate the reinforced concrete beam-column elements in the FAM. The load versus deformation comparison shows that the proposed algorithm gives results very similar to experimental data. Additionally, the P-Delta effect also has been considered in this study. Li *et al.* (2013a) implemented an existing brace physical theory model for use in the FAM. In the procedure, the physical theory model developed by Dicleli and Calik (2008) is chosen for implementation in the FAM, because it is a relatively simple and efficient model that has been shown to provide reasonable accuracy. Two sliding plastic mechanisms, which simulate axial displacements produced by transverse brace displacement and the so-called growth effect, are used to represent the inelastic brace behavior. The resulting model is shown to provide good agreement with experimental data. Moreover, this brace model is implemented in a frame where inelastic response occurs in both the frame and braces to demonstrate the value of the brace model and the potential for simulating complex inelastic dynamic behavior of concentric braced frames with the FAM. The model is validated against prior experimental results to be an accurate, efficient, and stable algorithm for conducting dynamic analysis when coupled with the state space formulation.

In addition, Li and Zhang (2013b) developed a framework for the seismic damage analysis of reinforced concrete frame structures considering the stiffness degradation based on the FAM. A damage hinge model, which is located at the ends of columns and beams, is proposed for modeling damage behavior due to concrete cracking. As a damage effect is implemented by introducing the damage indices as internal variables, the real-time structural performance and damage level can be evaluated during the computation process. The damage hinge, together with the plastic hinge arising from structural materials, forms a complete inelastic mechanism including stiffness degradation behavior for reinforced concrete frame structures. Since only initial stiffness is used throughout the dynamic computation analysis, the usage of the state space formulation, as an outstanding advantage of the FAM, is retained and makes the real-time damage analysis more accurate, efficient, and stable. As for the reinforced concrete shear wall member, a procedure for modeling the hysteretic response of reinforced concrete shear wall members based on the existing models in the FAM was established and will be discussed in Chapter 6. An reinforced concrete (RC) flexural member model, where the strength deterioration and stiffness degradation effect due to increasing loading cycles, and the pinching behavior that mainly roots in the crack opening and closing during loading reversals are considered, was established and incorporated in the FAM. The methodology will be presented in Chapter 4 together with several examples.

1.2 Applications of the Force Analogy Method

Because the FAM has two outstanding benefits in terms of computation efficiency and stability, it has the advantage over other analysis tools for the following applications:

1.2.1 Structural Vibration Control

Since the concept of structural vibration control in civil engineering was proposed by J.T.P Yao in 1972, it has made considerable progress in the development of theoretical and experimental researches. A number of structural control techniques and strategies have been developed and applied in practices, specifically in seismic regions. The structural vibration control began in the mechanical engineering in the early 20th century and the majority of control theories,

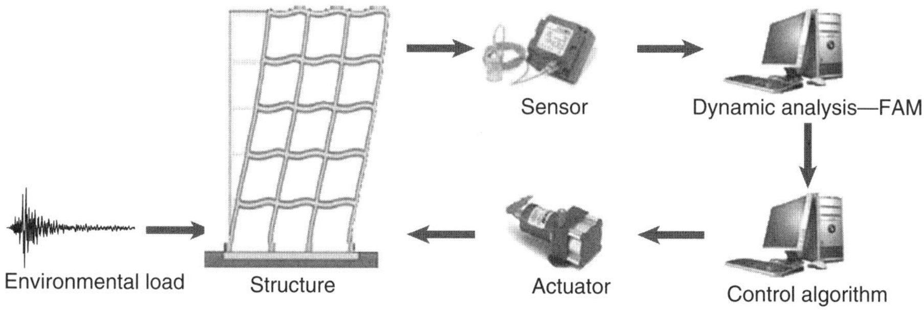

Figure 1.3 The framework for the combination of FAM and control algorithm.

includes the linear quadratic regulator, modal control, smart control, H2 control, H∞ control, etc., and algorithms were applied to elastic systems and have been matured. These control algorithms together with dynamic analysis procedure run together for determining controlling force of actuators, as shown in Figure 1.3. However, structural members in civil engineering buildings will always experience inelastic deformation when the buildings are subjected to excessive loadings. This causes significant problems, such as time delay, incompatible program, etc., during the combination of inelastic computation procedures and control algorithms.

The emergence of the FAM provides a way to solve this type of problem because the left-hand side of the equation of motion of nonlinear systems retains the linear properties of corresponding elastic system. Thus, many problems, which are relatively difficult to answer while applying traditional control algorithms to inelastic systems, have been solved to some degrees using the FAM. Wong and Yang (2003) and Wong (2005) proposed inelastic structural control algorithms, which compensates for the time delay that happens in practical control systems, through incorporating the FAM with the predictive instantaneous optimal control algorithm and the predictive instantaneous optimal control algorithm, respectively. Moreover, since the earthquake ground velocity is not at high frequency as compared with the ground acceleration, it can be predicted at certain time steps beforehand in the real-time domain with higher accuracy. Thus, Pang and Wong (2006) proposed a simple control algorithm expressed using the input ground velocity, namely the Predictive Instantaneous Optimal Control algorithm.

To capture the damaging effects during earthquake ground motions, the FAM is used to characterize structures responding in the inelastic domain. Li and Li (2011a) developed an approach based on the FAM to analyze the dynamic response of structure with energy-dissipation devices. The proposed algorithm is applicable to a variety of energy-dissipation devices by turning them to the equivalent force applied at the joints of the frame. Wong (2008) and Wong and Johnson (2009) presented studies on the use of tuned mass dampers as a passive energy-dissipation device to investigate the benefits of using such devices in reducing the inelastic structural responses. In addition, Wong (2011a) presented a simple numerical algorithm based on the combination of the state space method and FAM to calculate the inelastic dynamic analysis of structures with nonlinear fluid viscous dampers. Finally, Li *et al.* (2011b) proposed a control algorithm for inelastic structures through combining the market-based control strategy and force analogy method. The framework of this work will be discussed in Chapter 9.

1.2.2 Modal Dynamic Analysis Method

Since each term on the left-hand side of the equation of motion for nonlinear MDOF systems is feasible for modal decomposition like elastic systems, it suggests that the FAM is probably a good baseline for applying the modal dynamic analysis method to solve nonlinear MDOF system problems. Wong (2011b) extended the modal superposition to the nonlinear domain by using the FAM to address material nonlinearity. In addition, because linear modal superposition has found great acceptance in performance-based seismic engineering, geometric nonlinearity is incorporated into the analysis using stability functions. Through the combination of FAM, stability functions, the state space method, and modal superposition, numerical simulations are performed and results are demonstrated to be both accurate and efficient. Moreover, a simple analysis tool for capturing the effect of rigid-end offsets in framed structures under earthquake excitation has been incorporated into the above nonlinear modal analysis methodology by Wong (2012). Author also demonstrated that the equation of motion for nonlinear MDOF systems in the FAM can be uncoupled, but two other governing equations in the FAM relating the internal force, such as the moment and force of structural members are not decomposable. However, uncoupled modal SDOF system responses can be determined by incorporating the FAM with the modal pushover analysis method such that the modal superposition method is suitable for the solution of the nonlinear MDOF system. Although the procedure presented is still an approximation method due to the modal pushover analysis method application, its value and potential for the maximum displacement estimation of the nonlinear MDOF system based on the FAM were validated. The procedure will be discussed in Chapter 8 along with examples.

1.2.3 Other Design and Analysis Areas

Wong and Yang (2002) derived the plastic energy dissipation of structures based on the FAM and used the energy as the response parameter in evaluating the performance of the structure, and Wong and Wang (2003) extended the energy-balance equation to include control energy as an addition form of energy dissipation to resist earthquake inputs. In these studies, the FAM was modified and extended to analyze real moment-resisting frames with zero rotational mass moment of inertia using the method of static condensation. The static condensation method in the FAM will be discussed in Chapter 2 for static analysis and Chapter 3 for dynamic analysis.

Wang and Wong (2007) introduced the FAM for the first time into the field of stochastic dynamic analysis for inelastic structures. This stochastic FAM maintains the advantage of high efficiency in the numerical computation of the FAM in dynamic analysis. According to the stochastic FAM, the variance covariance functions of inelastic dynamic responses, such as displacement, velocity, inelastic displacement of the entire moment-resisting framed structures, and plastic rotation at individual plastic hinge location, can be produced for structures subjected to random excitation.

1.3 Background of the Force Analogy Method

The first step in learning the force analogy method for solving nonlinear structure problems is to understand the matrix method of structural analysis. Because understanding each term in the stiffness matrix (i.e. $12EI/L^3$, $6EI/L^2$, $4EI/L$, and $2EI/L$) is so important

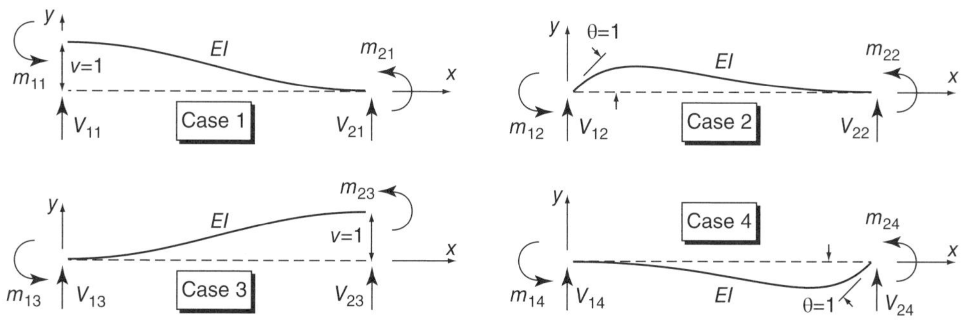

Figure 1.4 Displacement patterns and the corresponding fixed-end forces.

to the presentations in the subsequent chapters, it is appropriate and worthwhile to derive the elastic stiffness matrix for bending in this section for the completeness of the book.

Consider a beam of length L with uniform elastic modulus E and moment of inertia I that is subjected to loadings at the two ends. Due to the loadings, the deformation at the two ends (i.e. translations and rotations) of the beam can be related to the amount of shear and moments at the two ends through a stiffness matrix expressed in the following form:

$$\begin{bmatrix} V_1 \\ m_1 \\ V_2 \\ m_2 \end{bmatrix} = \begin{bmatrix} k_{11} & k_{12} & k_{13} & k_{14} \\ k_{21} & k_{22} & k_{23} & k_{24} \\ k_{31} & k_{32} & k_{33} & k_{34} \\ k_{41} & k_{42} & k_{43} & k_{44} \end{bmatrix} \begin{bmatrix} v_1 \\ \theta_1 \\ v_2 \\ \theta_2 \end{bmatrix} \tag{1.1}$$

where V is the shear, m is the moment, v is the transverse displacement, and θ is the rotation. The subscript '1' represents the quantities at the 1-end, and the subscript '2' represents the quantities at the 2-end. Finally, k represents the entries in the stiffness matrix. To determine the stiffness matrix, four cases of a beam deflection are separated as shown in Figure 1.4 using the unit displacement method. Here, V_{1i}, m_{1i}, V_{2i}, and m_{2i} represent the fixed-end shears and moments of the beam, and $i = 1, \ldots, 4$ represents the four cases of unit displacement patterns of beam deflection.

Using the classical Bernoulli–Euler beam theory with "plane sections remain plane", where the moment is proportional to the curvature, the governing equilibrium equation describing the deflection of the beam member can be written as

$$(EIv'')'' = 0 \tag{1.2}$$

By assuming EI is constant along the member, the solution to the fourth-order ordinary differential equation is:

$$v = Ax^3 + Bx^2 + Cx + D \tag{1.3}$$

and the corresponding, rotation, moment, and shear equations become:

$$\theta(x) = v' = 3Ax^2 + 2Bx + C \tag{1.4a}$$

$$m(x) = EIv'' = EI(6Ax + 2B) \tag{1.4b}$$

$$V(x) = EIv''' = EI(6A) \tag{1.4c}$$

In order to solve for the constants in Eq. (1.3), the following four cases of boundary conditions are considered.

Case 1: Imposing the boundary conditions $v(0) = 1$, $v'(0) = 0$, $v(L) = 0$, and $v'(L) = 0$ gives

$$v(0) = 1: \quad D = 1 \tag{1.5a}$$

$$v'(0) = 0: \quad C = 0 \tag{1.5b}$$

$$v(L) = 0: \quad AL^3 + BL^2 + CL + D = 0 \tag{1.5c}$$

$$v'(L) = 0: \quad 3AL^2 + 2BL + C = 0 \tag{1.5d}$$

Solving simultaneously for the constants in Eq. (1.5) gives

$$A = 2/L^3, \quad B = -3/L^2, \quad C = 0, \quad D = 1 \tag{1.6}$$

Now substituting the constants in Eq. (1.6) into the shear equation in Eq. (1.4c) and the moment equation in Eq. (1.4b) evaluated at appropriate end points gives the fixed-end forces as labeled in Figure 1.4 (Case 1) as:

$$V_{11} = EIv'''(0) = EI(6A) = 12EI/L^3 \tag{1.7a}$$

$$m_{11} = -EIv''(0) = -EI(2B) = 6EI/L^2 \tag{1.7b}$$

$$V_{21} = -EIv'''(L) = -EI(6A) = -12EI/L^3 \tag{1.7c}$$

$$m_{21} = EIv''(L) = EI(6AL + 2B) = 6EI/L^2 \tag{1.7d}$$

In matrix form, this is given as

$$\begin{bmatrix} V_{11} \\ m_{11} \\ V_{21} \\ m_{21} \end{bmatrix} = \begin{bmatrix} 12EI/L^3 & \times & \times & \times \\ 6EI/L^2 & \times & \times & \times \\ -12EI/L^3 & \times & \times & \times \\ 6EI/L^2 & \times & \times & \times \end{bmatrix} \begin{bmatrix} 1 \\ 0 \\ 0 \\ 0 \end{bmatrix} \tag{1.8}$$

Note that the minus signs in front of the calculations for m_{1i} and V_{2i} are used because of the differences in sign convention between the classical beam theory and the theory for stiffness method of structural analysis.

Case 2: Imposing the boundary conditions $v(0)=0$, $v'(0)=1$, $v(L)=0$, and $v'(L)=0$ gives

$$v(0)=0: \quad D=0 \tag{1.9a}$$

$$v'(0)=1: \quad C=1 \tag{1.9b}$$

$$v(L)=0: \quad AL^3+BL^2+CL=0 \tag{1.9c}$$

$$v'(L)=0: \quad 3AL^2+2BL+C=0 \tag{1.9d}$$

Solving simultaneously for the constants in Eq. (1.9) gives

$$A=1/L^2, \quad B=-2/L, \quad C=1, \quad D=0 \tag{1.10}$$

Now substituting the constants in Eq. (1.10) into the shear equation in Eq. (1.4c) and the moment equation in Eq. (1.4b) evaluated at appropriate end points gives the fixed-end forces as labeled in Figure 1.4 (Case 2) as:

$$V_{12}=EIv'''(0)=EI(6A)=6EI/L^2 \tag{1.11a}$$

$$m_{12}=-EIv''(0)=-EI(2B)=4EI/L \tag{1.11b}$$

$$V_{22}=-EIv'''(L)=-EI(6A)=-6EI/L^2 \tag{1.11c}$$

$$m_{22}=EIv''(L)=EI(6AL+2B)=2EI/L \tag{1.11d}$$

In matrix form, this is given as

$$\begin{bmatrix} V_{12} \\ m_{12} \\ V_{22} \\ m_{22} \end{bmatrix} = \begin{bmatrix} \times & 6EI/L^2 & \times & \times \\ \times & 4EI/L & \times & \times \\ \times & -6EI/L^2 & \times & \times \\ \times & 2EI/L & \times & \times \end{bmatrix} \begin{bmatrix} 0 \\ 1 \\ 0 \\ 0 \end{bmatrix} \tag{1.12}$$

Case 3: Imposing the boundary conditions $v(0)=0$, $v'(0)=0$, $v(L)=1$, and $v'(L)=0$ gives

$$v(0)=0: \quad D=0 \tag{1.13a}$$

$$v'(0)=0: \quad C=0 \tag{1.13b}$$

$$v(L)=1: \quad AL^3+BL^2+CL=1 \tag{1.13c}$$

$$v'(L)=0: \quad 3AL^2+2BL+C=0 \tag{1.13d}$$

Solving simultaneously for the constants in Eq. (1.13) gives

$$A=-2/L^3, \quad B=3/L^2, \quad C=0, \quad D=1 \tag{1.14}$$

Now substituting the constants in Eq. (1.14) into the shear equation in Eq. (1.4c) and the moment equation in Eq. (1.4b) evaluated at appropriate end points gives the fixed-end forces as labeled in Figure 1.4 (Case 3) as:

$$V_{13} = EIv'''(0) = EI(6A) = -12EI/L^3 \tag{1.15a}$$

$$m_{13} = -EIv''(0) = -EI(2B) = -6EI/L^2 \tag{1.15b}$$

$$V_{23} = -EIv'''(L) = -EI(6A) = 12EI/L^3 \tag{1.15c}$$

$$m_{23} = EIv''(L) = EI(6AL + 2B) = -6EI/L^2 \tag{1.15d}$$

In matrix form, this is given as

$$\begin{bmatrix} V_{13} \\ m_{13} \\ V_{23} \\ m_{23} \end{bmatrix} = \begin{bmatrix} \times & \times & -12EI/L^3 & \times \\ \times & \times & -6EI/L^2 & \times \\ \times & \times & 12EI/L^3 & \times \\ \times & \times & -6EI/L^2 & \times \end{bmatrix} \begin{bmatrix} 0 \\ 0 \\ 1 \\ 0 \end{bmatrix} \tag{1.16}$$

Case 4: Finally, imposing the boundary conditions $v(0)=0$, $v'(0)=0$, $v(L)=0$, and $v'(L)=1$ gives

$$v(0)=0: \quad D=0 \tag{1.17a}$$

$$v'(0)=0: \quad C=0 \tag{1.17b}$$

$$v(L)=0: \quad AL^3 + BL^2 + CL = 0 \tag{1.17c}$$

$$v'(L)=1: \quad 3AL^2 + 2BL + C = 1 \tag{1.17d}$$

Solving simultaneously for the constants in Eq. (1.17) gives

$$A = 1/L^2, \quad B = -1/L, \quad C = 1, \quad D = 0 \tag{1.18}$$

Now substituting the constants in Eq. (1.18) into the shear equation in Eq. (1.4c) and the moment equation in Eq. (1.4b) evaluated at appropriate end points gives the fixed-end forces as labeled in Figure 1.4 (Case 4) as:

$$V_{14} = EIv'''(0) = EI(6A) = 6EI/L^2 \tag{1.19a}$$

$$m_{14} = -EIv''(0) = -EI(2B) = 2EI/L \tag{1.19b}$$

$$V_{24} = -EIv'''(L) = -EI(6A) = -6EI/L^2 \tag{1.19c}$$

$$m_{24} = EIv''(L) = EI(6AL + 2B) = 4EI/L \tag{1.19d}$$

In matrix form, this is given as

$$\begin{bmatrix} V_{14} \\ m_{14} \\ V_{24} \\ m_{24} \end{bmatrix} = \begin{bmatrix} \times & \times & \times & 6EI/L^2 \\ \times & \times & \times & 2EI/L \\ \times & \times & \times & -6EI/L^2 \\ \times & \times & \times & 4EI/L \end{bmatrix} \begin{bmatrix} 0 \\ 0 \\ 0 \\ 1 \end{bmatrix} \tag{1.20}$$

In summary, based on the construction of the stiffness matrix using above four cases as shown in Eqs. (1.8), (1.12), (1.16), and (1.20), the stiffness equation of the ith beam member becomes:

$$\begin{bmatrix} V_1 \\ m_1 \\ V_2 \\ M_2 \end{bmatrix} = \begin{bmatrix} 12EI/L^3 & 6EI/L^2 & -12EI/L^3 & 6EI/L^2 \\ 6EI/L^2 & 4EI/L & -6EI/L^2 & 2EI/L \\ -12EI/L^3 & -6EI/L^2 & 12EI/L^3 & -6EI/L^2 \\ 6EI/L^2 & 2EI/L & -6EI/L^2 & 4EI/L \end{bmatrix} \begin{bmatrix} v_1 \\ \theta_1 \\ v_2 \\ \theta_2 \end{bmatrix} \tag{1.21}$$

From Eq. (1.21), it can be seen that:

- The stiffness relating the transverse displacement at one end of the beam with end shears is $12EI/L^3$.
- The stiffness relating the transverse displacement at one end of the beam with end moments is $6EI/L^2$.
- The stiffness relating the rotation at one end of the beam with end shears is $6EI/L^2$.
- The stiffness relating the rotation at one end of the beam with end moment at the same end is $4EI/L$.
- The stiffness relating the rotation at one end of the beam with end moment at the opposite end is $2EI/L$.

Example 1.1 One-Story One-Bay Frame

Consider a one-story one-bay frame as shown in Figure 1.5(a). Assume the members are axially rigid, this results in a three degrees of freedom system, one floor translation and two joint rotations, as labeled in Figure 1.5(a) as v_1, θ_2, and θ_3. Also assume that the beam and the two columns are of the same length L and elastic modulus E, but the moment of inertias of each member are as labeled in the figure. The global stiffness matrix relates v_1, θ_2, and θ_3 at the degrees of freedom with the corresponding applied forces, i.e.

$$\begin{bmatrix} F_1 \\ m_2 \\ m_3 \end{bmatrix} = \begin{bmatrix} K_{11} & K_{12} & K_{13} \\ K_{21} & K_{22} & K_{23} \\ K_{31} & K_{32} & K_{33} \end{bmatrix} \begin{bmatrix} v_1 \\ \theta_2 \\ \theta_3 \end{bmatrix} \tag{1.22}$$

where F_1, m_2, and m_3 are the global applied force and moments at the degrees of freedom as shown in Figure 1.5(b).

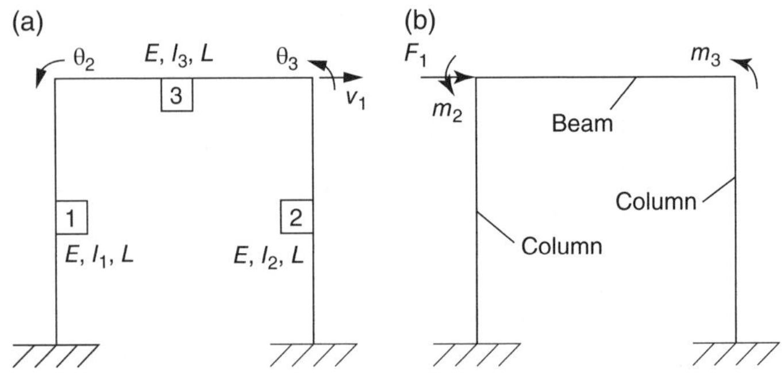

Figure 1.5 One-story one-bay moment-resisting frame: (a) Three degrees of freedom system; (b) Applied forces at the degrees of freedom.

The objective here is to construct the global stiffness matrix in Eq. (1.22). Three displacement patterns are used to construct this stiffness matrix as follows:

Case 1: Imposing the boundary conditions $v_1 = 1$, $\theta_2 = 0$, and $\theta_3 = 0$
Figure 1.6 shows the displacement pattern and the corresponding applied forces required to induce such displacement pattern. These applied forces give the first column of the stiffness matrix as

$$\begin{bmatrix} F_1 \\ m_2 \\ m_3 \end{bmatrix} = \begin{bmatrix} 12EI_1/L^3 + 12EI_2/L^3 & \times & \times \\ 6EI_1/L^2 & \times & \times \\ 6EI_2/L^2 & \times & \times \end{bmatrix} \begin{bmatrix} 1 \\ 0 \\ 0 \end{bmatrix} \tag{1.23}$$

Case 2: Imposing the boundary conditions $v_1 = 0$, $\theta_2 = 1$, and $\theta_3 = 0$
Figure 1.7 shows the displacement pattern and the corresponding applied forces required to induce such displacement pattern. These applied forces give the second column of the stiffness matrix as

$$\begin{bmatrix} F_1 \\ m_2 \\ m_3 \end{bmatrix} = \begin{bmatrix} \times & 6EI_1/L^2 & \times \\ \times & 4EI_1/L + 4EI_3/L & \times \\ \times & 2EI_3/L & \times \end{bmatrix} \begin{bmatrix} 0 \\ 1 \\ 0 \end{bmatrix} \tag{1.24}$$

Case 3: Imposing the boundary conditions $v_1 = 0$, $\theta_2 = 0$, and $\theta_3 = 1$
Figure 1.8 shows the displacement pattern and the corresponding applied forces required to induce such displacement pattern. These applied forces give the third column of the stiffness matrix as

$$\begin{bmatrix} F_1 \\ m_2 \\ m_3 \end{bmatrix} = \begin{bmatrix} \times & \times & 6EI_2/L^2 \\ \times & \times & 2EI_3/L \\ \times & \times & 4EI_2/L + 4EI_3/L \end{bmatrix} \begin{bmatrix} 0 \\ 0 \\ 1 \end{bmatrix} \tag{1.25}$$

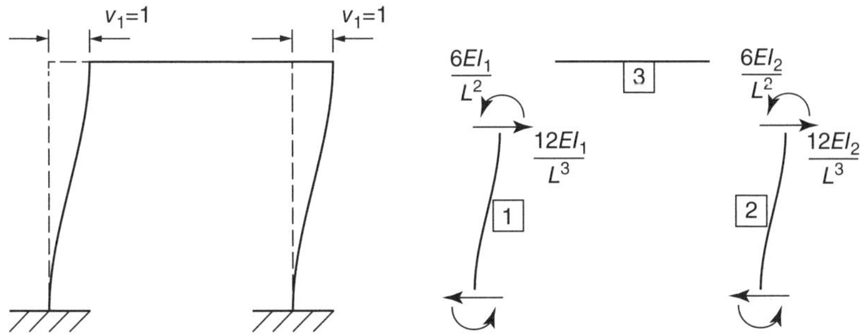

Figure 1.6 Displacement pattern using unit displacement at v_1.

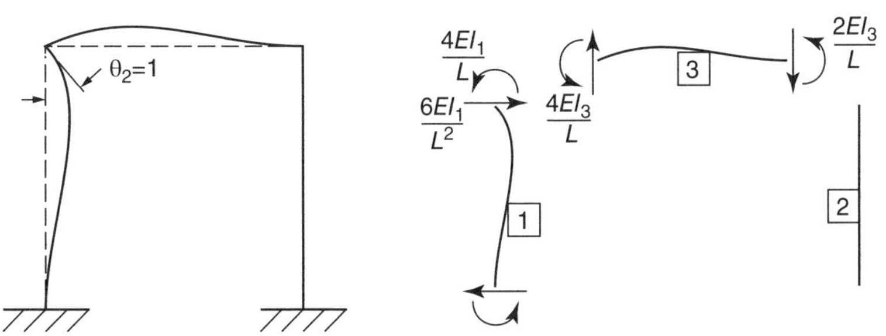

Figure 1.7 Displacement pattern using unit displacement at θ_2.

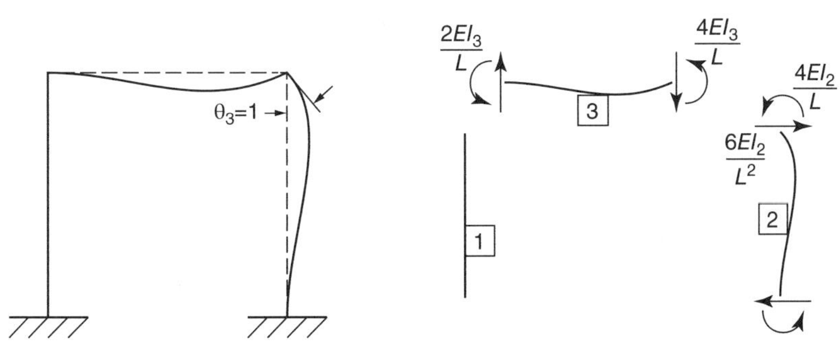

Figure 1.8 Displacement pattern using unit displacement at θ_3.

Based on Eqs. (1.23), (1.24), and (1.25), the global stiffness matrix is therefore constructed as:

$$\mathbf{K}=\begin{bmatrix} 12EI_1/L^3+12EI_2/L^3 & 6EI_1/L^2 & 6EI_2/L^2 \\ 6EI_1/L^2 & 4EI_1/L+4EI_3/L & 2EI_3/L \\ 6EI_2/L^2 & 2EI_3/L & 4EI_2/L+4EI_3/L \end{bmatrix} \begin{matrix} \leftarrow v_1 \\ \leftarrow \theta_2 \\ \leftarrow \theta_3 \end{matrix} \tag{1.26}$$

Note that the global stiffness matrix **K** in Eq. (1.26) is symmetric – an important property that is observed throughout this book.

References

Chao S-H. and Loh C-H. (2007). Inelastic response analysis of reinforced concrete structures using modified force analogy method. *Earthquake Engineering and Structural Dynamics*, 12(36): 1659–1683.

Dicleli M. and Calik E.E. (2008). Physical theory hysteretic model for steel braces. *Journal of Structural Engineering ASCE*, 134(7): 1215–1228.

Georgoussis G.K. (2008). Approximate modal analysis of multistory symmetrical buildings with restricted inelasticity. *The Structural Design of Tall and Special Buildings*, 17(2): 313–330.

Li G. and Li H.N. (2011a). Seismic response analysis of structure with energy-dissipation devices using force analogy method. *The Structural Design of Tall and Special Buildings*, 20(3): 291–313.

Li G., Liu Q. and Li H.N. (2011b). Inelastic structural control based on MBC and FAM. *Mathematical Problems in Engineering*, Volume 2011, Article ID 460731, 18 pages doi:10.1155/2011/460731.

Li G., Fahnestock L.A. and Li H.N. (2013a). Simulation of steel brace hysteretic response using the force analogy method. *Journal of Structural Engineering ASCE*, 139(4): 526–536.

Li G., Zhang Y., and Li H.N. (2013b). Seismic damage analysis of reinforced concrete frame using the force analogy method. *Journal of Engineering Mechanics ASCE*, 139(12): 1780–1789.

Lin T.H. (1968). *Theory of inelastic structures, Wiley, New York, USA.*

Newmark N.M. (1970). "Current trends in the seismic analysis and design of high-rise structures" Chapter 16 in *Earthquake Engineering*, Prentice-Hall, New Jersey.

Pang M. and Wong K.K.F. (2006). Predictive instantaneous optimal control of inelastic structures based on ground velocity. *The Structural Design of Tall and Special Buildings*, 15: 307–324.

Villaverde R. (1998). Modal superposition method for seismic design of non-linear multistory structures. *Earthquake Engineering and Structural Dynamics*, 16: 691–704.

Villaverde R. (1996). Simplified response-spectrum seismic analysis of nonlinear structures. *Journal of Engineering Mechanics ASCE*, 122(3): 282–285.

Wang Z. and Wong K.K.F. (2007). Stochastic dynamic analysis of inelastic structures using force analogy method. *Journal of Engineering Mechanics ASCE*, 133(4): 414–421.

Wong K.K.F. (1996). Optimal linear control of inelastic building response during earthquake. *Ph.D. Thesis*, University of California at Los Angeles.

Wong K.K.F. and Yang R. (1999). Inelastic dynamic response of structures using force analogy method. *Journal of Engineering Mechanics ASCE*, 125(10): 1190–1200.

Wong K.K.F. and Yang R. (2002). Earthquake response and energy evaluation of inelastic structures. *Journal of Engineering Mechanics ASCE*, 128(3): 308–318.

Wong K.K.F. and Yang R. (2003). Predictive instantaneous optimal control of inelastic structures during earthquakes. *Earthquake Engineering and Structural Dynamics*, (32): 2179–2195.

Wong K.K.F. and Wang Y. (2003). Energy-based design of structures using modified force analogy method. *The Structural Design of Tall and Special Building*. 12: 393–407.

Wong K.K.F. (2005). Predictive optimal linear control of inelastic structures during earthquake. Part II. *Journal of Engineering Mechanics ASCE*, 131(2): 142–152.

Wong K.K.F. and Zhao D. (2007). Uncoupling of potential energy in nonlinear seismic analysis of framed structures. *Journal of Engineering Mechanics ASCE*, 133(10): 1061–1071.

Wong K.K.F. (2008). Seismic energy dissipation of inelastic structures with tuned mass dampers. *Journal of Engineering Mechanics ASCE*, 134(2): 163–172.

Wong K.K.F. and Johnson J. (2009). Seismic energy dissipation of inelastic structures with multiple tuned mass dampers," *Journal of Engineering Mechanic ASCE*, 135(4): 265–275.

Wong K.K.F. (2011a). Seismic energy analysis of structures with nonlinear fluid viscous dampers – Algorithm and numerical verification. *The Structural Design of Tall and Special Buildings*, 20: 482–496.

Wong K.K.F. (2011b). Nonlinear dynamic analysis of structures using modal superposition, *Proceedings of the 2011 SEI Structures Congress*, *Paper* 336: pp. 770–781.

Wong K.K.F. (2012). Nonlinear modal analysis of structures with rigid-end offsets. *Proceedings of the 2012 SEI Structures Congress*, 20th Analysis and Computation Specialty Conference, Paper 185: 174–184.

Wong K.K.F. (2013). Evaluation of computational tools for performing nonlinear seismic analyses of structural collapse. *Proceedings of the 2013 SEI Structures Congress*, Paper 201: 2106–2117.

Zhao D. and Wong K.K.F. (2006). New approach for seismic nonlinear analysis of inelastic framed structures. *Journal of Engineering Mechanics ASCE*, 132(9): 959–966.

2

Nonlinear Static Analysis

The theory of nonlinear analysis is first presented in static problems using the Force Analogy Method (FAM). This method treats material nonlinearity through an equivalent force that is applied to the structure, while maintaining equilibrium and satisfying compatibility conditions. In this chapter, the FAM is first derived using single-degree-of-freedom systems, and it is then extended to analyzing multi-degree-of-freedom moment-resisting frames. Static condensation is included in the presentation to demonstrate that the FAM can also be characterized in a reduced form, which is an important part of the nonlinear dynamic analysis. A vast number of examples is presented to demonstrate the procedure for solving nonlinear static problems using the FAM.

2.1 Plastic Rotation

In order to develop the FAM for analyzing moment-resisting frame, the definition of plastic rotation must first be clearly explained. Some documents define plastic rotation based on chord rotations (e.g., ASCE 2006), while other documents define plastic rotation based on the Bernoulli–Euler beam theory (e.g., Powell 2010). Here, plastic rotation is defined as the measurement of the amount of concentrated rotation at the plastic hinge that is caused by the moment demand exceeding its yield moment m_y, and this moment demand is based on the structural analysis. Consider Figure 2.1(a) where a beam in a typical moment-resisting frame is subjected to moments at both ends. Here, the moments are drawn with a positive direction for performing structural analysis (i.e. counter-clockwise as positive). Note that in general,

Theory of Nonlinear Structural Analysis: The Force Analogy Method for Earthquake Engineering, First Edition. Gang Li and Kevin K.F. Wong.

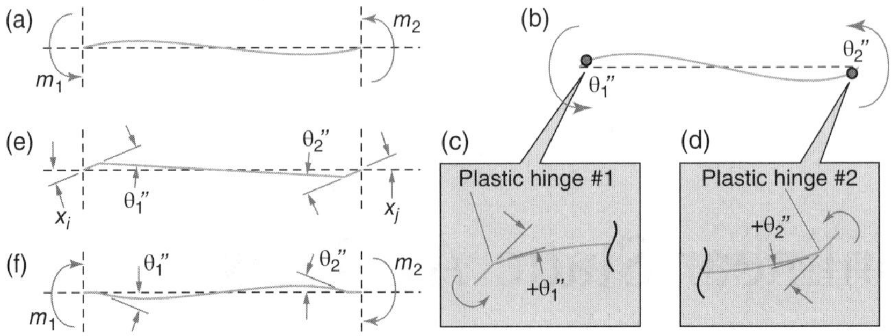

Figure 2.1 Definition of positive plastic rotations: (a) Beam with positive moments; (b) Beam with positive plastic rotations due to positive moments; (c) Positive plastic rotation at the left end. (d) Positive plastic rotation at the right end; (e) Beam member satisfying compatibility condition; (f) Beam member satisfying equilibrium condition.

$m_1 \neq -m_2$, and therefore shear forces at the two ends of the member are required for equilibrium. But these shear forces are not shown in the figure for simplicity.

It is important to note that positive moment causes positive plastic rotation, and negative moment causes negative plastic rotation. This means when the moments in Figure 2.1(a) exceed the corresponding yield moments, plastic hinges form with positive plastic rotations, as shown in Figure 2.1(b). Therefore, as shown in Figure 2.1(c), a positive plastic rotation at the '1'-end (or the near-end) of the beam is defined as a concentrated 'clockwise' rotation when one moves from the left side of the hinge to the right side of the hinge. Similarly, as shown in Figure 2.1(d), a positive plastic rotation at the '2'-end (or the far-end) of the beam is defined as a concentrated 'counter-clockwise' rotation when one moves from the left side of the hinge to the right side of the hinge.

After yielding, the moment versus plastic rotation relationship typically follows a nonlinear function, denoted here as $f(\cdot)$. It is important to specify the nonlinear function for each and every plastic hinge prior to performing the nonlinear structural analysis. The hardening rule for reversed loading (such as kinematic hardening or isotropic hardening) also needs to be specified, which is particularly important for capturing the cyclic response in a nonlinear dynamic analysis. Once the nonlinear behavior is defined, the moment versus plastic rotation relationship can be expressed in mathematical form as:

$$\text{if} \begin{cases} -m_y < m \\ -m_y \le m \le m_y, \\ m > m_y \end{cases} \quad \text{then} \begin{cases} m = f_-(\theta'') \\ \theta'' = 0 \\ m = f_+(\theta'') \end{cases} \tag{2.1}$$

where $f_+(\cdot)$ denotes the relationship between positive plastic rotation and positive moment, and $f_-(\cdot)$ denotes the relationship between negative plastic rotation and negative moment. For symmetric material (i.e. the magnitude of yielding is the same in positive and negative directions), $|f_+(\cdot)| = |f_-(\cdot)| = f(\cdot)$, then Eq. (2.1) can often be simplified as

$$\text{if} \begin{cases} m \le m_y \\ m > m_y \end{cases}, \quad \text{then} \begin{cases} \theta'' = 0 \\ m = f(\theta'') \end{cases} \tag{2.2}$$

Consider that after the beam yielded as shown in Figure 2.1(b), the applied moments are removed. Figure 2.1(e) shows the beam with plastic rotations remaining at the two ends. The beam returns to a straightened position after the applied moments have been removed, and therefore the beam is in equilibrium within itself. However, with the plastic rotations at the plastic hinges, it requires that Joint '*i*' and Joint '*j*' be rotated with x_i and x_j (i.e. the *i*th and *j*th degrees of freedom are rotations) while the connecting members of these joints require that these joint rotations be zero. Therefore, Figure 2.1(e) violates the compatibility condition when equilibrium is maintained.

On the other hand, Figure 2.1(f) shows the beam with plastic rotations at the two ends while imposing zero joint rotations (i.e. $x_i = x_j = 0$). This satisfies the compatibility condition because the connecting members at Joint '*i*' and Joint '*j*' also require that these joint rotations be zero. However, because the beam is now bent, it induces moments m_1 and m_2 at the two ends while the connecting members remain straight and induce no moment at their ends. Therefore, Figure 2.1(f) violates the equilibrium when the compatibility condition is satisfied.

It is therefore the objective of the force analogy method and this chapter to discuss the technique of analyzing the nonlinear structure while maintaining equilibrium and satisfying the compatibility condition.

2.2 Force Analogy Method for Static Single-Degree-of-Freedom Systems

2.2.1 *Inelastic Displacement*

The FAM begins with the concept of inelastic displacement. Consider a single degree of freedom (SDOF) system with a force versus displacement curve shown in Figure 2.2. In this figure, K_e denotes the initial stiffness, K_t denotes the varying post-yield stiffness, and F_y and x_y represents the yield force and yield displacement, respectively. Assume that a static force of F_s is applied at the only degree of freedom, the system reaches a displacement at Point C as shown in Figure 2.2. Let x represents and total displacement in the system, and note in the figure that $x > x_y$ (i.e. the system has yielded).

The concept of the FAM is to extend the initial stiffness line OA until it reaches the force F_s at Point B. By defining the displacement with respect to Point B as the elastic displacement x', the difference between the total displacement x and the elastic displacement x' is the inelastic

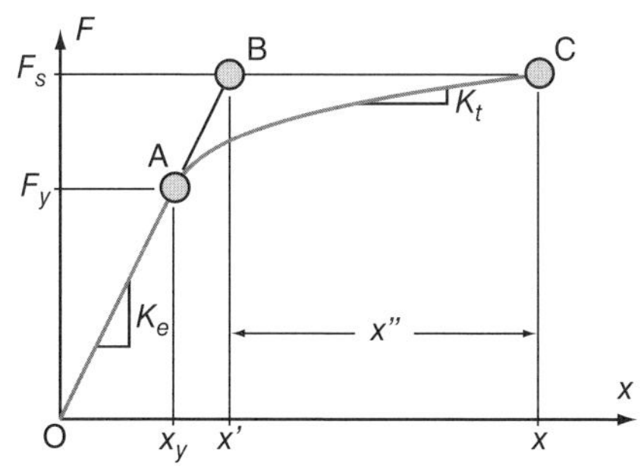

Figure 2.2 Inelastic displacement in the force–displacement relationship.

displacement x'', as shown in Figure 2.2. The elastic displacement x' represents the displacement that will be recovered after unloading, while the inelastic displacement x'' represents the permanent deformation that will not be recovered after unloading (or sometimes known as the "residual displacement"). Representing this concept in mathematical form for any value of total displacement x gives

$$x = x' + x'' \tag{2.3}$$

$$\text{if} \begin{cases} x \le x_y \\ x > x_y \end{cases}, \quad \text{then} \begin{cases} x' = F_s/K_e, & x'' = 0 \\ x' = F_s/K_e, & x'' > 0 \end{cases} \tag{2.4}$$

Therefore, the force in the system F_s at any displacement x is always equal to the elastic displacement multiplied by the initial stiffness, i.e.

$$F_s = K_e x' \tag{2.5}$$

Solving for x' in Eq. (2.3) and substituting the result into Eq. (2.5) gives

$$F_s = K_e(x - x'') \tag{2.6}$$

Equation (2.6) shows the effect of changing the displacement of the system on the force level. The elastic displacement x' is not a constant and changes as the applied force F_s changes, and therefore this elastic displacement is not equal to the yield displacement, i.e. $x' \neq x_y$, as presented in Figure 2.2 where Point A is at yield displacement x_y and Point B is at elastic displacement x'. Note that no assumption has been made on whether the material has hardening behavior (i.e. $K_t > 0$) or softening behavior (i.e. $K_t < 0$).

In summary, the FAM focuses on a change in displacement, not in stiffness, to give the same level of force.

2.2.2 *Application of the FAM on SDOF System*

A SDOF system as shown in Figure 2.3(a) is used to demonstrate the concept of the FAM in terms of how it satisfies the equilibrium and compatibility conditions at the system level. The

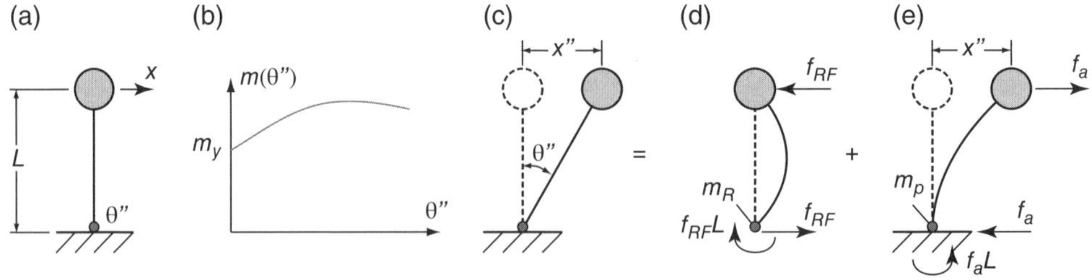

Figure 2.3 Force analogy method applied on a SDOF system: (a) A single-degree-of-freedom system; (b) Moment versus plastic rotation relationship; (c) Relationship between plastic rotation and inelastic displacement; (d) Force applied to satisfy compatibility condition; (e) Equivalent force applied to satisfy equilibrium condition.

total displacement is the only degree of freedom (DOF) of the system and is labeled as x, and one plastic hinge location (PHL) with the plastic rotation labeled as θ''. Let the length of the column be denoted as L. The displacement and moment relationships for a SDOF system with a single plastic hinge based on the FAM are

$$x = x' + x'' \tag{2.7}$$

$$m = m' + m'' \tag{2.8}$$

where m is the total moment at the PHL, m' is the elastic moment at the PHL due to the elastic displacement x' at the DOF, and m'' is the inelastic moment at the PHL due to the inelastic displacement x'' at the DOF. A typical moment versus plastic rotation relationship is presented in Figure 2.3(b), and a plastic rotation at the PHL causing a permanent deformation in the column is shown in Figure 2.3(c). Preliminary observation suggests that the top of the column is free to move to any position after yielding at the PHL, and no lateral fore is required to prevent any lateral movement of the column to remain straight. Therefore, Figure 2.3(c) already satisfies both the compatibility and equilibrium conditions, and it is predictable that the relationships between the displacement, moment, and plastic rotations are

$$x'' = L\theta'', \qquad m'' = 0 \tag{2.9}$$

Here, the stage is set to prove Eq. (2.9) through a detailed derivation of the FAM that is similar to a multi-degree-of-freedom system approach. Consider that the analysis of the purely inelastic behavior (i.e. when plastic rotation occurs after the loading is removed), as shown in Figure 2.3(c), is separated into two parts as shown in Figure 2.3(d) and Figure 2.3(e). First in Figure 2.3(d), applying a restoring force f_{RF} on the column to give a zero displacement restores the column back to its original position. If the column is connected to other parts of the structure, such as in a multi-degree-of-freedom system, applying the restoring force f_{RF} would result in satisfying the compatibility condition. This applied restoring force (see Figure 2.3(d)) is

$$f_{RF} = -\left(\frac{3EI}{L^2}\right)\theta'' = -K'\theta'' \tag{2.10}$$

where E is the Young's modulus, I is the moment of inertia of the column, and K' denotes the stiffness that relates the plastic rotation with the restoring force f_{RF}. Also, applying this restoring force f_{RF} induces a residual moment m_R at the plastic hinge:

$$m_R = -\left(\frac{3EI}{L}\right)\theta'' = -K''\theta'' \tag{2.11}$$

where K'' denotes the stiffness that relates the plastic rotation with the residual moment m_R.

The restoring force in Eq. (2.10) is actually not present in the structure (since the applied lateral force is not present in the analysis of a purely inelastic behavior), and therefore an equal and opposite force f_a must be applied to the structure in order to satisfy the equilibrium condition of the column in its inelastic state, as shown in Figure 2.3(e), i.e.

$$f_a = -f_{RF} = \left(\frac{3EI}{L^2}\right)\theta'' \tag{2.12}$$

When the equivalent force f_a is applied, the displacement x'' at the top of the column (i.e. the only DOF) is computed through the use of traditional stiffness method of structural analysis, or simply by the formula

$$f_a = \left(\frac{3EI}{L^3}\right)x'' = Kx'' \tag{2.13}$$

where K is the global stiffness of the structure. Equating f_a in Eqs. (2.12) and (2.13) gives

$$x'' = L\theta'' \tag{2.14}$$

which is exactly the kinematic relation as given in the first part of Eq. (2.9). In addition, applying the equivalent force f_a causes an induced moment m_p at the PHL, which can be determined by performing simple equilibrium together with Eq. (2.12):

$$m_p = Lf_a = \left(\frac{3EI}{L}\right)\theta'' \tag{2.15}$$

Finally, the inelastic moment m'' shown in Figure 2.3(c) can be computed by summing the residual moment m_R shown in Figure 2.3(d) and the induced moment m_p shown in Figure 2.3(e). Doing so and substituting Eqs. (2.11) and (2.15) into the resulting equation gives:

$$m'' = m_R + m_p = -\left(\frac{3EI}{L}\right)\theta'' + \left(\frac{3EI}{L}\right)\theta'' = 0 \tag{2.16}$$

This means that plastic rotation θ'' will not induce any inelastic moment m'' on this column, since the column is free to displace laterally. This proves that the second part of Eq. (2.9) is correct.

A similar relationship can be obtained between the elastic moment m', the elastic displacement x', and the external applied force F_a. Through the traditional stiffness method of structural analysis with the use of stiffness matrix K (see Eq. (2.13)):

$$F_a = Kx' = \left(\frac{3EI}{L^3}\right)x', \quad m' = K'x' = \left(\frac{3EI}{L^2}\right)x' \tag{2.17}$$

2.2.3 Nonlinear Analysis Using FAM

In summary, it can be seen that the force analogy method divides the nonlinear structural analysis problem into (1) the traditional stiffness method of analysis of the elastic portion, and (2) the traditional stiffness method of analysis of the inelastic portion. For the elastic portion, the applied force F_a cause elastic displacement x' and elastic moment m' as shown in Eq. (2.17). For the inelastic portion, the plastic rotation θ'' causes inelastic displacement x'' and inelastic moment m'' as shown in Eq. (2.9).

The objective now is to combine Eqs. (2.9) and (2.17) and represent the analytical solution using the total displacement x and the total moments m, which are typically the quantities of interest when solving the structural analysis problem. Here, Eq. (2.7) is first considered by solving for the elastic displacement x', and substituting the result into Eq. (2.17) gives

$$F_a = \left(\frac{3EI}{L^3}\right)x' = \left(\frac{3EI}{L^3}\right)[x - x''], \quad m' = \left(\frac{3EI}{L^2}\right)x' = \left(\frac{3EI}{L^2}\right)[x - x''] \tag{2.18}$$

Then substituting the inelastic displacement x'' in Eq. (2.9) into Eq. (2.18) gives

$$F_a = \left(\frac{3EI}{L^3}\right)x' = \left(\frac{3EI}{L^3}\right)[x - L\theta''], \quad m' = \left(\frac{3EI}{L^2}\right)x' = \left(\frac{3EI}{L^2}\right)[x - L\theta''] \tag{2.19}$$

Simplifying the first equation of Eq. (2.19) gives

$$\left(\frac{3EI}{L^3}\right)x - \left(\frac{3EI}{L^2}\right)\theta'' = F_a \tag{2.20}$$

which is the first important equation of the FAM.

Now, based on Eq. (2.8), the total moment m at the PHL can be calculated by substituting the elastic moments m' in Eq. (2.19) and the inelastic moments m'' in Eq. (2.9). Doing so gives

$$m = m' + m'' = \left(\frac{3EI}{L^2}\right)[x - L\theta''] + 0 = \left(\frac{3EI}{L^2}\right)x + \left(\frac{3EI}{L}\right)\theta'' \tag{2.21}$$

which is the second important equation of the FAM. Combining Eqs. (2.20) and (2.21) gives the governing equation of the force analogy method in matrix form for solving nonlinear static problems:

$$\begin{bmatrix} 3EI/L^3 & 3EI/L^2 \\ 3EI/L & 3EI/L \end{bmatrix} \begin{Bmatrix} x \\ -\theta'' \end{Bmatrix} = \begin{Bmatrix} F_a \\ m \end{Bmatrix} \tag{2.22}$$

or, according to the definitions of stiffnesses in Eqs. (2.10), (2.11), and (2.13), Eq. (2.22) can simply be written as

$$\begin{bmatrix} K & K' \\ K' & K'' \end{bmatrix} \begin{Bmatrix} x \\ -\theta'' \end{Bmatrix} = \begin{Bmatrix} F_a \\ m \end{Bmatrix} \tag{2.23}$$

For a given external applied force F_a, there are three unknowns according to Eq. (2.22), namely x, m, and θ''. The matrix equation in Eq. (2.22) provides two equations for the solution. The moment versus plastic rotation relationship of the plastic hinge as given in Figure 2.3(b) provides the third equation to solve for the three unknowns. This moment versus plastic rotation relationship comes from the mechanics of material behavior of the plastic hinge, which can be written in the form as shown in Eq. (2.1) or Eq. (2.2).

Example 2.1 SDOF System with Linear Response

Consider the SDOF column shown in Figure 2.4(a) subjected to an applied force F_o. The moment versus plastic rotation relationship is shown in Figure 2.4(b), with a yield moment of $m_y = 2F_oL$ and a post-yield stiffness of EI/L. It can foreseeable that the column will remain linear because the moment demand at the plastic hinge is only $m = F_oL$, which is less than the yield moment of $2F_oL$. But let us investigate how Eq. (2.22) can be used to solve the problem.

Just like any nonlinear static analysis, the first step is to check whether the load is large enough to cause any yielding. Therefore, assume that the structure remains linear and the plastic rotation θ'' is zero, then Eq. (2.22) becomes

$$\begin{bmatrix} 3EI/L^3 & 3EI/L^2 \\ 3EI/L^2 & 3EI/L \end{bmatrix} \begin{Bmatrix} x \\ 0 \end{Bmatrix} = \begin{Bmatrix} F_o \\ m \end{Bmatrix} \tag{2.24}$$

Using the first equation of Eq. (2.24) and solving for the total displacement x gives

$$x = \frac{F_oL^3}{3EI} \tag{2.25}$$

Now substituting Eq. (2.25) back into the second equation of Eq. (2.24) and solving for the moment m gives

$$m = \left(\frac{3EI}{L^2}\right)\left(\frac{F_oL^3}{3EI}\right) = F_oL \tag{2.26}$$

Since $m = F_oL$ is less than the yield moment $m_y = 2F_oL$, the assumption made is correct, and therefore the total displacement x given in Eq. (2.25) and the moment m in Eq. (2.26) represent the final answer.

Example 2.2 SDOF System with Nonlinear Response

Consider the SDOF column shown in Figure 2.4(a) is now subjected to an applied force of $3F_o$. At a yield moment of $m_y = 2F_oL$, it can be seen now that the column will behave nonlinearly because the moment demand at the plastic hinge of $m = 3F_oL$ exceeds the yield moment of $2F_oL$. Let us investigate how Eq. (2.22) can be used to solve the problem.

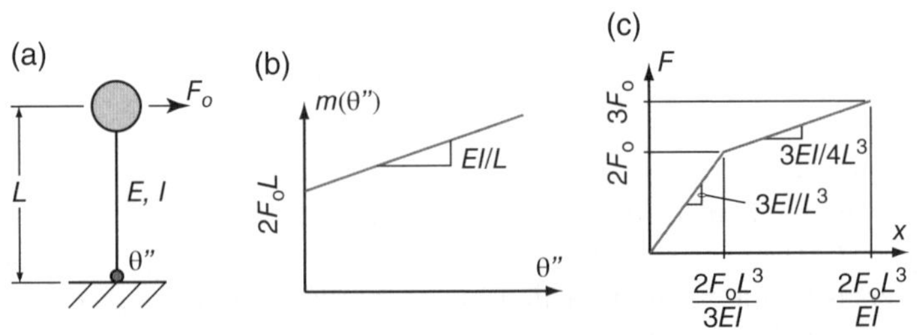

Figure 2.4 Single-degree-of-freedom system: (a) A single-degree-of-freedom column; (b) Moment versus plastic rotation relationship; (c) System pushover curve.

Similar to Example 2.1 and just like any nonlinear static analysis, the first step is to check whether the load is large enough to cause any yielding. Therefore, assume that the plastic rotation θ'' is zero, then Eq. (2.22) becomes

$$\left[\begin{array}{c:c} 3EI/L^3 & 3EI/L^2 \\ \hdashline 3EI/L^2 & 3EI/L \end{array}\right]\left\{\begin{array}{c} x \\ \hdashline 0 \end{array}\right\}=\left\{\begin{array}{c} 3F_o \\ \hdashline m \end{array}\right\} \tag{2.27}$$

Using the first equation of Eq. (2.27) and solving for the total displacement x gives

$$x=\frac{F_oL^3}{EI} \tag{2.28}$$

Now substituting Eq. (2.28) back into the second equation of Eq. (2.27) to solve for the moment m gives

$$m=\left(\frac{3EI}{L^2}\right)\left(\frac{F_aL^3}{EI}\right)=3F_oL \tag{2.29}$$

which exceeds the yield moment of $m_y=2F_oL$. Therefore, yielding occurs and the assumption that plastic rotation is zero is incorrect. But because the plastic hinge has yielded, the moment m in Eq. (2.22) can be expressed in terms of the plastic rotation θ'' based on the moment versus plastic rotation relationship given in Figure 2.4(b), i.e.

$$m=2F_oL+\frac{EI}{L}\theta'' \tag{2.30}$$

Substituting Eq. (2.30) into Eq. (2.22) gives

$$\left[\begin{array}{c:c} 3EI/L^3 & 3EI/L^2 \\ \hdashline 3EI/L^2 & 3EI/L+EI/L \end{array}\right]\left\{\begin{array}{c} x \\ \hdashline -\theta'' \end{array}\right\}=\left\{\begin{array}{c} 3F_o \\ \hdashline 2F_oL \end{array}\right\} \tag{2.31}$$

Note that the post-yield stiffness EI/L in Eq. (2.30) enters into Eq. (2.31) as an addition to the diagonal term in the stiffness matrix. Now solving for Eq. (2.31) gives

$$\left\{\begin{array}{c} x \\ \hdashline -\theta'' \end{array}\right\}=\frac{L^4}{3E^2I^2}\begin{bmatrix} 4EI/L & -3EI/L^2 \\ -3EI/L^2 & 3EI/L^3 \end{bmatrix}\left\{\begin{array}{c} 3F_o \\ \hdashline 2F_oL \end{array}\right\}=\left\{\begin{array}{c} 2F_oL^3/EI \\ \hdashline -F_oL^2/EI \end{array}\right\} \tag{2.32}$$

With the minus sign in front of θ'' in Eq. (2.32), the solution becomes

$$x=\frac{2F_oL^3}{EI},\qquad \theta''=\frac{F_oL^2}{EI} \tag{2.33}$$

and substituting Eq. (2.33) into Eq. (2.30) gives the moment m:

$$m=2F_oL+\frac{EI}{L}\left(\frac{F_oL^2}{EI}\right)=3F_oL \tag{2.34}$$

which is exactly equal to the applied force of $3F_o$ multiplied by the length of the column L.

The resulting force–displacement relationship of the column (or the pushover curve) is shown in Figure 2.4(c).

2.3 Nonlinear Structural Analysis of Moment-Resisting Frames

For a multi-degree-of-freedom (MDOF) system, the derivation of the FAM also begins with a similar concept of inelastic displacements while representing the force and displacement quantities in vector form. Consider a structure having n degrees of freedom (DOFs), the displacement can be written in vector form as

$$\mathbf{x} = \mathbf{x}' + \mathbf{x}'' = \begin{Bmatrix} x_1' \\ x_2' \\ \vdots \\ x_n' \end{Bmatrix} + \begin{Bmatrix} x_1'' \\ x_2'' \\ \vdots \\ x_n'' \end{Bmatrix} \tag{2.35}$$

where $\mathbf{x}$ represents the total displacement vector, $\mathbf{x}'$ is the elastic displacement vector, and $\mathbf{x}''$ is the inelastic displacement vector.

Consider a moment-resisting framed structures, the moments at the two ends of the members are typically the critical points where yielding occurs. Denote these yielding locations as plastic hinge locations (PHLs). Let the total moment vector $\mathbf{m}$ at the locations where plastic hinges may form be described by the equation

$$\mathbf{m} = \mathbf{m}' + \mathbf{m}'' = \begin{Bmatrix} m_1' \\ m_2' \\ \vdots \\ m_q' \end{Bmatrix} + \begin{Bmatrix} m_1'' \\ m_2'' \\ \vdots \\ m_q'' \end{Bmatrix} \tag{2.36}$$

where $\mathbf{m}'$ is the elastic moment vector due to elastic displacement $\mathbf{x}'$, and $\mathbf{m}''$ is the inelastic moment vector due to inelastic displacement $\mathbf{x}''$. The value q represents the total number of potential PHLs in the moment-resisting frame.

Consider first the inelastic moment vector, $\mathbf{m}''$. When plastic rotations occur at certain PHLs in the structure, these plastic rotations are replaced with a set of fictitious forces. Define the plastic rotation vector, $\boldsymbol{\Theta}''$, as

$$\boldsymbol{\Theta}'' = \begin{Bmatrix} \theta_1'' \\ \theta_2'' \\ \vdots \\ \theta_q'' \end{Bmatrix} \tag{2.37}$$

An example of having plastic rotations developed at two PHLs in the ith beam member is shown in Figure 2.5(a). This state of the structure can never exist because this ith beam member violates the compatibility condition. Without any force applied to the ith member, it should remain straight, yet the plastic rotations at the two ends restrict the member from being straight. To ensure the structure deforms in a compatible mode, the member with plastic rotations $\boldsymbol{\Theta}''$ is first isolated from the structure and restoring forces are applied to restore this member back to the original undeformed shape, as shown in Figure 2.5(b). This induces internal restoring forces $\mathbf{f}_{RF}$ as the fixed-end forces on the member. At the global degree of freedom level, the restoring force is an $n \times 1$ vector of the form:

$$\mathbf{f}_{RF} = \begin{Bmatrix} F_{RF1} \\ F_{RF2} \\ \vdots \\ F_{RFn} \end{Bmatrix} = -\mathbf{K}'\boldsymbol{\Theta}'' \tag{2.38}$$

where $\mathbf{K}'$ is the assembled stiffness matrix that relates the plastic rotation at the PHLs with the applied forces at the global DOFs. This $\mathbf{K}'$ matrix is the collection of individual $\mathbf{K}'_i$ matrices for each beam member appropriately assembled, where $\mathbf{K}'_i$ associated with the ith beam member is of the form

$$\mathbf{f}_{RF(i)} = \begin{Bmatrix} V_{RF1}^{(i)} \\ M_{RF1}^{(i)} \\ V_{RF2}^{(i)} \\ M_{RF2}^{(i)} \end{Bmatrix} = -\begin{bmatrix} 6EI/L^2 & 6EI/L^2 \\ 4EI/L & 2EI/L \\ -6EI/L^2 & -6EI/L^2 \\ 2EI/L & 4EI/L \end{bmatrix} \begin{Bmatrix} \theta''_{1i} \\ \theta''_{2i} \end{Bmatrix} = -\mathbf{K}'_i\boldsymbol{\Theta}''_i \tag{2.39}$$

where E is the Young's modulus, I is the moment of inertia, and L is the length of the beam member, subscripts '1' and '2' denote the '1'-end and the '2'-end of the beam member, respectively, and $V^{(i)}$ denotes fixed-end shear and $M^{(i)}$ denotes fixed-end moment of the ith beam member. In addition to the restoring forces $\mathbf{f}_{RF}$ that are applied at the global DOFs, plastic rotations $\boldsymbol{\Theta}''$ induce residual moments $\mathbf{m}_R$ at the PHLs as shown in Figure 2.5(b). At the local PHL level, the residual moment is a $q \times 1$ vector in the form:

$$\mathbf{m}_R = \begin{Bmatrix} m_{R,1} \\ m_{R,2} \\ \vdots \\ m_{R,q} \end{Bmatrix} = -\mathbf{K}''\boldsymbol{\Theta}'' \tag{2.40}$$

where $\mathbf{K}''$ is a matrix that relates the plastic rotations at the PHLs with the moments at these PHLs. This $\mathbf{K}''$ matrix is the collection of individual $\mathbf{K}''_i$ matrices for each beam member appropriately assembled, where $\mathbf{K}''_i$ associated with the ith beam member is

$$\mathbf{m}_{R(i)} = \begin{Bmatrix} m_{R,1}^{(i)} \\ m_{R,2}^{(i)} \end{Bmatrix} = -\begin{bmatrix} 4EI/L & 2EI/L \\ 2EI/L & 4EI/L \end{bmatrix} \begin{Bmatrix} \theta''_{1i} \\ \theta''_{2i} \end{Bmatrix} = -\mathbf{K}''_i\boldsymbol{\Theta}''_i \tag{2.41}$$

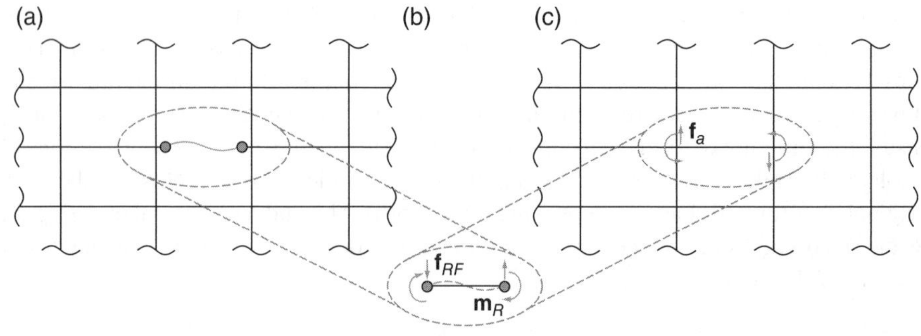

Figure 2.5 Satisfying compatibility and equilibrium conditions with the FAM: (a) A multi-degree-of-freedom system with plastic hinges; (b) Force applied to satisfy compatibility condition; (c) Equivalent force applied to satisfy equilibrium condition.

The minus signs appear in Eqs. (2.38), (2.39), (2.40), and (2.41) because negative restoring forces must be applied in order to rotate a positive plastic rotation back to zero, which results in negative restoring forces and negative residual moments.

Now the member is assembled back into the structure and the deformation of this ith beam member is compatible with that of the joint. However, the restoring forces $\mathbf{f}_{RF}$ as shown in Figure 2.5(b) are actually not present globally as shown in Figure 2.5(a), and as a result equal and opposite forces $\mathbf{f}_a$ must be applied to the structure's degrees of freedom to cancel these restoring forces, as shown in Figure 2.5(c). These induced equivalent forces $\mathbf{f}_a$ are described by the equation

$$\mathbf{f}_a = -\mathbf{f}_{RF} = \mathbf{K}'\boldsymbol{\Theta}'' \tag{2.42}$$

Applying the equivalent forces $\mathbf{f}_a$ results in a permanent deformation of the structure, which is exactly $\mathbf{x}''$. The relationship can be obtained through conducting the traditional stiffness method of structural analysis for MDOF systems (which can be found in many structural analysis textbooks, e.g., Hibbeler (2012), Leet *et al.* (2010)) by using the global structural stiffness matrix $\mathbf{K}$, i.e.

$$\mathbf{f}_a = \mathbf{K}\mathbf{x}'' \tag{2.43}$$

where the global structural stiffness matrix $\mathbf{K}$ is the collection of individual $\mathbf{K}_i$ matrices for each beam member appropriately assembled. The $\mathbf{K}_i$ associated with the ith beam member is

$$\mathbf{f}_{a(i)} = \begin{Bmatrix} V_1^{(i)} \\ M_1^{(i)} \\ V_2^{(i)} \\ M_2^{(i)} \end{Bmatrix} = \begin{bmatrix} 12EI/L^3 & 6EI/L^2 & -12EI/L^3 & 6EI/L^2 \\ 6EI/L^2 & 4EI/L & -6EI/L^2 & 2EI/L \\ -12EI/L^3 & -6EI/L^2 & 12EI/L^3 & -6EI/L^2 \\ 6EI/L^2 & 2EI/L & -6EI/L^2 & 4EI/L \end{bmatrix} \begin{Bmatrix} v_1^{(i)} \\ \theta_1^{(i)} \\ v_2^{(i)} \\ \theta_2^{(i)} \end{Bmatrix} = \mathbf{K}_i\mathbf{x}_i \tag{2.44}$$

where $v_1^{(i)}$, $\theta_1^{(i)}$, $v_2^{(i)}$, and $\theta_2^{(i)}$ are the lateral displacements and rotations at the two ends of the beam member in the local coordinate system. Through this process, the structure remains in equilibrium in addition to being compatible. Equating $\mathbf{f}_a$ in both Eqs. (2.43) and (2.44) and solving for the inelastic displacements $\mathbf{x}''$ gives

$$\mathbf{x}'' = \mathbf{K}^{-1}\mathbf{K}'\boldsymbol{\Theta}'' \tag{2.45}$$

Due to the induced equivalent forces $\mathbf{f}_a$ which produce inelastic displacements $\mathbf{x}''$ in the structure, additional moments are also induced at the PHLs. Denoting this induced moment vector as $\mathbf{m}_p$, it is related to the inelastic displacement $\mathbf{x}''$ by the equation

$$\mathbf{m}_p = \mathbf{K}'^T\mathbf{x}'' \tag{2.46}$$

Then substituting Eq. (2.45) into Eq. (2.46) gives

$$\mathbf{m}_p = \mathbf{K}'^T\mathbf{K}^{-1}\mathbf{K}'\boldsymbol{\Theta}'' \tag{2.47}$$

Note that the $\mathbf{K}'^T$ matrix is used in Eq. (2.46). Recall from Eq. (2.38) that the $\mathbf{K}'$ matrix relates the plastic rotation at the PHLs with the applied force at the DOFs. Therefore, from the theory of reciprocity, it follows that $\mathbf{K}'^T$ relates the inelastic displacement at the DOFs with the moments at the PHLs, i.e.

$$\begin{Bmatrix} m_{p1} \\ m_{p2} \\ \vdots \\ m_{pq} \end{Bmatrix} = \begin{bmatrix} K'_{11} & K'_{21} & \cdots & K'_{n1} \\ K'_{12} & K'_{22} & \cdots & K'_{n2} \\ \vdots & \vdots & \ddots & \vdots \\ K'_{1q} & K'_{2q} & \cdots & K_{nq} \end{bmatrix} \begin{Bmatrix} x''_1 \\ x''_2 \\ \vdots \\ x''_n \end{Bmatrix} \quad \text{and} \quad \begin{Bmatrix} f_{a1} \\ f_{a2} \\ \vdots \\ f_{an} \end{Bmatrix} = \begin{bmatrix} K'_{11} & K'_{12} & \cdots & K'_{1q} \\ K'_{21} & K'_{22} & \cdots & K'_{2q} \\ \vdots & \vdots & \ddots & \vdots \\ K'_{n1} & K'_{n2} & \cdots & K_{nq} \end{bmatrix} \begin{Bmatrix} \theta''_1 \\ \theta''_2 \\ \vdots \\ \theta''_q \end{Bmatrix} \tag{2.48}$$

Finally, the inelastic moments $\mathbf{m}''$ at the PHLs in Figure 2.5(a) can be determined by summing the residual moments $\mathbf{m}_R$ at the PHLs shown in Figure 2.5(b) and the induced moment $\mathbf{m}_p$ shown in Figure 2.5(c), i.e.

$$\mathbf{m}'' = \mathbf{m}_R + \mathbf{m}_p \tag{2.49}$$

Substituting Eqs. (2.40) and (2.47) in Eq. (2.49) gives the equation of inelastic moments $\mathbf{m}''$ as a function of plastic rotations $\boldsymbol{\Theta}''$ as:

$$\mathbf{m}'' = -\left(\mathbf{K}'' - \mathbf{K}'^T\mathbf{K}^{-1}\mathbf{K}'\right)\boldsymbol{\Theta}'' \tag{2.50}$$

Equation (2.50) represents inelastic moment vector due to the plastic rotations within the structure with no external applied force. For example, if an earthquake causes plastic rotations within the structure, then the inelastic moments represent the forces remaining in the members after the earthquake motion subsides.

Now consider the relationship between the elastic moments $\mathbf{m}'$ and the elastic displacements $\mathbf{x}'$ due to the external applied static load $\mathbf{F}_a$. Similar to Eq. (2.43) where the inelastic displacements $\mathbf{x}''$ are due to the application of the induced equivalent loads $\mathbf{f}_a$, the elastic displacements $\mathbf{x}'$ of the structure are the result of applying the external static loads $\mathbf{F}_a$. Again by using the traditional method of structural analysis, the external static loads $\mathbf{F}_a$ can be related to the elastic displacements $\mathbf{x}'$ through the global stiffness matrix $\mathbf{K}$, i.e.

$$\mathbf{F}_a = \mathbf{K}\mathbf{x}' \tag{2.51}$$

Similarly from Eq. (2.46), where the induced moments $\mathbf{m}_p$ are related to the inelastic displacements $\mathbf{x}''$ due to the application of the induced equivalent load $\mathbf{f}_a$, the elastic moments $\mathbf{m}'$ are related to the elastic displacements $\mathbf{x}'$, which can be obtained by inverting the stiffness matrix $\mathbf{K}$ in Eq. (2.51). This equation for computing the elastic moments $\mathbf{m}'$ is:

$$\mathbf{m}' = \mathbf{K}'^{T}\mathbf{x}' \tag{2.52}$$

Example 2.3 Element Stiffness Matrices of One-Story Frame

Consider a one-story one-bay frame as shown in Figure 2.6. Assume the members are axially rigid, this results in a three degrees of freedom system ($n = 3$) with six plastic hinge locations ($q = 6$). Assume also that the beam is of different length and size from the two columns. This gives element stiffness matrices of the form:

$$\mathbf{K}_1 = \mathbf{K}_2 = \begin{bmatrix} 12EI_c/L_c^3 & 6EI_c/L_c^2 & -12EI_c/L_c^3 & 6EI_c/L_c^2 \\ 6EI_c/L_c^2 & 4EI_c/L_c & -6EI_c/L_c^2 & 2EI_c/L_c \\ -12EI_c/L_c^3 & -6EI_c/L_c^2 & 12EI_c/L_c^3 & -6EI_c/L_c^2 \\ 6EI_c/L_c^2 & 2EI_c/L_c & -6EI_c/L_c^2 & 4EI_c/L_c \end{bmatrix} \tag{2.53a}$$

$$\mathbf{K}_3 = \begin{bmatrix} 12EI_b/L_b^3 & 6EI_b/L_b^2 & -12EI_b/L_b^3 & 6EI_b/L_b^2 \\ 6EI_b/L_b^2 & 4EI_b/L_b & -6EI_b/L_b^2 & 2EI_b/L_b \\ -12EI_b/L_b^3 & -6EI_b/L_b^2 & 12EI_b/L_b^3 & -6EI_b/L_b^2 \\ 6EI_b/L_b^2 & 2EI_b/L_b & -6EI_b/L_b^2 & 4EI_b/L_b \end{bmatrix} \tag{2.53b}$$

$$\mathbf{K}_1' = \mathbf{K}_2' = \begin{bmatrix} 6EI_c/L_c^2 & 6EI_c/L_c^2 \\ 4EI_c/L_c & 2EI_c/L_c \\ -6EI_c/L_c^2 & -6EI_c/L_c^2 \\ 2EI_c/L_c & 4EI_c/L_c \end{bmatrix}, \quad \mathbf{K}_3' = \begin{bmatrix} 6EI_b/L_b^2 & 6EI_b/L_b^2 \\ 4EI_b/L_b & 2EI_b/L_b \\ -6EI_b/L_b^2 & -6EI_b/L_b^2 \\ 2EI_b/L_b & 4EI_b/L_b \end{bmatrix} \tag{2.53c}$$

$$\mathbf{K}_1'' = \mathbf{K}_2'' = \begin{bmatrix} 4EI_c/L_c & 2EI_c/L_c \\ 2EI_c/L_c & 4EI_c/L_c \end{bmatrix}, \quad \mathbf{K}_3'' = \begin{bmatrix} 4EI_b/L_b & 2EI_b/L_b \\ 2EI_b/L_b & 4EI_b/L_b \end{bmatrix} \tag{2.53d}$$

Now, assembling each of the element stiffness matrices into the global stiffness matrices gives:

$$\mathbf{K} = \begin{bmatrix} 12EI_c/L_c^3 + 12EI_c/L_c^3 & 6EI_c/L_c^2 & 6EI_c/L_c^2 \\ 6EI_c/L_c^2 & 4EI_c/L_c + 4EI_b/L_b & 2EI_b/L_b \\ 6EI_c/L_c^2 & 2EI_b/L_b & 4EI_c/L_c + 4EI_b/L_b \end{bmatrix} \begin{matrix} \leftarrow x_1 \\ \leftarrow x_2 \\ \leftarrow x_3 \end{matrix} \tag{2.54a}$$

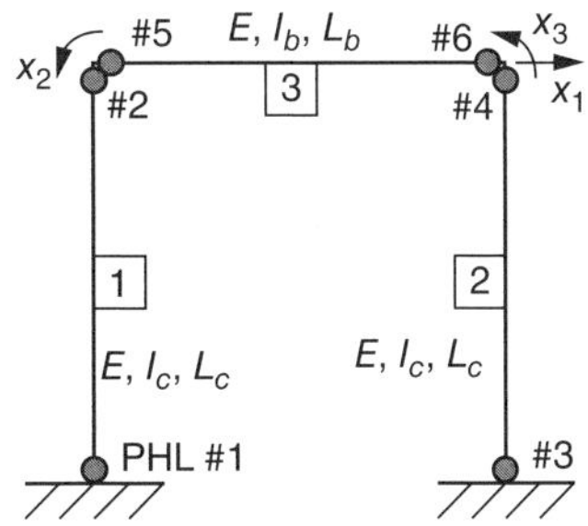

Figure 2.6 One-story one-bay moment-resisting frame

$$\mathbf{K}' = \left[\begin{array}{cc|cc|cc} 6EI_c/L_c^2 & 6EI_c/L_c^2 & 6EI_c/L_c^2 & 6EI_c/L_c^2 & 0 & 0 \\ 2EI_c/L_c & 4EI_c/L_c & 0 & 0 & 4EI_b/L_b & 2EI_b/L_b \\ 0 & 0 & 2EI_c/L_c & 4EI_c/L_c & 2EI_b/L_b & 4EI_b/L_b \end{array}\right] \begin{array}{l} \leftarrow x_1 \\ \leftarrow x_2 \\ \leftarrow x_3 \end{array} \tag{2.54b}$$

$$\mathbf{K}'' = \left[\begin{array}{cc|cc|cc} 4EI_c/L_c & 2EI_c/L_c & 0 & 0 & 0 & 0 \\ 2EI_c/L_c & 4EI_c/L_c & 0 & 0 & 0 & 0 \\ \hline 0 & 0 & 4EI_c/L_c & 2EI_c/L_c & 0 & 0 \\ 0 & 0 & 2EI_c/L_c & 4EI_c/L_c & 0 & 0 \\ \hline 0 & 0 & 0 & 0 & 4EI_b/L_b & 2EI_b/L_b \\ 0 & 0 & 0 & 0 & 2EI_b/L_b & 4EI_b/L_b \end{array}\right] \begin{array}{l} \leftarrow \theta_1'' \\ \leftarrow \theta_2'' \\ \leftarrow \theta_3'' \\ \leftarrow \theta_4'' \\ \leftarrow \theta_5'' \\ \leftarrow \theta_6'' \end{array} \tag{2.54c}$$

2.4 Force Analogy Method for Static Multi-Degree-of-Freedom Systems

At this stage, similar to the SDOF system, it can be seen that the FAM has divided the nonlinear MDOF structural analysis problem into (1) the traditional stiffness method of analysis of the elastic portion, and (2) the traditional stiffness method of analysis of the inelastic portion. For the elastic portion, the applied forces $\mathbf{F}_a$ cause elastic displacements $\mathbf{x}'$ (see Eq. (2.51)) and elastic moments $\mathbf{m}'$ (see Eq. (2.52)) of the form:

$$\mathbf{F}_a = \mathbf{K}\mathbf{x}', \quad \mathbf{m}' = \mathbf{K}'^T\mathbf{x}' \tag{2.55}$$

For the inelastic portion, the plastic rotations $\mathbf{\Theta}''$ cause inelastic displacements $\mathbf{x}''$ (see Eq. (2.45)) and inelastic moments $\mathbf{m}''$ (see Eq. (2.50)) of the form:

$$\mathbf{x}'' = \mathbf{K}^{-1}\mathbf{K}'\mathbf{\Theta}'', \quad \mathbf{m}'' = -\left(\mathbf{K}'' - \mathbf{K}'^T\mathbf{K}^{-1}\mathbf{K}'\right)\mathbf{\Theta}'' \tag{2.56}$$

The objective now is to combine Eqs. (2.55) and (2.56) and represent the analytical solution using the total displacements $\mathbf{x}$ and the total moments $\mathbf{m}$, where according to Eqs. (2.35) and (2.36), they are

$$\mathbf{x}=\mathbf{x}'+\mathbf{x}'' \tag{2.57}$$

$$\mathbf{m}=\mathbf{m}'+\mathbf{m}'' \tag{2.58}$$

Equation (2.57) is first considered by solving for the elastic displacements $\mathbf{x}'$, and substituting the result into Eq. (2.55) gives

$$\mathbf{F}_a=\mathbf{K}[\mathbf{x}-\mathbf{x}''], \quad \mathbf{m}'=\mathbf{K}'^T[\mathbf{x}-\mathbf{x}''] \tag{2.59}$$

Then substituting the inelastic displacements $\mathbf{x}''$ in Eq. (2.56) into Eq. (2.59) gives

$$\mathbf{F}_a=\mathbf{K}\left[\mathbf{x}-\mathbf{K}^{-1}\mathbf{K}'\mathbf{\Theta}''\right], \quad \mathbf{m}'=\mathbf{K}'^T\left[\mathbf{x}-\mathbf{K}^{-1}\mathbf{K}'\mathbf{\Theta}''\right] \tag{2.60}$$

Simplifying the first equation of Eq. (2.60) gives

$$\mathbf{F}_a=\mathbf{K}\mathbf{x}-\mathbf{K}'\mathbf{\Theta}'' \tag{2.61}$$

which is the first important equation of the FAM in matrix form. Now based on Eq. (2.58), the total moments $\mathbf{m}$ at all the PHLs can be calculated by substituting the elastic moments $\mathbf{m}'$ in Eq. (2.60) and the inelastic moments $\mathbf{m}''$ in Eq. (2.56). Doing so gives

$$\mathbf{m}=\mathbf{m}'+\mathbf{m}''=\mathbf{K}'^T\left[\mathbf{x}-\mathbf{K}^{-1}\mathbf{K}'\mathbf{\Theta}''\right]-[\mathbf{K}''-\mathbf{K}'^T\mathbf{K}^{-1}\mathbf{K}']\mathbf{\Theta}''=\mathbf{K}'^T\mathbf{x}-\mathbf{K}''\mathbf{\Theta}'' \tag{2.62}$$

which is the second important equation of the FAM in matrix form. Combining Eqs. (2.61) and (2.62) gives the governing equation of the force analogy method for solving nonlinear static problems:

$$\begin{bmatrix} \mathbf{K} & \mathbf{K}' \\ \mathbf{K}'^T & \mathbf{K}'' \end{bmatrix} \begin{Bmatrix} \mathbf{x} \\ -\mathbf{\Theta}'' \end{Bmatrix} = \begin{Bmatrix} \mathbf{F}_a \\ \mathbf{m} \end{Bmatrix} \tag{2.63}$$

Equation (2.63) shows both the total displacement vector $\mathbf{x}$ (n unknowns) and the plastic rotation vector $\mathbf{\Theta}''$ (q unknowns) make up the unknown vector that is typically required to be solved in the nonlinear structural analysis problem. This means that in the FAM, this unknown vector is comprised of the quantities in both the DOFs and PHLs. It should be pointed out that this treatment of displacements and plastic rotations is similar to that of the fast nonlinear analysis (FNA) presented in Wilson (2010), where in FNA both displacements and plastic rotations are labeled as degrees of freedom.

The solution to Eq. (2.63) is a simple iterative procedure. For any applied static force pattern $\mathbf{F}_a$, the unknowns are $\mathbf{x}$, $\mathbf{\Theta}''$, and $\mathbf{m}$. This gives a total of $n+2q$ unknowns. Equation (2.63) by itself gives $n+q$ independent equations. Also, each plastic hinge contains its own moment versus plastic rotation relationship, which gives additional q equations. Using these $n+2q$ equations, the $n+2q$ unknowns in Eq. (2.63) can be solved uniquely. Once the plastic rotations $\mathbf{\Theta}''$ are calculated, the inelastic displacements $\mathbf{x}''$ can be determined using the first equation shown in Eq. (2.56).

Example 2.4 Global Stiffness Matrices of the One-Story Frame

For the one-story one-bay moment-resisting frame shown in Figure 2.6, the global matrices are presented in Eq. (2.54). Therefore, the governing equation of the FAM for this one-story frame becomes:

$$\left[\begin{array}{ccc|cccccc}
\frac{24EI_c}{L_c^3} & \frac{6EI_c}{L_c^2} & \frac{6EI_c}{L_c^2} & \frac{6EI_c}{L_c^2} & \frac{6EI_c}{L_c^2} & \frac{6EI_c}{L_c^2} & \frac{6EI_c}{L_c^2} & 0 & 0 \\
\frac{6EI_c}{L_c^2} & \frac{4EI_c}{L_c}+\frac{4EI_b}{L_b} & \frac{2EI_b}{L_b} & \frac{2EI_c}{L_c} & \frac{4EI_c}{L_c} & 0 & 0 & \frac{4EI_b}{L_b} & \frac{2EI_b}{L_b} \\
\frac{6EI_c}{L_c^2} & \frac{2EI_b}{L_b} & \frac{4EI_c}{L_c}+\frac{4EI_b}{L_b} & 0 & 0 & \frac{2EI_c}{L_c} & \frac{4EI_c}{L_c} & \frac{2EI_b}{L_b} & \frac{4EI_b}{L_b} \\
\hline
\frac{6EI_c}{L_c^2} & \frac{2EI_c}{L_c} & 0 & \frac{4EI_c}{L_c} & \frac{2EI_c}{L_c} & 0 & 0 & 0 & 0 \\
\frac{6EI_c}{L_c^2} & \frac{4EI_c}{L_c} & 0 & \frac{2EI_c}{L_c} & \frac{4EI_c}{L_c} & 0 & 0 & 0 & 0 \\
\frac{6EI_c}{L_c^2} & 0 & \frac{2EI_c}{L_c} & 0 & 0 & \frac{4EI_c}{L_c} & \frac{2EI_c}{L_c} & 0 & 0 \\
\frac{6EI_c}{L_c^2} & 0 & \frac{4EI_c}{L_c} & 0 & 0 & \frac{2EI_c}{L_c} & \frac{4EI_c}{L_c} & 0 & 0 \\
0 & \frac{4EI_b}{L_b} & \frac{2EI_b}{L_b} & 0 & 0 & 0 & 0 & \frac{4EI_b}{L_b} & \frac{2EI_b}{L_b} \\
0 & \frac{2EI_b}{L_b} & \frac{4EI_b}{L_b} & 0 & 0 & 0 & 0 & \frac{2EI_b}{L_b} & \frac{4EI_b}{L_b}
\end{array}\right]
\left\{\begin{array}{c} x_1 \\ x_2 \\ x_3 \\ \hline -\theta_1'' \\ -\theta_2'' \\ -\theta_3'' \\ -\theta_4'' \\ -\theta_5'' \\ -\theta_6'' \end{array}\right\}
=
\left\{\begin{array}{c} F_1 \\ F_2 \\ F_3 \\ \hline m_1 \\ m_2 \\ m_3 \\ m_4 \\ m_5 \\ m_6 \end{array}\right\}
\tag{2.64}$$

Example 2.5 Linear Solution to the One-Story Frame

Consider the one-story one-bay moment-resisting frame in Example 2.3 with the figure shown in Figure 2.6. The governing equation of the FAM is presented in Eq. (2.64). For simplicity, let $I_c = I_b = I$ and $L_c = L_b = L$. Assume that only a lateral force of F_o is applied at the horizontal degree of freedom x_1, this gives $F_1 = F_o$ and $F_2 = F_3 = 0$. Also, assume that the columns yield at $m_{yc} = F_oL/2$ with a post-yield stiffness of EI/L (i.e. hardening behavior), and the beam yields at $m_{yb} = F_oL/3$ with a post-yield stiffness of zero (i.e. elastic–plastic). This gives the following two moment versus plastic rotation relationships:

$$if \begin{cases} m_i \le m_{yc} \\ m_i > m_{yc} \end{cases}, \quad then \begin{cases} \theta_i'' = 0 \\ m_i = F_oL/2 + (EI/L)\,\theta_i'' \end{cases} \quad i = 1,2,3,4 \tag{2.65a}$$

$$if \begin{cases} m_i \le m_{yb} \\ m_i > m_{yb} \end{cases}, \quad then \begin{cases} \theta_i'' = 0 \\ m_i = F_oL/3 \end{cases} \quad i = 5,6 \tag{2.65b}$$

To analyze the structural behavior subjected to the applied lateral load, the first step is to assume that the response is linear, i.e. $\theta_1'' = \theta_2'' = \theta_3'' = \theta_4'' = \theta_5'' = \theta_6'' = 0$. As a result, Eq. (2.64) becomes

$$\begin{bmatrix} 24EI/L^3 & 6EI/L^2 & 6EI/L^2 \\ 6EI/L^2 & 8EI/L & 2EI/L \\ 6EI/L^2 & 2EI/L & 8EI/L \end{bmatrix} \begin{Bmatrix} x_1 \\ x_2 \\ x_3 \end{Bmatrix} = \begin{Bmatrix} F_o \\ 0 \\ 0 \end{Bmatrix} \tag{2.66}$$

Solving for the displacements gives

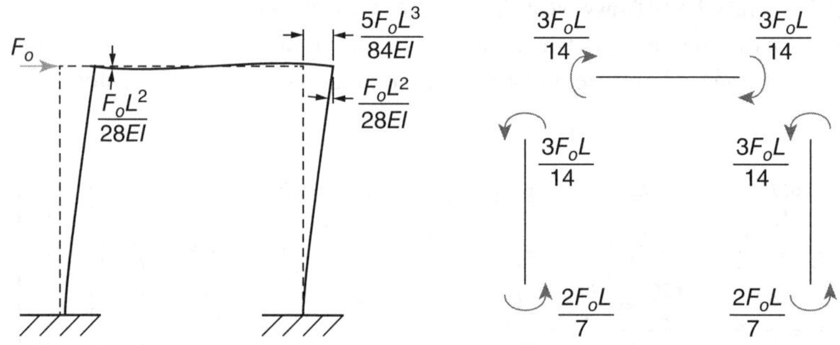

Figure 2.7 Graphical illustration of the response of linear one-story frame.

$$\begin{Bmatrix} x_1 \\ x_2 \\ x_3 \end{Bmatrix} = \begin{Bmatrix} 5F_oL^3/84EI \\ -F_oL^2/28EI \\ -F_oL^2/28EI \end{Bmatrix} \tag{2.67}$$

Substituting Eq. (2.67) back in Eq. (2.64) and solving for the moments gives

$$\begin{Bmatrix} m_1 \\ m_2 \\ m_3 \\ m_4 \\ m_5 \\ m_6 \end{Bmatrix} = \begin{bmatrix} 6EI/L^2 & 2EI/L & 0 \\ 6EI/L^2 & 4EI/L & 0 \\ 6EI/L^2 & 0 & 2EI/L \\ 6EI/L^2 & 0 & 4EI/L \\ 0 & 4EI/L & 2EI/L \\ 0 & 2EI/L & 4EI/L \end{bmatrix} \begin{Bmatrix} 5F_oL^3/84EI \\ -F_oL^2/28EI \\ -F_oL^2/28EI \end{Bmatrix} = \begin{Bmatrix} 2F_oL/7 \\ 3F_oL/14 \\ 2F_oL/7 \\ 3F_oL/14 \\ -3F_oL/14 \\ -3F_oL/14 \end{Bmatrix} \tag{2.68}$$

Since the moment demands in Eq. (2.68) are less than the corresponding yield moments, the assumption that all plastic rotations are zero in correct. Therefore, Eqs. (2.67) and (2.68) represent the solution to the problem, and Figure 2.7 summarizes these results.

Example 2.6 Nonlinear Solution to the One-Story Frame

Consider the same one-story one-bay moment-resisting frame used in Example 2.5 with the only difference being that the structure is now subjected to a lateral force of $2F_o$. This gives $F_1 = 2F_o$ and $F_2 = F_3 = 0$. Again, the first step is to assume that the response is linear, i.e. $\theta_1'' = \theta_2'' = \theta_3'' = \theta_4'' = \theta_5'' = \theta_6'' = 0$. As a result, Eq. (2.64) becomes

$$\begin{bmatrix} 24EI/L^3 & 6EI/L^2 & 6EI/L^2 \\ 6EI/L^2 & 8EI/L & 2EI/L \\ 6EI/L^2 & 2EI/L & 8EI/L \end{bmatrix} \begin{Bmatrix} x_1 \\ x_2 \\ x_3 \end{Bmatrix} = \begin{Bmatrix} 2F_o \\ 0 \\ 0 \end{Bmatrix} \tag{2.69}$$

Solving for the displacements gives

$$\begin{Bmatrix} x_1 \\ x_2 \\ x_3 \end{Bmatrix} = \begin{Bmatrix} 5F_oL^3/42EI \\ -F_oL^2/14EI \\ -F_oL^2/14EI \end{Bmatrix} \tag{2.70}$$

Substituting Eq. (2.70) back in Eq. (2.64) and solving for the moments gives

$$\begin{Bmatrix} m_1 \\ m_2 \\ m_3 \\ m_4 \\ m_5 \\ m_6 \end{Bmatrix} = \begin{bmatrix} 6EI/L^2 & 2EI/L & 0 \\ 6EI/L^2 & 4EI/L & 0 \\ 6EI/L^2 & 0 & 2EI/L \\ 6EI/L^2 & 0 & 4EI/L \\ 0 & 4EI/L & 2EI/L \\ 0 & 2EI/L & 4EI/L \end{bmatrix} \begin{Bmatrix} 5F_oL^3/42EI \\ -F_oL^2/14EI \\ -F_oL^2/14EI \end{Bmatrix} = \begin{Bmatrix} 4F_oL/7 \\ 3F_oL/7 \\ 4F_oL/7 \\ 3F_oL/7 \\ -3F_oL/7 \\ -3F_oL/7 \end{Bmatrix} \tag{2.71}$$

Comparing the moment demands in Eq. (2.71) with the corresponding yield moments show that m_1, m_3, m_5, and m_6 have all exceeded their corresponding yield moment. Therefore, the second step is to assume that plastic hinges are formed at these locations, and the resulting moment versus plastic rotation relationships follow from Eq. (2.65) as

$$m_1 = F_oL/2 + (EI/L)\,\theta_1'' \tag{2.72a}$$

$$m_3 = F_oL/2 + (EI/L)\,\theta_3'' \tag{2.72b}$$

$$m_5 = m_6 = -F_oL/3 \tag{2.72c}$$

$$\theta_2 = \theta_4 = 0 \tag{2.72d}$$

where the minus sign in front of $F_oL/3$ in Eq. (2.72c) denotes that the negative yield moment has been exceeded. Substituting Eq. (2.72) into Eq. (2.64) gives

$$\left[\begin{array}{ccc|cccc} 24EI/L^3 & 6EI/L^2 & 6EI/L^2 & 6EI/L^2 & 6EI/L^2 & 0 & 0 \\ 6EI/L^2 & 8EI/L & 2EI/L & 2EI/L & 0 & 4EI/L & 2EI/L \\ 6EI/L^2 & 2EI/L & 8EI/L & 0 & 2EI/L & 2EI/L & 4EI/L \\ \hline 6EI/L^2 & 2EI/L & 0 & 5EI/L & 0 & 0 & 0 \\ 6EI/L^2 & 0 & 2EI/L & 0 & 5EI/L & 0 & 0 \\ 0 & 4EI/L & 2EI/L & 0 & 0 & 4EI/L & 2EI/L \\ 0 & 2EI/L & 4EI/L & 0 & 0 & 2EI/L & 4EI/L \end{array}\right] \left\{\begin{array}{c} x_1 \\ x_2 \\ x_3 \\ \hline -\theta_1'' \\ -\theta_3'' \\ -\theta_5'' \\ \theta_6'' \end{array}\right\} = \left\{\begin{array}{c} 2F_o \\ 0 \\ 0 \\ \hline F_oL/2 \\ F_oL/2 \\ -F_oL/3 \\ F_oL/3 \end{array}\right\} \tag{2.73}$$

Note that the hardening stiffness of *EI*/*L* for PHLs #1 and #3 have been incorporated into Eq. (2.73) through the addition of stiffness in the corresponding diagonal terms of the stiffness matrix. Now solving for the displacements and plastic rotations in Eq. (2.73) gives

$$\left\{\begin{array}{c} x_1 \\ x_2 \\ x_3 \\ \hline -\theta_1'' \\ -\theta_3'' \\ -\theta_5'' \\ -\theta_6'' \end{array}\right\} = \left\{\begin{array}{c} F_oL^3/3EI \\ -F_oL^2/3EI \\ -F_oL^2/3EI \\ \hline -F_oL^2/6EI \\ -F_oL^2/6EI \\ 5F_oL^2/18EI \\ 5F_oL^2/18EI \end{array}\right\} \tag{2.74}$$

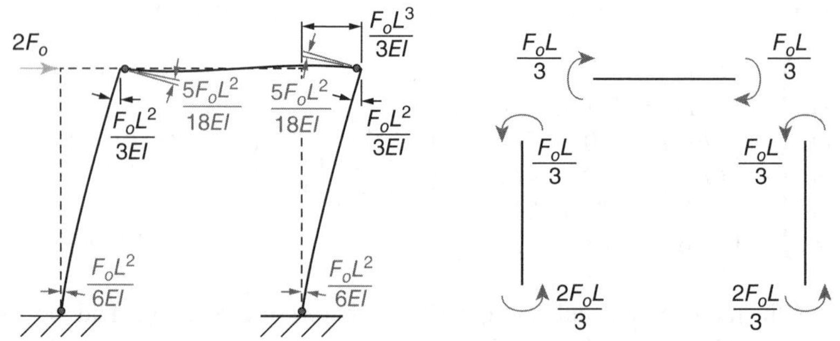

Figure 2.8 Graphical illustration of the response of nonlinear one-story frame.

Substituting Eq. (2.74) back in Eq. (2.64) and solving for the moments gives

$$\begin{Bmatrix} m_1 \\ m_2 \\ m_3 \\ m_4 \\ m_5 \\ m_6 \end{Bmatrix} = \left[\begin{array}{ccc:cccc} 6EI/L^2 & 2EI/L & 0 & 4EI/L & 0 & 0 & 0 \\ 6EI/L^2 & 4EI/L & 0 & 2EI/L & 0 & 0 & 0 \\ 6EI/L^2 & 0 & 2EI/L & 0 & 4EI/L & 0 & 0 \\ 6EI/L^2 & 0 & 4EI/L & 0 & 2EI/L & 0 & 0 \\ 0 & 4EI/L & 2EI/L & 0 & 0 & 4EI/L & 2EI/L \\ 0 & 2EI/L & 4EI/L & 0 & 0 & 2EI/L & 4EI/L \end{array}\right] \left\{\begin{array}{c} F_oL^3/3EI \\ -F_oL^2/3EI \\ -F_oL^2/3EI \\ \hdashline -F_oL^2/6EI \\ -F_oL^2/6EI \\ 5F_oL^2/18EI \\ 5F_oL^2/18EI \end{array}\right\} = \begin{Bmatrix} 2F_oL/3 \\ F_oL/3 \\ 2F_oL/3 \\ F_oL/3 \\ -F_oL/3 \\ -F_oL/3 \end{Bmatrix} \tag{2.75}$$

Since both moment demands for m_2 and m_4 are less than the corresponding yield moment, the assumption that PHLs #2 and #4 remain elastic as indicated in Eq. (2.72d) is correct. Therefore, in summary,

$$\begin{Bmatrix} x_1 \\ x_2 \\ x_3 \end{Bmatrix} = \begin{Bmatrix} F_oL^3/3EI \\ -F_oL^2/3EI \\ -F_oL^2/3EI \end{Bmatrix}, \begin{Bmatrix} \theta_1'' \\ \theta_2'' \\ \theta_3'' \\ \theta_4'' \\ \theta_5'' \\ \theta_6'' \end{Bmatrix} = \begin{Bmatrix} F_oL^2/6EI \\ 0 \\ F_oL^2/6EI \\ 0 \\ -5F_oL^2/18EI \\ -5F_oL^2/18EI \end{Bmatrix}, \begin{Bmatrix} m_1 \\ m_2 \\ m_3 \\ m_4 \\ m_5 \\ m_6 \end{Bmatrix} = \begin{Bmatrix} 2F_oL/3 \\ F_oL/3 \\ 2F_oL/3 \\ F_oL/3 \\ -F_oL/3 \\ -F_oL/3 \end{Bmatrix} \tag{2.76}$$

Note in Eq. (2.76) that positive moments m_1 and m_3 cause positive plastic rotations θ_1'' and θ_3'', and negative moments m_5 and m_6 cause negative plastic rotations θ_5'' and θ_6''. Also, note that joint equilibriums are satisfied when $m_2 = -m_5$ and $m_4 = -m_6$. Figure 2.8 summarizes these results.

2.5 Nonlinear Static Examples

Additional examples are here presented to demonstrate the use of the FAM for solving nonlinear static problems.

Example 2.7 Linear Column with Two Plastic Hinges

Consider that the SDOF column with two PHLs shown in Figure 2.9(a) is subjected to a lateral applied force F_o. Let the moment versus plastic rotation relationships of the two plastic hinges follow those plotted in Figure 2.9(b), where $m_{y1} = 2F_oL$, $m_{y2} = F_oL$, $K_{t1} = EI/L$, and $K_{t2} = 2EI/L$. These relationships can be summarized as:

$$\text{if } \begin{cases} m_1 \le 2F_oL \\ m_1 > 2F_oL \end{cases}, \quad \text{then } \begin{cases} \theta_1'' = 0 \\ m_1 = 2F_oL + (EI/L)\,\theta_1'' \end{cases} \tag{2.77a}$$

$$\text{if } \begin{cases} m_2 \le F_oL \\ m_2 > F_oL \end{cases}, \quad \text{then } \begin{cases} \theta_2'' = 0 \\ m_2 = F_oL + (2EI/L)\,\theta_2'' \end{cases} \tag{2.77b}$$

Based on the structure's configuration shown in Figure 2.9(a), the stiffness matrices are:

$$\mathbf{K} = \left[\frac{12EI}{L^3}\right], \quad \mathbf{K}' = \left[\frac{6EI}{L^2} \quad \frac{6EI}{L^2}\right], \quad \mathbf{K}'' = \begin{bmatrix} 4EI/L & 2EI/L \\ 2EI/L & 4EI/L \end{bmatrix} \tag{2.78}$$

and therefore,

$$\left[\begin{array}{c|cc} 12EI/L^3 & 6EI/L^2 & 6EI/L^2 \\ \hline 6EI/L^2 & 4EI/L & 2EI/L \\ 6EI/L^2 & 2EI/L & 4EI/L \end{array}\right] \left\{\begin{array}{c} x \\ \hline -\theta_1'' \\ -\theta_2'' \end{array}\right\} = \left\{\begin{array}{c} F_o \\ \hline m_1 \\ m_2 \end{array}\right\} \tag{2.79}$$

First assume that the structure is linear, this gives $\theta_1'' = \theta_2'' = 0$. Then using the first equation of Eq. (2.79) gives

$$\frac{12EI}{L^3} x = F_o \tag{2.80}$$

Solving for the displacement x in Eq. (2.80) gives

$$x = F_oL^3/12EI \tag{2.81}$$

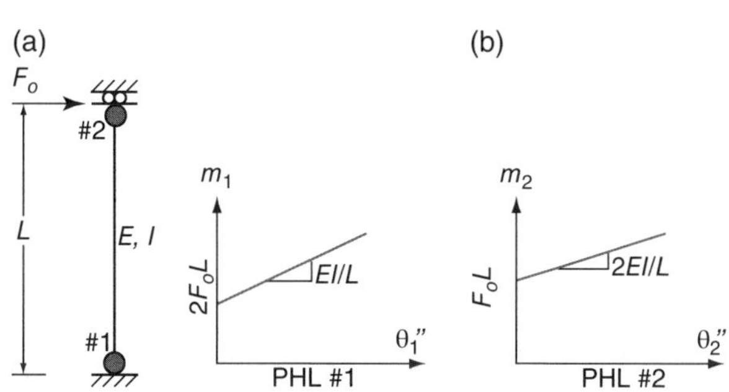

Figure 2.9 A SDOF system with two PHLs: (a) A single-degree-of-freedom system; (b) Moment versus plastic rotation relationships of the plastic hinges.

Then substituting Eq. (2.81) into the second and third equations of Eq. (2.79), the moments are calculated as

$$\begin{Bmatrix} m_1 \\ m_2 \end{Bmatrix} = \begin{bmatrix} 6EI/L^2 \\ 6EI/L^2 \end{bmatrix} \left(\frac{F_oL^3}{12EI} \right) = \begin{Bmatrix} F_oL/2 \\ F_oL/2 \end{Bmatrix} \tag{2.82}$$

Comparing these moment demands in Eq. (2.82) with the yield moments in Eq. (2.77) shows that the plastic hinges have not yielded, which means the linear structure assumption is correct. Therefore, the final displacement and moments are presented in Eqs. (2.81) and (2.82), respectively.

Example 2.8 Nonlinear Column with Yielding in One Plastic Hinge

Consider again the SDOF column as shown in Figure 2.9(a) with the moment versus plastic rotation relationships shown in Figure 2.9(b). Now let the lateral applied force be equal to $3F_o$. The matrix equation used for solving the problem is similar to Eq. (2.79) except for the applied load, i.e.

$$\left[\begin{array}{c|cc} 12EI/L^3 & 6EI/L^2 & 6EI/L^2 \\ \hline 6EI/L^2 & 4EI/L & 2EI/L \\ 6EI/L^2 & 2EI/L & 4EI/L \end{array} \right] \left\{ \begin{array}{c} x \\ \hline -\theta_1'' \\ -\theta_2'' \end{array} \right\} = \left\{ \begin{array}{c} 3F_o \\ \hline m_1 \\ m_2 \end{array} \right\} \tag{2.83}$$

Again, the structure is first assumed to be linear, this gives $\theta_1'' = \theta_2'' = 0$. Then using the first equation of Eq. (2.83) gives

$$\frac{12EI}{L^3} x = 3F_o \tag{2.84}$$

Solving for the displacement x in Eq. (2.84) gives

$$x = F_oL^3/4EI \tag{2.85}$$

Then substituting Eq. (2.85) into the second and third equations of Eq. (2.83), the moments are calculated as

$$\begin{Bmatrix} m_1 \\ m_2 \end{Bmatrix} = \begin{bmatrix} 6EI/L^2 \\ 6EI/L^2 \end{bmatrix} \left(\frac{F_oL^3}{4EI} \right) = \begin{Bmatrix} 3F_oL/2 \\ 3F_oL/2 \end{Bmatrix} \tag{2.86}$$

Comparing these moment demands in Eq. (2.86) with the yield moments in Eq. (2.77) shows that PHL #2 has yielded, but PHL #1 remains elastic. This means the original assumption that the structure is linear is incorrect.

Now assume that PHL #2 has yielded while PHL #1 remains elastic. This gives

$$\theta_1'' = 0, \quad m_2 = F_oL + (2EI/L)\,\theta_2'' \tag{2.87}$$

Substituting Eq. (2.87) into Eq. (2.83) and extracting the first and third equations of the matrix equation gives

$$\begin{bmatrix} 12EI/L^3 & 6EI/L^2 \\ 6EI/L^2 & 4EI/L+2EI/L \end{bmatrix} \begin{Bmatrix} x \\ -\theta_2'' \end{Bmatrix} = \begin{Bmatrix} 3F_o \\ F_oL \end{Bmatrix} \tag{2.88}$$

Solving for Eq. (2.88) gives

$$\begin{bmatrix} x \\ -\theta_2'' \end{bmatrix} = \begin{bmatrix} 12EI/L^3 & 6EI/L^2 \\ 6EI/L^2 & 6EI/L \end{bmatrix}^{-1} \begin{bmatrix} 3F_o \\ F_oL \end{bmatrix} = \begin{Bmatrix} F_oL^3/3EI \\ -F_oL^2/6EI \end{Bmatrix} \tag{2.89}$$

Then substituting Eq. (2.89) into the second and third equations of Eq. (2.83), the moments are calculated as

$$\begin{Bmatrix} m_1 \\ m_2 \end{Bmatrix} = \begin{bmatrix} 6EI/L^2 & 2EI/L \\ 6EI/L^2 & 4EI/L \end{bmatrix} \begin{Bmatrix} F_oL^3/3EI \\ -F_oL^2/6EI \end{Bmatrix} = \begin{Bmatrix} 5F_oL/3 \\ 4F_oL/3 \end{Bmatrix} \tag{2.90}$$

Comparing the moment m_1 with the corresponding yield moment in Eq. (2.77) shows that PHL #1 remains elastic. This means the original assumption that PHL #2 has yielded and PHL #1 remains elastic is correct. Therefore, in summary,

$$x = \frac{F_oL^3}{3EI}, \quad \begin{bmatrix} \theta_1'' \\ \theta_2'' \end{bmatrix} = \begin{Bmatrix} 0 \\ F_oL^2/6EI \end{Bmatrix}, \quad \begin{Bmatrix} m_1 \\ m_2 \end{Bmatrix} = \begin{Bmatrix} 5F_oL/3 \\ 4F_oL/3 \end{Bmatrix} \tag{2.91}$$

Example 2.9 Nonlinear Column with Yielding in All Plastic Hinges
Consider again the SDOF column as shown in Figure 2.9(a) with the moment versus plastic rotation relationships shown in Figure 2.9(b). Now let the lateral applied force be equal to $5F_o$. The matrix equation used for solving the problem is similar to Eq. (2.79) except for the applied load, i.e.

$$\left[\begin{array}{c:cc} 12EI/L^3 & 6EI/L^2 & 6EI/L^2 \\ \hdashline 6EI/L^2 & 4EI/L & 2EI/L \\ 6EI/L^2 & 2EI/L & 4EI/L \end{array}\right] \left[\begin{array}{c} x \\ \hdashline -\theta_1'' \\ -\theta_2'' \end{array}\right] = \left\{\begin{array}{c} 5F_o \\ \hdashline m_1 \\ m_2 \end{array}\right\} \tag{2.92}$$

Again, the structure is first assumed to be linear, this gives $\theta_1'' = \theta_2'' = 0$. Then using the first equation of Eq. (2.92) gives

$$\frac{12EI}{L^3}x = 5F_o \tag{2.93}$$

Solving for the displacement x in Eq. (2.93) gives

$$x = 5F_oL^3/12EI \tag{2.94}$$

Then substituting Eq. (2.94) into the second and third equations of Eq. (2.92), the moments are calculated as

$$\begin{Bmatrix} m_1 \\ m_2 \end{Bmatrix} = \begin{bmatrix} 6EI/L^2 \\ 6EI/L^2 \end{bmatrix} \left(\frac{5F_oL^3}{12EI} \right) = \begin{Bmatrix} 5F_oL/2 \\ 5F_oL/2 \end{Bmatrix} \tag{2.95}$$

Comparing these moment demands in Eq. (2.95) with the yield moments in Eq. (2.77) shows that both PHLs #1 and #2 have yielded. This means the original assumption that the structure is linear is incorrect.

Now assume that both PHLs #1 and #2 have yielded. This gives

$$m_1 = 2F_oL + (EI/L)\,\theta_1'', \quad m_2 = F_oL + (2EI/L)\,\theta_2'' \tag{2.96}$$

Substituting Eq. (2.96) into Eq. (2.92) gives

$$\begin{bmatrix} 12EI/L^3 & 6EI/L^2 & 6EI/L^2 \\ 6EI/L^2 & 4EI/L + EI/L & 2EI/L \\ 6EI/L^2 & 2EI/L & 4EI/L + 2EI/L \end{bmatrix} \begin{Bmatrix} x \\ -\theta_1'' \\ -\theta_2'' \end{Bmatrix} = \begin{Bmatrix} 5F_o \\ 2F_oL \\ F_oL \end{Bmatrix} \tag{2.97}$$

Solving for Eq. (2.97) gives

$$\begin{Bmatrix} x \\ -\theta_1'' \\ -\theta_2'' \end{Bmatrix} = \begin{bmatrix} 12EI/L^3 & 6EI/L^2 & 6EI/L^2 \\ 6EI/L^2 & 5EI/L & 2EI/L \\ 6EI/L^2 & 2EI/L & 6EI/L \end{bmatrix}^{-1} \begin{Bmatrix} 5F_o \\ 2F_oL \\ F_oL \end{Bmatrix} = \begin{Bmatrix} 16F_oL^3/15EI \\ -3F_oL^2/5EI \\ -7F_oL^2/10EI \end{Bmatrix} \tag{2.98}$$

Then substituting Eq. (2.98) into the second and third equations of Eq. (2.92), the moments are calculated as

$$\begin{Bmatrix} m_1 \\ m_2 \end{Bmatrix} = \begin{bmatrix} 6EI/L^2 & 4EI/L & 2EI/L \\ 6EI/L^2 & 2EI/L & 4EI/L \end{bmatrix} \begin{Bmatrix} 16F_oL^3/15EI \\ -3F_oL^2/5EI \\ -7F_oL^2/10EI \end{Bmatrix} = \begin{Bmatrix} 13F_oL/5 \\ 12F_oL/5 \end{Bmatrix} \tag{2.99}$$

These moments can similarly be obtained using Eq. (2.96), i.e.

$$m_1 = 2F_oL + \left(\frac{EI}{L}\right)\left(\frac{3F_oL^2}{5EI}\right) = \frac{13F_oL}{5} \tag{2.100a}$$

$$m_2 = F_oL + \left(\frac{2EI}{L}\right)\left(\frac{7F_oL^2}{10EI}\right) = \frac{12F_oL}{5} \tag{2.100b}$$

Comparing the moment demands m_1 and m_2 with the corresponding yield moments in Eq. (2.77) shows that both PHLs #1 and #2 have yielded. This means the original assumption is correct. Therefore, in summary,

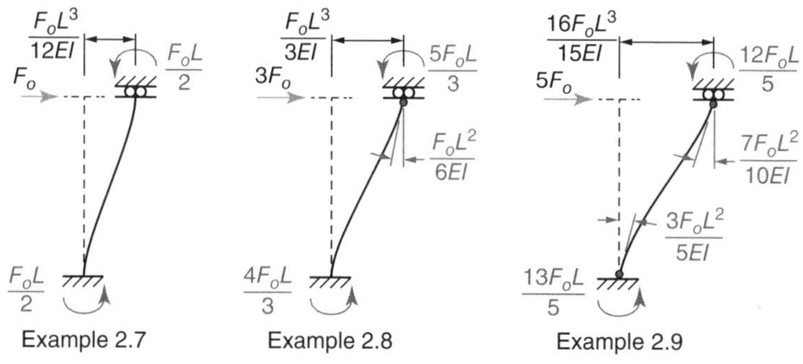

Figure 2.10 Graphical illustration of the responses of the SDOF system.

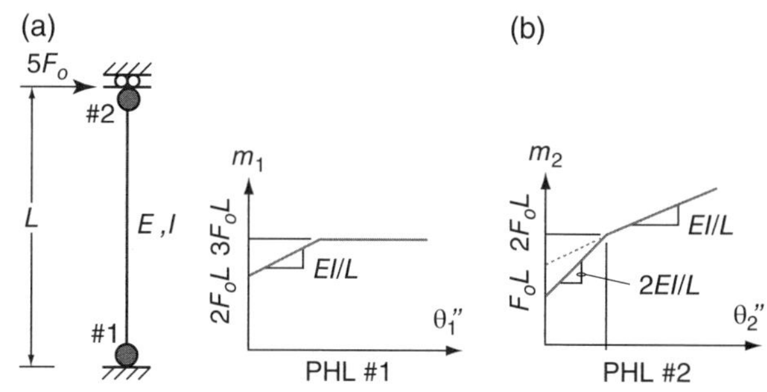

Figure 2.11 A SDOF system with two PHLs: (a) A single-degree-of-freedom system; (b) Moment versus plastic rotation relationships of the plastic hinges.

$$x = \frac{16F_oL^3}{15EI}, \quad \begin{bmatrix} \theta_1'' \\ \theta_2'' \end{bmatrix} = \begin{Bmatrix} 3F_oL^2/5EI \\ 7F_oL^2/10EI \end{Bmatrix}, \quad \begin{Bmatrix} m_1 \\ m_2 \end{Bmatrix} = \begin{Bmatrix} 13F_oL/5 \\ 12F_oL/5 \end{Bmatrix} \tag{2.101}$$

Figure 2.10 shows a summary of the results in Examples 2.7, 2.8, and 2.9.

Example 2.10 Nonlinear Column with Tri-linear Plastic Rotation Behavior
Consider that the SDOF column with two PHLs shown in Figure 2.11(a) is subjected to a lateral applied force $5F_o$. Let the moment versus plastic rotation relationships of the plastic hinges follow the tri-linear relationships plotted in Figure 2.11(b). These relationships can be expressed in equation forms as:

$$\text{if} \begin{cases} m_1 \le 2F_oL \\ 2F_oL < m_1 \le 3F_oL, \\ m_1 > 3F_oL \end{cases} \quad \text{then} \begin{cases} \theta_1'' = 0 \\ m_1 = 2F_oL + (EI/L)\,\theta_1'' \\ m_1 = 3F_oL \end{cases} \tag{2.102a}$$

$$\text{if} \begin{cases} m_2 \le F_oL \\ F_oL < m_2 \le 2F_oL, \\ m_2 > 2F_oL \end{cases} \quad \text{then} \begin{cases} \theta_2'' = 0 \\ m_2 = F_oL + (2EI/L)\,\theta_2'' \\ m_2 = 3F_oL/2 + (EI/L)\,\theta_2'' \end{cases} \tag{2.102b}$$

The matrix equation used for solving the problem can be obtained, just like the one in Eq. (2.92), as:

$$\left[\begin{array}{c|cc} 12EI/L^3 & 6EI/L^2 & 6EI/L^2 \\ \hline 6EI/L^2 & 4EI/L & 2EI/L \\ 6EI/L^2 & 2EI/L & 4EI/L \end{array}\right]\left\{\begin{array}{c} x \\ \hline -\theta_1'' \\ -\theta_2'' \end{array}\right\}=\left\{\begin{array}{c} 5F_o \\ \hline m_1 \\ m_2 \end{array}\right\} \tag{2.103}$$

The structure is first assumed to be linear, this gives $\theta_1''=\theta_2''=0$. Then using the first equation of Eq. (2.103) gives

$$\frac{12EI}{L^3}x=5F_o \tag{2.104}$$

Solving for the displacement x in Eq. (2.104) gives

$$x=5F_oL^3/12EI \tag{2.105}$$

Then substituting Eq. (2.105) into the second and third equations of Eq. (2.103), the moments are calculated as

$$\begin{Bmatrix} m_1 \\ m_2 \end{Bmatrix}=\begin{bmatrix} 6EI/L^2 \\ 6EI/L^2 \end{bmatrix}\left(\frac{5F_oL^3}{12EI}\right)=\begin{Bmatrix} 5F_oL/2 \\ 5F_oL/2 \end{Bmatrix} \tag{2.106}$$

Comparing these moment demands in Eq. (2.106) with the yield moments in Eq. (2.102) shows that both PHLs #1 and #2 have yielded. This means the original assumption that the structure is linear is incorrect.

Now assume that both PHLs #1 and #2 have yielded. Based on the moment demands in Eq. (2.106), it is also assumed that PHL #1 is at the first nonlinear slope and PHL #2 is at the second nonlinear slope. According to Eq. (2.102), this gives

$$m_1=2F_oL+(EI/L)\,\theta_1'', \qquad m_2=3F_oL/2+(EI/L)\,\theta_2'' \tag{2.107}$$

Substituting Eq. (2.107) into Eq. (2.103) gives

$$\begin{bmatrix} 12EI/L^3 & 6EI/L^2 & 6EI/L^2 \\ 6EI/L^2 & 4EI/L+EI/L & 2EI/L \\ 6EI/L^2 & 2EI/L & 4EI/L+EI/L \end{bmatrix}\begin{Bmatrix} x \\ -\theta_1'' \\ -\theta_2'' \end{Bmatrix}=\begin{Bmatrix} 5F_o \\ 2F_oL \\ 3F_oL/2 \end{Bmatrix} \tag{2.108}$$

Solving for Eq. (2.108) gives

$$\begin{Bmatrix} x \\ -\theta_1'' \\ -\theta_2'' \end{Bmatrix}=\begin{bmatrix} 12EI/L^3 & 6EI/L^2 & 6EI/L^2 \\ 6EI/L^2 & 5EI/L & 2EI/L \\ 6EI/L^2 & 2EI/L & 5EI/L \end{bmatrix}^{-1}\begin{Bmatrix} 5F_o \\ 2F_oL \\ 3F_oL/2 \end{Bmatrix}=\begin{Bmatrix} 7F_oL^3/6EI \\ -2F_oL^2/3EI \\ -5F_oL^2/6EI \end{Bmatrix} \tag{2.109}$$

Then substituting Eq. (2.109) into the second and third equations of Eq. (2.103), the moments are calculated as

$$\begin{Bmatrix} m_1 \\ m_2 \end{Bmatrix} = \begin{bmatrix} 6EI/L^2 & 4EI/L & 2EI/L \\ 6EI/L^2 & 2EI/L & 4EI/L \end{bmatrix} \begin{Bmatrix} 7F_oL^3/6EI \\ -2F_oL^2/3EI \\ -5F_oL^2/6EI \end{Bmatrix} = \begin{Bmatrix} 8F_oL/3 \\ 7F_oL/3 \end{Bmatrix} \tag{2.110}$$

These moments can similarly be obtained using Eq. (2.107), i.e.

$$m_1 = 2F_oL + \left(\frac{EI}{L}\right)\left(\frac{2F_oL^2}{3EI}\right) = \frac{8F_oL}{3} \tag{2.111a}$$

$$m_2 = \frac{3F_oL}{2} + \left(\frac{EI}{L}\right)\left(\frac{5F_oL^2}{6EI}\right) = \frac{7F_oL}{3} \tag{2.111b}$$

Comparing the moment demands m_1 and m_2 with the corresponding yield moments in Eq. (2.102) shows that PHL #1 is responding at the first nonlinear slope and PHL #2 is responding at the second nonlinear slope. This means the original assumption is correct. Therefore, in summary,

$$x = \frac{7F_oL^3}{6EI}, \quad \begin{bmatrix} \theta_1'' \\ \theta_2'' \end{bmatrix} = \begin{Bmatrix} 2F_oL^2/3EI \\ 5F_oL^2/6EI \end{Bmatrix}, \quad \begin{Bmatrix} m_1 \\ m_2 \end{Bmatrix} = \begin{Bmatrix} 8F_oL/3 \\ 7F_oL/3 \end{Bmatrix} \tag{2.112}$$

Example 2.11 Nonlinear Column with Softening Behavior

Consider again that the SDOF column with two PHLs shown in Figure 2.11(a) is subjected to a lateral applied force $5F_o$. But now, let the moment versus plastic rotation relationships of the plastic hinges follow:

$$if \begin{cases} m_1 \le 2F_oL \\ m_1 > 2F_oL \end{cases}, \quad then \begin{cases} \theta_1'' = 0 \\ m_1 = 2F_oL + (2EI/L)\,\theta_1'' \end{cases} \tag{2.113a}$$

$$if \begin{cases} m_2 \le F_oL \\ F_oL < m_2 \le 2F_oL \\ m_2 > 2F_oL \end{cases}, \quad then \begin{cases} \theta_2'' = 0 \\ m_2 = F_oL + (2EI/L)\,\theta_2'' \\ m_2 = 9F_oL/4 - (EI/2L)\,\theta_2'' \end{cases} \tag{2.113b}$$

Note that PHL #1 exhibits a bilinear hardening behavior, while PHL #2 exhibits a tri-linear softening behavior with a slope of $-EI/2L$ in the second nonlinear slope.

The matrix equation used for solving the problem can again be obtained as:

$$\left[\begin{array}{c|cc} 12EI/L^3 & 6EI/L^2 & 6EI/L^2 \\ \hline 6EI/L^2 & 4EI/L & 2EI/L \\ 6EI/L^2 & 2EI/L & 4EI/L \end{array}\right] \begin{Bmatrix} x \\ \hline -\theta_1'' \\ -\theta_2'' \end{Bmatrix} = \begin{Bmatrix} 5F_o \\ \hline m_1 \\ m_2 \end{Bmatrix} \tag{2.114}$$

The structure is first assumed to be linear, this gives $\theta_1'' = \theta_2'' = 0$. Then using the first equation of Eq. (2.114) gives

$$\frac{12EI}{L^3}x = 5F_o \tag{2.115}$$

Solving for the displacement x gives

$$x = 5F_oL^3/12EI \tag{2.116}$$

Then substituting Eq. (2.116) into the second and third equations of Eq. (2.114), the moments are calculated as

$$\begin{Bmatrix} m_1 \\ m_2 \end{Bmatrix} = \begin{bmatrix} 6EI/L^2 \\ 6EI/L^2 \end{bmatrix} \left(\frac{5F_oL^3}{12EI}\right) = \begin{Bmatrix} 5F_oL/2 \\ 5F_oL/2 \end{Bmatrix} \tag{2.117}$$

Comparing these moment demands in Eq. (2.117) with the yield moments in Eq. (2.113) shows that both PHLs #1 and #2 have yielded. This means the original assumption that the structure is linear is incorrect.

Now assume that both PHLs #1 and #2 have yielded. Based on the moment demands in Eq. (2.117), it is seen that the moment demand at PHL #2 has exceeded its maximum moment capacity, and therefore PHL #2 is assumed to reach its softening stage (i.e. the second nonlinear slope). According to Eq. (2.113), this gives

$$m_1 = 2F_oL + (2EI/L)\,\theta_1'', \quad m_2 = 9F_oL/4 - (EI/2L)\,\theta_2'' \tag{2.118}$$

Substituting Eq. (2.118) into Eq. (2.114) gives

$$\begin{bmatrix} 12EI/L^3 & 6EI/L^2 & 6EI/L^2 \\ 6EI/L^2 & 4EI/L + 2EI/L & 2EI/L \\ 6EI/L^2 & 2EI/L & 4EI/L - EI/2L \end{bmatrix} \begin{Bmatrix} x \\ -\theta_1'' \\ -\theta_2'' \end{Bmatrix} = \begin{Bmatrix} 5F_o \\ 2F_oL \\ 9F_oL/4 \end{Bmatrix} \tag{2.119}$$

Solving for Eq. (2.119) gives

$$\begin{Bmatrix} x \\ -\theta_1'' \\ -\theta_2'' \end{Bmatrix} = \begin{bmatrix} 12EI/L^3 & 6EI/L^2 & 6EI/L^2 \\ 6EI/L^2 & 6EI/L & 2EI/L \\ 6EI/L^2 & 2EI/L & 7EI/2L \end{bmatrix}^{-1} \begin{Bmatrix} 5F_o \\ 2F_oL \\ 9F_oL/4 \end{Bmatrix} = \begin{Bmatrix} 13F_oL^3/6EI \\ -F_oL^2/EI \\ -5F_oL^2/2EI \end{Bmatrix} \tag{2.120}$$

Then substituting Eq. (2.120) into the second and third equations of Eq. (2.114), the moments are calculated as

$$\begin{Bmatrix} m_1 \\ m_2 \end{Bmatrix} = \begin{bmatrix} 6EI/L^2 & 4EI/L & 2EI/L \\ 6EI/L^2 & 2EI/L & 4EI/L \end{bmatrix} \begin{Bmatrix} 13F_oL^3/6EI \\ -F_oL^2/EI \\ -5F_oL^2/2EI \end{Bmatrix} = \begin{Bmatrix} 4F_oL \\ F_oL \end{Bmatrix} \tag{2.121}$$

These moments can similarly be obtained using Eq. (2.118), i.e.

$$m_1 = 2F_oL + \left(\frac{2EI}{L}\right)\left(\frac{F_oL^2}{EI}\right) = 4F_oL \tag{2.122a}$$

$$m_2 = \frac{9F_oL}{4} - \left(\frac{EI}{2L}\right)\left(\frac{5F_oL^2}{2EI}\right) = F_oL \tag{2.122b}$$

Comparing the moment demand for m_1 with the corresponding yield moment in Eq. (2.113) shows that PHL #1 has yielded. In addition, the combination of plastic rotation θ_2'' in Eq. (2.120) and moment demand m_2 in Eq. (2.121) indicates that the plastic hinge is responding in the softening stage and has not reached the limiting value of zero moment. Therefore, this means the original assumption is correct. In summary,

$$x = \frac{13F_oL^3}{6EI}, \quad \begin{bmatrix}\theta_1''\\ \theta_2''\end{bmatrix} = \begin{Bmatrix} F_oL^2/EI \\ 5F_oL^2/2EI\end{Bmatrix}, \quad \begin{Bmatrix} m_1 \\ m_2 \end{Bmatrix} = \begin{Bmatrix} 4F_oL \\ F_oL \end{Bmatrix} \tag{2.123}$$

Example 2.12 Nonlinear Two-Story Frame

Consider the two-story moment-resisting frame as shown in Figure 2.12. Assume that all the members are axially rigid, which therefore gives a total of 6 DOFs (i.e. $n = 6$, with 2 translations and 4 rotations) and 12 PHLs (i.e. $q = 12$). All the beams and columns have the same elastic modulus E. All columns have length L and moment of inertia I, while the beams have length $2L$ and moment of inertia $2I$. The frame is subjected to a lateral force F_o on the second floor and another lateral force F_o on the third floor.

For this two-story frame, the displacement, force, moment, and plastic rotation vectors are:

$$\mathbf{x} = \{x_1\ x_2\ x_3\ x_4\ x_5\ x_6\}^T \tag{2.124a}$$

$$\mathbf{F} = \{F_o\ F_o\ 0\ 0\ 0\ 0\}^T \tag{2.124b}$$

$$\mathbf{\Theta}'' = \{\theta_1''\ \theta_2''\ \theta_3''\ \theta_4''\ \theta_5''\ \theta_6''\ \theta_7''\ \theta_8''\ \theta_9''\ \theta_{10}''\ \theta_{11}''\ \theta_{12}''\}^T \tag{2.124c}$$

$$\mathbf{m} = \{m_1\ m_2\ m_3\ m_4\ m_5\ m_6\ m_7\ m_8\ m_9\ m_{10}\ m_{11}\ m_{12}\}^T \tag{2.124d}$$

The 6×6 stiffness matrix $\mathbf{K}$, the 6×12 stiffness matrix $\mathbf{K}'$, and the 12×12 stiffness matrix $\mathbf{K}''$ are:

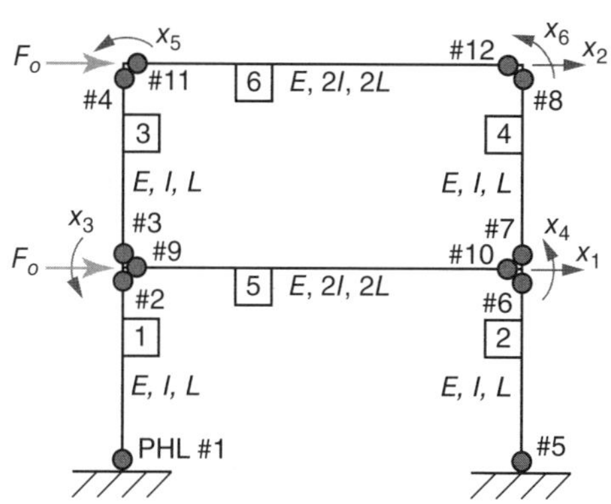

Figure 2.12 A two-story frame with 12 PHLs.

$$
\mathbf{K}=\begin{bmatrix}
48EI/L^3 & -24EI/L^3 & 0 & 0 & -6EI/L^2 & -6EI/L^2 \\
-24EI/L^3 & 24EI/L^3 & 6EI/L^2 & 6EI/L^2 & 6EI/L^2 & 6EI/L^2 \\
0 & 6EI/L^2 & 12EI/L & 2EI/L & 2EI/L & 0 \\
0 & 6EI/L^2 & 2EI/L & 12EI/L & 0 & 2EI/L \\
-6EI/L^2 & 6EI/L^2 & 2EI/L & 0 & 8EI/L & 2EI/L \\
-6EI/L^2 & 6EI/L^2 & 0 & 2EI/L & 2EI/L & 8EI/L
\end{bmatrix} \tag{2.125a}
$$

$$
\mathbf{K}'=\begin{bmatrix}
\frac{6EI}{L^2} & \frac{6EI}{L^2} & -\frac{6EI}{L^2} & -\frac{6EI}{L^2} & \frac{6EI}{L^2} & \frac{6EI}{L^2} & -\frac{6EI}{L^2} & -\frac{6EI}{L^2} & 0 & 0 & 0 & 0 \\
0 & 0 & \frac{6EI}{L^2} & \frac{6EI}{L^2} & 0 & 0 & \frac{6EI}{L^2} & \frac{6EI}{L^2} & 0 & 0 & 0 & 0 \\
\frac{2EI}{L} & \frac{4EI}{L} & \frac{4EI}{L} & \frac{2EI}{L} & 0 & 0 & 0 & 0 & \frac{4EI}{L} & \frac{2EI}{L} & 0 & 0 \\
0 & 0 & 0 & 0 & \frac{2EI}{L} & \frac{4EI}{L} & \frac{4EI}{L} & \frac{2EI}{L} & \frac{2EI}{L} & \frac{4EI}{L} & 0 & 0 \\
0 & 0 & \frac{2EI}{L} & \frac{4EI}{L} & 0 & 0 & 0 & 0 & 0 & 0 & \frac{4EI}{L} & \frac{2EI}{L} \\
0 & 0 & 0 & 0 & 0 & 0 & \frac{2EI}{L} & \frac{4EI}{L} & 0 & 0 & \frac{2EI}{L} & \frac{4EI}{L}
\end{bmatrix} \tag{2.125b}
$$

$$
\mathbf{K}''=\left[\begin{array}{cc:cc:cc:cc:cc:cc}
\frac{4EI}{L} & \frac{2EI}{L} & 0 & 0 & 0 & 0 & 0 & 0 & 0 & 0 & 0 & 0 \\
\frac{2EI}{L} & \frac{4EI}{L} & 0 & 0 & 0 & 0 & 0 & 0 & 0 & 0 & 0 & 0 \\
\hdashline
0 & 0 & \frac{4EI}{L} & \frac{2EI}{L} & 0 & 0 & 0 & 0 & 0 & 0 & 0 & 0 \\
0 & 0 & \frac{2EI}{L} & \frac{4EI}{L} & 0 & 0 & 0 & 0 & 0 & 0 & 0 & 0 \\
\hdashline
0 & 0 & 0 & 0 & \frac{4EI}{L} & \frac{2EI}{L} & 0 & 0 & 0 & 0 & 0 & 0 \\
0 & 0 & 0 & 0 & \frac{2EI}{L} & \frac{4EI}{L} & 0 & 0 & 0 & 0 & 0 & 0 \\
\hdashline
0 & 0 & 0 & 0 & 0 & 0 & \frac{4EI}{L} & \frac{2EI}{L} & 0 & 0 & 0 & 0 \\
0 & 0 & 0 & 0 & 0 & 0 & \frac{2EI}{L} & \frac{4EI}{L} & 0 & 0 & 0 & 0 \\
\hdashline
0 & 0 & 0 & 0 & 0 & 0 & 0 & 0 & \frac{4EI}{L} & \frac{2EI}{L} & 0 & 0 \\
0 & 0 & 0 & 0 & 0 & 0 & 0 & 0 & \frac{2EI}{L} & \frac{4EI}{L} & 0 & 0 \\
\hdashline
0 & 0 & 0 & 0 & 0 & 0 & 0 & 0 & 0 & 0 & \frac{4EI}{L} & \frac{2EI}{L} \\
0 & 0 & 0 & 0 & 0 & 0 & 0 & 0 & 0 & 0 & \frac{2EI}{L} & \frac{4EI}{L}
\end{array}\right] \tag{2.125c}
$$

and finally the governing equation of the FAM can be written as:

$$
\left[\begin{array}{c:c}
\mathbf{K}_{6\times6} & \mathbf{K}'_{6\times12} \\
\hdashline
\mathbf{K}'^{T}_{12\times6} & \mathbf{K}''_{12\times12}
\end{array}\right]
\left\{\begin{array}{c}
\mathbf{x}_{6\times1} \\
\hdashline
-\boldsymbol{\Theta}''_{12\times1}
\end{array}\right\}
=
\left\{\begin{array}{c}
\mathbf{F}_{6\times1} \\
\hdashline
\mathbf{m}_{12\times1}
\end{array}\right\} \tag{2.126}
$$

Assume that the plastic hinges for the beams and columns take the following moment versus plastic rotation relationships:

$$if\ \begin{cases} m_i \le F_oL/2 \\ m_i > F_oL/2 \end{cases},\quad then\ \begin{cases} \theta_i'' = 0 \\ m_i = F_oL/2 + (2EI/L)\,\theta_i'' \end{cases}\quad i = 1,\ldots,8 \tag{2.127a}$$

$$if\ \begin{cases} m_i \le 2F_oL/5 \\ m_i > 2F_oL/5 \end{cases},\quad then\ \begin{cases} \theta_i'' = 0 \\ m_i = 2F_oL/5 + (EI/L)\,\theta_i'' \end{cases}\quad i = 9,\ldots,12 \tag{2.127b}$$

where $i = 1, \ldots, 8$ denotes PHLs #1 to #8 in the columns, and $i = 9, \ldots, 12$ denotes PHLs #9 to #12 in the beams.

The structure is first assumed to be linear, this gives $\boldsymbol{\Theta}'' = \mathbf{0}$. Then using the first matrix equation of Eq. (2.126) gives

$$\begin{bmatrix} 48EI/L^3 & -24EI/L^3 & 0 & 0 & -6EI/L^2 & -6EI/L^2 \\ -24EI/L^3 & 24EI/L^3 & 6EI/L^2 & 6EI/L^2 & 6EI/L^2 & 6EI/L^2 \\ 0 & 6EI/L^2 & 12EI/L & 2EI/L & 2EI/L & 0 \\ 0 & 6EI/L^2 & 2EI/L & 12EI/L & 0 & 2EI/L \\ -6EI/L^2 & 6EI/L^2 & 2EI/L & 0 & 8EI/L & 2EI/L \\ -6EI/L^2 & 6EI/L^2 & 0 & 2EI/L & 2EI/L & 8EI/L \end{bmatrix} \begin{Bmatrix} x_1 \\ x_2 \\ x_3 \\ x_4 \\ x_5 \\ x_6 \end{Bmatrix} = \begin{Bmatrix} F_o \\ F_o \\ 0 \\ 0 \\ 0 \\ 0 \end{Bmatrix} \tag{2.128}$$

Solving for the displacements $\mathbf{x}$ gives

$$\begin{Bmatrix} x_1 \\ x_2 \\ x_3 \\ x_4 \\ x_5 \\ x_6 \end{Bmatrix} = \begin{Bmatrix} 2F_oL^3/15EI \\ F_oL^3/4EI \\ -F_oL^2/10EI \\ -F_oL^2/10EI \\ -F_oL^2/20EI \\ -F_oL^2/20EI \end{Bmatrix} \tag{2.129}$$

Then substituting Eq. (2.129) into the second matrix equation of Eq. (2.126), the moments are calculated as

$$\begin{Bmatrix} m_1 \\ m_2 \\ m_3 \\ m_4 \\ m_5 \\ m_6 \\ m_7 \\ m_8 \\ m_9 \\ m_{10} \\ m_{11} \\ m_{12} \end{Bmatrix} = F_oL \begin{Bmatrix} 0.6 \\ 0.4 \\ 0.2 \\ 0.3 \\ 0.6 \\ 0.4 \\ 0.2 \\ 0.3 \\ -0.6 \\ -0.6 \\ -0.3 \\ -0.3 \end{Bmatrix},\quad \begin{Bmatrix} m_{y1} \\ m_{y2} \\ m_{y3} \\ m_{y4} \\ m_{y5} \\ m_{y6} \\ m_{y7} \\ m_{y8} \\ m_{y9} \\ m_{y10} \\ m_{y11} \\ m_{y12} \end{Bmatrix} = F_oL \begin{Bmatrix} 0.5 \\ 0.5 \\ 0.5 \\ 0.5 \\ 0.5 \\ 0.5 \\ 0.5 \\ 0.5 \\ -0.4 \\ -0.4 \\ -0.4 \\ -0.4 \end{Bmatrix} \tag{2.130}$$

The corresponding yield moments of the 12 PHLs are also presented in Eq. (2.130) as a comparison. It can be seen that the moment demands m_1, m_5, m_9, and m_{10} exceed their corresponding yield moments. This means the original assumption that the structure is linear is incorrect.

Now assume that PHLs #1, #5, #9, and #10 have yielded. This gives

$$\theta_2''=\theta_3''=\theta_4''=\theta_6''=\theta_7''=\theta_8''=\theta_{11}''=\theta_{12}''=0 \tag{2.131a}$$

$$m_1=F_oL/2+(2EI/L)\,\theta_1'',\quad m_5=F_oL/2+(2EI/L)\,\theta_5'' \tag{2.131b}$$

$$m_9=-2F_oL/5+(EI/L)\,\theta_9'',\quad m_{10}=-2F_oL/5+(EI/L)\,\theta_{10}'' \tag{2.131c}$$

where the minus signs in front of $2F_oL/5$ in Eq. (2.131c) denote that the negative yield moments have been exceeded. From Eq. (2.126), it then follows that

$$\left[\begin{array}{cccccc|cccc}
\frac{48EI}{L^3} & -\frac{24EI}{L^3} & 0 & 0 & -\frac{6EI}{L^2} & -\frac{6EI}{L^2} & \frac{6EI}{L^2} & \frac{6EI}{L^2} & 0 & 0\\
-\frac{24EI}{L^3} & \frac{24EI}{L^3} & \frac{6EI}{L^2} & \frac{6EI}{L^2} & \frac{6EI}{L^2} & \frac{6EI}{L^2} & 0 & 0 & 0 & 0\\
0 & \frac{6EI}{L^2} & \frac{12EI}{L} & \frac{2EI}{L} & \frac{2EI}{L} & 0 & \frac{2EI}{L} & 0 & \frac{4EI}{L} & \frac{2EI}{L}\\
0 & \frac{6EI}{L^2} & \frac{2EI}{L} & \frac{12EI}{L} & 0 & \frac{2EI}{L} & 0 & \frac{2EI}{L} & \frac{2EI}{L} & \frac{4EI}{L}\\
-\frac{6EI}{L^2} & \frac{6EI}{L^2} & \frac{2EI}{L} & 0 & \frac{8EI}{L} & \frac{2EI}{L} & 0 & 0 & 0 & 0\\
-\frac{6EI}{L^2} & \frac{6EI}{L^2} & 0 & \frac{2EI}{L} & \frac{2EI}{L} & \frac{8EI}{L} & 0 & 0 & 0 & 0\\
\hline
\frac{6EI}{L^2} & 0 & \frac{2EI}{L} & 0 & 0 & 0 & \frac{6EI}{L} & 0 & 0 & 0\\
\frac{6EI}{L^2} & 0 & 0 & \frac{2EI}{L} & 0 & 0 & 0 & \frac{6EI}{L} & 0 & 0\\
0 & 0 & \frac{4EI}{L} & \frac{2EI}{L} & 0 & 0 & 0 & 0 & \frac{5EI}{L} & \frac{2EI}{L}\\
0 & 0 & \frac{2EI}{L} & \frac{4EI}{L} & 0 & 0 & 0 & 0 & \frac{2EI}{L} & \frac{5EI}{L}
\end{array}\right]
\left\{\begin{array}{c} x_1\\ x_2\\ x_3\\ x_4\\ x_5\\ x_6\\ \hline -\theta_1''\\ -\theta_5''\\ -\theta_9''\\ -\theta_{10}'' \end{array}\right\}
=
\left\{\begin{array}{c} F_o\\ F_o\\ 0\\ 0\\ 0\\ 0\\ \hline F_oL/2\\ F_oL/2\\ -2F_oL/2\\ -2F_oL/5 \end{array}\right\} \tag{2.132}$$

Solving for Eq. (2.132) gives

$$\left\{\begin{array}{c} x_1\\ x_2\\ x_3\\ x_4\\ x_5\\ x_6\\ \hline -\theta_1''\\ -\theta_5''\\ -\theta_9''\\ -\theta_{10}'' \end{array}\right\}
=
\left\{\begin{array}{c} 0.2073\times F_oL^3/EI\\ 0.3731\times F_oL^3/EI\\ -0.1860\times F_oL^2/EI\\ -0.1860\times F_oL^2/EI\\ -0.0623\times F_oL^2/EI\\ -0.0623\times F_oL^2/EI\\ \hline -0.0620\times F_oL^2/EI\\ -0.0620\times F_oL^2/EI\\ 0.1023\times F_oL^2/EI\\ 0.1023\times F_oL^2/EI \end{array}\right\} \tag{2.133}$$

Then substituting Eq. (2.133) into the second matrix equation of Eq. (2.126), the moments are calculated as

$$\begin{Bmatrix} m_1 \\ m_2 \\ m_3 \\ m_4 \\ m_5 \\ m_6 \\ m_7 \\ m_8 \\ m_9 \\ m_{10} \\ m_{11} \\ m_{12} \end{Bmatrix} = F_oL \begin{Bmatrix} 0.6240 \\ 0.3760 \\ 0.1263 \\ 0.3737 \\ 0.6240 \\ 0.3760 \\ 0.1263 \\ 0.3737 \\ -0.5023 \\ -0.5023 \\ -0.3737 \\ -0.3737 \end{Bmatrix}, \begin{Bmatrix} m_{y1} \\ m_{y2} \\ m_{y3} \\ m_{y4} \\ m_{y5} \\ m_{y6} \\ m_{y7} \\ m_{y8} \\ m_{y9} \\ m_{y10} \\ m_{y11} \\ m_{y12} \end{Bmatrix} = F_oL \begin{Bmatrix} 0.5 \\ 0.5 \\ 0.5 \\ 0.5 \\ 0.5 \\ 0.5 \\ 0.5 \\ 0.5 \\ -0.4 \\ -0.4 \\ -0.4 \\ -0.4 \end{Bmatrix} \tag{2.134}$$

The corresponding yield moments of the 12 PHLs are also presented in Eq. (2.134) as a comparison. The moment demands in Eq. (2.134) can similarly be obtained using Eqs. (2.131b) and (2.131c), i.e.

$$m_1 = \frac{F_oL}{2} + \left(\frac{2EI}{L}\right)\left(0.0620 \times \frac{F_oL^2}{EI}\right) = 0.6240 \times F_oL \tag{2.135a}$$

$$m_5 = \frac{F_oL}{2} + \left(\frac{2EI}{L}\right)\left(0.0620 \times \frac{F_oL^2}{EI}\right) = 0.6240 \times F_oL \tag{2.135b}$$

$$m_9 = -\frac{2F_oL}{5} + \left(\frac{EI}{L}\right)\left(-0.1023 \times \frac{F_oL^2}{EI}\right) = -0.5023 \times F_oL \tag{2.135c}$$

$$m_{10} = -\frac{2F_oL}{5} + \left(\frac{EI}{L}\right)\left(-0.1023 \times \frac{F_oL^2}{EI}\right) = -0.5023 \times F_oL \tag{2.135d}$$

Comparing the moment demands m_2, m_3, m_4, m_6, m_7, m_8, m_{11}, and m_{12} with the corresponding yield moments in Eq. (2.134) shows that all these moment demands are less than the corresponding yield moments. This means the original assumption is correct. Therefore, in summary,

$$\begin{Bmatrix} x_1 \\ x_2 \\ x_3 \\ x_4 \\ x_5 \\ x_6 \end{Bmatrix} = \begin{Bmatrix} 0.2073 \times F_oL^3/EI \\ 0.3731 \times F_oL^3/EI \\ -0.1860 \times F_oL^2/EI \\ -0.1860 \times F_oL^2/EI \\ -0.0623 \times F_oL^2/EI \\ -0.0623 \times F_oL^2/EI \end{Bmatrix}, \begin{Bmatrix} \theta''_1 \\ \theta''_2 \\ \theta''_3 \\ \theta''_4 \\ \theta''_5 \\ \theta''_6 \\ \theta''_7 \\ \theta''_8 \\ \theta''_9 \\ \theta''_{10} \\ \theta''_{11} \\ \theta''_{12} \end{Bmatrix} = \frac{F_oL^2}{EI} \begin{Bmatrix} 0.0620 \\ 0 \\ 0 \\ 0 \\ 0.0620 \\ 0 \\ 0 \\ 0 \\ -0.1023 \\ -0.1023 \\ 0 \\ 0 \end{Bmatrix}, \begin{Bmatrix} m_1 \\ m_2 \\ m_3 \\ m_4 \\ m_5 \\ m_6 \\ m_7 \\ m_8 \\ m_9 \\ m_{10} \\ m_{11} \\ m_{12} \end{Bmatrix} = F_oL \begin{Bmatrix} 0.6240 \\ 0.3760 \\ 0.1263 \\ 0.3737 \\ 0.6240 \\ 0.3760 \\ 0.1263 \\ 0.3737 \\ -0.5023 \\ -0.5023 \\ -0.3737 \\ -0.3737 \end{Bmatrix} \tag{2.136}$$

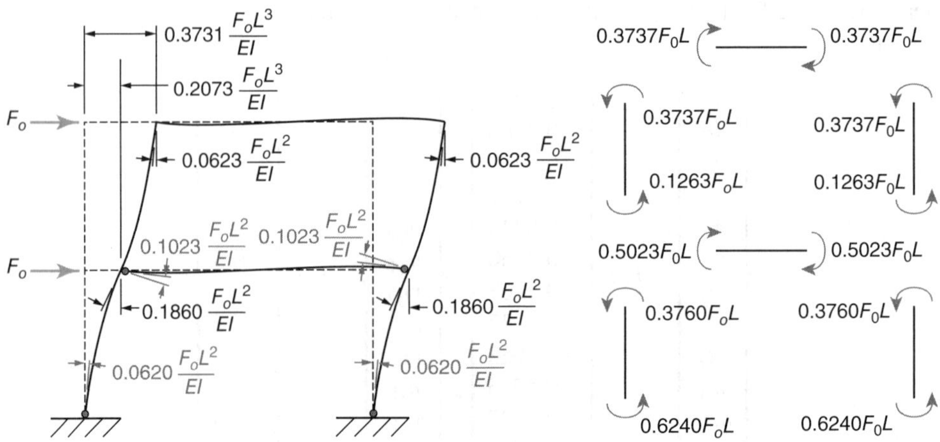

Figure 2.13 Graphical illustration of the responses of the two-story frame.

Figure 2.13 summarizes of these results in graphical form.

Example 2.13 Nonlinear Two-Story Frame with Loads Removed

Consider the two-story moment-resisting frame as shown in Figure 2.12 with the results obtained in Eq. (2.136) and shown in Figure 2.13. Now assume the applied loads of F_o on both stories are removed. The result is that both displacements and moments change, but the plastic rotations remain in the members, i.e.

$$\begin{Bmatrix} \theta_1'' \\ \theta_5'' \\ \theta_9'' \\ \theta_{10}'' \end{Bmatrix} = \frac{F_oL^2}{EI}\begin{Bmatrix} 0.0620 \\ 0.0620 \\ -0.1023 \\ -0.1023 \end{Bmatrix} \tag{2.137}$$

The stiffness is assumed to be elastic during unloading (i.e. with no stiffness degradation), and therefore Eq. (2.126) is applicable. From the first matrix equation of Eq. (2.126), it follows that

$$\begin{bmatrix} \frac{48EI}{L^3} & -\frac{24EI}{L^3} & 0 & 0 & -\frac{6EI}{L^2} & -\frac{6EI}{L^2} \\ -\frac{24EI}{L^3} & \frac{24EI}{L^3} & \frac{6EI}{L^2} & \frac{6EI}{L^2} & \frac{6EI}{L^2} & \frac{6EI}{L^2} \\ 0 & \frac{6EI}{L^2} & \frac{12EI}{L} & \frac{2EI}{L} & \frac{2EI}{L} & 0 \\ 0 & \frac{6EI}{L^2} & \frac{2EI}{L} & \frac{12EI}{L} & 0 & \frac{2EI}{L} \\ -\frac{6EI}{L^2} & \frac{6EI}{L^2} & \frac{2EI}{L} & 0 & \frac{8EI}{L} & \frac{2EI}{L} \\ -\frac{6EI}{L^2} & \frac{6EI}{L^2} & 0 & \frac{2EI}{L} & \frac{2EI}{L} & \frac{8EI}{L} \end{bmatrix} \begin{Bmatrix} x_1 \\ x_2 \\ x_3 \\ x_4 \\ x_5 \\ x_6 \end{Bmatrix} + \begin{bmatrix} \frac{6EI}{L^2} & \frac{6EI}{L^2} & 0 & 0 \\ 0 & 0 & 0 & 0 \\ \frac{2EI}{L} & 0 & \frac{4EI}{L} & \frac{2EI}{L} \\ 0 & \frac{2EI}{L} & \frac{2EI}{L} & \frac{4EI}{L} \\ 0 & 0 & 0 & 0 \\ 0 & 0 & 0 & 0 \end{bmatrix} \begin{Bmatrix} -\theta_1'' \\ -\theta_5'' \\ -\theta_9'' \\ -\theta_{10}'' \end{Bmatrix} = \begin{Bmatrix} 0 \\ 0 \\ 0 \\ 0 \\ 0 \\ 0 \end{Bmatrix} \tag{2.138}$$

Substituting Eq. (2.137) into Eq. (2.138) and solving for the displacements give

$$\begin{Bmatrix} x_1 \\ x_2 \\ x_3 \\ x_4 \\ x_5 \\ x_6 \end{Bmatrix} = \begin{Bmatrix} 0.0740 \times F_o L^3/EI \\ 0.1231 \times F_o L^3/EI \\ -0.0860 \times F_o L^2/EI \\ -0.0860 \times F_o L^2/EI \\ -0.0123 \times F_o L^2/EI \\ -0.0123 \times F_o L^2/EI \end{Bmatrix} \tag{2.139}$$

Then substituting Eqs. (2.137) and (2.139) into the second matrix equation of Eq. (2.126) gives

$$\begin{Bmatrix} m_1 \\ m_2 \\ m_3 \\ m_4 \\ m_5 \\ m_6 \\ m_7 \\ m_8 \\ m_9 \\ m_{10} \\ m_{11} \\ m_{12} \end{Bmatrix} = = \begin{bmatrix} \frac{6EI}{L^2} & 0 & \frac{2EI}{L} & 0 & 0 & 0 \\ \frac{6EI}{L^2} & 0 & \frac{4EI}{L} & 0 & 0 & 0 \\ -\frac{6EI}{L^2} & \frac{6EI}{L^2} & \frac{4EI}{L} & 0 & \frac{2EI}{L} & 0 \\ -\frac{6EI}{L^2} & \frac{6EI}{L^2} & \frac{2EI}{L} & 0 & \frac{4EI}{L} & 0 \\ \frac{6EI}{L^2} & 0 & 0 & \frac{2EI}{L} & 0 & 0 \\ \frac{6EI}{L^2} & 0 & 0 & \frac{4EI}{L} & 0 & 0 \\ -\frac{6EI}{L^2} & \frac{6EI}{L^2} & 0 & \frac{4EI}{L} & 0 & \frac{2EI}{L} \\ -\frac{6EI}{L^2} & \frac{6EI}{L^2} & 0 & \frac{2EI}{L} & 0 & \frac{4EI}{L} \\ 0 & 0 & \frac{4EI}{L} & \frac{2EI}{L} & 0 & 0 \\ 0 & 0 & \frac{2EI}{L} & \frac{4EI}{L} & 0 & 0 \\ 0 & 0 & 0 & 0 & \frac{4EI}{L} & \frac{2EI}{L} \\ 0 & 0 & 0 & 0 & \frac{2EI}{L} & \frac{4EI}{L} \end{bmatrix} \begin{Bmatrix} x_1 \\ x_2 \\ x_3 \\ x_4 \\ x_5 \\ x_6 \end{Bmatrix} + \begin{bmatrix} \frac{4EI}{L} & 0 & 0 & 0 \\ \frac{2EI}{L} & 0 & 0 & 0 \\ 0 & 0 & 0 & 0 \\ 0 & 0 & 0 & 0 \\ 0 & \frac{4EI}{L} & 0 & 0 \\ 0 & \frac{2EI}{L} & 0 & 0 \\ 0 & 0 & 0 & 0 \\ 0 & 0 & 0 & 0 \\ 0 & 0 & \frac{4EI}{L} & \frac{2EI}{L} \\ 0 & 0 & \frac{2EI}{L} & \frac{4EI}{L} \\ 0 & 0 & 0 & 0 \\ 0 & 0 & 0 & 0 \end{bmatrix} \begin{Bmatrix} -\theta_1'' \\ -\theta_5'' \\ -\theta_9'' \\ -\theta_{10}'' \end{Bmatrix} = F_o L \begin{Bmatrix} 0.0240 \\ -0.0240 \\ -0.0737 \\ 0.0737 \\ 0.0240 \\ -0.0240 \\ -0.0737 \\ 0.0737 \\ 0.0977 \\ 0.0977 \\ -0.0737 \\ -0.0737 \end{Bmatrix} \tag{2.140}$$

Note that when the plastic rotation is nonzero but the applied load is zero, the displacements in Eq. (2.139) are also the inelastic displacements (i.e. $\mathbf{x}'' = \mathbf{x}$ and $\mathbf{x}' = \mathbf{0}$) and the moments in Eq. (2.140) are also the inelastic moments (i.e. $\mathbf{m}'' = \mathbf{m}$ and $\mathbf{m}' = \mathbf{0}$).

Figure 2.14 summarizes these results in graphical form.

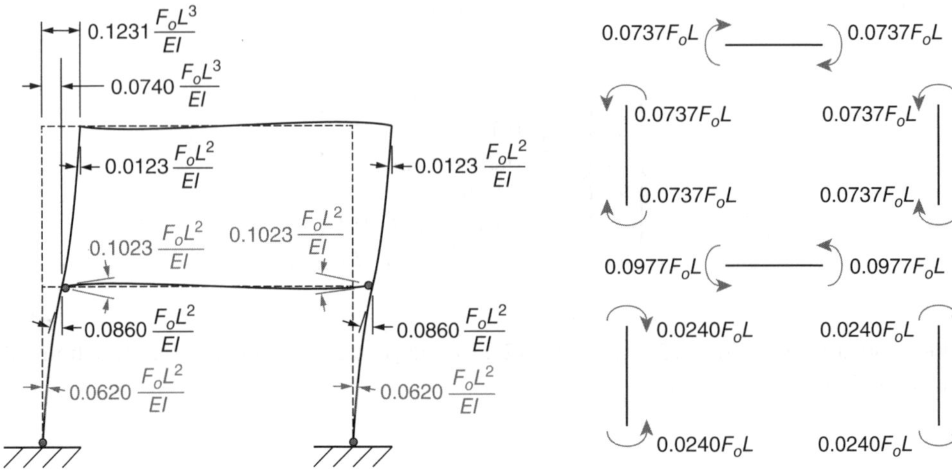

Figure 2.14 Graphical illustration of the responses of the two-story frame after loads removed.

2.6 Static Condensation

One important aspect of matrix structural analysis is its ability to reduce the size of the problem through static condensation, which is a method that is commonly used to compress the stiffness matrix when the applied forces at certain DOFs are zero. Consider Eq. (2.63), which is repeated here as follows:

$$\begin{bmatrix} \mathbf{K} & \mathbf{K}' \\ \mathbf{K}'^T & \mathbf{K}'' \end{bmatrix} \begin{Bmatrix} \mathbf{x} \\ -\mathbf{\Theta}'' \end{Bmatrix} = \begin{Bmatrix} \mathbf{F}_a \\ \mathbf{m} \end{Bmatrix} \tag{2.141}$$

Some entries of $\mathbf{F}_a$ may be zero, such as when no moment is applied at the rotational degrees of freedom. Let d denotes the number of degrees of freedom that have nonzero applied forces, and r denotes the number of degrees of freedom that have zero applied forces. This gives $n = d + r$ in an n-DOF system. The matrices and vectors in Eq. (2.141) are now partitioned as follow:

$$\mathbf{x} = \begin{Bmatrix} \mathbf{x}_d \\ \mathbf{x}_r \end{Bmatrix}, \qquad \mathbf{F}_a = \begin{Bmatrix} \mathbf{F}_{ad} \\ \mathbf{0} \end{Bmatrix}, \qquad \mathbf{K} = \begin{bmatrix} \mathbf{K}_{dd} & \mathbf{K}_{dr} \\ \mathbf{K}_{rd} & \mathbf{K}_{rr} \end{bmatrix}, \qquad \mathbf{K}' = \begin{bmatrix} \mathbf{K}'_d \\ \mathbf{K}'_r \end{bmatrix} \tag{2.142}$$

where the subscript 'd' denotes the quantities associated with the DOFs with nonzero applied forces, and the subscript 'r' denotes the quantities associated with the DOFs with zero applied forces. Substituting Eq. (2.142) into Eq. (2.141) gives

$$\begin{bmatrix} \mathbf{K}_{dd} & \mathbf{K}_{dr} & \mathbf{K}'_d \\ \mathbf{K}_{rd} & \mathbf{K}_{rr} & \mathbf{K}'_r \\ \mathbf{K}'^T_d & \mathbf{K}'^T_r & \mathbf{K}'' \end{bmatrix} \begin{Bmatrix} \mathbf{x}_d \\ \mathbf{x}_r \\ -\mathbf{\Theta}'' \end{Bmatrix} = \begin{Bmatrix} \mathbf{F}_{ad} \\ \mathbf{0} \\ \mathbf{m} \end{Bmatrix} \tag{2.143}$$

By performing static condensation, the second equation of Eq. (2.143) is first extracted and written in the long form:

$$\mathbf{K}_{rd}\mathbf{x}_d + \mathbf{K}_{rr}\mathbf{x}_r - \mathbf{K}'_r\mathbf{\Theta}'' = \mathbf{0} \tag{2.144}$$

Solving for $\mathbf{x}_r$ in Eq. (2.144) gives

$$\mathbf{x}_r = \mathbf{K}_{rr}^{-1}\mathbf{K}'_r\mathbf{\Theta}'' - \mathbf{K}_{rr}^{-1}\mathbf{K}_{rd}\mathbf{x}_d \tag{2.145}$$

Now substituting Eq. (2.145) back into the first and third equations of Eq. (2.143) gives

$$\mathbf{K}_{dd}\mathbf{x}_d + \mathbf{K}_{dr}\left(\mathbf{K}_{rr}{}^{-1}\mathbf{K}'_r\mathbf{\Theta}'' - \mathbf{K}_{rr}^{-1}\mathbf{K}_{rd}\mathbf{x}_d\right) - \mathbf{K}'_d\mathbf{\Theta}'' = \mathbf{F}_{ad} \tag{2.146a}$$

$$\mathbf{K}'^T_d\mathbf{x}_d + \mathbf{K}'^T_r\left(\mathbf{K}_{rr}{}^{-1}\mathbf{K}'_r\mathbf{\Theta}'' - \mathbf{K}_{rr}^{-1}\mathbf{K}_{rd}\mathbf{x}_d\right) - \mathbf{K}''\mathbf{\Theta}'' = \mathbf{m} \tag{2.146b}$$

and rearranging the terms and simplifying gives

$$\left(\mathbf{K}_{dd} - \mathbf{K}_{dr}\mathbf{K}_{rr}^{-1}\mathbf{K}_{rd}\right)\,\mathbf{x}_d - \left(\mathbf{K}'_d - \mathbf{K}_{dr}\mathbf{K}_{rr}^{-1}\mathbf{K}'_r\right)\,\mathbf{\Theta}'' = \mathbf{F}_{ad} \tag{2.147a}$$

$$\left(\mathbf{K}'^T_d - \mathbf{K}'^T_r\mathbf{K}_{rr}{}^{-1}\mathbf{K}_{rd}\right)\,\mathbf{x}_d - \left(\mathbf{K}'' - \mathbf{K}'^T_r\mathbf{K}_{rr}{}^{-1}\mathbf{K}'_r\right)\,\mathbf{\Theta}'' = \mathbf{m} \tag{2.147b}$$

Finally, representing Eq. (2.147) in matrix form gives:

$$\begin{bmatrix} \mathbf{K}_{dd} - \mathbf{K}_{dr}\mathbf{K}_{rr}^{-1}\mathbf{K}_{rd} & \mathbf{K}'_d - \mathbf{K}_{dr}\mathbf{K}_{rr}^{-1}\mathbf{K}'_r \\ \mathbf{K}'^T_d - \mathbf{K}'^T_r\mathbf{K}_{rr}^{-1}\mathbf{K}_{rd} & \mathbf{K}'' - \mathbf{K}'^T_r\mathbf{K}_{rr}^{-1}\mathbf{K}'_r \end{bmatrix} \begin{Bmatrix} \mathbf{x}_d \\ -\mathbf{\Theta}'' \end{Bmatrix} = \begin{Bmatrix} \mathbf{F}_{ad} \\ \mathbf{m} \end{Bmatrix} \tag{2.148}$$

For simplicity, define

$$\bar{\mathbf{K}} = \mathbf{K}_{dd} - \mathbf{K}_{dr}\mathbf{K}_{rr}^{-1}\mathbf{K}_{rd}, \quad \bar{\mathbf{K}}' = \mathbf{K}'_d - \mathbf{K}_{dr}\mathbf{K}_{rr}^{-1}\mathbf{K}'_r, \quad \bar{\mathbf{K}}'' = \mathbf{K}'' - \mathbf{K}'^T_r\mathbf{K}_{rr}^{-1}\mathbf{K}'_r \tag{2.149}$$

Then, Eq. (2.148) becomes

$$\begin{bmatrix} \bar{\mathbf{K}} & \bar{\mathbf{K}}' \\ \bar{\mathbf{K}}'^T & \bar{\mathbf{K}}'' \end{bmatrix} \begin{Bmatrix} \mathbf{x}_d \\ -\mathbf{\Theta}'' \end{Bmatrix} = \begin{Bmatrix} \mathbf{F}_{ad} \\ \mathbf{m} \end{Bmatrix} \tag{2.150}$$

Example 2.14 Nonlinear Solution to the One-Story Frame with Static Condensation
Consider the same one-story one-bay moment-resisting frame used in Example 2.3 with $I_c = I_b = I$, $L_c = L_b = L$, and an applied lateral force of $2F_o$ at the degree of freedom labeled x_1. This gives $F_1 = 2F_o$ and $F_2 = F_3 = 0$. For the moment versus plastic rotation relationships of the six PHLs, let

$$\text{if } \begin{cases} m_i \le F_o L/2 \\ m_i > F_o L/2 \end{cases}, \quad \text{then } \begin{cases} \theta''_i = 0 \\ m_i = F_o L/2 + (EI/L)\,\theta''_i \end{cases} \quad i = 1,2,3,4 \tag{2.151a}$$

$$if \begin{cases} m_i \le F_o L/3 \\ m_i > F_o L/3 \end{cases}, \qquad then \begin{cases} \theta_i'' = 0 \\ m_i = F_o L/3 \end{cases} \qquad i = 5,6 \tag{2.151b}$$

The stiffness matrices were presented in Eq. (2.54). Recognizing that $F_2 = F_3 = 0$, static condensation can be performed with the stiffness matrices partitioned as follows:

$$\begin{Bmatrix} \mathbf{x}_d \\ \hdashline \mathbf{x}_r \end{Bmatrix} = \begin{Bmatrix} x_1 \\ \hdashline x_2 \\ x_3 \end{Bmatrix}, \qquad \begin{Bmatrix} \mathbf{F}_{ad} \\ \hdashline \mathbf{0} \end{Bmatrix} = \begin{Bmatrix} 2F_o \\ \hdashline 0 \\ 0 \end{Bmatrix} \tag{2.152a}$$

$$\begin{bmatrix} \mathbf{K}_{dd} & \mathbf{K}_{dr} \\ \mathbf{K}_{rd} & \mathbf{K}_{rr} \end{bmatrix} = \begin{bmatrix} 24EI/L^3 & 6EI/L^2 & 6EI/L^2 \\ 6EI/L^2 & 8EI/L & 2EI/L \\ 6EI/L^2 & 2EI/L & 8EI/L \end{bmatrix} \tag{2.152b}$$

$$\begin{bmatrix} \mathbf{K}_d' \\ \hdashline \mathbf{K}_r' \end{bmatrix} = \begin{bmatrix} 6EI/L^2 & 6EI/L^2 & 6EI/L^2 & 6EI/L^2 & 0 & 0 \\ \hdashline 2EI/L & 4EI/L & 0 & 0 & 4EI/L & 2EI/L \\ 0 & 0 & 2EI/L & 4EI/L & 2EI/L & 4EI/L \end{bmatrix} \tag{2.152c}$$

$$\mathbf{K}'' = \begin{bmatrix} 4EI/L & 2EI/L & 0 & 0 & 0 & 0 \\ 2EI/L & 4EI/L & 0 & 0 & 0 & 0 \\ 0 & 0 & 4EI/L & 2EI/L & 0 & 0 \\ 0 & 0 & 2EI/L & 4EI/L & 0 & 0 \\ 0 & 0 & 0 & 0 & 4EI/L & 2EI/L \\ 0 & 0 & 0 & 0 & 2EI/L & 4EI/L \end{bmatrix} \tag{2.152d}$$

From Eq. (2.149), it follows that

$$\bar{\mathbf{K}} = \mathbf{K}_{dd} - \mathbf{K}_{dr}\mathbf{K}_{rr}^{-1}\mathbf{K}_{rd} = \frac{24EI}{L^3} - \begin{bmatrix} \frac{6EI}{L^2} & \frac{6EI}{L^2} \end{bmatrix} \begin{bmatrix} \frac{8EI}{L} & \frac{2EI}{L} \\ \frac{2EI}{L} & \frac{8EI}{L} \end{bmatrix}^{-1} \begin{bmatrix} \frac{6EI}{L^2} \\ \frac{6EI}{L^2} \end{bmatrix} = \frac{84EI}{5L^3} \tag{2.153a}$$

$$\bar{\mathbf{K}}' = \mathbf{K}_d' - \mathbf{K}_{dr}\mathbf{K}_{rr}^{-1}\mathbf{K}_r' = \begin{bmatrix} \frac{24EI}{5L^2} & \frac{18EI}{5L^2} & \frac{24EI}{5L^2} & \frac{18EI}{5L^2} & -\frac{18EI}{5L^2} & -\frac{18EI}{5L^2} \end{bmatrix} \tag{2.153b}$$

$$\bar{\mathbf{K}}'' = \mathbf{K}'' - \mathbf{K}_r'^T\mathbf{K}_{rr}^{-1}\mathbf{K}_r' = \frac{EI}{15L} \begin{bmatrix} 52 & 14 & 2 & 4 & -14 & -4 \\ 14 & 28 & 4 & 8 & -28 & -8 \\ 2 & 4 & 52 & 14 & -4 & -14 \\ 4 & 8 & 14 & 28 & -8 & -28 \\ -14 & -28 & -4 & -8 & 28 & 8 \\ -4 & -8 & -14 & -28 & 8 & 28 \end{bmatrix} \tag{2.153c}$$

To analyze the structural response, the first step is to assume that the response is linear, i.e. $\theta_1'' = \theta_2'' = \theta_3'' = \theta_4'' = \theta_5'' = \theta_6'' = 0$. As a result, the first matrix equation of Eq. (2.150) becomes

$$\frac{84EI}{5L^3} x_1 = 2F_o \tag{2.154}$$

Solving for the displacement x_1 in Eq. (2.154) gives

$$x_1 = 5F_oL^3/42EI \tag{2.155}$$

Substituting Eq. (2.155) back into the second matrix equation of Eq. (2.150) and solving for the moments gives

$$\begin{Bmatrix} m_1 \\ m_2 \\ m_3 \\ m_4 \\ m_5 \\ m_6 \end{Bmatrix} = \begin{bmatrix} 24EI/5L^2 \\ 18EI/5L^2 \\ 24EI/5L^2 \\ 18EI/5L^2 \\ -18EI/5L^2 \\ -18EI/5L^2 \end{bmatrix} \frac{5F_oL^3}{42EI} = \begin{Bmatrix} 4F_oL/7 \\ 3F_oL/7 \\ 4F_oL/7 \\ 3F_oL/7 \\ -3F_oL/7 \\ -3F_oL/7 \end{Bmatrix} \tag{2.156}$$

Comparing the moment demands in Eq. (2.156) with the corresponding yield moments in Eq. (2.151) show that m_1, m_3, m_5, and m_6 have all exceeded their corresponding yield moment. This means the original assumption that the structure is linear is incorrect. Therefore, the second step is to assume that plastic hinges are formed at these locations, and the resulting moment versus plastic rotation relationships become

$$m_1 = F_oL/2 + (EI/L)\,\theta_1'' \tag{2.157a}$$

$$m_3 = F_oL/2 + (EI/L)\,\theta_3'' \tag{2.157b}$$

$$m_5 = m_6 = -F_oL/3 \tag{2.157c}$$

$$\theta_2 = \theta_4 = 0 \tag{2.157d}$$

where the minus sign in front of $F_oL/3$ in Eq. (2.157c) denotes that the negative yield moment has been exceeded. Substituting Eq. (2.157) into Eq. (2.150) gives

$$\left[\begin{array}{c|cccc} 84EI/5L^3 & 24EI/5L^2 & 24EI/5L^2 & -18EI/5L^2 & -18EI/5L^2 \\ \hline 24EI/5L^2 & 67EI/15L & 2EI/15L & -14EI/15L & -4EI/15L \\ 24EI/5L^2 & 2EI/15L & 67EI/15L & -4EI/15L & -14EI/15L \\ -18EI/5L^2 & -14EI/15L & -4EI/15L & 28EI/15L & 8EI/15L \\ -18EI/5L^2 & -4EI/15L & -14EI/15L & 8EI/15L & 28EI/15L \end{array}\right] \begin{Bmatrix} x_1 \\ \hline -\theta_1'' \\ -\theta_3'' \\ -\theta_5'' \\ -\theta_6'' \end{Bmatrix} = \begin{Bmatrix} 2F_o \\ \hline F_oL/2 \\ F_oL/2 \\ -F_oL/3 \\ -F_oL/3 \end{Bmatrix} \tag{2.158}$$

Note that the hardening stiffness of EI/L for PHLs #1 and #3 have been incorporated into Eq. (2.158) through the addition of stiffness in the corresponding diagonal terms of the stiffness matrix. Now solving for the displacements and plastic rotations in Eq. (2.158) gives

$$\begin{Bmatrix} x_1 \\ \hline -\theta_1'' \\ -\theta_3'' \\ -\theta_5'' \\ -\theta_6'' \end{Bmatrix} = \begin{Bmatrix} F_oL^3/3EI \\ \hline -F_oL^2/6EI \\ -F_oL^2/6EI \\ 5F_oL^2/18EI \\ 5F_oL^2/18EI \end{Bmatrix} \tag{2.159}$$

Substituting Eq. (2.159) back into the second matrix equation of Eq. (2.150) and solving for the moments gives

$$\begin{Bmatrix} m_1 \\ m_2 \\ m_3 \\ m_4 \\ m_5 \\ m_6 \end{Bmatrix} = \left[\begin{array}{c|cccc} \frac{24EI}{5L^2} & \frac{52EI}{15L} & \frac{2EI}{15L} & -\frac{14EI}{15L} & -\frac{4EI}{15L} \\ \frac{18EI}{5L^2} & \frac{14EI}{15L} & \frac{4EI}{15L} & -\frac{28EI}{15L} & -\frac{8EI}{15L} \\ \frac{24EI}{5L^2} & \frac{2EI}{15L} & \frac{52EI}{15L} & -\frac{4EI}{15L} & -\frac{14EI}{15L} \\ \frac{18EI}{5L^2} & \frac{4EI}{15L} & \frac{14EI}{15L} & -\frac{8EI}{15L} & -\frac{28EI}{15L} \\ -\frac{18EI}{5L^2} & -\frac{14EI}{15L} & -\frac{4EI}{15L} & \frac{28EI}{15L} & \frac{8EI}{15L} \\ -\frac{18EI}{5L^2} & -\frac{4EI}{15L} & -\frac{14EI}{15L} & \frac{8EI}{15L} & \frac{28EI}{15L} \end{array}\right] \begin{Bmatrix} \frac{F_oL^3}{3EI} \\ \hline -\frac{F_oL^3}{6EI} \\ -\frac{F_oL^3}{6EI} \\ \frac{5F_oL^3}{18EI} \\ \frac{5F_oL^3}{18EI} \end{Bmatrix} = \begin{Bmatrix} 2F_oL/3 \\ F_oL/3 \\ 2F_oL/3 \\ F_oL/3 \\ -F_oL/3 \\ -F_oL/3 \end{Bmatrix} \tag{2.160}$$

Since both moment demands for m_2 and m_4 are less than the corresponding yield moment, the assumption that PHLs #2 and #4 remain elastic as indicated in Eq. (2.157d) is correct.

Finally, the DOFs that were statically condensed, $\mathbf{x}_r$, can be calculated using Eq. (2.145), which is

$$\begin{aligned} \begin{Bmatrix} x_2 \\ x_3 \end{Bmatrix} &= \begin{bmatrix} \frac{8EI}{L} & \frac{2EI}{L} \\ \frac{2EI}{L} & \frac{8EI}{L} \end{bmatrix}^{-1} \begin{bmatrix} 2EI/L & 0 \\ 4EI/L & 0 \\ 0 & 2EI/L \\ 0 & 4EI/L \\ 4EI/L & 2EI/L \\ 2EI/L & 4EI/L \end{bmatrix}^T \begin{Bmatrix} F_oL^2/6EI \\ 0 \\ F_oL^2/6EI \\ 0 \\ -5F_oL^2/18EI \\ -5F_oL^2/18EI \end{Bmatrix} \\ &\quad - \begin{bmatrix} \frac{8EI}{L} & \frac{2EI}{L} \\ \frac{8EI}{L} & \frac{8EI}{L} \end{bmatrix}^{-1} \begin{bmatrix} \frac{6EI}{L^2} \\ \frac{6EI}{L^2} \end{bmatrix} \begin{Bmatrix} \frac{F_oL^3}{3EI} \end{Bmatrix} = \begin{Bmatrix} -\frac{F_oL^2}{3EI} \\ -\frac{F_oL^2}{3EI} \end{Bmatrix} \end{aligned} \tag{2.161}$$

Therefore, in summary,

$$\begin{Bmatrix} x_1 \\ \hdashline x_2 \\ x_3 \end{Bmatrix} = \begin{Bmatrix} F_oL^3/3EI \\ \hdashline -F_oL^2/3EI \\ -F_oL^2/3EI \end{Bmatrix}, \quad \begin{Bmatrix} \theta_1'' \\ \theta_2'' \\ \theta_3'' \\ \theta_4'' \\ \theta_5'' \\ \theta_6'' \end{Bmatrix} = \begin{Bmatrix} F_oL^2/6EI \\ 0 \\ F_oL^2/6EI \\ 0 \\ -5F_oL^2/18EI \\ -5F_oL^2/18EI \end{Bmatrix}, \quad \begin{Bmatrix} m_1 \\ m_2 \\ m_3 \\ m_4 \\ m_5 \\ m_6 \end{Bmatrix} = \begin{Bmatrix} 2F_oL/3 \\ F_oL/3 \\ 2F_oL/3 \\ F_oL/3 \\ -F_oL/3 \\ -F_oL/3 \end{Bmatrix} \tag{2.162}$$

which are the same as those obtained in Eq. (2.76) of Example 2.6.

Example 2.15 Nonlinear Solution to the Two-Story Frame with Static Condensation
Consider the two-story moment-resisting frame used in Example 2.12 with the same applied loads, i.e. $F_1 = F_2 = F_o$ and $F_3 = F_4 = F_5 = F_6 = 0$. Assume that the moment versus plastic rotation relationships are:

$$if \begin{cases} m_i \le F_oL/2 \\ m_i > F_oL/2 \end{cases}, \quad then \begin{cases} \theta_i'' = 0 \\ m_i = F_oL/2 + (2EI/L)\,\theta_i'' \end{cases} \quad i = 1, \ldots, 8 \tag{2.163a}$$

$$if \begin{cases} m_i \le 2F_oL/5 \\ m_i > 2F_oL/5 \end{cases}, \quad then \begin{cases} \theta_i'' = 0 \\ m_i = 2F_oL/5 + (EI/L)\,\theta_i'' \end{cases} \quad i = 9, \ldots, 12 \tag{2.163b}$$

The stiffness matrices were presented in Eq. (2.125). Recognizing $F_3 = F_4 = F_5 = F_6 = 0$, static condensation can be performed with the matrices and vectors being partitioned as follows:

$$\begin{Bmatrix} \mathbf{x}_d \\ \hdashline \mathbf{x}_r \end{Bmatrix} = \begin{Bmatrix} x_1 \\ x_2 \\ \hdashline x_3 \\ x_4 \\ x_5 \\ x_6 \end{Bmatrix}, \quad \begin{Bmatrix} \mathbf{F}_{ad} \\ \hdashline \mathbf{0} \end{Bmatrix} = \begin{Bmatrix} F_o \\ F_o \\ \hdashline 0 \\ 0 \\ 0 \\ 0 \end{Bmatrix} \tag{2.164a}$$

$$\begin{bmatrix} \mathbf{K}_{dd} & \mathbf{K}_{dr} \\ \mathbf{K}_{rd} & \mathbf{K}_{rr} \end{bmatrix} = \begin{bmatrix} 48EI/L^3 & -24EI/L^3 & 0 & 0 & -6EI/L^2 & -6EI/L^2 \\ -24EI/L^3 & 24EI/L^3 & 6EI/L^2 & 6EI/L^2 & 6EI/L^2 & 6EI/L^2 \\ \hdashline 0 & 6EI/L^2 & 12EI/L & 2EI/L & 2EI/L & 0 \\ 0 & 6EI/L^2 & 2EI/L & 12EI/L & 0 & 2EI/L \\ -6EI/L^2 & 6EI/L^2 & 2EI/L & 0 & 8EI/L & 2EI/L \\ -6EI/L^2 & 6EI/L^2 & 0 & 2EI/L & 2EI/L & 8EI/L \end{bmatrix} \tag{2.164b}$$

$$\begin{bmatrix} \mathbf{K}'_d \\ \hdashline \mathbf{K}'_r \end{bmatrix} = \left[\begin{array}{cccccccccccc} \frac{6EI}{L^2} & \frac{6EI}{L^2} & -\frac{6EI}{L^2} & -\frac{6EI}{L^2} & \frac{6EI}{L^2} & \frac{6EI}{L^2} & -\frac{6EI}{L^2} & -\frac{6EI}{L^2} & 0 & 0 & 0 & 0 \\ 0 & 0 & \frac{6EI}{L^2} & \frac{6EI}{L^2} & 0 & 0 & \frac{6EI}{L^2} & \frac{6EI}{L^2} & 0 & 0 & 0 & 0 \\ \hdashline \frac{2EI}{L} & \frac{4EI}{L} & \frac{4EI}{L} & \frac{2EI}{L} & 0 & 0 & 0 & 0 & \frac{4EI}{L} & \frac{2EI}{L} & 0 & 0 \\ 0 & 0 & 0 & 0 & \frac{2EI}{L} & \frac{4EI}{L} & \frac{4EI}{L} & \frac{2EI}{L} & \frac{2EI}{L} & \frac{4EI}{L} & 0 & 0 \\ 0 & 0 & \frac{2EI}{L} & \frac{4EI}{L} & 0 & 0 & 0 & 0 & 0 & 0 & \frac{4EI}{L} & \frac{2EI}{L} \\ 0 & 0 & 0 & 0 & 0 & 0 & \frac{2EI}{L} & \frac{4EI}{L} & 0 & 0 & \frac{2EI}{L} & \frac{4EI}{L} \end{array}\right] \tag{2.164c}$$

$$\mathbf{K}'' = \begin{bmatrix} \frac{4EI}{L} & \frac{2EI}{L} & 0 & 0 & 0 & 0 & 0 & 0 & 0 & 0 & 0 & 0 \\ \frac{2EI}{L} & \frac{4EI}{L} & 0 & 0 & 0 & 0 & 0 & 0 & 0 & 0 & 0 & 0 \\ 0 & 0 & \frac{4EI}{L} & \frac{2EI}{L} & 0 & 0 & 0 & 0 & 0 & 0 & 0 & 0 \\ 0 & 0 & \frac{2EI}{L} & \frac{4EI}{L} & 0 & 0 & 0 & 0 & 0 & 0 & 0 & 0 \\ 0 & 0 & 0 & 0 & \frac{4EI}{L} & \frac{2EI}{L} & 0 & 0 & 0 & 0 & 0 & 0 \\ 0 & 0 & 0 & 0 & \frac{2EI}{L} & \frac{4EI}{L} & 0 & 0 & 0 & 0 & 0 & 0 \\ 0 & 0 & 0 & 0 & 0 & 0 & \frac{4EI}{L} & \frac{2EI}{L} & 0 & 0 & 0 & 0 \\ 0 & 0 & 0 & 0 & 0 & 0 & \frac{2EI}{L} & \frac{4EI}{L} & 0 & 0 & 0 & 0 \\ 0 & 0 & 0 & 0 & 0 & 0 & 0 & 0 & \frac{4EI}{L} & \frac{2EI}{L} & 0 & 0 \\ 0 & 0 & 0 & 0 & 0 & 0 & 0 & 0 & \frac{2EI}{L} & \frac{4EI}{L} & 0 & 0 \\ 0 & 0 & 0 & 0 & 0 & 0 & 0 & 0 & 0 & 0 & \frac{4EI}{L} & \frac{2EI}{L} \\ 0 & 0 & 0 & 0 & 0 & 0 & 0 & 0 & 0 & 0 & \frac{2EI}{L} & \frac{4EI}{L} \end{bmatrix} \tag{2.164d}$$

From Eq. (2.149), it follows that

$$\bar{\mathbf{K}} = \mathbf{K}_{dd} - \mathbf{K}_{dr}\mathbf{K}_{rr}^{-1}\mathbf{K}_{rd} = \frac{EI}{17L^3}\begin{bmatrix} 690 & -300 \\ -300 & 228 \end{bmatrix} \tag{2.165a}$$

$$\bar{\mathbf{K}}' = \frac{EI}{17L^2}\begin{bmatrix} 99 & 96 & -87 & -63 & 99 & 96 & -87 & -63 & -9 & -9 & 63 & 63 \\ -12 & -24 & 60 & 54 & -12 & -24 & 60 & 54 & -36 & -36 & -54 & -54 \end{bmatrix} \tag{2.165b}$$

$$\bar{\mathbf{K}}'' = \frac{EI}{119L}\begin{bmatrix} 433 & 152 & -74 & -19 & 8 & 16 & 11 & -2 & -78 & -27 & 19 & 2 \\ 152 & 304 & -148 & -38 & 16 & 32 & 22 & -4 & -156 & -54 & 38 & 4 \\ -74 & -148 & 285 & 78 & 11 & 22 & 30 & 27 & -137 & -52 & -78 & -27 \\ -19 & -38 & 78 & 213 & -2 & -4 & 27 & 60 & -40 & -23 & -213 & -60 \\ 8 & 16 & 11 & -2 & 433 & 152 & -74 & -19 & -27 & -78 & 2 & 19 \\ 16 & 32 & 22 & -4 & 152 & 304 & -148 & -38 & -54 & -156 & 4 & 38 \\ 11 & 22 & 30 & 27 & -74 & -148 & 285 & 78 & -52 & -137 & -27 & -78 \\ -2 & -4 & 27 & 60 & -19 & -38 & 78 & 213 & -23 & -40 & -60 & -213 \\ -78 & -156 & -137 & -40 & -27 & -54 & -52 & -23 & 293 & 106 & 40 & 23 \\ -27 & -54 & -52 & -23 & -78 & -156 & -137 & -40 & 106 & 293 & 23 & 40 \\ 19 & 38 & -78 & -213 & 2 & 4 & -27 & -60 & 40 & 23 & 213 & 60 \\ 2 & 4 & -27 & -60 & 19 & 38 & -78 & -213 & 23 & 40 & 60 & 213 \end{bmatrix} \tag{2.165c}$$

The structure is first assumed to be linear, this gives $\mathbf{\Theta}'' = \mathbf{0}$. Then using the first matrix equation of Eq. (2.150) gives

$$\frac{EI}{17L^3}\begin{bmatrix} 690 & -300 \\ -300 & 228 \end{bmatrix}\begin{Bmatrix} x_1 \\ x_2 \end{Bmatrix} = \begin{Bmatrix} F_o \\ F_o \end{Bmatrix} \tag{2.166}$$

Solving for the displacements in Eq. (2.166) gives

$$\begin{Bmatrix} x_1 \\ x_2 \end{Bmatrix} = \begin{Bmatrix} F_oL^3/6EI \\ F_oL^3/4EI \end{Bmatrix} \tag{2.167}$$

Then substituting Eq. (2.167) into the second matrix equation of Eq. (2.150), the moments are calculated as

$$\begin{Bmatrix} m_1 \\ m_2 \\ m_3 \\ m_4 \\ m_5 \\ m_6 \\ m_7 \\ m_8 \\ m_9 \\ m_{10} \\ m_{11} \\ m_{12} \end{Bmatrix} = F_oL\begin{Bmatrix} 0.6 \\ 0.4 \\ 0.2 \\ 0.3 \\ 0.6 \\ 0.4 \\ 0.2 \\ 0.3 \\ -0.6 \\ -0.6 \\ -0.3 \\ -0.3 \end{Bmatrix}, \quad \begin{Bmatrix} m_{y1} \\ m_{y2} \\ m_{y3} \\ m_{y4} \\ m_{y5} \\ m_{y6} \\ m_{y7} \\ m_{y8} \\ m_{y9} \\ m_{y10} \\ m_{y11} \\ m_{y12} \end{Bmatrix} = F_oL\begin{Bmatrix} 0.5 \\ 0.5 \\ 0.5 \\ 0.5 \\ 0.5 \\ 0.5 \\ 0.5 \\ 0.5 \\ -0.4 \\ -0.4 \\ -0.4 \\ -0.4 \end{Bmatrix} \tag{2.168}$$

The corresponding yield moments of the 12 PHLs are also presented in Eq. (2.168) as a comparison. It can be seen in Eq. (2.168) that the moment demands m_1, m_5, m_9, and m_{10} exceed their corresponding yield moments. This means the original assumption that the structure is linear is incorrect.

Now assume that PHLs #1, #5, #9, and #10 have yielded. According to Eq. (2.163), this gives

$$\theta_2'' = \theta_3'' = \theta_4'' = \theta_6'' = \theta_7'' = \theta_8'' = \theta_{11}'' = \theta_{12}'' = 0 \tag{2.169a}$$

$$m_1 = F_oL/2 + (2EI/L)\,\theta_1'', \qquad m_5 = F_oL/2 + (2EI/L)\,\theta_5'' \tag{2.169b}$$

$$m_9 = -2F_oL/5 + (EI/L)\,\theta_9'', \qquad m_{10} = -2F_oL/5 + (EI/L)\,\theta_{10}'' \tag{2.169c}$$

where the minus signs in front of $2F_oL/5$ in Eq. (2.169c) denote that the negative yield moments have been exceeded. From Eq. (2.150), it then follows that

$$\begin{bmatrix} \frac{690EI}{17L^3} & -\frac{300EI}{17L^3} & \frac{99EI}{17L^2} & \frac{99EI}{17L^2} & -\frac{9EI}{17L^2} & -\frac{9EI}{17L^2} \\ -\frac{300EI}{17L^3} & \frac{228EI}{17L^3} & -\frac{12EI}{17L^2} & -\frac{12EI}{17L^2} & -\frac{36EI}{17L^2} & -\frac{36EI}{17L^2} \\ \frac{99EI}{17L^2} & -\frac{12EI}{17L^2} & \frac{433EI}{119L}+\frac{2EI}{L} & \frac{8EI}{119L} & -\frac{78EI}{119L} & -\frac{27EI}{119L} \\ \frac{99EI}{17L^2} & -\frac{12EI}{17L^2} & \frac{8EI}{119L} & \frac{433EI}{119L}+\frac{2EI}{L} & -\frac{27EI}{119L} & -\frac{78EI}{119L} \\ -\frac{9EI}{17L^2} & -\frac{36EI}{17L^2} & -\frac{78EI}{119L} & -\frac{27EI}{119L} & \frac{293EI}{119L}+\frac{EI}{L} & \frac{106EI}{119L} \\ -\frac{9EI}{17L^2} & -\frac{36EI}{17L^2} & -\frac{27EI}{119L} & -\frac{78EI}{119L} & \frac{106EI}{119L} & \frac{293EI}{119L}+\frac{EI}{L} \end{bmatrix} \begin{Bmatrix} x_1 \\ x_2 \\ -\theta_1'' \\ -\theta_5'' \\ -\theta_9'' \\ -\theta_{10}'' \end{Bmatrix} = \begin{Bmatrix} F_o \\ F_o \\ F_oL/2 \\ F_oL/2 \\ -2F_oL/5 \\ -2F_oL/5 \end{Bmatrix} \tag{2.170}$$

Solving for Eq. (2.170) gives

$$\begin{Bmatrix} x_1 \\ x_2 \\ -\theta_1'' \\ -\theta_5'' \\ -\theta_9'' \\ -\theta_{10}'' \end{Bmatrix} = \begin{Bmatrix} 0.2073\times F_oL^3/EI \\ 0.3731\times F_oL^3/EI \\ -0.0620\times F_oL^2/EI \\ -0.0620\times F_oL^2/EI \\ 0.1023\times F_oL^2/EI \\ 0.1023\times F_oL^2/EI \end{Bmatrix} \tag{2.171}$$

Then substituting Eq. (2.171) into the second matrix equation of Eq. (2.150), the moments are calculated as

$$\begin{Bmatrix} m_1 \\ m_2 \\ m_3 \\ m_4 \\ m_5 \\ m_6 \\ m_7 \\ m_8 \\ m_9 \\ m_{10} \\ m_{11} \\ m_{12} \end{Bmatrix} = F_oL \begin{Bmatrix} 0.6240 \\ 0.3760 \\ 0.1263 \\ 0.3737 \\ 0.6240 \\ 0.3760 \\ 0.1263 \\ 0.3737 \\ -0.5023 \\ -0.5023 \\ -0.3737 \\ -0.3737 \end{Bmatrix}, \quad \begin{Bmatrix} m_{y1} \\ m_{y2} \\ m_{y3} \\ m_{y4} \\ m_{y5} \\ m_{y6} \\ m_{y7} \\ m_{y8} \\ m_{y9} \\ m_{y10} \\ m_{y11} \\ m_{y12} \end{Bmatrix} = F_oL \begin{Bmatrix} 0.5 \\ 0.5 \\ 0.5 \\ 0.5 \\ 0.5 \\ 0.5 \\ 0.5 \\ 0.5 \\ -0.4 \\ -0.4 \\ -0.4 \\ -0.4 \end{Bmatrix} \tag{2.172}$$

The corresponding yield moments of the 12 PHLs are also presented in Eq. (2.172) as a comparison. Since the moment demands m_2, m_3, m_4, m_6, m_7, m_8, m_{11}, and m_{12} are less than the corresponding yield moments, it means the original assumption is correct.

Finally, the DOFs that were statically condensed, $\mathbf{x}_r$, can be calculated from Eq. (2.145):

$$\begin{Bmatrix} x_3 \\ x_4 \\ x_5 \\ x_6 \end{Bmatrix} = \begin{bmatrix} \frac{43}{238} & -\frac{4}{119} & \frac{39}{119} & \frac{27}{238} \\ -\frac{4}{119} & \frac{43}{238} & \frac{27}{238} & \frac{39}{119} \\ -\frac{6}{119} & \frac{5}{238} & -\frac{19}{238} & -\frac{1}{119} \\ \frac{5}{238} & -\frac{6}{119} & -\frac{1}{119} & -\frac{19}{238} \end{bmatrix} \begin{Bmatrix} \theta_1'' \\ \theta_5'' \\ \theta_9'' \\ \theta_{10}'' \end{Bmatrix} - \begin{bmatrix} \frac{3}{34L} & \frac{6}{17L} \\ \frac{3}{34L} & \frac{6}{17L} \\ -\frac{21}{34L} & \frac{9}{17L} \\ -\frac{21}{34L} & \frac{9}{17L} \end{bmatrix} \begin{Bmatrix} x_1 \\ x_2 \end{Bmatrix} = \frac{F_oL^2}{EI} \begin{Bmatrix} -0.1860 \\ -0.1860 \\ -0.0623 \\ -0.0623 \end{Bmatrix} \tag{2.173}$$

Therefore, in summary,

$$\begin{Bmatrix} x_1 \\ x_2 \\ x_3 \\ x_4 \\ x_5 \\ x_6 \end{Bmatrix} = \begin{Bmatrix} 0.2073 \times F_oL^3/EI \\ 0.3731 \times F_oL^3/EI \\ -0.1860 \times F_oL^2/EI \\ -0.1860 \times F_oL^2/EI \\ -0.0623 \times F_oL^2/EI \\ -0.0623 \times F_oL^2/EI \end{Bmatrix}, \begin{Bmatrix} \theta_1'' \\ \theta_2'' \\ \theta_3'' \\ \theta_4'' \\ \theta_5'' \\ \theta_6'' \\ \theta_7'' \\ \theta_8'' \\ \theta_9'' \\ \theta_{10}'' \\ \theta_{11}'' \\ \theta_{12}'' \end{Bmatrix} = \frac{F_oL^2}{EI} \begin{Bmatrix} 0.0620 \\ 0 \\ 0 \\ 0 \\ 0.0620 \\ 0 \\ 0 \\ 0 \\ -0.1023 \\ -0.1023 \\ 0 \\ 0 \end{Bmatrix}, \begin{Bmatrix} m_1 \\ m_2 \\ m_3 \\ m_4 \\ m_5 \\ m_6 \\ m_7 \\ m_8 \\ m_9 \\ m_{10} \\ m_{11} \\ m_{12} \end{Bmatrix} = F_oL \begin{Bmatrix} 0.6240 \\ 0.3760 \\ 0.1263 \\ 0.3737 \\ 0.6240 \\ 0.3760 \\ 0.1263 \\ 0.3737 \\ -0.5023 \\ -0.5023 \\ -0.3737 \\ -0.3737 \end{Bmatrix} \tag{2.174}$$

which are the same as those obtained in Eq. (2.136) of Example 2.12, with results presented using graphical form in Figure 2.13.

References

American Society of Civil Engineers (2006). ASCE Standard ASCE/SEI 41–06, *Seismic Rehabilitation of Existing Building*, Reston, Virginia, USA.

Hibbeler R.C. (2012). *Structural Analysis, 8th Edition*, Prentice Hall, New Jersey, USA.

Leet K.M., Uang C.M. and Gilbert A.M. (2010). *Fundamentals of Structural Analysis, 4th Edition*, McGraw Hill, New York, USA.

Powell G. (2010). *Modeling for Structural Analysis*, Computers and Structures, Inc., Berkeley, California, USA.

Wilson E.L. (2010). *Static and Dynamic Analysis of Structures*, Computers and Structures, Inc., Berkeley, California, USA.

3

Nonlinear Dynamic Analysis

The theory of nonlinear static analysis of moment-resisting frames using the Force Analogy Method (FAM) was presented in Chapter 2. In this chapter, the FAM is combined with the state space method of linear dynamic analysis to form the nonlinear dynamic analysis technique that will be used for solving structural dynamic problems in the remainder of this book. Because the FAM uses initial stiffness in the formulation to quantify the stiffness force in the structure, it works harmoniously with the state space method that requires stiffness matrix to remain constant during the analysis. At the same time, the state space method is an explicit algorithm that is very efficient in the time-stepping calculations, and the outputs from the state space method feed directly to the FAM for evaluating the internal forces of the structure. Therefore, the end result of combining the force analogy method and the state space method is an accurate and efficient algorithm for performing nonlinear dynamic analysis of structures.

In this chapter, the state space method is first presented for performing linear dynamic analysis in Section 3.1. It is followed by its combination with the FAM for solving nonlinear dynamic problems in Section 3.2. In Section 3.3, static condensation is included in the derivation of the nonlinear dynamic analysis procedure for solving moment-resisting frame problems. Finally, in Section 3.4, two numerical examples are presented to demonstrate the use of the FAM for solving large moment-resisting frame problems.

3.1 State Space Method for Linear Dynamic Analysis

Before nonlinear dynamic analysis based on the force analogy method (FAM) is presented in detail, a quick overview on the theory of linear dynamic analysis is crucial to the understanding

Theory of Nonlinear Structural Analysis: The Force Analogy Method for Earthquake Engineering, First Edition. Gang Li and Kevin K.F. Wong.

of dynamic response calculations. Today, linear dynamic analysis for multi-degree-of-freedom (MDOF) systems is a well-established subject, and it has been discussed in detail in many structural dynamic textbooks (e.g., Chopra 2011). Therefore, the goal of this section is not to focus on the general linear dynamic analysis, but rather to focus on a specific dynamic analysis tool – the state space method – that works together harmoniously with the force analogy method on solving nonlinear dynamic problems. To date, only one textbook talks about the state space method for earthquake engineering purposes (Hart and Wong 2000), though the method is widely used in control engineering. Therefore, it is worthwhile for this chapter to begin with the discussion on the state space method for solving linear dynamic problems.

3.1.1 Equation of Motion

One important aspect of performing a dynamic analysis is that the displacement response changes with respect to time. The time variable t is typically inserted to the displacement vector in the form $\mathbf{x}(t)$ to denote that the displacement vector is a function of time. When the displacement is differentiated with respect to time, velocity $\dot{\mathbf{x}}(t)$ is obtained, i.e.

$$\dot{\mathbf{x}}(t) = \frac{d\mathbf{x}(t)}{dt} \tag{3.1}$$

where a dot above the variable represents differentiation of the variable with respect to time. Differentiating the velocity with respect to time once again gives the acceleration $\ddot{\mathbf{x}}(t)$:

$$\ddot{\mathbf{x}}(t) = \frac{d\dot{\mathbf{x}}(t)}{dt} = \frac{d^2\mathbf{x}(t)}{dt^2} \tag{3.2}$$

where, for an n-degree-of-freedom (n-DOF) system, these $n \times 1$ displacement, velocity, and acceleration vectors are of the form:

$$\mathbf{x}(t) = \begin{Bmatrix} x_1(t) \\ x_2(t) \\ \vdots \\ x_n(t) \end{Bmatrix}, \quad \dot{\mathbf{x}}(t) = \begin{Bmatrix} \dot{x}_1(t) \\ \dot{x}_2(t) \\ \vdots \\ \dot{x}_n(t) \end{Bmatrix}, \quad \ddot{\mathbf{x}}(t) = \begin{Bmatrix} \ddot{x}_1(t) \\ \ddot{x}_2(t) \\ \vdots \\ \ddot{x}_n(t) \end{Bmatrix} \tag{3.3}$$

Consider the equation of motion for an n-DOF system subjected to the $n \times 1$ earthquake ground acceleration vector $\ddot{\mathbf{g}}(t)$:

$$\mathbf{M}\ddot{\mathbf{x}}(t) + \mathbf{C}\dot{\mathbf{x}}(t) + \mathbf{K}\mathbf{x}(t) = -\mathbf{M}\ddot{\mathbf{g}}(t) \tag{3.4}$$

where $\mathbf{M}$ is the $n \times n$ mass matrix, $\mathbf{C}$ is the $n \times n$ damping matrix, and $\mathbf{K}$ is the $n \times n$ stiffness matrix. The stiffness matrix $\mathbf{K}$ for an n-DOF system is the same as what is assembled in the classical stiffness method of analysis, similar to what was obtained in Section 1.3 and throughout Chapter 2. The earthquake ground acceleration vector $\ddot{\mathbf{g}}(t)$ corresponds to the effect of ground motion at each degree of freedom (DOF).

The mass matrix of a discrete n-DOF system is usually lumped into an $n \times n$ diagonal matrix of the form:

$$\mathbf{M} = \begin{bmatrix} M_1 & 0 & \cdots & 0 \\ 0 & M_2 & \ddots & \vdots \\ \vdots & \ddots & \ddots & 0 \\ 0 & \cdots & 0 & M_n \end{bmatrix} \tag{3.5}$$

However, when certain degrees of freedom are representing the rotation of joints, the mass moment of inertia at these joints may typically be ignored by setting the value equal to zero. In this case, the mass matrix becomes

$$\mathbf{M} = \left[\begin{array}{ccc|ccc} M_1 & 0 & 0 & 0 & \cdots & 0 \\ 0 & \ddots & 0 & \vdots & \ddots & \vdots \\ 0 & 0 & M_d & 0 & \cdots & 0 \\ \hline 0 & \cdots & 0 & 0 & \cdots & 0 \\ \vdots & \ddots & \vdots & \vdots & \ddots & \vdots \\ 0 & \cdots & 0 & 0 & \cdots & 0 \end{array}\right] = \left[\begin{array}{c|c} \mathbf{M}_{dd} & \mathbf{0} \\ \hline \mathbf{0} & \mathbf{0} \end{array}\right] \tag{3.6}$$

where the subscript 'd' represents the total number of translational degrees of freedom with mass, $d \leq n$, and $\mathbf{M}_{dd}$ is an invertible mass matrix in its condensed form with only the translational degrees of freedom.

The damping matrix for an n-DOF system is typically a fully populated $n \times n$ matrix of the form:

$$\mathbf{C} = \begin{bmatrix} C_{11} & C_{12} & \cdots & C_{1n} \\ C_{21} & C_{22} & \ddots & \vdots \\ \vdots & \ddots & \ddots & C_{n-1,n} \\ C_{n1} & \cdots & C_{n,n-1} & C_{nn} \end{bmatrix} \tag{3.7}$$

The exact formulation of the damping matrix has never been developed in theory, so it is often assumed to be proportional to the mass matrix $\mathbf{M}$ and stiffness matrix $\mathbf{K}$ of the form:

$$\mathbf{C} = a_1\mathbf{M} + a_2\mathbf{K} \tag{3.8}$$

where a_1 and a_2 are the proportional damping constants. This type of proportional damping is called Rayleigh damping. However, if some degrees of freedom correspond to the rotation of joints with the mass moment of inertia being ignored, the damping matrix can also be condensed in the form:

$$\mathbf{C} = \left[\begin{array}{c|c} \mathbf{C}_{dd} & \mathbf{0} \\ \hline \mathbf{0} & \mathbf{0} \end{array}\right] \tag{3.9}$$

where $\mathbf{C}_{dd}$ is an invertible $d \times d$ damping matrix corresponding to the translational degrees of freedom only. The condensed damping matrix $\mathbf{C}_{dd}$ can then be assumed to be proportional to the condensed mass matrix $\mathbf{M}_{dd}$ and the condensed stiffness matrix $\bar{\mathbf{K}}$ of the form:

$$\mathbf{C}_{dd} = a_1 \mathbf{M}_{dd} + a_2 \bar{\mathbf{K}} \tag{3.10}$$

where $\bar{\mathbf{K}}$ is the stiffness matrix after static condensation is performed (see Eq. (2.149) in Section 2.6). In any event, by assuming that the structural damping is equal to certain percentages of critical damping for any two modes of vibration (e.g., 5% in the first two modes), the corresponding a_1 and a_2 values can be calculated and therefore the damping matrix $\mathbf{C}$ can be evaluated using either Eq. (3.8) or Eq. (3.10).

3.1.2 State Space Solution

Once the mass matrix $\mathbf{M}$, damping matrix $\mathbf{C}$, and stiffness matrix $\mathbf{K}$ (or $\mathbf{M}_{dd}$, $\mathbf{C}_{dd}$, and $\bar{\mathbf{K}}$ in the condensed form) are determined, the state space method can be used to perform the linear dynamic analysis via the equation of motion given in Eq. (3.4). To represent this equation in state space form, let the state vector $\mathbf{z}(t)$ be defined as

$$\mathbf{z}(t) = \begin{Bmatrix} \mathbf{x}(t) \\ \dot{\mathbf{x}}(t) \end{Bmatrix} \tag{3.11}$$

which is a $2n \times 1$ vector with a collection of displacements and velocities (i.e. the states) of the responses. It follows from Eq. (3.4) that

$$\dot{\mathbf{z}}(t) = \begin{Bmatrix} \dot{\mathbf{x}}(t) \\ \ddot{\mathbf{x}}(t) \end{Bmatrix} = \begin{bmatrix} \mathbf{0} & \mathbf{I} \\ -\mathbf{M}^{-1}\mathbf{K} & -\mathbf{M}^{-1}\mathbf{C} \end{bmatrix} \begin{Bmatrix} \mathbf{x}(t) \\ \dot{\mathbf{x}}(t) \end{Bmatrix} + \begin{Bmatrix} \mathbf{0} \\ -\ddot{\mathbf{g}}(t) \end{Bmatrix} \tag{3.12}$$

The ground acceleration vector $\ddot{\mathbf{g}}(t)$ is an $n \times 1$ vector with entries that correspond to each degree of freedom. To express this ground motion in terms of the three translational components of a typical earthquake ground motion in the global x-, y-, and z-directions, namely $\ddot{g}_x(t)$, $\ddot{g}_y(t)$, and $\ddot{g}_z(t)$, let

$$\ddot{\mathbf{g}}(t) = \mathbf{h} \begin{Bmatrix} \ddot{g}_x(t) \\ \ddot{g}_y(t) \\ \ddot{g}_z(t) \end{Bmatrix} = \mathbf{h}\mathbf{a}(t) \tag{3.13}$$

where $\mathbf{h}$ is an $n \times 3$ matrix that relates the directions of each DOF with the global x-, y-, and z-directions (i.e. a collection of 0s and 1s in all entries), and $\mathbf{a}(t)$ is the 3×1 ground acceleration vector in the three global directions. When only a two-dimensional analysis is performed with one component of earthquake ground motion where $a(t) = \ddot{g}_x(t)$, the $\mathbf{h}$ matrix becomes an $n \times 1$ matrix that relates the directions of each DOF with respect to the global x-direction.

To simplify Eq. (3.12), let

$$\mathbf{A} = \begin{bmatrix} \mathbf{0} & \mathbf{I} \\ -\mathbf{M}^{-1}\mathbf{K} & -\mathbf{M}^{-1}\mathbf{C} \end{bmatrix}, \qquad \mathbf{H} = \begin{bmatrix} \mathbf{0} \\ -\mathbf{h} \end{bmatrix} \tag{3.14}$$

where $\mathbf{A}$ is the $2n \times 2n$ state transition matrix in the continuous form, and $\mathbf{H}$ is the $2n \times 3$ ground motion transition matrix in the continuous form. Then Eq. (3.12) becomes

$$\dot{\mathbf{z}}(t) = \mathbf{A}\mathbf{z}(t) + \mathbf{H}\mathbf{a}(t) \tag{3.15}$$

Solving for the first-order linear differential equation in Eq. (3.15) gives

$$\mathbf{z}(t) = \mathbf{e}^{\mathbf{A}(t-t_o)}\mathbf{z}(t_o) + \mathbf{e}^{\mathbf{A}t}\int_{t_o}^{t} \mathbf{e}^{-\mathbf{A}s}\mathbf{H}\mathbf{a}(s)\,ds \tag{3.16}$$

where t_o is the time of reference when the integration begins, which is typically the time when the states $\mathbf{z}(t_o)$ are known, such as the initial conditions.

In order to integrate the term on the right-hand side of Eq. (3.16), the ground acceleration vector $\mathbf{a}(s)$ should be expressed in a form of a continuous function. However, this ground acceleration vector is typically digitized numerically in a short time interval for the duration of the actual earthquake ground motion. Therefore, the integral needs to be evaluated numerically. Let $t_{k+1} = t$, $t_k = t_o$, and $\Delta t = t_{k+1} - t_k$, and the subscript k denotes the kth time step, then it follows from Eq. (3.16) that

$$\mathbf{z}_{k+1} = \mathbf{e}^{\mathbf{A}\Delta t}\mathbf{z}_k + \mathbf{e}^{\mathbf{A}t_{k+1}}\int_{t_k}^{t_{k+1}} \mathbf{e}^{-\mathbf{A}s}\mathbf{H}\mathbf{a}(s)\,ds \tag{3.17}$$

Note that the objective of integration in Eq. (3.17) is to compute the area underneath the $\mathbf{e}^{-\mathbf{A}s}\mathbf{H}\mathbf{a}(s)$ curve from t_k to t_{k+1}, with a small time step. Because it is the area underneath the curve that is of interest, any reasonable method of approximating the curve for $\mathbf{a}(s)$ between the two time steps (e.g., constant acceleration or linear acceleration) will give a reasonable approximation to the calculation of the enclosed area. In this presentation, the delta forcing function approximation is used, where the ground acceleration vector takes the form:

$$\mathbf{a}(s) = \mathbf{a}_k\delta(s - t_k)\Delta t, \qquad t_k \le s < t_{k+1} \tag{3.18}$$

Substituting Eq. (3.18) into Eq. (3.17) and performing the integration gives

$$\begin{aligned} \mathbf{z}_{k+1} &= \mathbf{e}^{\mathbf{A}\Delta t}\mathbf{z}_k + \mathbf{e}^{\mathbf{A}t_{k+1}}\int_{t_k}^{t_{k+1}} \mathbf{e}^{-\mathbf{A}s}\mathbf{H}\mathbf{a}_k\delta(s-t_k)\Delta t\,ds \\ &= \mathbf{e}^{\mathbf{A}\Delta t}\mathbf{z}_k + \mathbf{e}^{\mathbf{A}t_{k+1}}\mathbf{e}^{-\mathbf{A}t_k}\mathbf{H}\mathbf{a}_k\Delta t \\ &= \mathbf{e}^{\mathbf{A}\Delta t}\mathbf{z}_k + \Delta t\,\mathbf{e}^{\mathbf{A}\Delta t}\mathbf{H}\mathbf{a}_k \end{aligned} \tag{3.19}$$

where $\mathbf{z}_k$ and $\mathbf{a}_k$ are the discretized forms of $\mathbf{z}(t)$ and $\mathbf{a}(t)$, respectively. Let

$$\mathbf{F}_d = \mathbf{e}^{\mathbf{A}\Delta t}, \qquad \mathbf{H}_d = \mathbf{e}^{\mathbf{A}\Delta t}\mathbf{H}\Delta t \tag{3.20}$$

Then, Eq. (3.19) becomes

$$\mathbf{z}_{k+1} = \mathbf{F}_d \mathbf{z}_k + \mathbf{H}_d \mathbf{a}_k \tag{3.21}$$

where $\mathbf{F}_d$ is the $2n \times 2n$ state transition matrix in discretized form and $\mathbf{H}_d$ is the $2n \times 3$ ground acceleration transition matrix in discretized form. Note that by representing the earthquake ground acceleration $\mathbf{a}(s)$ as a delta forcing function in Eq. (3.18), Eq. (3.21) becomes an explicit solution procedure.

3.1.3 Solution Procedure

The state transition matrix $\mathbf{F}_d = \mathbf{e}^{\mathbf{A}\Delta t}$ requires that an exponential power of a matrix be performed. For a single degree of freedom (SDOF) system, it has a closed-form solution:

$$\mathbf{e}^{\mathbf{A}\Delta t} = e^{-\zeta\omega\Delta t} \begin{bmatrix} \cos\omega_d\Delta t + \dfrac{\zeta}{\sqrt{1-\zeta^2}}\sin\omega_d\Delta t & \dfrac{1}{\omega\sqrt{1-\zeta^2}}\sin\omega_d\Delta t \\ -\dfrac{\omega}{\sqrt{1-\zeta^2}}\sin\omega_d\Delta t & \cos\omega_d\Delta t - \dfrac{\zeta}{\sqrt{1-\zeta^2}}\sin\omega_d\Delta t \end{bmatrix} \tag{3.22}$$

where ζ is the damping ratio, ω is the natural frequency, and ω_d is the damped natural frequency. These quantities can be computed for a SDOF system with mass M, damping C, and stiffness K as:

$$\omega = \sqrt{\frac{K}{M}}, \quad \zeta = \frac{C}{2\omega M}, \quad \omega_d = \omega\sqrt{1-\zeta^2} \tag{3.23}$$

However, for multi-degree-of-freedom system, the state transition matrix has no closed-form solution and therefore it generally needs to be evaluated numerically. One good method is to use the Taylor series expansion, i.e.

$$\mathbf{e}^{\mathbf{A}\Delta t} = \mathbf{I} + \mathbf{A}\Delta t + \frac{(\mathbf{A}\Delta t)^2}{2!} + \frac{(\mathbf{A}\Delta t)^3}{3!} + \ldots \tag{3.24}$$

However, if some entries in $\mathbf{A}\Delta t$ are much larger than 1.0, squaring and cubing this matrix can result in an even larger matrix that requires summing many terms before the factorial in the denominator takes over for convergence. This may also lead to numerical error in the evaluation due to multiplication, addition, and then division of large numbers. An improved method called the scaling and squaring method may be used to enhance the numerical accuracy of the calculation and increase the speed of convergence:

$$\mathbf{e}^{\mathbf{A}\Delta t} = \left(\mathbf{e}^{\mathbf{A}\Delta t/p}\right)^p = \left[\mathbf{I} + \frac{\mathbf{A}\Delta t}{p} + \frac{(\mathbf{A}\Delta t/p)^2}{2!} + \frac{(\mathbf{A}\Delta t/p)^3}{3!} + \ldots\right]^p \tag{3.25}$$

where p is a positive integer. By making p in the denominator sufficiently large, the squaring and cubing of a scaled $\mathbf{A}\Delta t/p$ matrix will quickly be overshadowed by the factorial in the denominator, which in turn requires fewer terms for convergence. Once the Taylor series converges, the resulting $\mathbf{e}^{\mathbf{A}\Delta t/p}$ matrix is then multiplied by itself p times to produce $\mathbf{e}^{\mathbf{A}\Delta t}$.

Once the calculation of the state transition matrix $\mathbf{e}^{\mathbf{A}\Delta t}$ is complete, the numerical solution to Eq. (3.21) is a recursive process. Assume that all the information at time step k is known, which include $\mathbf{x}_k$, $\dot{\mathbf{x}}_k$, $\ddot{\mathbf{x}}_k$, and the given input earthquake ground motion $\mathbf{a}_k$, Eq. (3.21) can be used to compute $\mathbf{z}_{k+1}$, which include $\mathbf{x}_{k+1}$ and $\dot{\mathbf{x}}_{k+1}$. Then, Eq. (3.4) can be used to calculate the acceleration response, i.e.

$$\ddot{\mathbf{x}}_{k+1} = -\mathbf{h}\mathbf{a}_{k+1} - \mathbf{M}^{-1}\mathbf{C}\dot{\mathbf{x}}_{k+1} - \mathbf{M}^{-1}\mathbf{K}\mathbf{x}_{k+1} \tag{3.26}$$

This completes the calculation of all the information at time step k+1, and the solution can proceed to the next time step.

One important acceleration quantity that often needs to be computed is the absolute acceleration of the masses. Let $\ddot{\mathbf{y}}(t)$ be denoted as the absolute acceleration with its discretized form be denoted as $\ddot{\mathbf{y}}_k$. The equation to calculate $\ddot{\mathbf{y}}_k$ can be written as

$$\ddot{\mathbf{y}}_k = \ddot{\mathbf{x}}_k + \ddot{\mathbf{g}}_k = \ddot{\mathbf{x}}_k + \mathbf{h}\mathbf{a}_k \tag{3.27}$$

where $\ddot{\mathbf{g}}_k$ is the discretized form of $\ddot{\mathbf{g}}(t)$. Substituting Eq. (3.26) into Eq. (3.27) gives

$$\ddot{\mathbf{y}}_{k+1} = -\mathbf{M}^{-1}\mathbf{C}\dot{\mathbf{x}}_{k+1} - \mathbf{M}^{-1}\mathbf{K}\mathbf{x}_{k+1} \tag{3.28}$$

Example 3.1 Linear Dynamic Response of a SDOF System

Consider the single degree of freedom (SDOF) system as shown in Figure 3.1. Assume the system has a mass of M = 1.0 Mg, a damping of C = 0.0 kN s/m, and a stiffness of K = 157.9 kN/m. This gives:

$$\omega = \sqrt{K/M} = \sqrt{157.9/1.0} = 12.566 \text{ rad/s} \tag{3.29a}$$

$$T = 2\pi\sqrt{M/K} = 2\pi\sqrt{1.0/157.9} = 0.5 \text{ s} \tag{3.29b}$$

$$\zeta = C/2\omega M = 0\% \tag{3.29c}$$

where T denotes the period of the SDOF system.

Assume the time step size of the earthquake ground motion is Δt = 0.01 s, the state transition matrix $\mathbf{F}_d$ as given in Eq. (3.22) and the ground acceleration transition matrix $\mathbf{H}_d$ defined in Eq. (3.20) can be evaluated as:

$$\mathbf{F}_d = \begin{bmatrix} 0.992115 & 0.009974 \\ -1.57498 & 0.992115 \end{bmatrix}, \quad \mathbf{H}_d = \begin{bmatrix} -0.000100 \\ -0.009921 \end{bmatrix} \tag{3.30}$$

Then, the recursive equation as given in Eq. (3.21) becomes

$$\begin{Bmatrix} x_{k+1} \\ \dot{x}_{k+1} \end{Bmatrix} = \begin{bmatrix} 0.992115 & 0.009974 \\ -1.57498 & 0.992115 \end{bmatrix} \begin{Bmatrix} x_k \\ \dot{x}_k \end{Bmatrix} + \begin{bmatrix} -0.000100 \\ -0.009921 \end{bmatrix} a_k \tag{3.31}$$

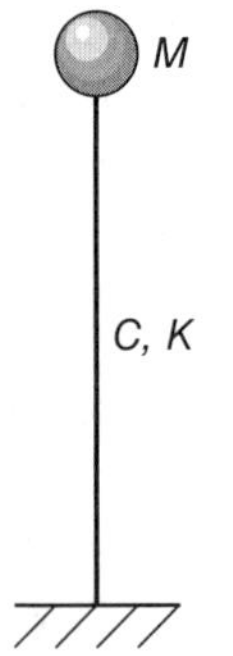

Figure 3.1 A SDOF system.

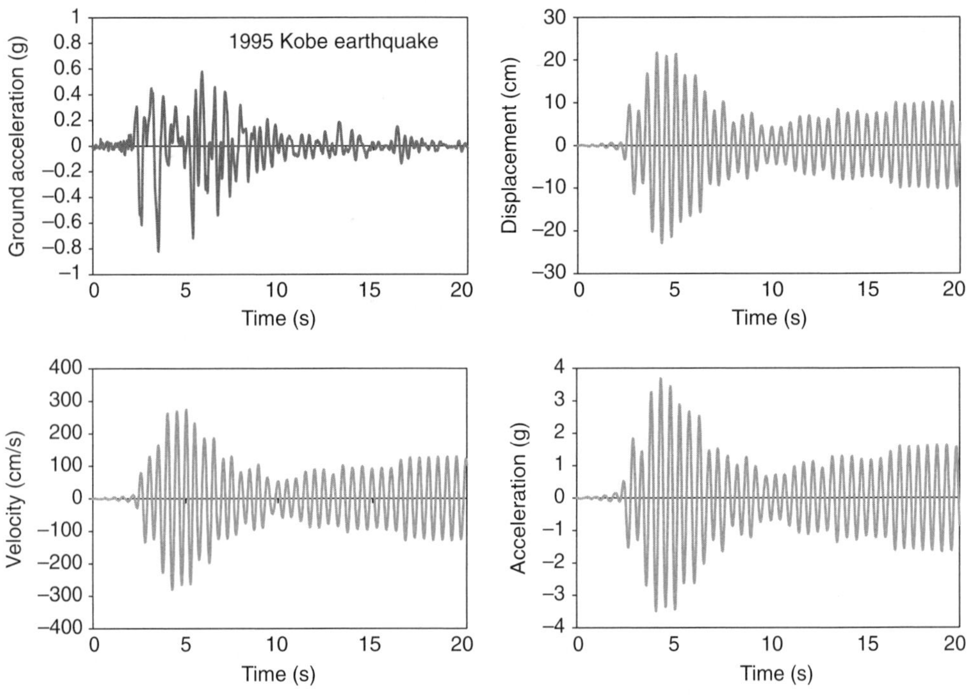

Figure 3.2 Response of a SDOF system with period of 0.5 s due to 1995 Kobe earthquake.

where a_k is the ground acceleration record in the digitized form. Assume that the earthquake ground acceleration a_k is the 1995 Kobe earthquake record as shown in Figure 3.2, with the initial displacement and velocity both set equal to zero (i.e. $x_0 = \dot{x}_0 = 0$), it follows that the displacement and velocity response histories can be evaluated using Eq. (3.31), and the results are presented in Figure 3.2.

The absolute acceleration $\ddot{y}_k$ of the SDOF system can be computed by using Eq. (3.28) rewritten in the form as:

$$\ddot{y}_{k+1} = 2\zeta\omega\dot{x}_{k+1} - \omega^2 x_{k+1} \tag{3.32}$$

Based on the calculation of the displacement and velocity responses using Eq. (3.31), the absolute acceleration response can be computed using Eq. (3.32) and the results are also presented in Figure 3.2.

Now assume that the SDOF system has a mass of $M = 1.0$ Mg, $C = 0.251$ kN s/m, and $K = 39.48$ kN/m. This gives:

$$\omega = \sqrt{K/M} = \sqrt{39.48/1.0} = 6.283 \ \text{rad/s} \tag{3.33a}$$

$$T = 2\pi\sqrt{M/K} = 2\pi\sqrt{1.0/39.48} = 1.0\,\text{s} \tag{3.33b}$$

$$\zeta = C/2\omega M = 0.251/(2\times 6.283\times 1.0) = 0.02 = 2\% \tag{3.33c}$$

Assume again that the time step size of the earthquake ground motion is $\Delta t = 0.01$ s, the state transition matrix $\mathbf{F}_d$ as given in Eq. (3.22) and the ground acceleration transition matrix $\mathbf{H}_d$ defined in Eq. (3.20) can be evaluated as:

$$\mathbf{F}_d = \begin{bmatrix} 0.998028 & 0.009981 \\ -0.39403 & 0.995520 \end{bmatrix}, \quad \mathbf{H}_d = \begin{bmatrix} -0.000100 \\ -0.009955 \end{bmatrix} \tag{3.34}$$

The recursive equation as given in Eq. (3.21) becomes

$$\begin{Bmatrix} x_{k+1} \\ \dot{x}_{k+1} \end{Bmatrix} = \begin{bmatrix} 0.998028 & 0.009981 \\ -0.39403 & 0.995520 \end{bmatrix} \begin{Bmatrix} x_k \\ \dot{x}_k \end{Bmatrix} + \begin{bmatrix} -0.000100 \\ -0.009955 \end{bmatrix} a_k \tag{3.35}$$

and Eq. (3.32) gives the calculation of the absolute acceleration response.

By subjecting the SDOF system to the 1994 Northridge earthquake record, as shown in Figure 3.3, with the initial displacement and velocity both set equal to zero (i.e. $x_0 = \dot{x}_0 = 0$), it follows that the displacement, velocity, and absolute acceleration response histories can be evaluated and the results are presented in Figure 3.3.

Finally, assume that the SDOF system has a mass of $M = 1.0$ Mg, $C = 0.419$ kN s/m, and $K = 17.55$ kN/m. This gives:

$$\omega = \sqrt{K/M} = \sqrt{17.55/1.0} = 4.189 \ \text{rad/s} \tag{3.36a}$$

$$T = 2\pi\sqrt{M/K} = 2\pi\sqrt{1.0/17.55} = 1.5\,\text{s} \tag{3.36b}$$

$$\zeta = C/2\omega M = 0.419/(2\times 4.189\times 1.0) = 0.05 = 5\% \tag{3.36c}$$

Assume again that the time step size of the earthquake ground motion is $\Delta t = 0.01$ s, the state transition matrix $\mathbf{F}_d$ as given in Eq. (3.22) and the ground acceleration transition matrix $\mathbf{H}_d$ defined in Eq. (3.20) can be evaluated as:

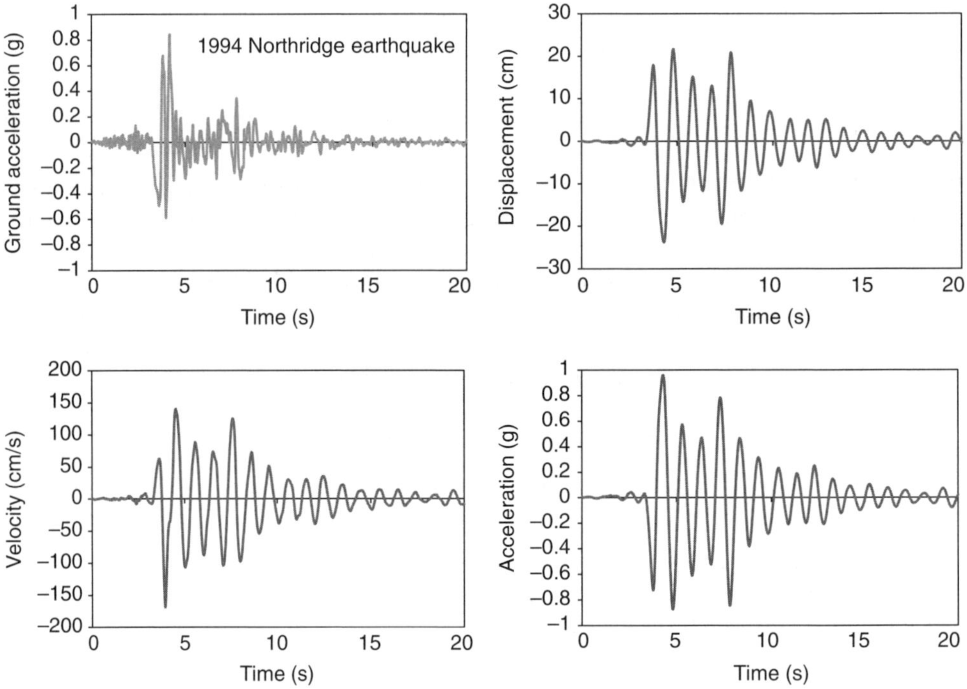

Figure 3.3 Response of a SDOF system with period of 1.0 s due to 1994 Northridge earthquake.

$$\mathbf{F}_d = \begin{bmatrix} 0.999124 & 0.009976 \\ -0.17504 & 0.994945 \end{bmatrix}, \quad \mathbf{H}_d = \begin{bmatrix} -0.000100 \\ -0.009949 \end{bmatrix} \tag{3.37}$$

The recursive equation as given in Eq. (3.21) becomes

$$\begin{Bmatrix} x_{k+1} \\ \dot{x}_{k+1} \end{Bmatrix} = \begin{bmatrix} 0.999124 & 0.009976 \\ -0.17504 & 0.994945 \end{bmatrix} \begin{Bmatrix} x_k \\ \dot{x}_k \end{Bmatrix} + \begin{bmatrix} -0.000100 \\ -0.009949 \end{bmatrix} a_k \tag{3.38}$$

and Eq. (3.32) gives the calculation of the absolute acceleration response.

By subjecting the SDOF system to the 1940 El-Centro earthquake record as shown in Figure 3.4, with the initial displacement and velocity both set equal to zero (i.e. $x_0 = \dot{x}_0 = 0$), it follows that the displacement, velocity, and absolute acceleration response histories can be evaluated and the results are presented in Figure 3.4.

3.2 Dynamic Analysis with Material Nonlinearity

3.2.1 Force Analogy Method

Changing the displacement, not the stiffness, to capture the yielding force is the basic concept of the FAM and was presented in detail for static analysis in Chapter 2. The details of the method are briefly summarized here with the representation of the equations in time domain for dynamic analysis.

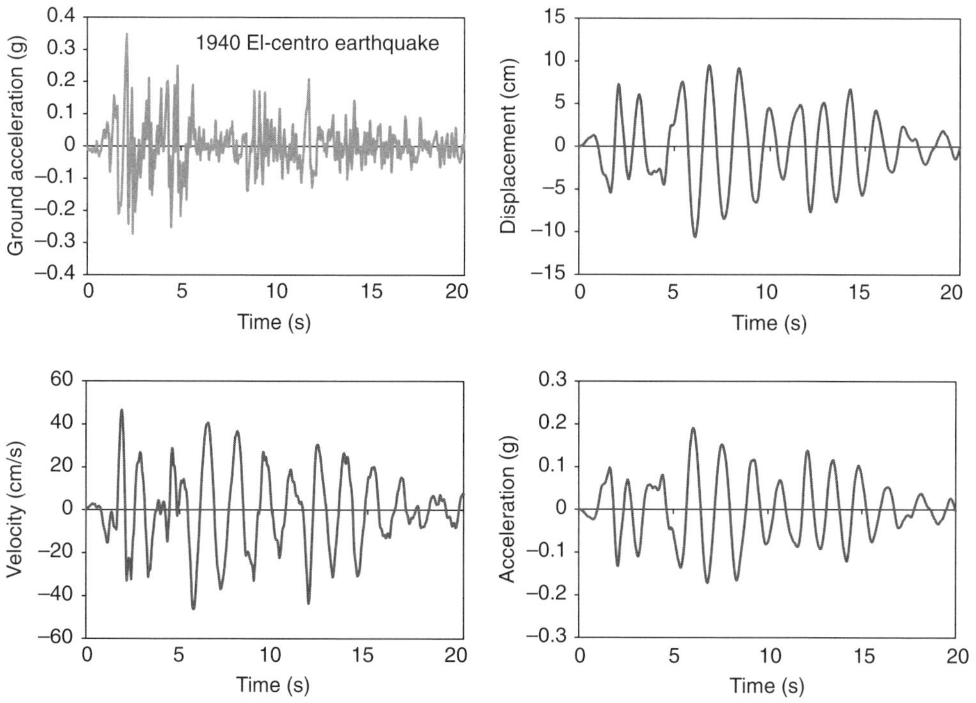

Figure 3.4 Response of a SDOF system with period of 1.5 s due to 1940 El-Centro earthquake.

The total displacement $\mathbf{x}(t)$ of a structure can be represented as the summation of the elastic displacement $\mathbf{x}'(t)$, which can be recovered after unloading, and the inelastic displacement $\mathbf{x}''(t)$, which represents permanent deformation once the structure exhibits inelastic behavior and is often called as the "residual displacement". In a moment-resisting frame with n degrees of freedom (DOFs) and q plastic hinge locations (PHLs), the displacement relationship can be expressed as

$$\mathbf{x}(t) = \mathbf{x}'(t) + \mathbf{x}''(t) \tag{3.39}$$

Similar to the total displacements in Eq. (3.39), the total moment $\mathbf{m}(t)$ at the PHLs of a typical moment-resisting frame can be separated into elastic moment and inelastic moment, i.e.

$$\mathbf{m}(t) = \mathbf{m}'(t) + \mathbf{m}''(t) \tag{3.40}$$

where $\mathbf{m}'(t)$ is the elastic moment due to the elastic displacement $\mathbf{x}'(t)$, and $\mathbf{m}''(t)$ is the inelastic moment due to the inelastic displacement $\mathbf{x}''(t)$. The displacements in Eq. (3.39) and the moments in Eq. (3.40) are related by the following equations:

$$\mathbf{m}'(t) = \mathbf{K}'^T \mathbf{x}'(t) \tag{3.41}$$

$$\mathbf{m}''(t) = -\left(\mathbf{K}'' - \mathbf{K}'^T\mathbf{K}^{-1}\mathbf{K}'\right)\mathbf{\Theta}''(t) \tag{3.42}$$

where $\mathbf{\Theta}''(t)$ is the plastic rotation at the PHLs, $\mathbf{K}$ is the $n \times n$ global stiffness matrix, $\mathbf{K}'$ is the $n \times q$ matrix that relates the plastic rotations at the PHLs and the forces at the DOFs, and $\mathbf{K}''$ is the $q \times q$ matrix that relates the plastic rotations with the corresponding moments at the PHLs.

Substituting Eqs. (3.41) and (3.42) into Eq. (3.40) and rearranging the terms, the first governing equation of the FAM for dynamic analysis is obtained as follows:

$$\mathbf{m}(t) + \mathbf{K}''\mathbf{\Theta}''(t) = \mathbf{K}'^T\mathbf{x}(t) \tag{3.43}$$

The second governing equation of the FAM relates the inelastic displacement $\mathbf{x}''(t)$ at the DOFs and the plastic rotation $\mathbf{\Theta}''(t)$ in the PHLs. This equation can be written as

$$\mathbf{x}''(t) = \mathbf{K}^{-1}\mathbf{K}'\mathbf{\Theta}''(t) \tag{3.44}$$

Note that the equation that relates the structural responses with the applied force, similar to the one shown in Eq. (2.61), is not used in the dynamic analysis.

3.2.2 State Space Analysis with the Force Analogy Method

When the FAM is used, the stiffness force in the equation of motion is calculated by multiplying the initial stiffness matrix $\mathbf{K}$ with the elastic displacement $\mathbf{x}'(t)$. Therefore, For an n-DOF system subjected to earthquake ground motions, this equation can be written as

$$\mathbf{M}\ddot{\mathbf{x}}(t) + \mathbf{C}\dot{\mathbf{x}}(t) + \mathbf{K}\mathbf{x}'(t) = -\mathbf{M}\ddot{\mathbf{g}}(t) \tag{3.45}$$

where the variables have been defined in Eq. (3.4). Replacing the elastic displacement $\mathbf{x}'(t)$ in Eq. (3.45) by the difference of total displacement $\mathbf{x}(t)$ and inelastic displacement $\mathbf{x}''(t)$ as obtained by rearranging the terms in Eq. (3.39), Eq. (3.45) becomes

$$\mathbf{M}\ddot{\mathbf{x}}(t) + \mathbf{C}\dot{\mathbf{x}}(t) + \mathbf{K}\mathbf{x}(t) = -\mathbf{M}\ddot{\mathbf{g}}(t) + \mathbf{K}\mathbf{x}''(t) \tag{3.46}$$

To represent Eq. (3.46) in state space form, let the state vector $\mathbf{z}(t)$ be again defined as

$$\mathbf{z}(t) = \begin{Bmatrix} \mathbf{x}(t) \\ \dot{\mathbf{x}}(t) \end{Bmatrix} \tag{3.47}$$

which is a $2n \times 1$ vector with a collection of states of the responses. It follows from Eq. (3.46) that

$$\dot{\mathbf{z}}(t) = \begin{Bmatrix} \dot{\mathbf{x}}(t) \\ \ddot{\mathbf{x}}(t) \end{Bmatrix} = \begin{bmatrix} \mathbf{0} & \mathbf{I} \\ -\mathbf{M}^{-1}\mathbf{K} & -\mathbf{M}^{-1}\mathbf{C} \end{bmatrix} \begin{Bmatrix} \mathbf{x}(t) \\ \dot{\mathbf{x}}(t) \end{Bmatrix} + \begin{bmatrix} \mathbf{0} \\ -\mathbf{h} \end{bmatrix} \mathbf{a}(t) + \begin{bmatrix} \mathbf{0} \\ \mathbf{M}^{-1}\mathbf{K} \end{bmatrix} \mathbf{x}''(t) \tag{3.48}$$

where the ground acceleration vector $\ddot{\mathbf{g}}(t)$ in Eq. (3.48) has been expressed in terms of the three earthquake ground motion components, i.e.

$$\ddot{\mathbf{g}}(t)=\mathbf{h}\begin{Bmatrix}\ddot{g}_x(t)\\ \ddot{g}_y(t)\\ \ddot{g}_z(t)\end{Bmatrix}=\mathbf{ha}(t) \tag{3.49}$$

similar to what was presented in Eq. (3.13).

To simplify Eq. (3.48), let

$$\mathbf{A}=\begin{bmatrix}\mathbf{0} & \mathbf{I}\\ -\mathbf{M}^{-1}\mathbf{K} & -\mathbf{M}^{-1}\mathbf{C}\end{bmatrix},\qquad \mathbf{H}=\begin{bmatrix}\mathbf{0}\\ -\mathbf{h}\end{bmatrix},\qquad \mathbf{G}=\begin{bmatrix}\mathbf{0}\\ \mathbf{M}^{-1}\mathbf{K}\end{bmatrix} \tag{3.50}$$

where $\mathbf{A}$ is again the $2n\times 2n$ state transition matrix in the continuous form, $\mathbf{H}$ is again the $2n\times 3$ ground motion transition matrix in the continuous form, and $\mathbf{G}$ is the $2n\times n$ inelastic displacement transition matrix in the continuous form. Then, Eq. (3.48) becomes

$$\dot{\mathbf{z}}(t)=\mathbf{Az}(t)+\mathbf{Ha}(t)+\mathbf{Gx}''(t) \tag{3.51}$$

Solving for the first-order linear differential equation in Eq. (3.51) gives

$$\mathbf{z}(t)=\mathbf{e}^{\mathbf{A}(t-t_o)}\mathbf{z}(t_o)+\mathbf{e}^{\mathbf{A}t}\int_{t_o}^{t}\mathbf{e}^{-\mathbf{A}s}[\mathbf{Ha}(s)+\mathbf{Gx}''(s)]ds \tag{3.52}$$

where t_o is the time of reference when the integration begins, which is typically the time when the states $\mathbf{z}(t_o)$ are known, such as the initial conditions.

To integrate Eq. (3.52) numerically, let $t_{k+1}=t$, $t_k=t_o$, and $\Delta t=t_{k+1}-t_k$, and the subscript k denotes the kth time step, then it follows from Eq. (3.52) that

$$\mathbf{z}_{k+1}=\mathbf{e}^{\mathbf{A}\Delta t}\mathbf{z}_k+\mathbf{e}^{\mathbf{A}t_{k+1}}\int_{t_k}^{t_{k+1}}\mathbf{e}^{-\mathbf{A}s}[\mathbf{Ha}(s)+\mathbf{Gx}''(s)]ds \tag{3.53}$$

Using the delta function approximation for the variables in the integral, where the ground acceleration vector and the inelastic displacement vector take the form:

$$\mathbf{a}(s)=\mathbf{a}_k\delta(s-t_k)\Delta t,\quad t_k\le s<t_{k+1} \tag{3.54a}$$

$$\mathbf{x}''(s)=\mathbf{x}_k''\delta(s-t_k)\Delta t,\quad t_k\le s<t_{k+1} \tag{3.54b}$$

Substituting Eq. (3.54) into Eq. (3.53) and performing the integration gives

$$\mathbf{z}_{k+1}=\mathbf{e}^{\mathbf{A}\Delta t}\mathbf{z}_k+\Delta t\mathbf{e}^{\mathbf{A}\Delta t}\mathbf{Ha}_k+\Delta t\mathbf{e}^{\mathbf{A}\Delta t}\mathbf{Gx}_k'' \tag{3.55}$$

where $\mathbf{z}_k$, $\mathbf{a}_k$, and $\mathbf{x}_k''$ are the discretized forms of $\mathbf{z}(t)$, $\mathbf{a}(t)$, and $\mathbf{x}''(t)$, respectively. Let

$$\mathbf{F}_d=\mathbf{e}^{\mathbf{A}\Delta t},\qquad \mathbf{H}_d=\mathbf{e}^{\mathbf{A}\Delta t}\mathbf{H}\Delta t,\qquad \mathbf{G}_d=\mathbf{e}^{\mathbf{A}\Delta t}\mathbf{G}\Delta t \tag{3.56}$$

Then Eq. (3.55) becomes

$$\mathbf{z}_{k+1} = \mathbf{F}_d \mathbf{z}_k + \mathbf{H}_d \mathbf{a}_k + \mathbf{G}_d \mathbf{x}''_k \tag{3.57}$$

3.2.3 Solution Procedure

Equation (3.57) is a recursive equation written in explicit form, which can be proven to be more computationally efficient than any implicit numerical algorithms. Based on all known information at time step k, both displacement $\mathbf{x}_{k+1}$ and velocity $\dot{\mathbf{x}}_{k+1}$, which are embedded in $\mathbf{z}_{k+1}$, can be calculated using Eq. (3.57). This completes the portion of using the state space method in the analysis, and the computed displacement $\mathbf{x}_{k+1}$ is passed to the FAM to compute the internal forces of the structure.

Consider the first governing equation of the FAM in Eq. (3.43) with the input displacement $\mathbf{x}_{k+1}$ coming from the state space method of analysis. To take care of the loading and unloading issues in a dynamic analysis, let the plastic rotation vector be written in an incremental form as:

$$\mathbf{\Theta}''_{k+1} = \mathbf{\Theta}''_k + \Delta\mathbf{\Theta}'' \tag{3.58}$$

where $\Delta\mathbf{\Theta}''$ is the change in plastic rotation from time step k to time step k+1, and $\mathbf{\Theta}''_k$ is the discretized form of $\mathbf{\Theta}''(t)$. Rewriting Eq. (3.43) in the discretized form at time step k+1 and substituting Eq. (3.58) into the resulting discretized equation gives

$$\mathbf{m}_{k+1} + \mathbf{K}''\Delta\mathbf{\Theta}'' = \mathbf{K}'^T \mathbf{x}_{k+1} - \mathbf{K}''\mathbf{\Theta}''_k \tag{3.59}$$

where $\mathbf{m}_{k+1}$ is the discretized form of $\mathbf{m}(t)$. Given the right-hand side of Eq. (3.59) is known, the objective is to calculate $\mathbf{m}_{k+1}$ and $\Delta\mathbf{\Theta}''$, a total of $2q$ unknowns, where q represents the total number of PHLs. Equation (3.59) provides q equations, along with q moment versus plastic rotation relationships from each plastic hinge, giving a total of $2q$ equations to solve for the $2q$ unknowns uniquely. Once $\mathbf{m}_{k+1}$ and $\Delta\mathbf{\Theta}''$ are solved, the plastic rotation at time step k+1 (i.e. $\mathbf{\Theta}''_{k+1}$) can be obtained using Eq. (3.58). Then, substituting the result into the second governing equation of the FAM in Eq. (3.44) gives the inelastic displacement $\mathbf{x}''_{k+1}$:

$$\mathbf{x}''_{k+1} = \mathbf{K}^{-1}\mathbf{K}'\mathbf{\Theta}''_{k+1} \tag{3.60}$$

This completes the portion of using the FAM in computing the internal forces of the structure, and the inelastic displacement $\mathbf{x}''_{k+1}$ is passed back to the state space method of Eq. (3.57) to compute the response at the next time step.

Finally, the absolute acceleration vector in Eq. (3.27) can be computed by rewriting Eq. (3.28) as

$$\ddot{\mathbf{y}}_k = -\mathbf{M}^{-1}\mathbf{C}\dot{\mathbf{x}}_k - \mathbf{M}^{-1}\mathbf{K}\mathbf{x}'_k \tag{3.61}$$

where again the stiffness force is equal to the stiffness matrix multiplied by the elastic displacement $\mathbf{x}'_k$, and $\mathbf{x}'_k$ is the discretized form of $\mathbf{x}'(t)$. Solving for $\mathbf{x}'_k$ in Eq. (3.39) and

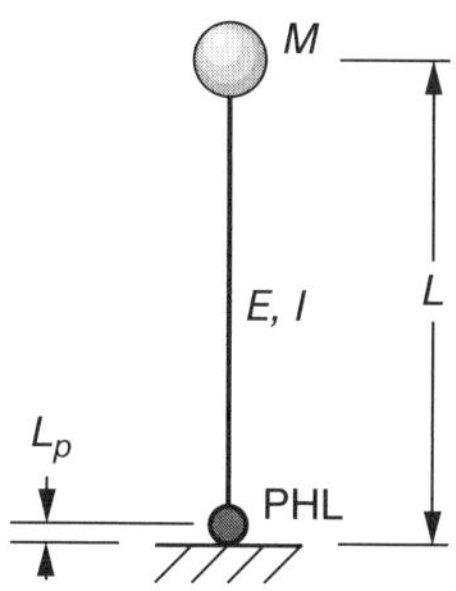

Figure 3.5 A SDOF system with one PHL.

substituting the result into Eq. (3.61) gives the equation for calculating the absolute acceleration response:

$$\ddot{\mathbf{y}}_k = -\mathbf{M}^{-1}\mathbf{C}\dot{\mathbf{x}}_k - \mathbf{M}^{-1}\mathbf{K}\left(\mathbf{x}_k - \mathbf{x}_k''\right) \tag{3.62}$$

This completes the calculation of all response quantities in a nonlinear dynamic analysis.

Example 3.2 Force Analogy Method with Input Displacement
Unlike static analysis in Chapter 2 where the input is the applied force, dynamic analysis requires an input displacement to the FAM as shown in Eq. (3.59). Consider a SDOF system as shown in Figure 3.5 with a plastic hinge at the base of the column that exhibits elastic–plastic behavior with a yield moment of m_p. The stiffness matrices for this column and the moment versus plastic rotation relationship can be calculated as:

$$K = \frac{3EI}{L^3}, \qquad K' = \frac{3EI}{L^3}\left(L - L_p\right), \qquad K'' = \frac{3EI}{L^3}\left(L - L_p\right)^2 \tag{3.63a}$$

$$if \begin{cases} m(t) \le m_p \\ m(t) > m_p \end{cases}, \qquad then \begin{cases} \theta''(t) = 0 \\ m(t) = m_p \end{cases} \tag{3.63b}$$

Now let the lateral displacement of the mass be monotonically applied (i.e. applying a displacement to the system rather than a force, or sometimes referred to as a displacement control type of loading) and be equal to $x_k = m_p L^2/4EI$ at time step k. Also for simplicity, let $L_p = 0$. It follows from Eqs. (3.43) and (3.63a) that

$$m_k + \frac{3EI}{L}\theta_k'' = \frac{3EI}{L^2}\left(\frac{m_p L^2}{4EI}\right) \tag{3.64}$$

First, assume that $\theta_k'' = 0$, i.e. the column remains linear. It follows from Eq. (3.64) that

$$m_k = 3m_p/4 \tag{3.65}$$

Since the calculated m_k is less than m_p, the assumption that the column remains linear is correct, and the moment is therefore given in Eq. (3.65).

Now instead, let the lateral displacement of the mass be monotonically applied and be equal to $m_pL^2/2EI$ at time step k. It follows from Eqs. (3.43) and (3.63a) that

$$m_k + \frac{3EI}{L}\theta_k'' = \frac{3EI}{L^2}\left(\frac{m_pL^2}{2EI}\right) \tag{3.66}$$

First, assume that $\theta_k''=0$, i.e. the column remains linear. It follows from Eq. (3.66) that

$$m_k = 3m_p/2 \tag{3.67}$$

Since the calculated m_k is greater than m_p, the assumption that the column remains linear is incorrect and the column has yielded. Now, let $m_k = m_p$ according to Eq. (3.63b). It follows from Eq. (3.43) that

$$m_p + \frac{3EI}{L}\theta_k'' = \frac{3EI}{L^2}\left(\frac{m_pL^2}{2EI}\right) \tag{3.68}$$

Solving for θ_k'' in Eq. (3.68) gives

$$\theta_k'' = \frac{m_pL}{6EI} \tag{3.69}$$

which is the final answer for θ_k''.

Note that a positive moment $m_k = m_p$ induces a positive plastic rotation $\theta_k'' = m_pL/6EI$. This is an important requirement in dynamic analysis for keeping track of the plastic hinge development, which will be seen in the next example.

Example 3.3 Force Analogy Method with Input Displacement after Yielding
Consider the SDOF system as shown in Figure 3.5 with $L_p = 0$ has reached a lateral displacement of $m_pL^2/2EI$ at time step k, where m_p is the yield moment of the plastic hinge. According to Example 3.2, the moment versus plastic rotation equation is

$$\text{if } \begin{cases} m(t) \le m_p \\ m(t) > m_p \end{cases}, \quad \text{then } \begin{cases} \theta''(t) = 0 \\ m(t) = m_p \end{cases} \tag{3.70}$$

and the plastic hinge response was calculated as

$$\theta_k'' = \frac{m_pL}{6EI}, \qquad m_k = m_p \tag{3.71}$$

Now let the lateral displacement of the mass at time step $k+1$ be equal to $x_{k+1} = 2m_pL^2/3EI$. Since the loading is an increase in displacement from the previous time step, incremental analysis presented in Eq. (3.59) must be used. Based on the new displacement at time step t_{k+1}, it follows from Eqs. (3.59) with stiffnesses given in Eq. (3.63a) that

$$m_{k+1}+\frac{3EI}{L}\Delta\theta''=\frac{3EI}{L^2}\left(\frac{2m_pL^2}{3EI}\right)-\frac{3EI}{L}\left(\frac{m_pL}{6EI}\right) \tag{3.72}$$

First, assume that $\Delta\theta''=0$, i.e. the column is at the unloading stage. It follows from Eq. (3.72) that

$$m_{k+1}=3m_p/2 \tag{3.73}$$

Since the calculated m_{k+1} is greater than m_p, the assumption that the column is at the unloading stage is incorrect. The displacement is actually getting larger and resulting plastic rotation must be increased as well. Now, let $m_{k+1}=m_p$ according to Eq. (3.70). It follows from Eq. (3.72) that

$$m_p+\frac{3EI}{L}\Delta\theta''=\frac{3EI}{L^2}\left(\frac{2m_pL^2}{3EI}\right)-\frac{3EI}{L}\left(\frac{m_pL}{6EI}\right) \tag{3.74}$$

Solving for $\Delta\theta''$ gives

$$\Delta\theta''=\frac{m_pL}{6EI} \tag{3.75}$$

and therefore, the total plastic rotation at time step k+1 becomes

$$\theta''_{k+1}=\theta''_k+\Delta\theta''=\frac{m_pL}{6EI}+\frac{m_pL}{6EI}=\frac{m_pL}{3EI} \tag{3.76}$$

This completes the calculation of response at time step t_{k+1}.

Now let the lateral displacement of the mass at time step $k+2$ be equal to $x_{k+2}=m_pL^2/2EI$ (i.e. a reduction in displacement from x_{k+1}). With the moment equal to m_p and plastic rotation equal to $m_pL/3EI$ at time step k+1, it follows from Eqs. (3.59) with stiffnesses given in Eq. (3.63a) that

$$m_{k+2}+\frac{3EI}{L}\Delta\theta''=\frac{3EI}{L^2}\left(\frac{m_pL^2}{2EI}\right)-\frac{3EI}{L}\left(\frac{m_pL}{3EI}\right) \tag{3.77}$$

First, assume that $\Delta\theta''=0$, i.e. the column is at the unloading stage. It follows from Eq. (3.77) that

$$m_{k+2}=m_p/2 \tag{3.78}$$

Since the calculated m_{k+2} is less than m_p, the assumption that the column is at the unloading stage is correct, and the moment is therefore given in Eq. (3.78). The plastic rotation is then calculated as

$$\theta''_{k+2}=\theta''_{k+1}+\Delta\theta''=\frac{m_pL}{3EI}+0=\frac{m_pL}{3EI} \tag{3.79}$$

Note that a positive moment $m_{k+1}=m_p$ induces a positive incremental plastic rotation $\Delta\theta''=m_pL/6EI$, and a positive reduced moment $m_{k+2}=m_p/2$ induces no additional plastic rotation at the plastic hinge (i.e. $\theta''_{k+2} = \theta''_{k+1}$ and $\Delta\theta'' = 0$). Typically, if a plastic hinge is assumed to yield in a positive direction with a positive moment but negative plastic rotation is obtained, it means that the assumption that the plastic hinge has further yielding is incorrect, and the moment should be calculated by assuming the plastic hinge is responding in the unloading path with $\Delta\theta'' = 0$.

Example 3.4 Force Analogy Method Using Hardening Model with Input Displacement

Consider the SDOF system used in Example 3.2 with $L_p = 0$ as shown in Figure 3.5 but with a moment versus plastic rotation relationship that exhibits a hardening behavior with post-yield stiffness K_t. Regardless of how the column responded to the input displacement prior to time t_k, assume that the responses at time t_k is on the hardening branch with the following values:

$$x_k = \frac{m_q L^2}{2EI}, \qquad \theta''_k = \frac{m_q L}{6EI}, \qquad m_k = m_q \tag{3.80}$$

where m_q is the yield moment value at time step k, which can be computed by adding the initial yield moment m_p with the post-yield stiffness of the plastic hinge K_t multiplied by the total plastic rotation θ''_k, i.e.

$$m_q = m_p + K_t \theta''_k \tag{3.81}$$

First, consider that the displacement at time step k+1 is increased to $x_{k+1} = 2m_q L^2/3EI$. Let the moment versus plastic rotation relationship be a bilinear model with a hardening stiffness of $K_t = EI/L$ after yielding. Then at this time step, the moment versus plastic rotation relationship can be written as

$$if \begin{cases} m_{k+1} \le m_q \\ m_{k+1} > m_q \end{cases}, \qquad then \begin{cases} \Delta\theta'' = 0 \\ m_{k+1} = m_q + (EI/L)\Delta\theta'' \end{cases} \tag{3.82}$$

Note the similarity between Eq. (3.81) and Eq. (3.82) in the calculation of moments by following the hysteresis curve. From Eq. (3.59),

$$m_{k+1} + \frac{3EI}{L}\Delta\theta'' = \frac{3EI}{L^2}\left(\frac{2m_q L^2}{3EI}\right) - \frac{3EI}{L}\left(\frac{m_q L}{6EI}\right) \tag{3.83}$$

First, assume that $\Delta\theta'' = 0$, i.e. the column is at the unloading stage. It follows from Eq. (3.83) that

$$m_{k+1} = 3m_q/2 \tag{3.84}$$

Since the calculated m_{k+1} is greater than m_q, the assumption that the column is at the unloading stage is incorrect according to Eq. (3.82). The displacement is actually getting larger and the resulting plastic rotation must be increased as well. Now, let $m_{k+1} = m_q + (EI/L)\Delta\theta''$ according to Eq. (3.82). It follows from Eq. (3.83) that

$$m_q + \left(\frac{3EI}{L} + \frac{EI}{L}\right)\Delta\theta'' = \frac{3EI}{L^2}\left(\frac{2m_qL^2}{3EI}\right) - \frac{3EI}{L}\left(\frac{m_qL}{6EI}\right) \tag{3.85}$$

Solving for $\Delta\theta''$ gives

$$\Delta\theta'' = \frac{m_qL}{8EI} \tag{3.86}$$

and therefore, the total plastic rotation and moment becomes

$$\theta''_{k+1} = \theta''_k + \Delta\theta'' = \frac{m_qL}{6EI} + \frac{m_qL}{8EI} = \frac{7m_qL}{24EI} \tag{3.87a}$$

$$m_{k+1} = m_q + \left(\frac{EI}{L}\right)\left(\frac{m_qL}{8EI}\right) = \frac{9m_q}{8} \tag{3.87b}$$

This completes the calculation of response at time step $k+1$. In summary,

$$x_{k+1} = \frac{2m_qL^2}{3EI}, \qquad \theta''_{k+1} = \frac{7m_qL}{24EI}, \qquad m_{k+1} = \frac{9m_q}{8} \tag{3.88}$$

and the moment versus plastic rotation are updated for the next time step, and it becomes

$$\text{if} \begin{cases} m_{k+2} \le 9m_q/8 \\ m_{k+2} > 9m_q/8 \end{cases}, \quad \text{then} \begin{cases} \Delta\theta'' = 0 \\ m_{k+2} = 9m_q/8 + (EI/L)\Delta\theta'' \end{cases} \tag{3.89}$$

Example 3.5 Force Analogy Method Using Softening Model with Input Displacement Now consider Example 3.4 again with the responses at time step k given in Eq. (3.80), which are

$$x_k = \frac{m_qL^2}{2EI}, \qquad \theta''_k = \frac{m_qL}{6EI}, \qquad m_k = m_q \tag{3.90}$$

Also, let the displacement be increased to $x_{k+1} = 2m_qL^2/3EI$ at time step $k+1$, but in this example let the moment versus plastic rotation relationship be a bilinear model with a softening stiffness of $K_t = -EI/L$ after yielding. Then at this time step, the moment versus plastic rotation relationship following Eq. (3.90) can be written as

$$\text{if} \begin{cases} m_{k+1} \le m_q \\ m_{k+1} > m_q \end{cases}, \quad \text{then} \begin{cases} \Delta\theta'' = 0 \\ m_{k+1} = m_q - (EI/L)\Delta\theta'' \end{cases} \tag{3.91}$$

From Eq. (3.59),

$$m_{k+1} + \frac{3EI}{L}\Delta\theta'' = \frac{3EI}{L^2}\left(\frac{2m_qL^2}{3EI}\right) - \frac{3EI}{L}\left(\frac{m_qL}{6EI}\right) \tag{3.92}$$

First, assume that $\Delta\theta'' = 0$, i.e. the column is at the unloading stage. It follows from Eq. (3.92) that

$$m_{k+1} = 3m_q/2 \tag{3.93}$$

Since the calculated m_{k+1} is greater than m_q, the assumption that the column is at the unloading stage is incorrect according to Eq. (3.91). Now, let $m_{k+1} = m_q - (EI/L)\Delta\theta''$ according to Eq. (3.91). It follows from Eq. (3.92) that

$$m_q + \left(\frac{3EI}{L} - \frac{EI}{L}\right)\Delta\theta'' = \frac{3EI}{L^2}\left(\frac{2m_qL^2}{3EI}\right) - \frac{3EI}{L}\left(\frac{m_qL}{6EI}\right) \tag{3.94}$$

Solving for $\Delta\theta''$ gives

$$\Delta\theta'' = \frac{m_qL}{4EI} \tag{3.95}$$

and therefore, the total plastic rotation and moment becomes

$$\theta''_{k+1} = \theta''_k + \Delta\theta'' = \frac{m_qL}{6EI} + \frac{m_qL}{4EI} = \frac{5m_qL}{12EI} \tag{3.96a}$$

$$m_{k+1} = m_q - \left(\frac{EI}{L}\right)\left(\frac{m_qL}{4EI}\right) = \frac{3m_q}{4} \tag{3.96b}$$

This completes the calculation of response at time step k+1. In summary,

$$x_k = \frac{2m_qL^2}{3EI}, \quad \theta''_k = \frac{5m_qL}{12EI}, \quad m_k = \frac{3m_q}{4} \tag{3.97}$$

and the moment versus plastic rotation at the next time step becomes

$$\text{if} \begin{cases} m_{k+2} \le 3m_q/4 \\ m_{k+2} > 3m_q/4 \end{cases}, \quad \text{then} \begin{cases} \Delta\theta'' = 0 \\ m_{k+2} = 3m_q/4 - (EI/L)\Delta\theta'' \end{cases} \tag{3.98}$$

Note from this example that the history of the response is not needed in the calculation from time step k to time step k+1. This means once the column yielded, the initial yield moment m_p can be discarded. Therefore, it is advisable to update the yield moment at every time step to keep track of the subsequent yielding behavior in a dynamic analysis.

Example 3.6 Dynamic Analysis of SDOF Systems with the Force Analogy Method
Consider the SDOF system as shown in Figure 3.5 with a plastic hinge at the base of the column that exhibits a kinematic hardening relationship. Let the properties of the column be $E = 1,000$ kN/m^2, $L = 10$ m, $L_p = 0$ m, and a mass of $M = 1.0$ Mg.

First, assume the column has other properties of $C = 0.0$ Mg s/m, $I = 52.64$ m^4, a yield displacement of $d_y = 15$ cm, and an elastic–plastic moment versus plastic rotation relationship. This gives the following structural properties:

$$K = 3EI/L^3 = 3\times 1000\times 52.64/(10)^3 = 157.92\ \text{kN/m} \tag{3.99a}$$

$$K' = 3EI/L^2 = 3\times 1000\times 52.64/(10)^2 = 1579.2\ \text{kN} \tag{3.99b}$$

$$K'' = 3EI/L = 3\times 1000\times 52.64/10 = 15,792\ \text{kN m} \tag{3.99c}$$

$$T = 2\pi\sqrt{M/K} = 2\pi\sqrt{1/157.92} = 0.5\ \text{s} \tag{3.99d}$$

$$m_p = d_y\times 3EI/L^2 = 0.15\times 1579.2 = 236.9\ \text{kN m} \tag{3.99e}$$

$$if \begin{cases} m_k \le 236.9 \\ m_k > 236.9 \end{cases}, \quad then \begin{cases} \Delta\theta'' = 0 \\ m_k = 236.9 \end{cases} \tag{3.99f}$$

With a time step size of the earthquake ground motion of $\Delta t = 0.01$ s, the matrices in Eq. (3.56) are evaluated as:

$$\mathbf{F}_d = \begin{bmatrix} 0.992115 & 0.009974 \\ -1.57498 & 0.992115 \end{bmatrix}, \quad \mathbf{H}_d = \begin{bmatrix} -0.000100 \\ -0.009921 \end{bmatrix}, \quad \mathbf{G}_d = \begin{bmatrix} 0.01575 \\ 1.5667 \end{bmatrix} \tag{3.100}$$

The recursive equation as given in Eq. (3.57) becomes

$$\begin{Bmatrix} x_{k+1} \\ \dot{x}_{k+1} \end{Bmatrix} = \begin{bmatrix} 0.992115 & 0.009974 \\ -1.57498 & 0.992115 \end{bmatrix} \begin{Bmatrix} x_k \\ \dot{x}_k \end{Bmatrix} + \begin{bmatrix} -0.000100 \\ -0.009921 \end{bmatrix} a_k + \begin{bmatrix} 0.01575 \\ 1.5667 \end{bmatrix} x''_k \tag{3.101}$$

and the governing equations of the FAM in Eqs. (3.59) and (3.60) are

$$m_{k+1} + (15792)\Delta\theta'' = (1579.2)x_{k+1} - (15792)\theta''_k \tag{3.102a}$$

$$x''_{k+1} = 10\left(\theta''_k + \Delta\theta''\right) \tag{3.102b}$$

Finally, the absolute acceleration response can be calculated using Eq. (3.62) as

$$\ddot{y}_{k+1} = -157.92\left(x_{k+1} - x''_{k+1}\right) \tag{3.103}$$

Assume that the earthquake ground acceleration record is the 1995 Kobe earthquake record as shown in Figure 3.2, and the initial displacement and velocity are both set equal to zero (i.e. $x_0 = \dot{x}_0 = 0$). It follows that the displacement, velocity, acceleration, moment, and plastic rotation response histories can be evaluated using Eqs. (3.101) through Eq. (3.103), and the results are presented in Figure 3.6. The responses are plotted along with the linear responses from Example 3.1 as comparisons (note that both the present example and Example 3.1 use the same period of 0.5 s, damping of 0%, and the same Kobe earthquake ground motion as the input). In addition, the yielding envelope and the backbone curve are plotted in the figure as appropriate.

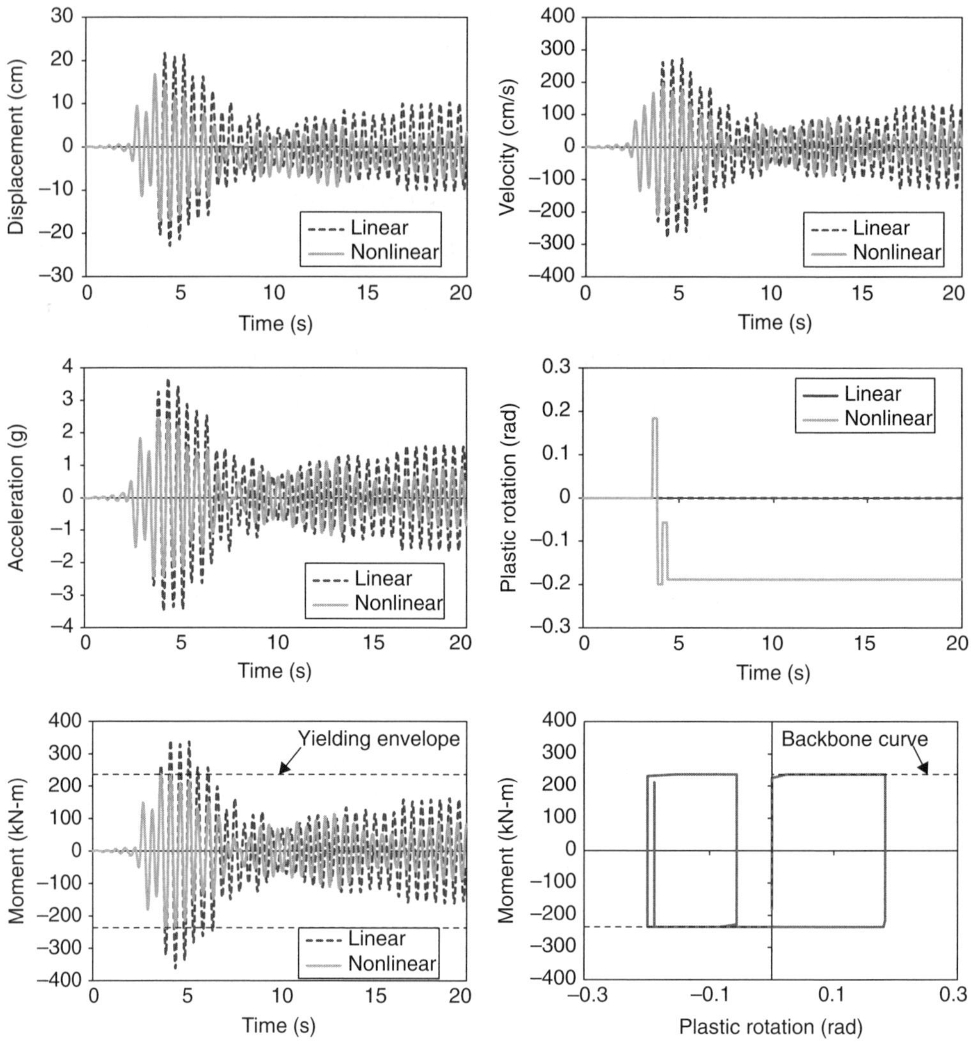

Figure 3.6 Comparison of 0% damped SDOF responses due to 1995 Kobe earthquake.

Now assume that the SDOF system has properties of $C = 0.251$ kN s/m, $I = 13.16\ \text{m}^4$, $d_y = 10$ cm, and a bilinear moment versus plastic rotation relationship with a kinematic hardening ratio of 8%. This gives the following structural properties:

$$K = 3EI/L^3 = 3 \times 1000 \times 13.16/(10)^3 = 39.48\ \text{kN/m} \tag{3.104a}$$

$$K' = 3EI/L^2 = 3 \times 1000 \times 13.16/(10)^2 = 394.8\ \text{kN} \tag{3.104b}$$

$$K'' = 3EI/L = 3 \times 1000 \times 13.16/10 = 3{,}948\ \text{kN m} \tag{3.104c}$$

$$T = 2\pi\sqrt{M/K} = 2\pi\sqrt{1/39.48} = 1.0\ \text{s} \tag{3.104d}$$

$$\zeta = C/2\omega M = 0.251/\left(2\times\sqrt{39.48}\times 1.0\right) = 0.02 = 2\% \tag{3.104e}$$

$$m_p = d_y \times 3EI/L^2 = 0.10\times 394.8 = 39.48\,\text{kN m} \tag{3.104f}$$

$$if \begin{cases} m_{k+1} \le 39.48 \\ m_{k+1} > 39.48 \end{cases}, \quad then \begin{cases} \Delta\theta'' = 0 \\ m_{k+1} = m_k + 0.08K''\Delta\theta'' \end{cases} \tag{3.104g}$$

The recursive equation as given in Eq. (3.57) becomes

$$\begin{Bmatrix} x_{k+1} \\ \dot{x}_{k+1} \end{Bmatrix} = \begin{bmatrix} 0.998028 & 0.009981 \\ -0.39403 & 0.995520 \end{bmatrix} \begin{Bmatrix} x_k \\ \dot{x}_k \end{Bmatrix} + \begin{bmatrix} -0.000100 \\ -0.009955 \end{bmatrix} a_k + \begin{bmatrix} 0.003940 \\ 0.39303 \end{bmatrix} x_k'' \tag{3.105}$$

and the governing equations of the FAM in Eqs. (3.59) and (3.60) are

$$m_{k+1} + (3948)\Delta\theta'' = (394.8)x_{k+1} - (3948)\theta_k'' \tag{3.106a}$$

$$x_{k+1}'' = 10\left(\theta_k'' + \Delta\theta''\right) \tag{3.106b}$$

Finally, the absolute acceleration response can be calculated using Eq. (3.62) as

$$\ddot{y}_{k+1} = -0.251\dot{x}_{k+1} - 39.48\left(x_{k+1} - x_{k+1}''\right) \tag{3.107}$$

Assume that the earthquake ground acceleration record is the 1994 Northridge earthquake record as shown in Figure 3.3, and the initial displacement and velocity are both set equal to zero (i.e. $x_0 = \dot{x}_0 = 0$). It follows that the displacement, velocity, acceleration, moment, and plastic rotation response histories can be evaluated using Eqs. (3.105) through Eq. (3.107), and the results are presented in Figure 3.7. The responses are plotted along with the linear responses from Example 3.1 as comparisons (note that both the present example and Example 3.1 use the same period of 1.0 s, damping of 2%, and the same Northridge earthquake ground motion as the input). In addition, the initial yielding moment and the backbone curve are plotted in the figure as appropriate.

Finally, assume that the SDOF column has properties of $C = 0.419$ kN s/m, $I = 5.85$ m^4, $d_y = 5$ cm, and a bilinear moment versus plastic rotation relationship with a kinematic softening ratio of –4%. This gives the following structural properties:

$$K = 3EI/L^3 = 3\times 1000\times 5.85/(10)^3 = 17.55\,\text{kN/m} \tag{3.108a}$$

$$K' = 3EI/L^2 = 3\times 1000\times 5.85/(10)^2 = 175.5\,\text{kN} \tag{3.108b}$$

$$K'' = 3EI/L = 3\times 1000\times 5.85/10 = 1{,}755\,\text{kN m} \tag{3.108c}$$

$$T = 2\pi\sqrt{M/K} = 2\pi\sqrt{1/17.55} = 1.5\,\text{s} \tag{3.108d}$$

$$\zeta = C/2\omega M = 0.419/\left(2\times\sqrt{17.55}\times 1.0\right) = 0.05 = 5\% \tag{3.108e}$$

$$m_p = d_y \times 3EI/L^2 = 0.05\times 175.5 = 8.775\,\text{kN m} \tag{3.108f}$$

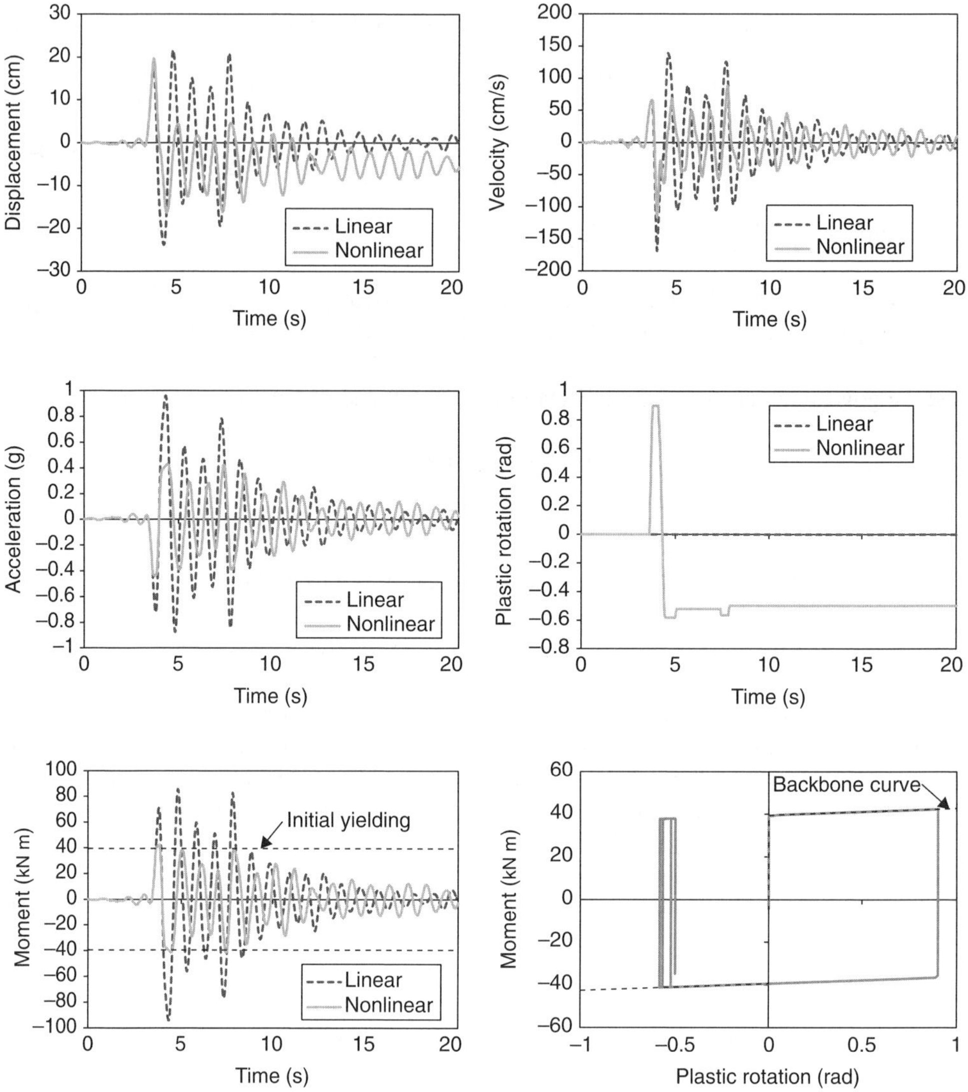

Figure 3.7 Comparison of 2% damped SDOF responses due to 1994 Northridge earthquake.

$$if \begin{cases} m_{k+1} \le 39.48 \\ m_{k+1} > 39.48 \end{cases}, \quad then \begin{cases} \Delta\theta'' = 0 \\ m_{k+1} = m_k - 0.04K''\Delta\theta'' \end{cases} \tag{3.108g}$$

The recursive equation as given in Eq. (3.57) becomes

$$\begin{Bmatrix} x_{k+1} \\ \dot{x}_{k+1} \end{Bmatrix} = \begin{bmatrix} 0.999124 & 0.009976 \\ -0.17504 & 0.994945 \end{bmatrix} \begin{Bmatrix} x_k \\ \dot{x}_k \end{Bmatrix} + \begin{bmatrix} -0.000100 \\ -0.009949 \end{bmatrix} a_k + \begin{bmatrix} 0.001751 \\ 0.17461 \end{bmatrix} x_k'' \tag{3.109}$$

The governing equations of the FAM in Eqs. (3.59) and (3.60) are

$$m_{k+1} + (1755)\Delta\theta'' = (175.5)x_{k+1} - (1755)\theta''_k \tag{3.110a}$$

$$x''_{k+1} = 10(\theta''_k + \Delta\theta'') \tag{3.110b}$$

Finally, the absolute acceleration response can be calculated using Eq. (3.62) as

$$\ddot{y}_{k+1} = -0.419\dot{x}_{k+1} - 17.55(x_{k+1} - x''_{k+1}) \tag{3.111}$$

Assume that the earthquake ground acceleration record is the 1940 El-Centro earthquake record as shown in Figure 3.4, and the initial displacement and velocity are both set equal to zero (i.e. $x_0 = \dot{x}_0 = 0$). It follows that the displacement, velocity, acceleration, moment, and plastic rotation response histories can be evaluated using Eqs. (3.109) through Eq. (3.111), and the results are presented in Figure 3.8. The responses are plotted along with the linear responses from Example 3.1 as comparisons (note that both the present example and Example 3.1 use the same period of 1.5 s, damping of 5%, and the same El-Centro earthquake ground motion as the input). In addition, the initial yielding moment and the backbone curve are plotted in the figure as appropriate.

3.3 Nonlinear Dynamic Analysis with Static Condensation

In Section 2.6, the static condensation was derived for the nonlinear static analysis of framed structures. In Section 3.1, the static condensation was briefly introduced in the dynamic analysis of linear systems. In this section, the knowledge gained from the previous sections is combined for the dynamic analysis of nonlinear systems. The goal is to condense the stiffness matrix when the mass moment of inertia is ignored at certain DOFs. But more importantly, it provides an avenue to invert the mass matrix embedded in the **A** matrix of Eq. (3.50) that would otherwise be undefined with a singular mass matrix.

Consider a moment-resisting frame with n DOFs and q PHLs, the equation of motion can be written in the matrix form as

$$\begin{bmatrix} \mathbf{M}_{dd} & \mathbf{0} \\ \mathbf{0} & \mathbf{0} \end{bmatrix} \begin{Bmatrix} \ddot{\mathbf{x}}_d(t) \\ \ddot{\mathbf{x}}_r(t) \end{Bmatrix} + \begin{bmatrix} \mathbf{C}_{dd} & \mathbf{0} \\ \mathbf{0} & \mathbf{0} \end{bmatrix} \begin{Bmatrix} \dot{\mathbf{x}}_d(t) \\ \dot{\mathbf{x}}_r(t) \end{Bmatrix} + \begin{bmatrix} \mathbf{K}_{dd} & \mathbf{K}_{dr} \\ \mathbf{K}_{rd} & \mathbf{K}_{rr} \end{bmatrix} \begin{Bmatrix} \mathbf{x}'_d(t) \\ \mathbf{x}'_r(t) \end{Bmatrix} = -\begin{bmatrix} \mathbf{M}_{dd} & \mathbf{0} \\ \mathbf{0} & \mathbf{0} \end{bmatrix} \begin{Bmatrix} \ddot{\mathbf{g}}(t) \\ \mathbf{0} \end{Bmatrix} \tag{3.112}$$

where the mass matrix **M** has been defined in Eq. (3.6), the damping matrix **C** has been defined in Eq. (3.9), and the stiffness matrix **K** has been defined in Eq. (2.142). The vector $\mathbf{x}'(t)$ is the elastic displacement response, $\dot{\mathbf{x}}(t)$ is the velocity response, $\ddot{\mathbf{x}}(t)$ is the acceleration response, and earthquake ground acceleration vector $\ddot{\mathbf{g}}(t)$ corresponds to the effect of ground motion at each DOF associated with nonzero mass. The subscript d denotes the number of degrees of freedom that have nonzero mass, and subscript r denotes the number of degrees of freedom that have zero mass moment of inertia. This gives $n = d + r$ in an n-DOF system.

The governing equations of the FAM presented in Eqs. (3.43) and (3.44) can similarly be written as:

$$\mathbf{m}(t) + \mathbf{K}''\mathbf{\Theta}''(t) = \begin{bmatrix} \mathbf{K}'^T_d & \mathbf{K}'^T_r \end{bmatrix} \begin{Bmatrix} \mathbf{x}_d(t) \\ \mathbf{x}_r(t) \end{Bmatrix} \tag{3.113}$$

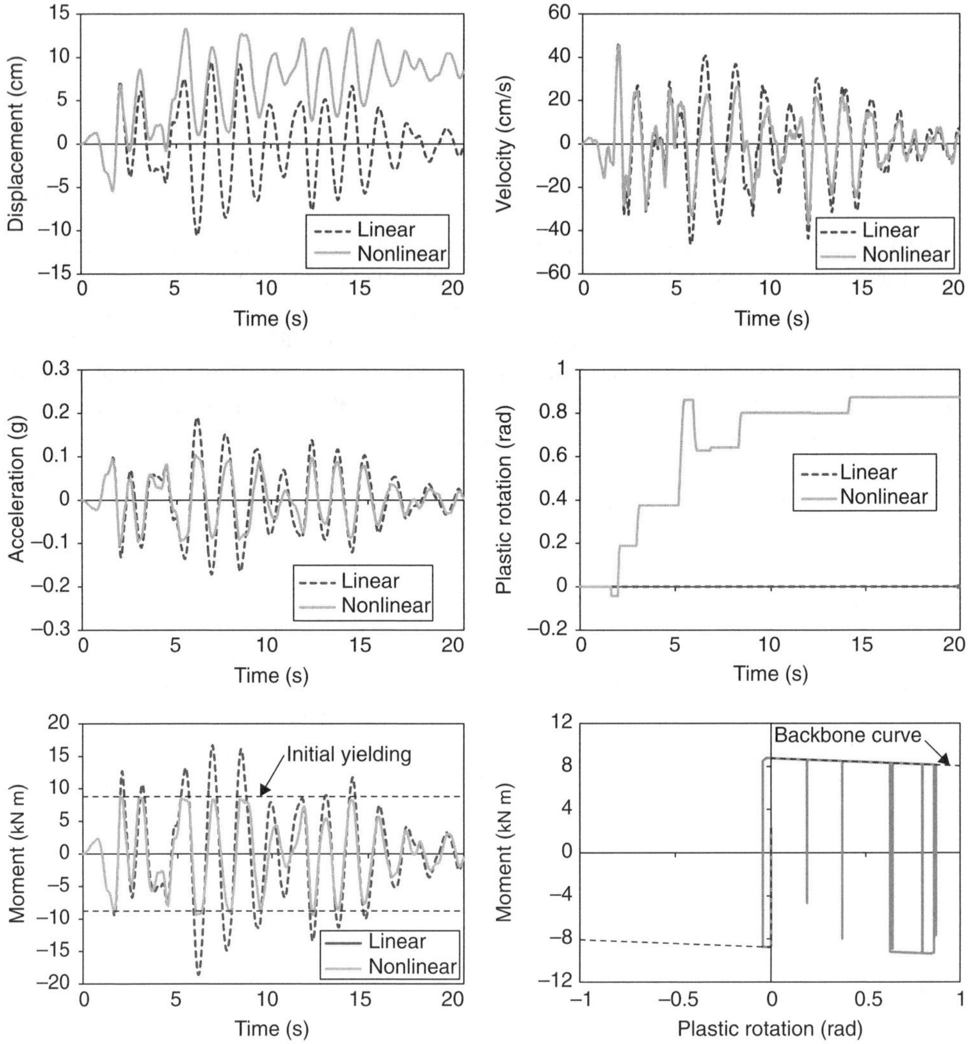

Figure 3.8 Comparison of 5% damped SDOF responses due to 1940 El-Centro earthquake.

$$\begin{Bmatrix} \mathbf{x}_d''(t) \\ \hline \mathbf{x}_r''(t) \end{Bmatrix} = \left[\begin{array}{c|c} \mathbf{K}_{dd} & \mathbf{K}_{dr} \\ \hline \mathbf{K}_{rd} & \mathbf{K}_{rr} \end{array} \right]^{-1} \begin{bmatrix} \mathbf{K}_d' \\ \hline \mathbf{K}_r' \end{bmatrix} \mathbf{\Theta}''(t) \tag{3.114}$$

where $\mathbf{m}(t)$ is the $q \times 1$ total moment vector of the PHLs, $\mathbf{\Theta}''(t)$ is the $q \times 1$ plastic rotation vector of the PHLs, and

$$\mathbf{K}' = \begin{bmatrix} \mathbf{K}_d' \\ \hline \mathbf{K}_r' \end{bmatrix}, \qquad \mathbf{x}''(t) = \begin{Bmatrix} \mathbf{x}_d''(t) \\ \hline \mathbf{x}_r''(t) \end{Bmatrix}, \qquad \begin{Bmatrix} \mathbf{x}_d(t) \\ \hline \mathbf{x}_r(t) \end{Bmatrix} = \begin{Bmatrix} \mathbf{x}_d'(t) \\ \hline \mathbf{x}_r'(t) \end{Bmatrix} + \begin{Bmatrix} \mathbf{x}_d''(t) \\ \hline \mathbf{x}_r''(t) \end{Bmatrix} \tag{3.115}$$

First, consider Eq. (3.112). By performing static condensation, the second equation of Eq. (3.112) is extracted and written in the long form as

$$\mathbf{K}_{rd}\mathbf{x}'_d(t) + \mathbf{K}_{rr}\mathbf{x}'_r(t) = \mathbf{0} \tag{3.116}$$

Solving for $\mathbf{x}'_r(t)$ in Eq. (3.116) gives

$$\mathbf{x}'_r(t) = -\mathbf{K}_{rr}^{-1}\mathbf{K}_{rd}\mathbf{x}'_d(t) \tag{3.117}$$

Now substituting Eq. (3.117) back into the first equation of Eq. (3.112) gives

$$\mathbf{M}_{dd}\ddot{\mathbf{x}}_d(t) + \mathbf{C}_{dd}\dot{\mathbf{x}}_d(t) + \mathbf{K}_{dd}\mathbf{x}'_d(t) - \mathbf{K}_{dr}\mathbf{K}_{rr}^{-1}\mathbf{K}_{rd}\mathbf{x}'_d(t) = -\mathbf{M}_{dd}\ddot{\mathbf{g}}(t) \tag{3.118}$$

As in Eq. (2.149), define

$$\bar{\mathbf{K}} = \mathbf{K}_{dd} - \mathbf{K}_{dr}\mathbf{K}_{rr}^{-1}\mathbf{K}_{rd} \tag{3.119}$$

Then, Eq. (3.118) becomes

$$\mathbf{M}_{dd}\ddot{\mathbf{x}}_d(t) + \mathbf{C}_{dd}\dot{\mathbf{x}}_d(t) + \bar{\mathbf{K}}\mathbf{x}'_d(t) = -\mathbf{M}_{dd}\ddot{\mathbf{g}}(t) \tag{3.120}$$

which represents the equation of motion in the statically condensed form. Similarities can be observed between Eq. (3.45) and Eq. (3.120). The $\mathbf{M}$, $\mathbf{C}$, and $\mathbf{K}$ matrices in Eq. (3.45) have been replaced by $\mathbf{M}_{dd}$, $\mathbf{C}_{dd}$, and $\bar{\mathbf{K}}$ matrices in Eq. (3.120), respectively, and the $\mathbf{x}'(t)$, $\dot{\mathbf{x}}(t)$, and $\ddot{\mathbf{x}}(t)$ vectors in Eq. (3.45) have been replaced by $\mathbf{x}'_d(t)$, $\dot{\mathbf{x}}_d(t)$, and $\ddot{\mathbf{x}}_d(t)$ vectors in Eq. (3.120), respectively.

Now consider Eq. (3.114). Pre-multiplying both sides of Eq. (3.114) by the stiffness matrix $\mathbf{K}$ gives

$$\left[\begin{array}{c|c} \mathbf{K}_{dd} & \mathbf{K}_{dr} \\ \hline \mathbf{K}_{rd} & \mathbf{K}_{rr} \end{array}\right]\left\{\begin{array}{c} \mathbf{x}''_d(t) \\ \hline \mathbf{x}''_r(t) \end{array}\right\} = \left[\begin{array}{c} \mathbf{K}'_d \\ \hline \mathbf{K}'_r \end{array}\right]\boldsymbol{\Theta}''(t) \tag{3.121}$$

Extracting the second equation of Eq. (3.121) and solving for $\mathbf{x}''_r(t)$ gives

$$\mathbf{x}''_r(t) = -\mathbf{K}_{rr}^{-1}\mathbf{K}_{rd}\mathbf{x}''_d(t) + \mathbf{K}_{rr}^{-1}\mathbf{K}'_r\boldsymbol{\Theta}''(t) \tag{3.122}$$

Now substituting Eq. (3.122) back into the first equation of Eq. (3.121) gives

$$\mathbf{K}_{dd}\mathbf{x}''_d(t) + \mathbf{K}_{dr}\left[-\mathbf{K}_{rr}^{-1}\mathbf{K}_{rd}\mathbf{x}''_d(t) + \mathbf{K}_{rr}^{-1}\mathbf{K}'_r\boldsymbol{\Theta}''(t)\right] = \mathbf{K}'_d\boldsymbol{\Theta}''(t) \tag{3.123}$$

and rearranging the terms gives

$$\left[\mathbf{K}_{dd} - \mathbf{K}_{dr}\mathbf{K}_{rr}^{-1}\mathbf{K}_{rd}\right]\ \mathbf{x}''_d(t) = \left[\mathbf{K}'_d - \mathbf{K}_{dr}\mathbf{K}_{rr}^{-1}\mathbf{K}'_r\right]\ \boldsymbol{\Theta}''(t) \tag{3.124}$$

As in Eq. (2.149), define

$$\bar{\mathbf{K}}' = \mathbf{K}'_d - \mathbf{K}_{dr}\mathbf{K}_{rr}^{-1}\mathbf{K}'_r \tag{3.125}$$

Substituting Eqs. (3.119) and (3.125) into Eq. (3.124) gives

$$\bar{\mathbf{K}}\mathbf{x}_d''(t) = \bar{\mathbf{K}}'\boldsymbol{\Theta}''(t) \tag{3.126}$$

Pre-multiplying both sides of Eq. (3.126) by the inverse of the condensed stiffness matrix (i.e. $\bar{\mathbf{K}}^{-1}$) gives

$$\mathbf{x}_d''(t) = \bar{\mathbf{K}}^{-1}\bar{\mathbf{K}}'\boldsymbol{\Theta}''(t) \tag{3.127}$$

which represents the second governing equation of the FAM in the statically condensed form. Similarities can also be observed between Eq. (3.44) and Eq. (3.127). The $\mathbf{K}$ and $\mathbf{K}'$ matrices in Eq. (3.44) have been replaced by $\bar{\mathbf{K}}$ and $\bar{\mathbf{K}}'$ matrices in Eq. (3.127), respectively, and the $\mathbf{x}''(t)$ vector in Eq. (3.44) have been replaced by $\mathbf{x}_d''(t)$ vector in Eq. (3.127).

Finally, consider Eq. (3.113). Expanding the right-hand side of Eq. (3.113) gives

$$\mathbf{m}(t) + \mathbf{K}''\boldsymbol{\Theta}''(t) = \mathbf{K}_d'^T\mathbf{x}_d(t) + \mathbf{K}_r'^T\mathbf{x}_r(t) \tag{3.128}$$

In Eq. (3.128), the term $\mathbf{x}_r(t)$ must be first calculated. Since $\mathbf{x}_r(t) = \mathbf{x}_r'(t) + \mathbf{x}_r''(t)$ according to Eq. (3.115), substituting Eqs. (3.117) and (3.122) into this equation gives

$$\mathbf{x}_r(t) = \mathbf{x}_r'(t) + \mathbf{x}_r''(t) = -\mathbf{K}_{rr}^{-1}\mathbf{K}_{rd}\mathbf{x}_d'(t) - \mathbf{K}_{rr}^{-1}\mathbf{K}_{rd}\mathbf{x}_d''(t) + \mathbf{K}_{rr}^{-1}\mathbf{K}_r'\boldsymbol{\Theta}''(t) \tag{3.129}$$

Since $\mathbf{x}_d'(t) + \mathbf{x}_d''(t) = \mathbf{x}_d(t)$ according to Eq. (3.115), substituting this result in Eq. (3.129) gives

$$\mathbf{x}_r(t) = -\mathbf{K}_{rr}^{-1}\mathbf{K}_{rd}\mathbf{x}_d(t) + \mathbf{K}_{rr}^{-1}\mathbf{K}_r'\boldsymbol{\Theta}''(t) \tag{3.130}$$

Now, substituting Eq. (3.130) into Eq. (3.128) and rearranging the terms gives

$$\mathbf{m}(t) + \left[\mathbf{K}'' - \mathbf{K}_r'^T\mathbf{K}_{rr}^{-1}\mathbf{K}_r'\right]\boldsymbol{\Theta}''(t) = \left[\mathbf{K}_d'^T - \mathbf{K}_r'^T\mathbf{K}_{rr}^{-1}\mathbf{K}_{rd}\right]\mathbf{x}_d(t) \tag{3.131}$$

As in Eq. (2.149), define

$$\bar{\mathbf{K}}'' = \mathbf{K}'' - \mathbf{K}_r'^T\mathbf{K}_{rr}^{-1}\mathbf{K}_r' \tag{3.132}$$

Substituting Eqs. (3.125) and (3.132) into Eq. (3.131) gives

$$\mathbf{m}(t) + \bar{\mathbf{K}}''\boldsymbol{\Theta}''(t) = \bar{\mathbf{K}}'^T\mathbf{x}_d(t) \tag{3.133}$$

which represents the first governing equation of the FAM in the statically condensed form. Similarities can again be observed between Eq. (3.43) and Eq. (3.133). The $\mathbf{K}'$ and $\mathbf{K}''$ matrices in Eq. (3.43) have been replaced by $\bar{\mathbf{K}}'$ and $\bar{\mathbf{K}}''$ matrices in Eq. (3.133), respectively, and the $\mathbf{x}(t)$ vector in Eq. (3.43) have been replaced by the $\mathbf{x}_d(t)$ vector in Eq. (3.133).

In summary, by applying static condensation to eliminate the DOFs associated with zero mass and zero mass moment of inertia, the resulting equations for nonlinear dynamic analysis with the force analogy method are presented in Eqs. (3.120), (3.133), and (3.127), which are rewritten here:

$$\mathbf{M}_{dd}\ddot{\mathbf{x}}_d(t)+\mathbf{C}_{dd}\dot{\mathbf{x}}_d(t)+\bar{\mathbf{K}}\mathbf{x}'_d(t)=-\mathbf{M}_{dd}\ddot{\mathbf{g}}(t) \tag{3.134a}$$

$$\mathbf{m}_k+\bar{\mathbf{K}}''\mathbf{\Theta}''_k=\bar{\mathbf{K}}'^T\mathbf{x}_{d,k} \tag{3.134b}$$

$$\mathbf{x}''_{d,k}=\bar{\mathbf{K}}^{-1}\bar{\mathbf{K}}'\mathbf{\Theta}''_k \tag{3.134c}$$

where

$$\bar{\mathbf{K}}=\mathbf{K}_{dd}-\mathbf{K}_{dr}\mathbf{K}_{rr}^{-1}\mathbf{K}_{rd},\quad \bar{\mathbf{K}}'=\mathbf{K}'_d-\mathbf{K}_{dr}\mathbf{K}_{rr}^{-1}\mathbf{K}'_r,\quad \bar{\mathbf{K}}''=\mathbf{K}''-\mathbf{K}'^T_r\mathbf{K}_{rr}^{-1}\mathbf{K}'_r \tag{3.135}$$

Note that Eqs. (3.134b) and (3.134c) are presented in discretized forms, where $\mathbf{m}_k$, $\mathbf{\Theta}''_k$, $\mathbf{x}_{d,k}$, and $\mathbf{x}''_{d,k}$ are the discretized forms of $\mathbf{m}(t)$, $\mathbf{\Theta}''(t)$, $\mathbf{x}_d(t)$, and $\mathbf{x}''_d(t)$, respectively. The displacement vector associated with the DOFs with zero mass moment of inertia (i.e. $\mathbf{x}_r(t)$) can be calculated using Eq. (3.130), which is rewritten here in discretized form:

$$\mathbf{x}_{r,k}=-\mathbf{K}_{rr}^{-1}\mathbf{K}_{rd}\mathbf{x}_{d,k}+\mathbf{K}_{rr}^{-1}\mathbf{K}'_r\mathbf{\Theta}''_k \tag{3.136}$$

where $\mathbf{x}_{r,k}$ is the discretized form of $\mathbf{x}_r(t)$. Note that Eq. (3.136) is consistent with Eq. (2.145).

Once the matrices are set up with the condensed mass matrix $\mathbf{M}_{dd}$ in Eq. (3.134a) becoming invertible, the nonlinear dynamic analysis can be performed using the combination of state space method and force analogy method presented in Eq. (3.57) of Section 3.2, i.e.

$$\mathbf{z}_{k+1}=\mathbf{F}_d\mathbf{z}_k+\mathbf{H}_d\mathbf{a}_k+\mathbf{G}\mathbf{x}''_{d,k} \tag{3.137}$$

where

$$\mathbf{F}_d=\mathbf{e}^{\mathbf{A}\Delta t},\quad \mathbf{H}_d=\mathbf{e}^{\mathbf{A}\Delta t}\mathbf{H}\Delta t,\quad \mathbf{G}_d=\mathbf{e}^{\mathbf{A}\Delta t}\mathbf{G}\Delta t \tag{3.138}$$

$$\mathbf{z}_k=\begin{Bmatrix}\mathbf{x}_{d,k}\\ \dot{\mathbf{x}}_{d,k}\end{Bmatrix},\quad \mathbf{A}=\begin{bmatrix}\mathbf{0} & \mathbf{I}\\ -\mathbf{M}_{dd}^{-1}\bar{\mathbf{K}} & -\mathbf{M}_{dd}^{-1}\mathbf{C}_{dd}\end{bmatrix},\quad \mathbf{H}=\begin{bmatrix}\mathbf{0}\\ -\mathbf{h}\end{bmatrix},\quad \mathbf{G}=\begin{bmatrix}\mathbf{0}\\ \mathbf{M}_{dd}^{-1}\bar{\mathbf{K}}\end{bmatrix} \tag{3.139}$$

and

$$\ddot{\mathbf{g}}_k=\mathbf{h}\mathbf{a}_k \tag{3.140}$$

Here, $\dot{\mathbf{x}}_{d,k}$, $\mathbf{a}_k$, and $\ddot{\mathbf{g}}_k$ are the discretized forms of $\dot{\mathbf{x}}_d(t)$, $\mathbf{a}(t)$, and $\ddot{\mathbf{g}}(t)$, respectively.

Finally, the absolute acceleration can be computed using Eq. (3.62) as:

$$\ddot{\mathbf{y}}_{d,k}=-\mathbf{M}_{dd}^{-1}\mathbf{C}_{dd}\dot{\mathbf{x}}_{d,k}-\mathbf{M}_{dd}^{-1}\bar{\mathbf{K}}\left(\mathbf{x}_{d,k}-\mathbf{x}''_{d,k}\right) \tag{3.141}$$

where $\ddot{\mathbf{y}}_{d,k}$ is the discretized form of $\ddot{\mathbf{y}}_d(t)$.

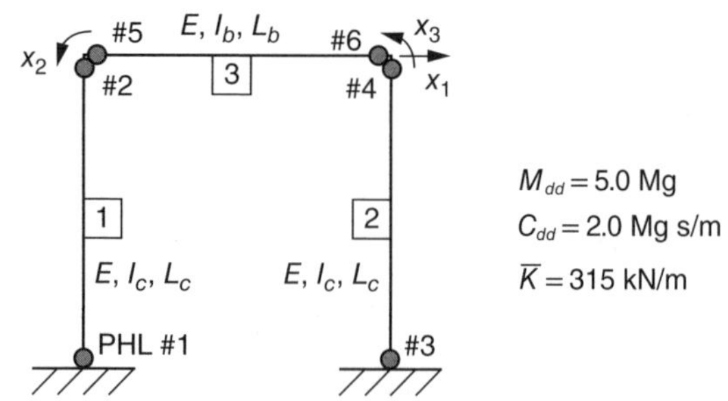

Figure 3.9 One-story one-bay moment-resisting frame as a SDOF system.

Example 3.7 One-story Moment-Resisting Frame

Consider a one-story one-bay moment-resisting frame as shown in Figure 3.9 with members assumed to be axially rigid. This frame has a total of 3 DOFs (i.e. $n=3$) and 6 PHLs (i.e. $q=6$) as labeled in the figure. The stiffness matrices $\mathbf{K}$, $\mathbf{K}'$, and $\mathbf{K}''$ were presented in Eq. (2.54) of Example 2.3. Assume that the frame can be reduced to a SDOF system with a mass of $M_{dd}=5.0$ Mg and a damping $C_{dd}=2.0$ Mg s/m at the DOF #1, while the mass moment of inertia at DOFs #2 and #3 are zero. Therefore, static condensation can be applied to eliminate the DOFs for x_2 and x_3, and the condensed stiffness matrices $\bar{\mathbf{K}}$, $\bar{\mathbf{K}}'$, and $\bar{\mathbf{K}}''$ were presented in Eq. (2.153) of Example 2.14 for the special case where $I_b=I_c=I$ and $L_b=L_c=L$.

Assume that the one-story frame has member properties of $E=200.0$ GPa, $I_b=I_c=6.0\times10^6$ mm^4, and $L_b=L_c=4.0$ m. Then, according to Eq. (2.153), the condensed stiffness matrices become

$$\bar{K}=\frac{84\times200\times3}{5(4)^3}=315\,\text{kN/m} \tag{3.142a}$$

$$\bar{\mathbf{K}}'=\mathbf{K}'_d-\mathbf{K}_{dr}\mathbf{K}_{rr}^{-1}\mathbf{K}'_r=[360\ \ 270\ \ 360\ \ 270\ \ -270\ \ -270]\,\text{kN} \tag{3.142b}$$

$$\bar{\mathbf{K}}''=\begin{bmatrix}1{,}040 & 280 & 40 & 80 & -280 & -80\\ 280 & 560 & 80 & 160 & -560 & -160\\ 40 & 80 & 1{,}040 & 280 & -80 & -280\\ 80 & 160 & 280 & 560 & -160 & -560\\ -280 & -560 & -80 & -160 & 560 & 160\\ -80 & -160 & -280 & -560 & 160 & 560\end{bmatrix}\text{kN m/rad} \tag{3.142c}$$

The period of vibration T and the damping ratio ζ can also be calculated:

$$T=2\pi\sqrt{M_{dd}/\bar{K}}=2\pi\sqrt{5.0/315}=0.792\,\text{s} \tag{3.143a}$$

$$\zeta=\frac{C}{2\sqrt{\bar{K}M_{dd}}}=\frac{2.0}{2\sqrt{315\times5.0}}=0.0252=2.52\% \tag{3.143b}$$

With a time step size of $\Delta t = 0.01$ s, the transition matrices are calculated as:

$$\mathbf{A} = \begin{bmatrix} 0 & 1 \\ -63 & -0.4 \end{bmatrix}, \quad \mathbf{H} = \begin{bmatrix} 0 \\ -1 \end{bmatrix}, \quad \mathbf{G} = \begin{bmatrix} 0 \\ 63 \end{bmatrix} \tag{3.144a}$$

$$\mathbf{F}_d = \mathbf{e}^{\mathbf{A}\Delta t} = \begin{bmatrix} 0.996856 & 0.009970 \\ -0.62808 & 0.992868 \end{bmatrix}, \quad \mathbf{H}_d = \begin{bmatrix} -0.00010 \\ -0.00993 \end{bmatrix}, \quad \mathbf{G}_d = \begin{bmatrix} 0.00628 \\ 0.62551 \end{bmatrix} \tag{3.144b}$$

and the equations for performing nonlinear dynamic analysis become

$$\begin{Bmatrix} x_1 \\ \dot{x}_1 \end{Bmatrix}_{k+1} = \begin{bmatrix} 0.996856 & 0.009970 \\ -0.62808 & 0.992868 \end{bmatrix} \begin{Bmatrix} x_1 \\ \dot{x}_1 \end{Bmatrix}_k + \begin{bmatrix} -0.00010 \\ -0.00993 \end{bmatrix} a_k + \begin{bmatrix} 0.00628 \\ 0.62551 \end{bmatrix} x''_{d,k} \tag{3.145a}$$

$$\begin{Bmatrix} m_1 \\ m_2 \\ m_3 \\ m_4 \\ m_5 \\ m_6 \end{Bmatrix}_k + \begin{bmatrix} 1{,}040 & 280 & 40 & 80 & -280 & -80 \\ 280 & 560 & 80 & 160 & -560 & -160 \\ 40 & 80 & 1{,}040 & 280 & -80 & -280 \\ 80 & 160 & 280 & 560 & -160 & -560 \\ -280 & -560 & -80 & -160 & 560 & 160 \\ -80 & -160 & -280 & -560 & 160 & 560 \end{bmatrix} \begin{Bmatrix} \theta''_1 \\ \theta''_2 \\ \theta''_3 \\ \theta''_4 \\ \theta''_5 \\ \theta''_6 \end{Bmatrix}_k = \begin{bmatrix} 360 \\ 270 \\ 360 \\ 270 \\ -270 \\ -270 \end{bmatrix} x_{1,k} \tag{3.145b}$$

$$x''_{1,k} = \begin{bmatrix} \frac{8}{7} & \frac{6}{7} & \frac{8}{7} & \frac{6}{7} & -\frac{6}{7} & -\frac{6}{7} \end{bmatrix} \begin{Bmatrix} \theta''_1 \\ \theta''_2 \\ \theta''_3 \\ \theta''_4 \\ \theta''_5 \\ \theta''_6 \end{Bmatrix}_k \tag{3.145c}$$

Assume that the plastic hinges exhibit elastic–plastic behavior with moment capacities of $m_b = 15.0$ kN m for the beam and $m_c = 20.0$ kN m for the two columns. This gives

$$if \begin{cases} m_i \le 20.0 \\ m_i > 20.0 \end{cases}, \quad then \begin{cases} \Delta\theta''_i = 0 \\ m_i = 20.0 \end{cases} \quad i = 1,2,3,4 \tag{3.146a}$$

$$if \begin{cases} m_i \le 15.0 \\ m_i > 15.0 \end{cases}, \quad then \begin{cases} \Delta\theta''_i = 0 \\ m_i = 15.0 \end{cases} \quad i = 5,6 \tag{3.146b}$$

Finally, the equations for the displacements at the DOFs #2 and #3 (i.e. $x_{2,k}$ and $x_{3,k}$), the elastic displacement at DOF #1 (i.e. $x'_{1,k}$), and the absolute acceleration responses at DOF #1 (i.e. $\ddot{y}_{1,k}$) can be calculated using Eqs. (3.136), (3.115), and (3.141), respectively:

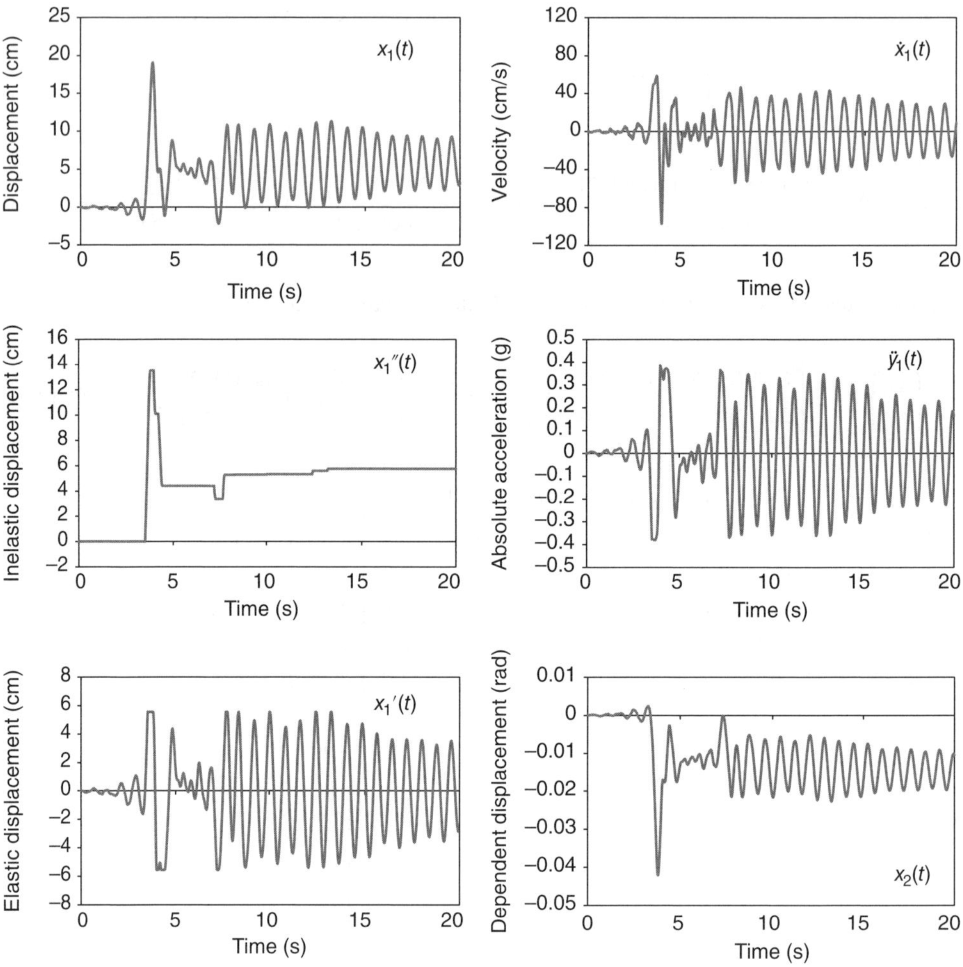

Figure 3.10 Global responses of the one-story one-bay moment-resisting frame.

$$\begin{Bmatrix} x_2 \\ x_3 \end{Bmatrix}_k = -\frac{3}{5\times 4}\begin{bmatrix} 1 \\ 1 \end{bmatrix} x_{1,k} + \frac{1}{15}\begin{bmatrix} 4 & 8 & -1 & -2 & 7 & 2 \\ -1 & -2 & 4 & 8 & 2 & 7 \end{bmatrix}\begin{Bmatrix} \theta_1'' \\ \theta_2'' \\ \theta_3'' \\ \theta_4'' \\ \theta_5'' \\ \theta_6'' \end{Bmatrix}_k \tag{3.147a}$$

$$x'_{1,k} = x_{1,k} - x''_{1,k} \tag{3.147b}$$

$$\ddot{y}_{1,k} = -0.4\times \dot{x}_{1,k} - 63\times x'_{1,k} \tag{3.147c}$$

Consider now the frame is subjected to the 1994 Northridge earthquake ground motion as shown in Figure 3.3. The nonlinear dynamic responses are calculated using Eqs. (3.145) and (3.147), and the global responses are plotted in Figure 3.10. These responses include the

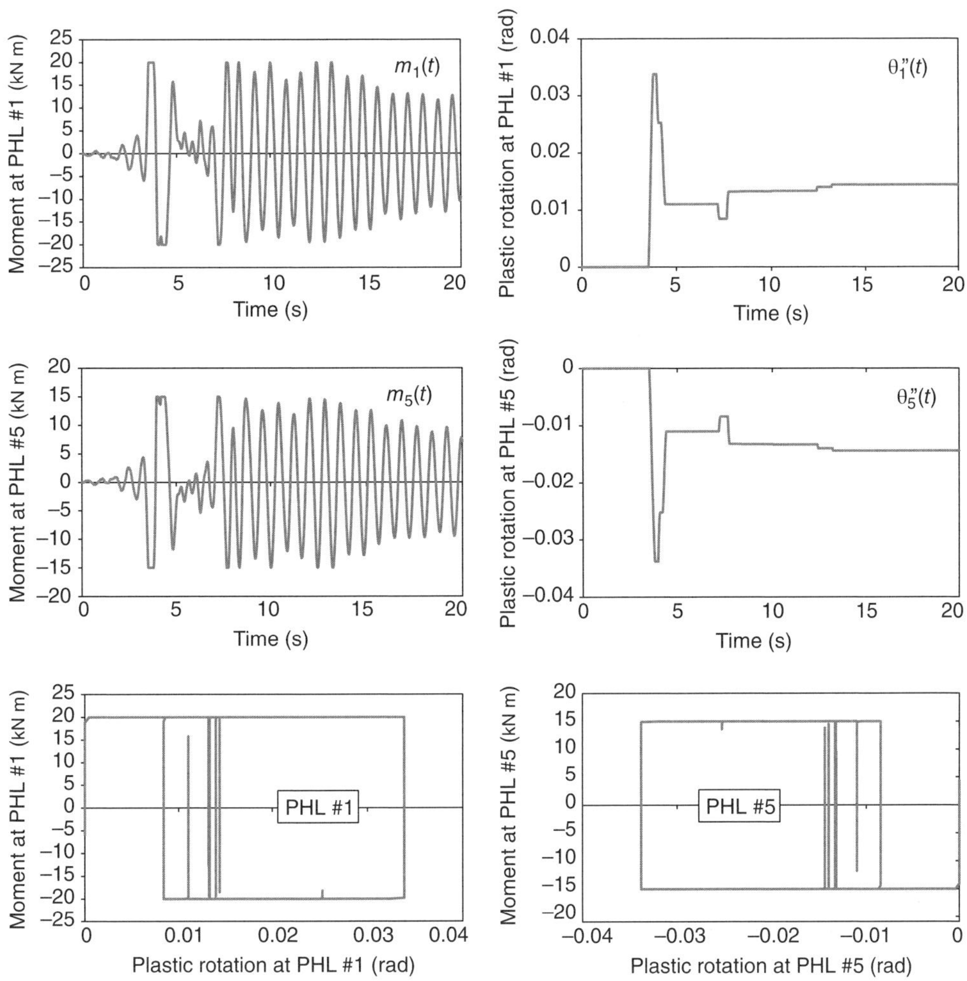

Figure 3.11 Local responses of the one-story one-bay moment-resisting frame.

displacement, velocity, acceleration, inelastic displacement, and elastic displacement at DOF #1, and the joint rotation at DOF #2. Due to symmetry, $x_2(t) = x_3(t)$, and therefore only the rotation at DOF #2 is plotted.

The local responses at selected plastic hinges are plotted in Figure 3.11. Only the responses for PHLs #1 and #5 are plotted, since due to symmetry:

$$m_1(t) = m_3(t),\quad \theta_1''(t) = \theta_3''(t),\quad m_5(t) = m_6(t),\quad \theta_5''(t) = \theta_6''(t) \tag{3.148}$$

In addition, PHLs #2 and #4 remain elastic due to the strong-column weak-beam design, and therefore the responses of these two plastic hinges are not plotted.

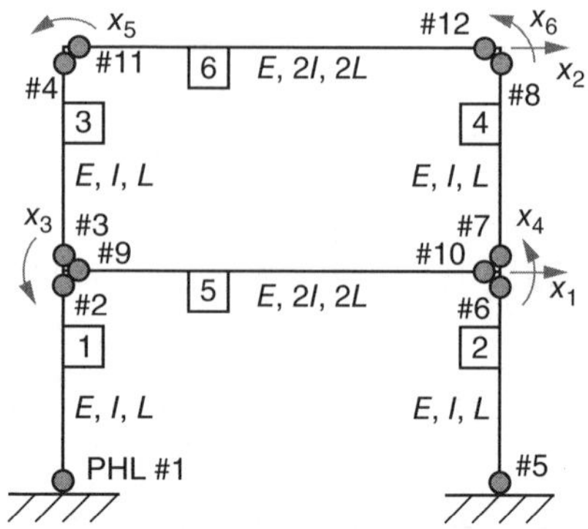

Figure 3.12 Two-story moment-resisting frame and a 2-DOF system.

Example 3.8 Two-story Moment-Resisting Frame

Consider a two-story one-bay moment-resisting frame as shown in Figure 3.12 with members assumed to be axially rigid. This frame has a total of 6 DOFs (i.e. $n = 6$) and 12 PHLs (i.e. $q = 12$) as labeled in the figure. The stiffness matrices $\mathbf{K}$, $\mathbf{K}'$, and $\mathbf{K}''$ were presented in Eq. (2.125) of Example 2.12. Assume that the mass moment of inertia are ignored for the rotational joints, the frame can then be statically condensed to a 2-DOF system. The condensed stiffness matrices $\bar{\mathbf{K}}$, $\bar{\mathbf{K}}'$, and $\bar{\mathbf{K}}''$ were presented in Eq. (2.165) of Example 2.15. Assume that the frame has member properties of $E = 200.0$ GPa, $I = 6.0 \times 10^6\,\text{mm}^4$, and $L = 4.0$ m. Then according to Eq. (2.165a), the condensed global stiffness matrix becomes

$$\bar{\mathbf{K}} = \frac{200 \times 6.0}{17(4)^3}\begin{bmatrix} 690 & -300 \\ -300 & 228 \end{bmatrix} = \begin{bmatrix} 761.03 & -330.88 \\ -330.88 & 251.47 \end{bmatrix} \text{kN/m} \tag{3.149}$$

and the $\bar{\mathbf{K}}'$ and $\bar{\mathbf{K}}''$ matrices can similarly be calculated using Eqs. (2.165b) and (2.165c).

Let the mass be 2.5 Mg on each floor and the damping be 2% for the two modes of vibration. This gives

$$\mathbf{M}_{dd} = \begin{bmatrix} 2.5 & 0 \\ 0 & 2.5 \end{bmatrix} \text{Mg}, \qquad \mathbf{C}_{dd} = \begin{bmatrix} 2.4955 & -0.7885 \\ -0.7885 & 1.2812 \end{bmatrix} \text{kN s/m} \tag{3.150}$$

and two periods of vibration of 1.055 s and 0.327 s. Using Eq. (3.139), the calculations of the transition matrices in the continuous form give

$$\mathbf{A} = \begin{bmatrix} 0.0 & 0.0 & 1.0 & 0.0 \\ 0.0 & 0.0 & 0.0 & 1.0 \\ -304.4 & 132.4 & -0.9982 & 0.3154 \\ 132.4 & -100.6 & 0.3154 & -0.5125 \end{bmatrix}, \quad \mathbf{H} = \begin{bmatrix} 0 \\ 0 \\ -1 \\ -1 \end{bmatrix}, \quad \mathbf{G} = \begin{bmatrix} 0.0 & 0.0 \\ 0.0 & 0.0 \\ 304.4 & -132.4 \\ -132.4 & 100.6 \end{bmatrix} \tag{3.151}$$

Then using Eq. (3.138) with a time step size of $\Delta t = 0.01$ s, the transition matrices in the discrete form are calculated as

$$\mathbf{F}_d = \begin{bmatrix} 0.984883 & 0.006568 & 0.009900 & 0.000038 \\ 0.006568 & 0.994998 & 0.000038 & 0.009958 \\ -3.00868 & 1.30649 & 0.975012 & 0.009671 \\ 1.30649 & -0.996660 & 0.009671 & 0.989906 \end{bmatrix} \tag{3.152a}$$

$$\mathbf{H}_d = \begin{bmatrix} -0.000099 \\ -0.000100 \\ -0.009847 \\ -0.009996 \end{bmatrix}, \quad \mathbf{G}_d = \begin{bmatrix} 0.030087 & -0.013065 \\ -0.013065 & 0.009967 \\ 2.95525 & -1.28072 \\ -1.28072 & 0.998293 \end{bmatrix} \tag{3.152b}$$

Therefore, using on the matrices presented in Eq. (3.152), Eq. (3.137) becomes

$$\begin{Bmatrix} x_1 \\ x_2 \\ \dot{x}_1 \\ \dot{x}_2 \end{Bmatrix}_{k+1} = \mathbf{F}_d \begin{Bmatrix} x_1 \\ x_2 \\ \dot{x}_1 \\ \dot{x}_2 \end{Bmatrix}_k + \mathbf{H}_d a_k + \mathbf{G}_d \begin{Bmatrix} x_1'' \\ x_2'' \end{Bmatrix}_k \tag{3.153}$$

Assume that the plastic hinges exhibit elastic–plastic behavior with moment capacities of $m_b = 15.0$ kN m for the two beams and $m_c = 20.0$ kN m for the four columns. This gives

$$if \begin{cases} m_i \le 20.0 \\ m_i > 20.0 \end{cases}, \quad then \begin{cases} \Delta\theta_i'' = 0 \\ m_i = 20.0 \end{cases} \quad i = 1, \ldots, 8 \tag{3.154a}$$

$$if \begin{cases} m_i \le 15.0 \\ m_i > 15.0 \end{cases}, \quad then \begin{cases} \Delta\theta_i'' = 0 \\ m_i = 15.0 \end{cases} \quad i = 9, \ldots, 12 \tag{3.154b}$$

and Eqs. (3.134b) and (3.134c) become

$$\begin{Bmatrix} m_1 \\ m_2 \\ \vdots \\ m_{12} \end{Bmatrix}_{k+1} + \bar{\mathbf{K}}'' \begin{Bmatrix} \Delta\theta_1'' \\ \Delta\theta_2'' \\ \vdots \\ \Delta\theta_{12}'' \end{Bmatrix} = \bar{\mathbf{K}}'^T \begin{Bmatrix} x_1 \\ x_2 \end{Bmatrix}_k - \bar{\mathbf{K}}'' \begin{Bmatrix} \theta_1'' \\ \theta_2'' \\ \vdots \\ \theta_{12}'' \end{Bmatrix}_k, \quad \begin{Bmatrix} x_1'' \\ x_2'' \end{Bmatrix}_k = \bar{\mathbf{K}}^{-1} \bar{\mathbf{K}}' \begin{Bmatrix} \theta_1'' \\ \theta_2'' \\ \vdots \\ \theta_{12}'' \end{Bmatrix}_k \tag{3.155}$$

Consider now the two-story moment-resisting frame is subjected to the 1994 Northridge earthquake ground motion as shown in Figure 3.3. The nonlinear dynamic responses are calculated using Eqs. (3.153) and (3.155), and the displacement, velocity, acceleration, and inelastic displacement responses are plotted in Figures 3.13 to 3.16, respectively.

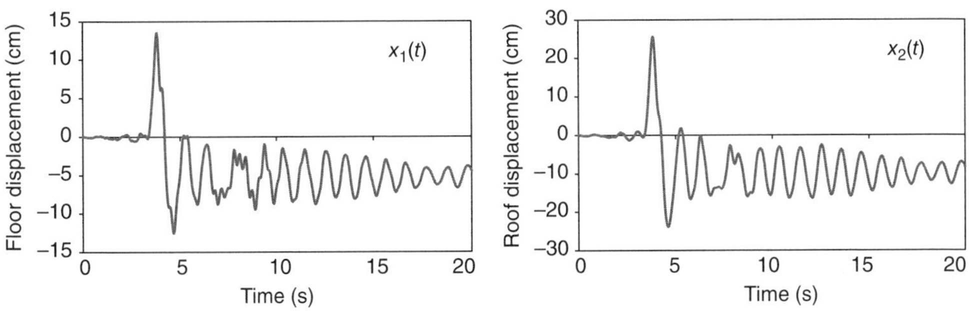

Figure 3.13 Global displacement responses of the two-story moment-resisting frame.

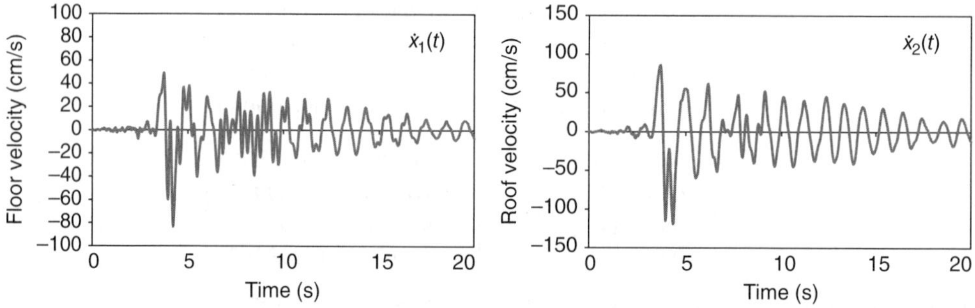

Figure 3.14 Global velocity responses of the two-story moment-resisting frame.

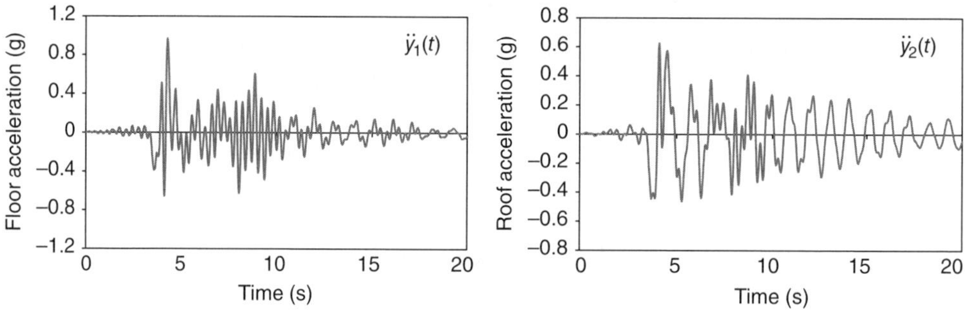

Figure 3.15 Global acceleration responses of the two-story moment-resisting frame.

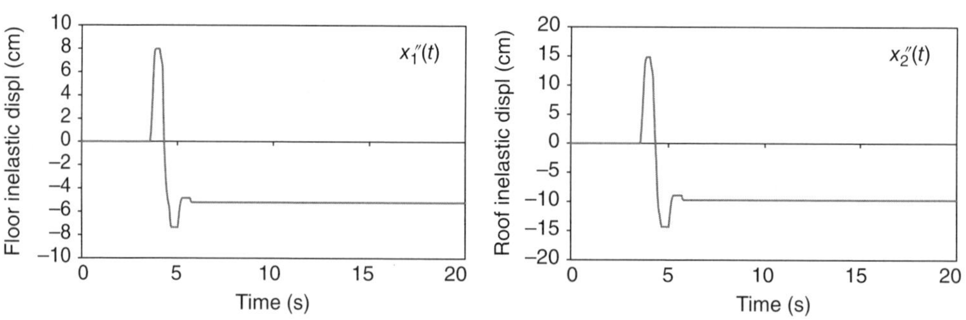

Figure 3.16 Global inelastic displacement responses of the two-story moment-resisting frame.

The local plastic hinge moment and plastic rotation responses at selected PHLs are plotted in Figures 3.17 and 3.18, respectively. These include PHLs #1 and #9, and the corresponding hysteresis loops are plotted in Figure 3.19.

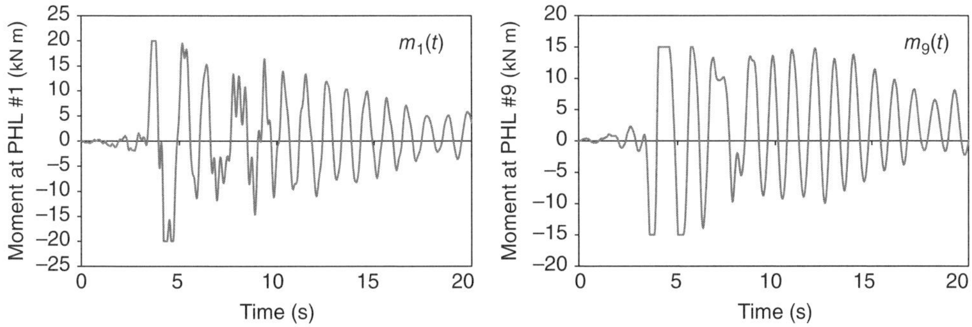

Figure 3.17 Moments at selected PHLs of the two-story moment-resisting frame.

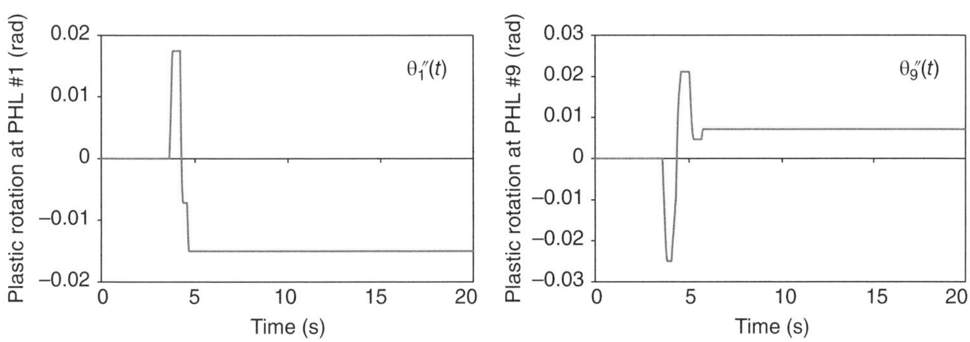

Figure 3.18 Plastic rotations at selected PHLs of the two-story moment-resisting frame.

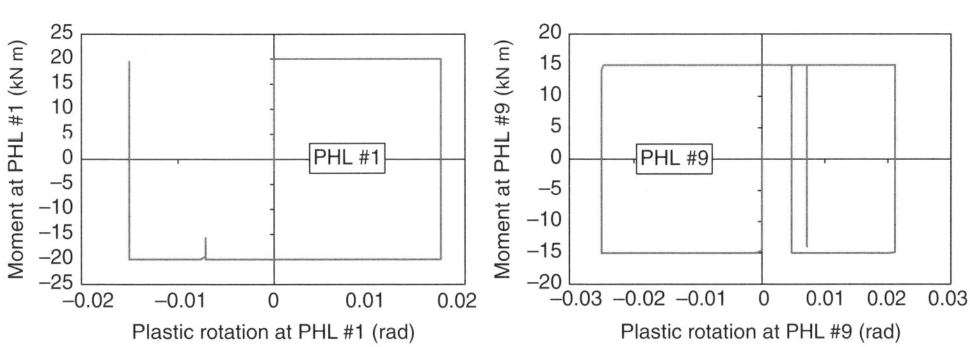

Figure 3.19 Hysteresis loops at selected PHLs of the two-story moment-resisting frame.

3.4 Nonlinear Dynamic Examples

Additional examples are presented here to demonstrate the use of the FAM with static condensation and the state space method for solving nonlinear dynamic problems of two-dimensional moment-resisting frames.

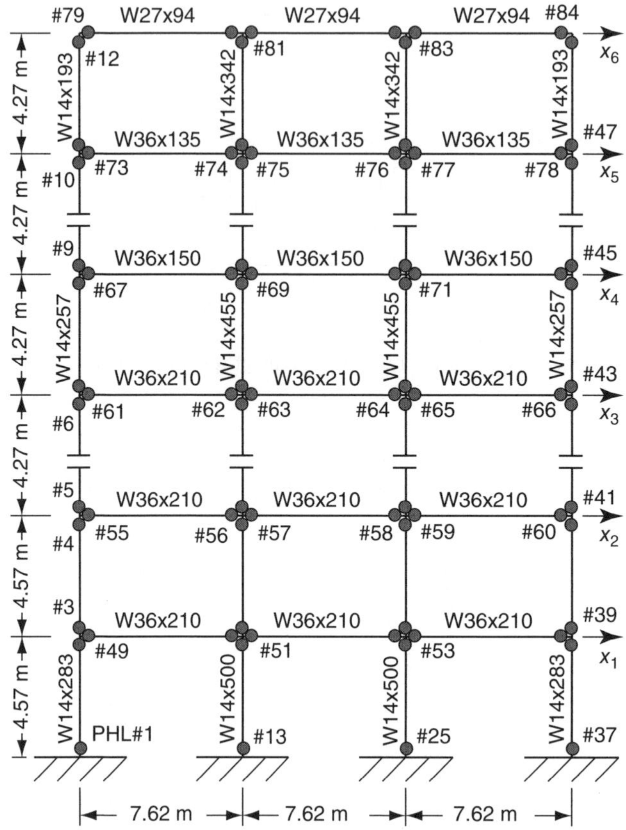

Mode	Period (s)	Damping
1	1.000	2%
2	0.362	2%
3	0.207	2%
4	0.148	2%
5	0.114	2%
6	0.093	2%

Figure 3.20 Six-story moment-resisting steel frame.

Example 3.9 Six-story Moment-Resisting Steel Frame

Consider the 6-story moment-resisting steel frame as shown in Figure 3.20. Assuming the members are axially rigid, this gives a total of 30 DOFs (i.e. $n = 30$) and 84 PHLs (i.e. $q = 84$). No mass moment of inertia is assumed at the rotational joints, and therefore static condensation is used to eliminate the 24 rotational DOFs (i.e. $r = 24$), resulting in only 6 translational DOFs for the frame (i.e. $d = 6$). The stiffness matrices are then formulated based on the standard properties of each member. The 30×30 $\mathbf{K}$ matrix, 30×84 $\mathbf{K}'$ matrix, and 84×84 $\mathbf{K}''$ matrix are first constructed by assembling the individual stiffness matrices of each member. Then static condensation using Eq. (3.135) is applied to obtain the 6×6 $\bar{\mathbf{K}}$ matrix, 6×84 $\bar{\mathbf{K}}'$ matrix, and 84×84 $\bar{\mathbf{K}}''$ matrix. As a numerical illustration, the resulting $\bar{\mathbf{K}}$ matrix is of the form

$$\bar{\mathbf{K}} = \begin{bmatrix} 464300 & -252200 & 39180 & -3715 & 345.8 & -29.4 \\ -252200 & 445700 & -266500 & 38160 & -3550 & 305.1 \\ 39180 & -266500 & 451000 & -252800 & 36180 & -3106 \\ -3715 & 38160 & -252800 & 384500 & -190900 & 24440 \\ 345.8 & -3550 & 36180 & -190900 & 282900 & -125000 \\ -29.4 & 305.1 & -3106 & 24440 & -125000 & 103300 \end{bmatrix} \text{kN/m} \tag{3.156}$$

Let the mass be 200.0 Mg on each floor, giving a total mass of 1,200 Mg for the entire frame, and the diagonal nonzero mass matrix $\mathbf{M}_{dd}$ can be written as

$$\mathbf{M}_{dd} = 200.0 \times \mathbf{I} = \begin{bmatrix} 200 & 0 & 0 & 0 & 0 & 0 \\ 0 & 200 & 0 & 0 & 0 & 0 \\ 0 & 0 & 200 & 0 & 0 & 0 \\ 0 & 0 & 0 & 200 & 0 & 0 \\ 0 & 0 & 0 & 0 & 200 & 0 \\ 0 & 0 & 0 & 0 & 0 & 200 \end{bmatrix} \text{Mg} \tag{3.157}$$

Using the resulting condensed $\bar{\mathbf{K}}$ stiffness matrix in Eq. (3.156) and the diagonal $\mathbf{M}_{dd}$ mass matrix in Eq. (3.157), the 6 periods of vibration are calculated and summarized in Figure 3.20. The damping is assumed to be 2% in all six modes of vibration. This gives a damping matrix of

$$\mathbf{C}_{dd} = \begin{bmatrix} 368 & -116 & -4.6 & -4.7 & -2.1 & -1.3 \\ -116 & 335 & -131 & -9.5 & -6.3 & -3.5 \\ -4.6 & -131 & 332 & -130 & -7.5 & -7.0 \\ -4.7 & -9.5 & -130 & 306 & -111 & -9.2 \\ -2.1 & -6.3 & -7.5 & -111 & 261 & -100 \\ -1.3 & -3.5 & -7.0 & -9.2 & -100 & 152 \end{bmatrix} \text{kN}\, s/\text{m} \tag{3.158}$$

Assume that all 84 plastic hinges exhibit elastic–plastic behavior with moment capacity of the ith plastic hinge, $m_{c,i}$, calculated as

$$m_{c,i} = f_y \times Z_i \quad i = 1, \ldots, 84 \tag{3.159}$$

where f_y is the yield stress of steel and Z_i is the plastic section modulus of the ith plastic hinge. This gives

$$if \begin{cases} m_i(t) \le m_{c,i} \\ m_i(t) > m_{c,i} \end{cases}, \; then \begin{cases} \Delta\theta_i''(t) = 0 \\ m_i(t) = m_{c,i} \end{cases} \quad i = 1, \ldots, 84 \tag{3.160}$$

By subjecting the 6-story frame to the 1995 Kobe earthquake ground motion as shown in Figure 3.2, Figures 3.21 to 3.23 show the displacement, velocity, and absolute acceleration responses, respectively, of selected floors. Figures 3.24 to 3.26 show the responses and hysteresis loops of selected plastic hinges. The plastic hinge locations include the base of the exterior column (PHL #1), base of the interior column (PHL #13), exterior hinge of the beam on the second floor (PHL #49), and interior hinge of the beam on the third floor (PHL #56).

Example 3.10 Ten-story Moment-Resisting Steel Frame

Consider the 10-story moment-resisting steel frame as shown in Figure 3.27. Assuming the members are axially rigid, this gives a total of 50 DOFs (i.e. $n = 50$) and 140 PHLs (i.e. $q = 140$). No mass moment of inertia is assumed at the rotational joints, and therefore

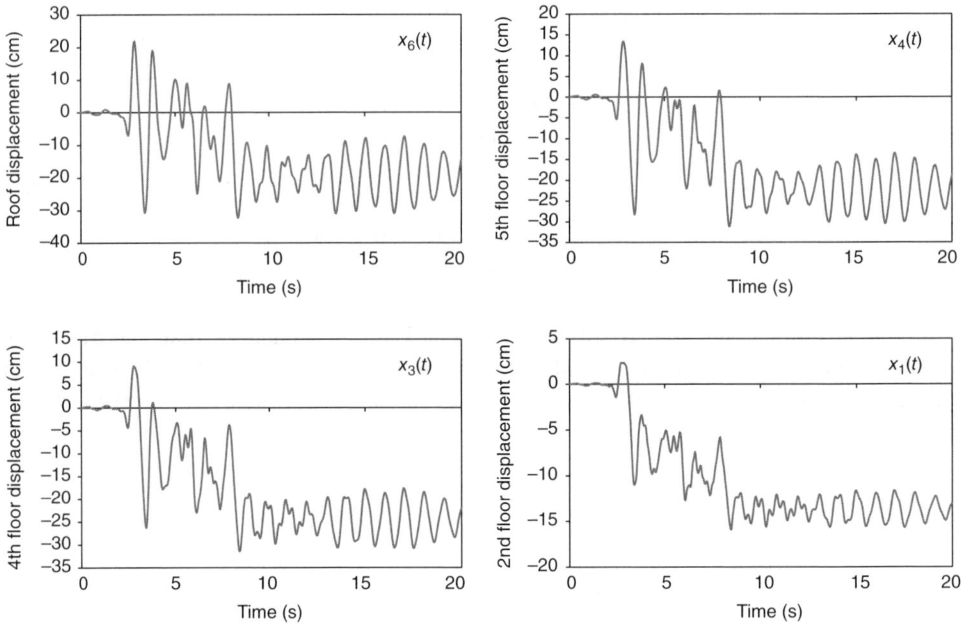

Figure 3.21 Global displacement responses of the 6-story moment-resisting frame.

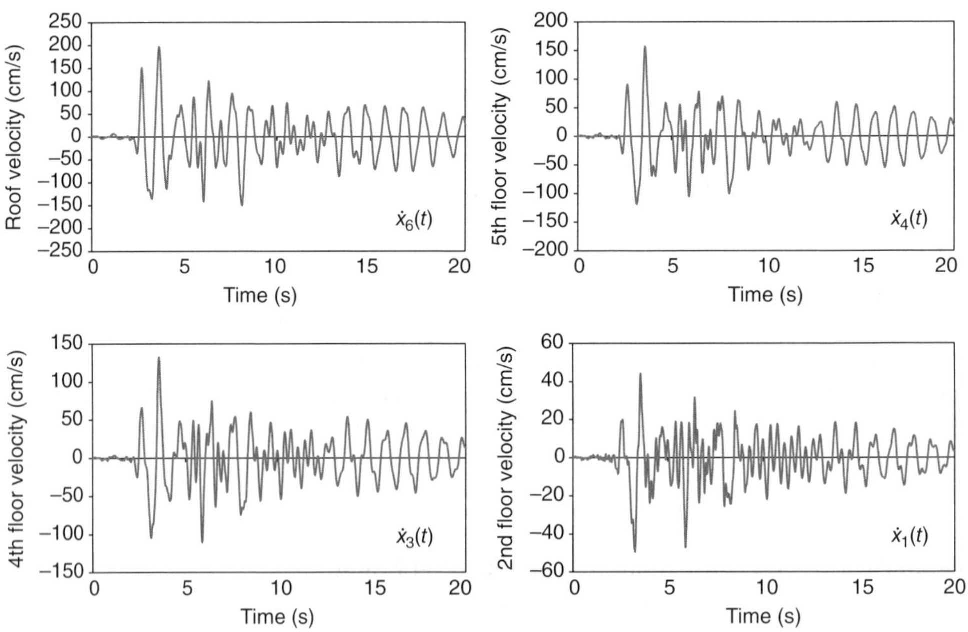

Figure 3.22 Global velocity responses of the 6-story moment-resisting frame.

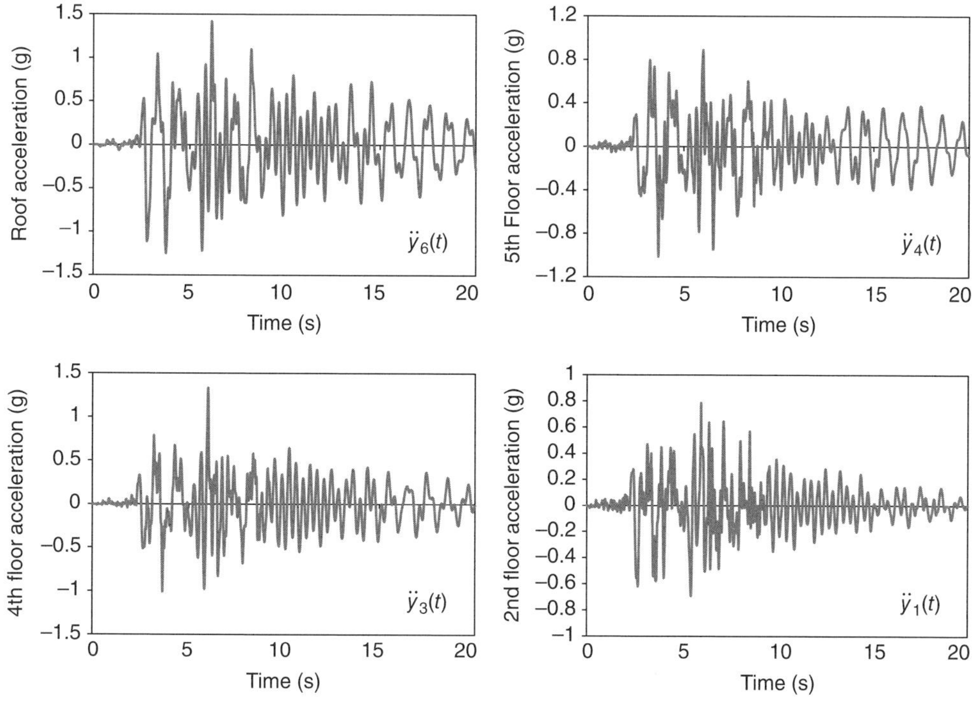

Figure 3.23 Global acceleration responses of the 6-story moment-resisting frame.

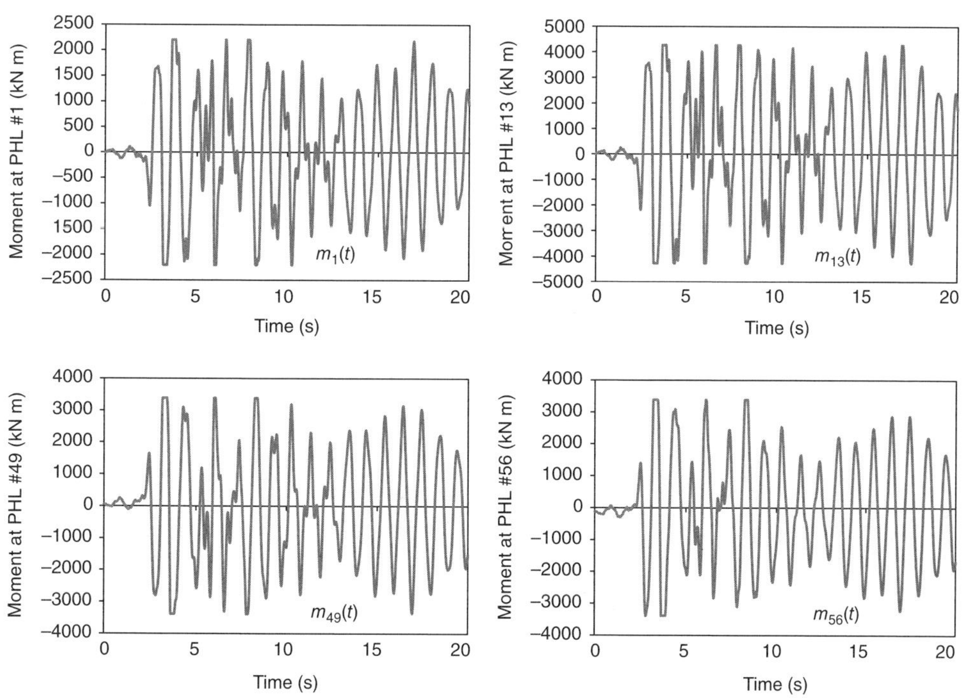

Figure 3.24 Local moment responses at selected PHLs of the 6-story frame.

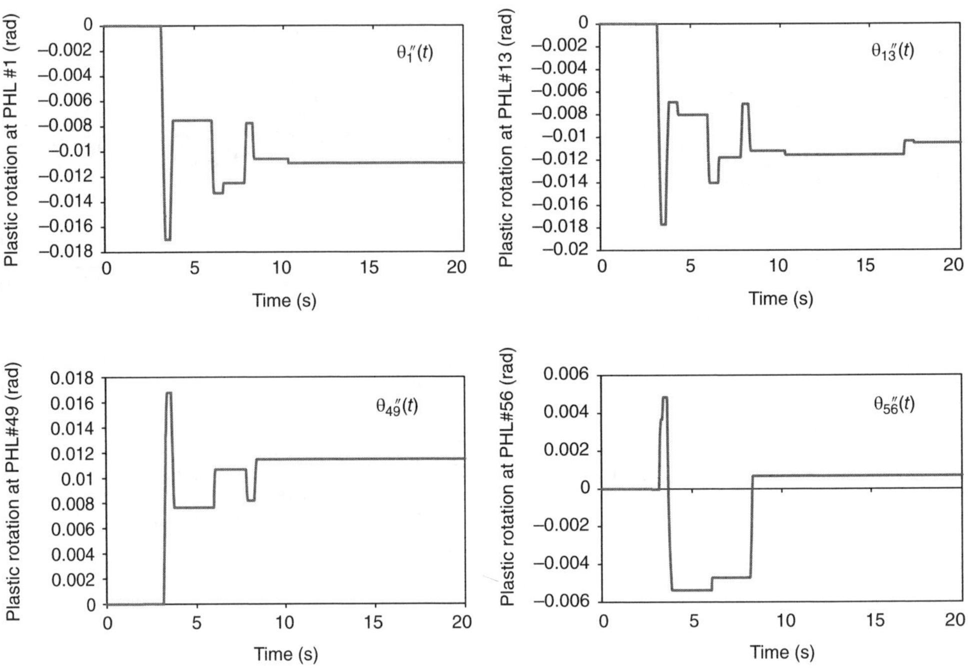

Figure 3.25 Local plastic rotation responses at selected PHLs of the 6-story frame.

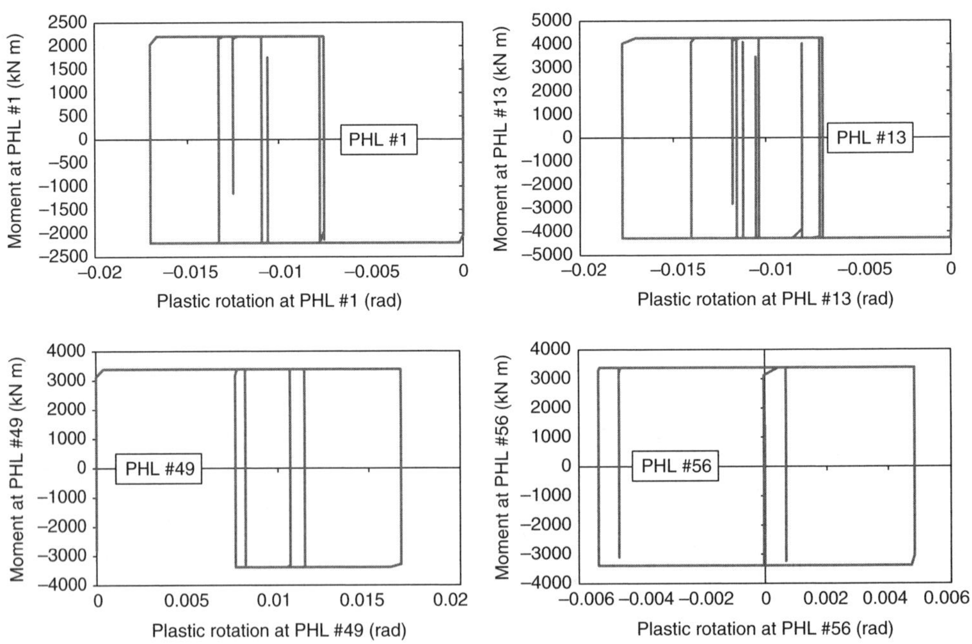

Figure 3.26 Hysteresis loops of selected PHLs of the 6-story frame.

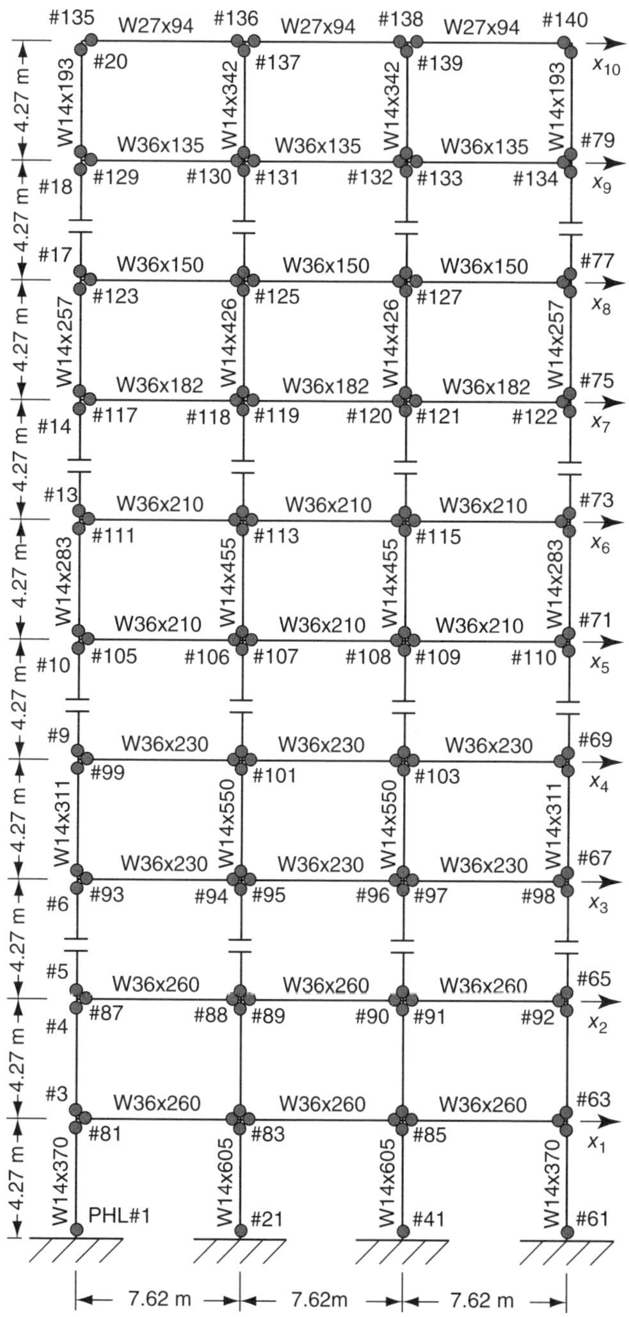

Mode	Period (s)	Damping
1	1.499	3%
2	0.546	3%
3	0.331	3%
4	0.232	3%
5	0.178	3%
6	0.145	3%
7	0.121	3%
8	0.104	3%
9	0.093	3%
10	0.081	3%

Figure 3.27 Ten-story moment-resisting steel frame.

static condensation is used to eliminate the 40 rotational DOFs (i.e. $r=40$), resulting in only 10 translational DOFs for the frame (i.e. $d=10$). Let the mass be 218.9 Mg on each floor, giving a total mass of 2,189 Mg for the entire frame, and the $\mathbf{M}_{dd}$ matrix can be written as

$$\mathbf{M}_{dd}=218.9\times\mathbf{I}=\begin{bmatrix}218.9 & 0 & \cdots & 0\\ 0 & 218.9 & \ddots & \vdots\\ \vdots & \ddots & \ddots & 0\\ 0 & \cdots & 0 & 218.9\end{bmatrix}\text{Mg} \qquad (3.161)$$

The damping is assumed to be 3% in all ten modes.

The stiffness matrices are then formulated based on the standard properties of each member. The 50×50 $\mathbf{K}$ matrix, 50×140 $\mathbf{K}'$ matrix, and 140×140 $\mathbf{K}''$ matrix are first constructed by assembling the individual stiffness matrices of each member. Then static condensation using Eq. (3.135) is applied to obtain the 10×10 $\bar{\mathbf{K}}$ matrix, 10×140 $\bar{\mathbf{K}}'$ matrix, and 140×140 $\bar{\mathbf{K}}''$ matrix. Using the resulting condensed $\bar{\mathbf{K}}$ stiffness matrix and the diagonal $\mathbf{M}_{dd}$ mass matrix, the 10 periods of vibration are calculated and summarized in Figure 3.27.

Assume that all 140 plastic hinges exhibit elastic–plastic behavior with moment capacity $m_{c,i}$ of the ith plastic hinge calculated as

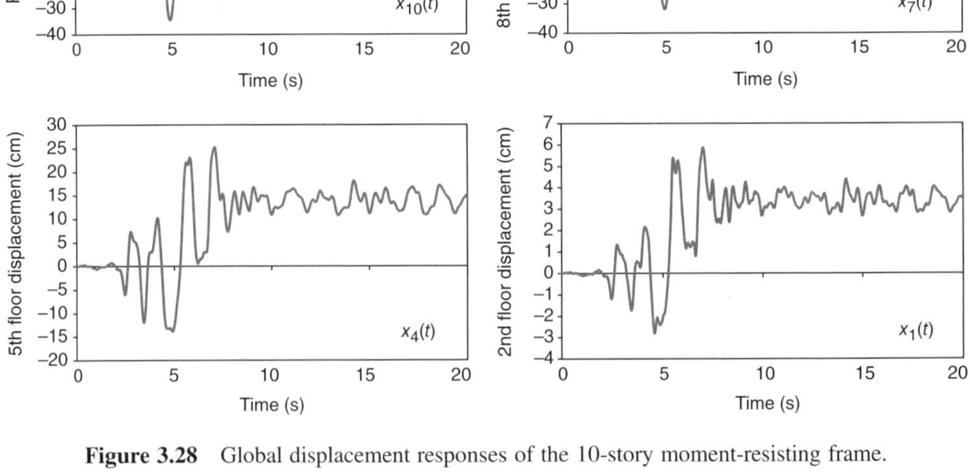

Figure 3.28 Global displacement responses of the 10-story moment-resisting frame.

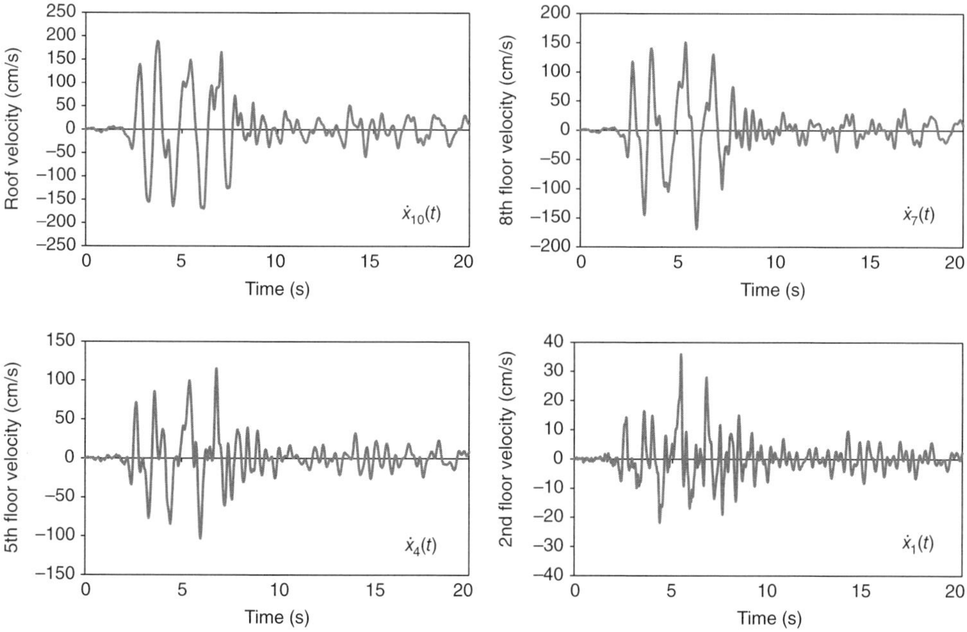

Figure 3.29 Global velocity responses of the 10-story moment-resisting frame.

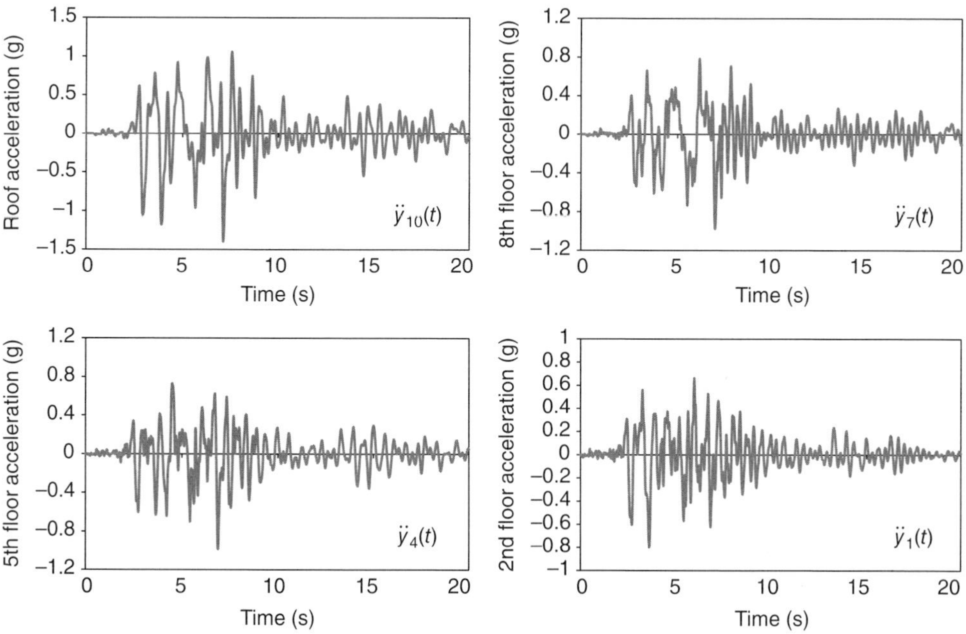

Figure 3.30 Global acceleration responses of the 10-story moment-resisting frame.

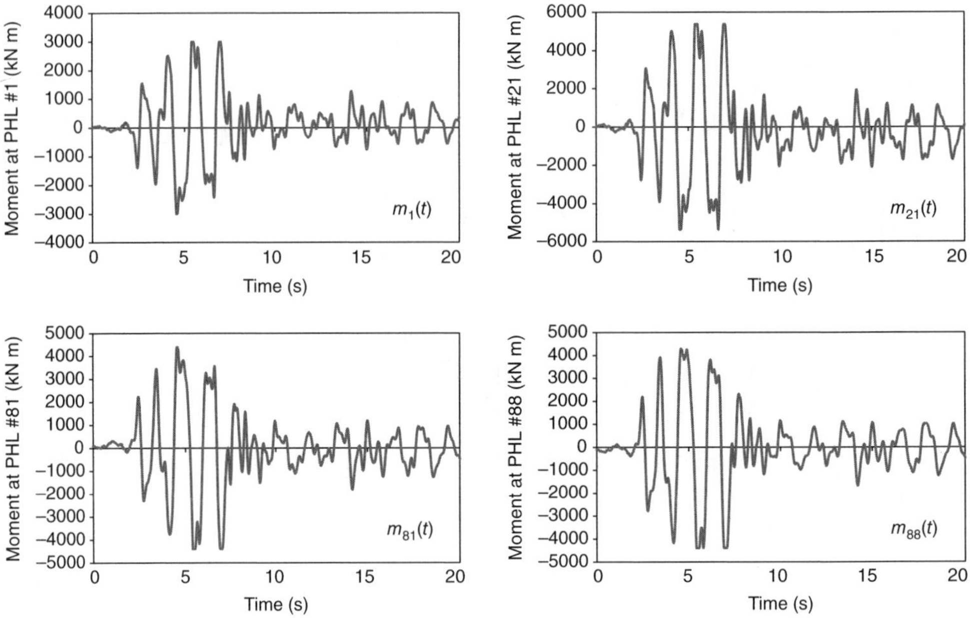

Figure 3.31 Local moment responses at selected PHLs of the 10-story frame.

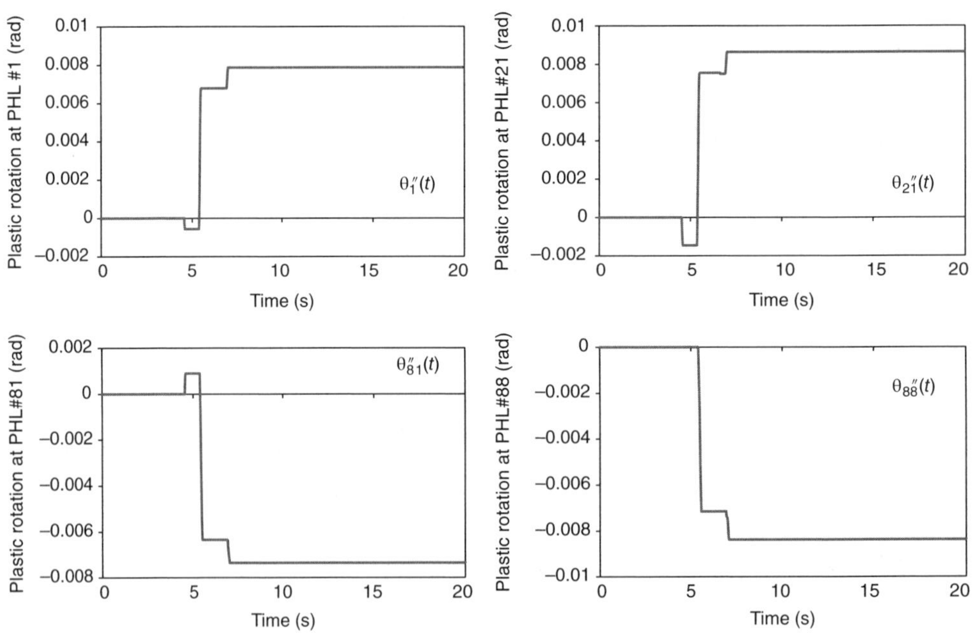

Figure 3.32 Local plastic rotation responses at selected PHLs of the 10-story frame.

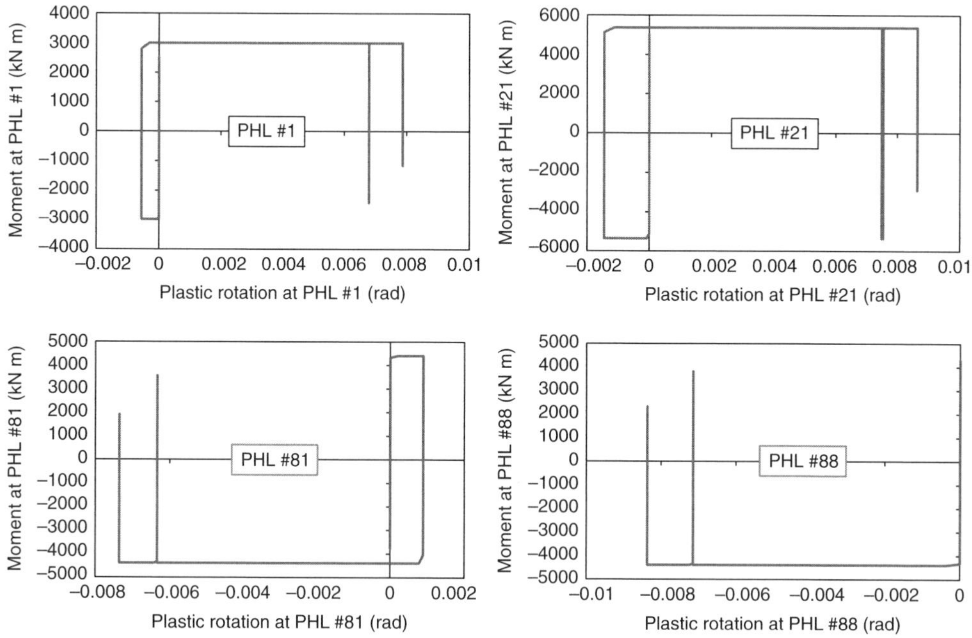

Figure 3.33 Hysteresis loops of selected PHLs of the 10-story frame.

$$m_{c,i} = f_y \times Z_i \quad i = 1, \ldots, 140 \tag{3.162}$$

where f_y is the yield stress of steel and Z_i is the plastic section modulus of the ith plastic hinge. This gives

$$\textit{if} \begin{cases} m_i(t) \le m_{c,i} \\ m_i(t) > m_{c,i} \end{cases}, \ \textit{then} \begin{cases} \Delta\theta_i''(t) = 0 \\ m_i(t) = m_{c,i} \end{cases} \quad i = 1, \ldots, 140 \tag{3.163}$$

By subjecting the 10-story frame to the 1995 Kobe earthquake ground motion as shown in Figure 3.2, Figures 3.28 to 3.30 show the displacement, velocity, and absolute acceleration responses, respectively, of selected floors. Figures 3.31 to 3.33 show the responses and hysteresis loops of selected plastic hinges. The plastic hinge locations include the base of the exterior column (PHL #1), base of the interior column (PHL #21), exterior hinge of the beam on the second floor (PHL #81), and interior hinge of the beam on the third floor (PHL #88).

References

Chopra A.K. (2011). *Dynamics of Structures*, 4th Edition, Prentice Hall, New Jersey, USA.

Hart G.C. and Wong K. (2000). *Structural Dynamics for Structural Engineers*, John Wiley & Sons, New York, USA.

4

Flexural Member

The flexural member is a commonly used structural component in structural engineering. The typical flexural members in building structures are the beams and columns that are mainly subjected to moments and shear forces. However, in practice, the shear effect of beams is traditionally neglected except for some special beams, such as the deep beam and the wall beam. In this chapter, the nonlinear behavior of reinforced concrete (RC) flexural members, including stiffness and strength degradation, is expressed with the FAM through local plastic mechanisms, which are the moment versus plastic rotation and shear force versus plastic deformation relationships, respectively. The local plastic mechanisms are established based on the existing hysteretic models, such as the models proposed by *Takeda et al.* (1970) and *Ozcebe* and *Saatcioglu* (1989), and their modified version established by *Sezen* and *Chowdhury* (2009) and *Xu* and *Zhang* (2011). The nonlinear analysis for the flexural member under static and dynamic loads in the FAM is interpreted through several single-degree-of-freedom (SDOF) systems. In addition, the procedure of performing nonlinear dynamic analyses for multi-degree-of-freedom (MDOF) systems with components of flexural members is verified against a RC framed structure.

4.1 Bending and Shear Behaviors

4.1.1 Hysteretic Models

The hysteretic model of the RC flexural member is composed by the total primary curve that defines the monotonic load–deformation response and the hysteretic rules that describe the loading, unloading, and reloading behaviors. The primary curve shown in Figure 4.1 consists

---

Theory of Nonlinear Structural Analysis: The Force Analogy Method for Earthquake Engineering, First Edition. Gang Li and Kevin K.F. Wong.

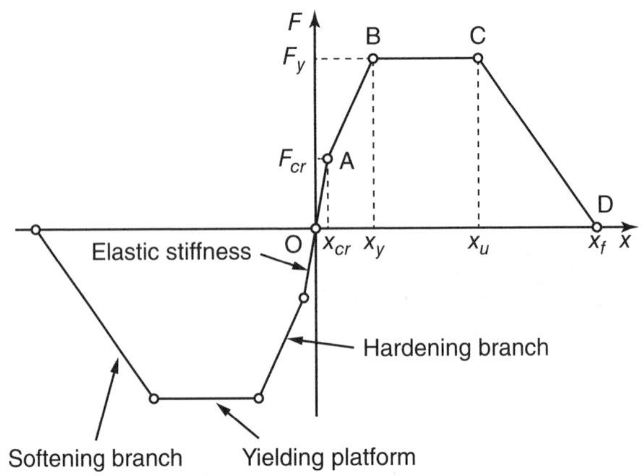

Figure 4.1 Primary curve of the flexural member.

of four segments: (1) line OA with cracking load capacity F_{cr} at point A; (2) line AB, which depicts the inelastic hardening behavior after cracking; (3) line BC, the yielding platform in which the yielding load capacity F_y is reached; (4) line CD, which is a softening branch with negative stiffness and the point D with displacement x_f depicts the failure of the flexural member.

The total monotonic behavior shown in Figure 4.1 can be decomposed into the bending and shear behaviors. In addition, separating the bending and shear behaviors is significant in theory, since the different behaviors can be tracked directly, which leads to a more detailed understanding of response according to the bending-shear interaction. The bending and shear primary curves can be derived empirically or calibrated from the envelope of the hysteretic loops obtained from the tests, or as an alternative, the analytical or theoretical method can be selected to obtain the primary curves, such as the modified compression field theory (MCFT) (*Vecchio* and *Collins* 1988) and the moment-curvature analysis (*Park* and *Paulay* 1975). However, in order to simulate the full-range nonlinearity of the RC flexural member, including the yielding platform and softening branch, empirical equations (*Sezen* and *Chowdhury* 2009) should also be incorporated. The decomposed primary curves for bending and shear behaviors are illustrated in Figure 4.2. As shown in Figure 4.2 (a), the bending primary curve represents the monotonic behavior of moment m versus rotation θ at the member's end, while the shear primary curve shown in Figure 4.2(b) expresses the monotonic behavior of the shear force V versus lateral deformation τ of the member.

The bending hysteretic model shown in Figure 4.2(a) was proposed by *Takeda et al.* (1970) and now is selected as a baseline model here. The hysteretic model for exhibiting the shear behavior in this chapter is shown in Figure 4.2(b), which is an improved model proposed by *Xu* and *Zhang* (2011) based on the work done by *Ozcebe* and *Saatcioglu* (1989). However, for the shear hysteretic model shown in Figure 4.2(b), the author developed it on the reloading branches beyond the cracking load capacity V_{cr} be targeted at the peak deformation and corresponding load reached during previous cycle, instead of a projected point as proposed in the original model. In general, the bending and shear hysteretic models used here can represent the strength deterioration, stiffness degradation, and pinching behaviors, which are the main characteristics of RC flexural members.

The nonlinear bending behavior of RC flexural members is associated with the following parameters: the cracking moment capacity m_{cr}, yielding moment capacity m_y, and their

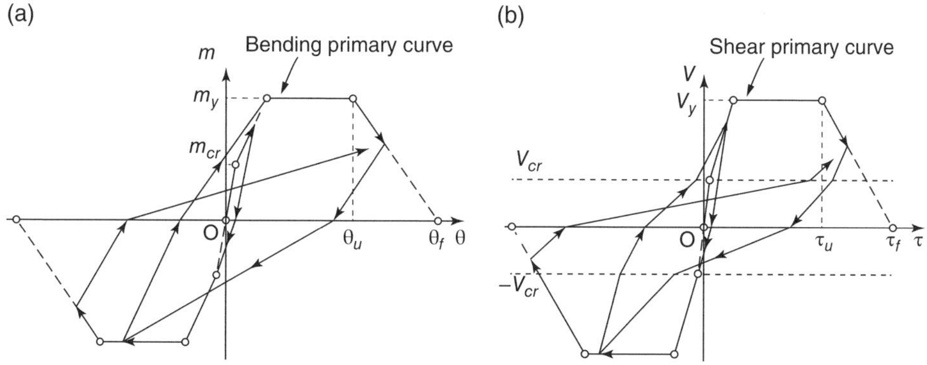

Figure 4.2 Hysteretic models: (a) Bending behavior; (b) Shear behavior.

corresponding rotations θ_{cr} and θ_y. Besides, the rotation θ_u shown in Figure 4.2(a) is the onset of strength degradation and the rotation θ_f on the horizontal axis denotes the bending failure of the member. Similarly, the parameters in the shear model of RC flexural members can be summarized as: the cracking shear capacity V_{cr}, yielding shear capacity V_y, and their corresponding lateral displacements τ_{cr} and τ_y. Besides, the lateral displacement τ_u shown in Figure 4.2(b) is the onset of strength degradation and the lateral displacement τ_f denotes the shear failure of the member. In addition, several basic parameters are also necessary in both the bending and shear models, i.e. the elastic modulus E, shear modulus G, section area A, moment of inertia for the section I, and the length of flexural members L.

4.1.2 Displacement Decomposition

In order to incorporate the FAM in the bending and shear models expressed in the previous section, let us investigate how to decompose the total displacements into the elastic and plastic components. First, take the bending hysteretic model, as shown in Figure 4.3(a) for instance. The point Q located on the hardening branch AB represents the rotation θ of the flexural member under the moment m. Extend line OA until it reaches the load m at point P, then the rotation at point P is the elastic rotation θ' that is equal to the difference between the total rotation θ and the plastic rotation θ'':

$$\theta' = \theta - \theta'' \tag{4.1}$$

The moment m at point Q or point P can be written by:

$$m = k_f \theta' \tag{4.2}$$

where k_f is the initial elastic stiffness of the bending hysteretic model. Substituting Eq. (4.1) to Eq. (4.2) yields:

$$m = k_f [\theta - \theta''] \tag{4.3}$$

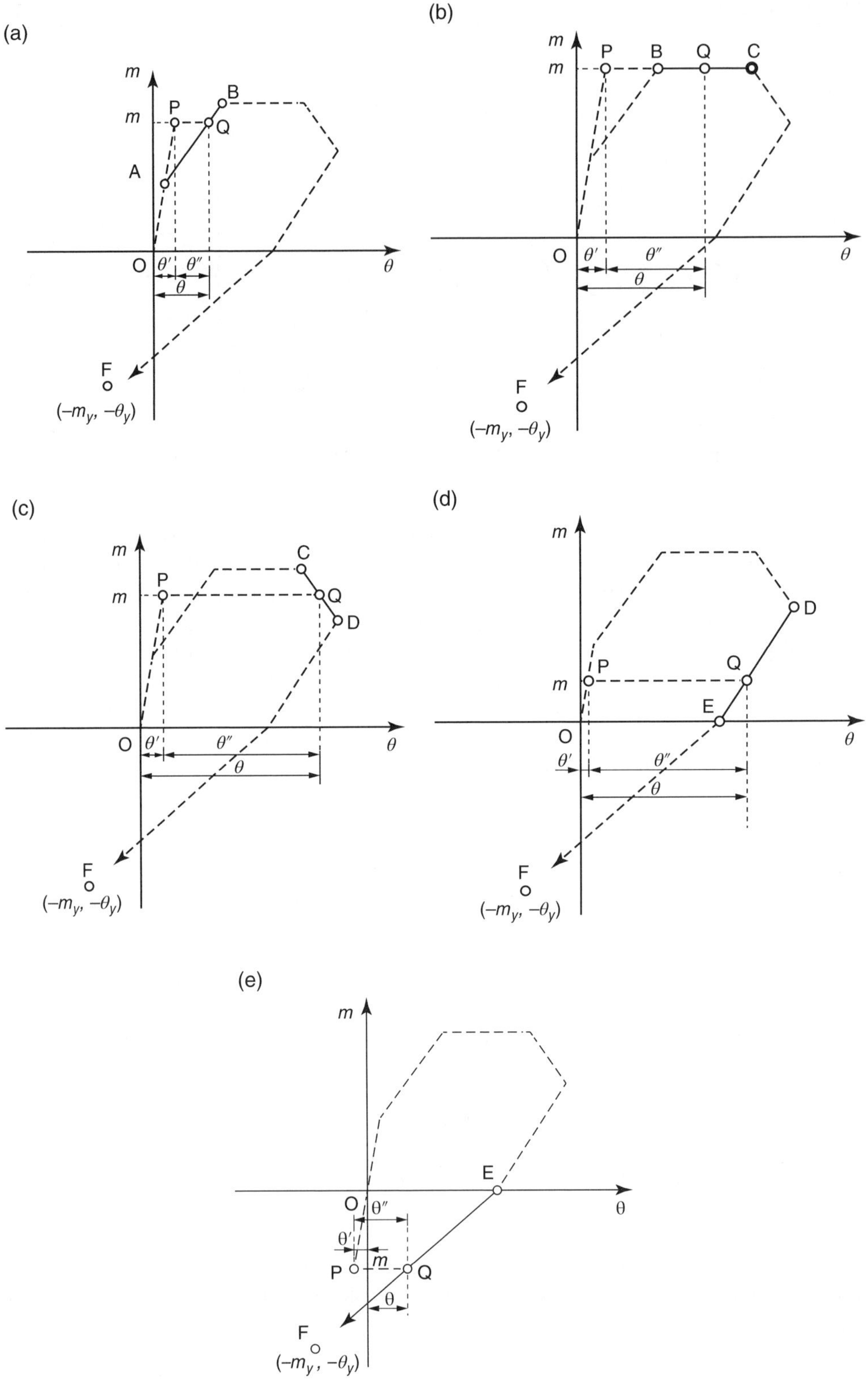

Figure 4.3 Decomposition of bending displacement: (a) Branch AB; (b) Branch BC; (c) Branch CD; (d) Branch DE; (e) Branch EF.

In addition, the total rotation at other branches in the bending model can also be composed into the elastic and plastic components, such as yielding platform branch BC, strength deterioration branch CD, unloading branch DE, and reloading branch EF shown in Figures 4.3(b) to 4.3(e).

Similarly, the above decomposition procedure can also be used to divide the total shear displacement into the elastic and plastic components. Because of the presence of the pinching effect, the decomposition of the shear model is more complicated in unloading and reloading branches. However, there is a similar relationship as:

$$V = k_s[\tau - \tau''] \tag{4.4}$$

where k_s is the initial elastic stiffness of the shear hysteretic model.

4.1.3 Local Plastic Mechanisms

The local plastic mechanisms (i.e. m–θ'' and V–τ'') in the FAM provide additional relationships for solving the nonlinear problems uniquely. When the flexural member is in the plastic domain, the bending and shear behaviors are governed by the inelastic branches. Define α_f the tangent stiffness of a branch in the bending hysteretic model shown in Figure 4.2(a), α_s the tangent stiffness of a branch in the shear hysteretic behavior shown in Figure 4.2(b). Then the local plastic mechanisms can be established as

$$\begin{aligned} m &= \Gamma(\theta'') = \alpha_f''\theta'' + m_{int} \\ V &= \Psi(\tau'') = \alpha_s''\tau'' + V_{int} \end{aligned} \tag{4.5}$$

in which

$$\begin{cases} \alpha_f'' = \dfrac{k_f\alpha_f}{k_f - \alpha_f} \\ \alpha_s'' = \dfrac{k_s\alpha_s}{k_s - \alpha_s} \end{cases} \tag{4.6}$$

where α_f'' is the tangent stiffness of the local bending plastic mechanism, m–θ''; α_s'' is the tangent stiffness of the local shear plastic mechanism, V–τ''; m_{int} is the intercept of m–θ'' relationship; V_{int} is the intercept of V–τ'' relationship; Γ and Ψ define the relationships of the local plastic mechanisms in terms of plastic rotation θ'' and plastic deformation τ'', respectively.

4.2 Inelastic Mechanisms of Flexural Members

A cantilever RC column subjected to lateral load F shown in Figure 4.4(a) is taken as an example to illustrate the inelastic mechanisms of flexural members. The total deformation of the RC column is the sum of bending and shear components. As shown in Figure 4.4(a), the bending

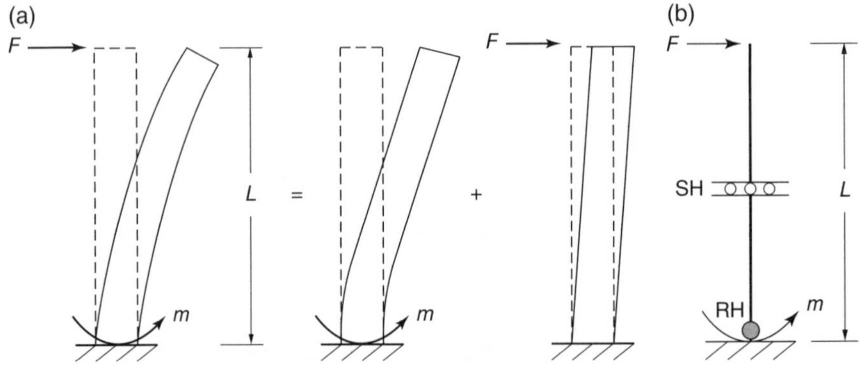

Figure 4.4 The flexural member: (a) Cantilever column; (b) Equivalent SDOF system.

deformation varies with the moment along the length of the column, and is usually assumed to be concentrated at the column end. However, the shear deformation, unlike the bending behavior, is generated along the column, so the deformation of the shear effect can be considered as a sliding mechanism. As a result, in the FAM, the flexural member shown in Figure 4.4(a) can be simplified to the system shown in Figure 4.4(b). As illustration, a rotation hinge (RH) is assembled at the column end to represent the plastic rotation due to the bending behavior and a slide hinge (SH) is assigned to the column to represent the lateral plastic deformation due to the shear behavior. The two hinges, RH and SH, are the local plastic mechanisms.

In addition, the total lateral displacement x of the flexural member is the sum of the elastic displacement x' and the inelastic displacement x'':

$$x = x' + x'' \tag{4.7}$$

where x'' is the sum of the plastic bending displacement x_1'' due to RH and plastic shear displacement x_2'' due to SH:

$$x'' = x_1'' + x_2'' \tag{4.8}$$

4.2.1 Elastic Displacement x'

In order to obtain the elastic displacement x', the RH and SH should be locked as shown in Figure 4.5(a), and the relationship between the lateral load F and the elastic displacement x' is satisfied by

$$F = k_0 x' \tag{4.9}$$

where k_0 is the initial elastic stiffness.

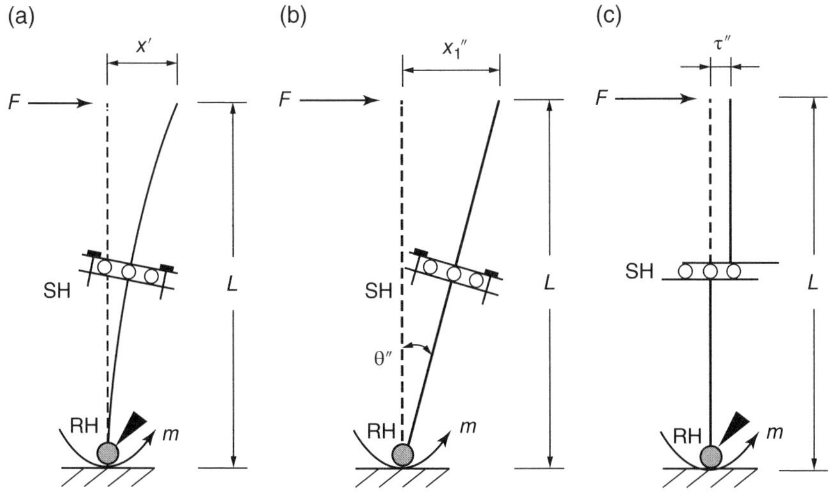

Figure 4.5 Determination of the displacements in the FAM: (a) Elastic displacement; (b) Plastic bending displacement; (c) Plastic shear displacement.

4.2.2 *Plastic Bending Displacement x_1''*

Assuming that there is no elastic displacement and the SH is locked, as shown in Figure 4.5(b), the RH rotates θ'', which is the plastic rotation due to bending behavior, while other parts of the system remain rigid. According to the kinematic relationship, the plastic bending displacement x_1'' can be written as:

$$x_1'' = \theta'' L \tag{4.10}$$

4.2.3 *Plastic Shear Displacement x_2''*

Similar to the procedure for x_1'', as shown in Figure 4.5(c), assume that there is no elastic displacement of the system and the RH is locked. The SH begins to slide due to the lateral load F, while other parts of the SDOF system remain rigid. Then the plastic shear displacement x_2'' equals to the sliding τ'' of SH, which is the plastic shear deformation due to shear behavior as:

$$x_2'' = \tau'' \tag{4.11}$$

4.2.4 *Combination of the Bending and Shear Behaviors*

The displacement decomposition and inelastic mechanisms of the SDOF system subjected to the lateral load F are illustrated in Figure 4.6, in which the displacement x contains the elastic

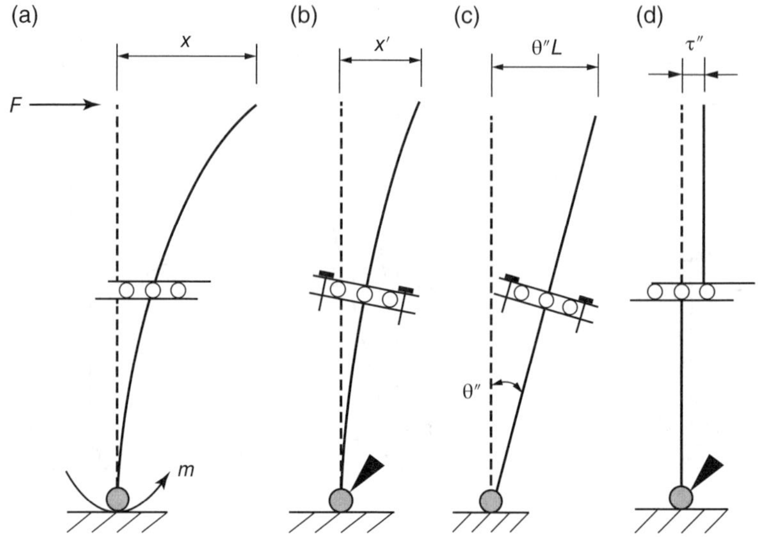

Figure 4.6 Schematic representation of the inelastic mechanisms for flexural members: (a) Total displacement; (b) Elastic component; (c) Plastic bending component; (d) Plastic shear component.

component and the plastic components generated from RH and SH. The applied lateral load F can be written by

$$F = k_0[x - (\theta''L + \tau'')] \tag{4.12}$$

4.3 Nonlinear Static Analysis of Structures with Flexural Members

4.3.1 Force Analogy Method for Static Single-Degree-of-Freedom Systems

Consider a SDOF system with a RC column and lump mass on its top. The RC column is a unique lateral-resistant member in the system, and two local plastic mechanisms, RH and SH shown in Figure 4.4(b), are used. Figure 4.7 shows the backbone curves of the bending and shear behaviors. For bending backbone curve shown in Figure 4.7(a), it can be seen that there are totally four categories: (1) elastic branch OA_1; (2) strength hardening branch A_1B_1; (3) yielding platform branch B_1C_1; (4) and strength softening branch C_1D_1. While branches with the subscripts 2 in Figure 4.7(b) express the corresponding categories of the shear backbone curve, which is the relationship of the shear force V versus shear deformation τ of the flexural member.

When the SDOF system subjected to the lateral load F remains elastic, there are following relations as:

$$\text{if} \quad \begin{cases} m \le m_{cr} \\ V \le V_{cr} \end{cases}, \quad \text{then} \quad F = k_0 x, \quad \begin{cases} \theta'' = 0 \\ \tau'' = 0 \end{cases} \tag{4.13}$$

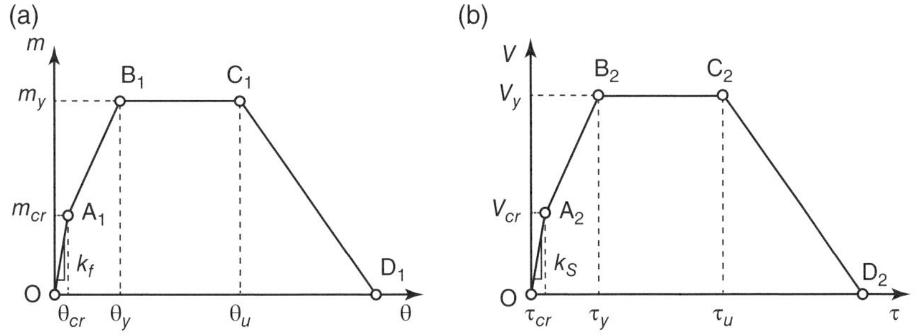

Figure 4.7 Backbone curves of the flexural member: (a) Bending backbone curve; (b) Shear backbone curve.

In addition, the critical points A_1, A_2, B_1, B_2, etc. in the primary curves can be derived according to the lateral load F, for example $V_{cr} = F_{cr}$ and $m_{cr} = V_{cr}L$, which also means that the bending and shear effects must behave on the same type of branches for the SDOF system. In the above equation, x is the total displacement; the initial elastic stiffness of the SDOF system k_0 is satisfied by

$$\frac{1}{k_0} = \frac{L^2}{k_f} + \frac{1}{k_s} \tag{4.14}$$

$$k_f = \frac{3EI}{L}, \quad k_s = \frac{GA}{L} \tag{4.15}$$

When the SDOF system is in the plastic domain and both the bending and shear behaviors are governed by the hardening branches A_1B_1 and A_2B_2, referring to Eq. (4.5), there are following relationships:

$$if \quad \begin{cases} m_{cr} < m < m_y \\ V_{cr} < V < V_y \end{cases}, \quad then \quad \begin{cases} m = \Gamma(\theta'') = \alpha_f''\theta'' + m_{int} \\ V = \Psi(\tau'') = \alpha_s''\tau'' + V_{int} \end{cases} \tag{4.16}$$

where α_f'' and α_s'' are the stiffnesses of the local plastic mechanisms, RH and SH, on branches A_1B_1 and A_2B_2; m_{int} and V_{int} are the intercepts of the m–θ'' and V–τ'' relationships, respectively.

Once the moment m at the RH and the shear force V of the SDOF system exceed or equal to the corresponding yielding capacities m_y and V_y, the bending and shear behaviors are represented by the yielding platforms B_1C_1 and B_2C_2. The local plastic mechanisms shown in Eq. (4.16) are still feasible by satisfying the conditions as:

$$\begin{cases} m \geq m_y \\ V \geq V_y \end{cases} \tag{4.17}$$

and α_f and α_s here are specified with smaller values (close to zero) to simulate the yielding platform behaviors.

When the deformations θ and τ of the SDOF system exceed θ_u and τ_u, the bending and shear behaviors are depicted by the softening branches C_1D_1 and C_2D_2. The tangent stiffnesses α_f and

α_s here are normally specified by negative values, and the local plastic mechanisms shown in Eq. (4.16) are still feasible by satisfying the conditions as

$$\begin{cases} \theta > \theta_u \\ \tau > \tau_u \end{cases} \tag{4.18}$$

Once the cracking capacities are exceeded, the RH and SH begin to deform and the applied load is written by

$$F = k_0[x - (\theta'' L + \tau'')] \tag{4.19}$$

Based on the discussion in Section 2.2, the governing equations of the FAM in matrix form for solving nonlinear static problems of the flexural members in the SDOF system can be expressed by

$$\left[\begin{array}{c|cc} k_0 & k_0 L & k_0 \\ \hline k_0 L & k_0 L^2 & k_0 L \\ k_0 & k_0 L & k_0 \end{array}\right] \left\{\begin{array}{c} x \\ \hline -\theta'' \\ -\tau'' \end{array}\right\} = \left\{\begin{array}{c} F \\ \hline m \\ V \end{array}\right\} \tag{4.20}$$

Equation (4.20) can be simplified by

$$\left[\begin{array}{c|c} \mathbf{K} & \mathbf{K}' \\ \hline \mathbf{K}'^T & \mathbf{K}'' \end{array}\right] \left\{\begin{array}{c} \mathbf{x} \\ \hline -\boldsymbol{\Lambda}'' \end{array}\right\} = \left\{\begin{array}{c} \mathbf{F} \\ \hline \mathbf{f} \end{array}\right\} \tag{4.21}$$

in which

$$\boldsymbol{\Lambda}'' = \begin{Bmatrix} \theta'' \\ \tau'' \end{Bmatrix}, \quad \mathbf{f} = \begin{Bmatrix} m \\ V \end{Bmatrix} \tag{4.22}$$

For a given external lateral force F, there are five unknowns x, m, V, θ'' and τ'' in Eq. (4.20), but only three equations are provided for the solution. Therefore, two relationships of RH and SH that expressed in Eq. (4.5) are introduced to provide five equations to solve for the five unknowns.

Example 4.1 SDOF System with Linear Response

Consider the SDOF column shown in Figure 4.8(a) which is subjected to an applied lateral force F_0. The backbone curves of the bending and shear behaviors of the column are shown in Figure 4.8(b). It can be forecasted that the column will remain linear because the moment and shear force demands at the RH and SH are $F_0 L$ and F_0, respectively. This example is to illustrate how to use Eq. (4.20) to solve the static problem when the SDOF system is in the elastic domain.

First, assume that the structure remains elastic and the plastic rotation θ'' at RH and plastic deformation τ'' at SH are zero. Therefore, Eq. (4.20) becomes:

$$\left[\begin{array}{c|cc} k_0 & k_0 L & k_0 \\ \hline k_0 L & k_0 L^2 & k_0 L \\ k_0 & k_0 L & k_0 \end{array}\right] \left\{\begin{array}{c} x \\ \hline 0 \\ 0 \end{array}\right\} = \left\{\begin{array}{c} F_0 \\ \hline m \\ V \end{array}\right\} \tag{4.23}$$

where k_0 can be computed by using Eqs. (4.14) and (4.15).

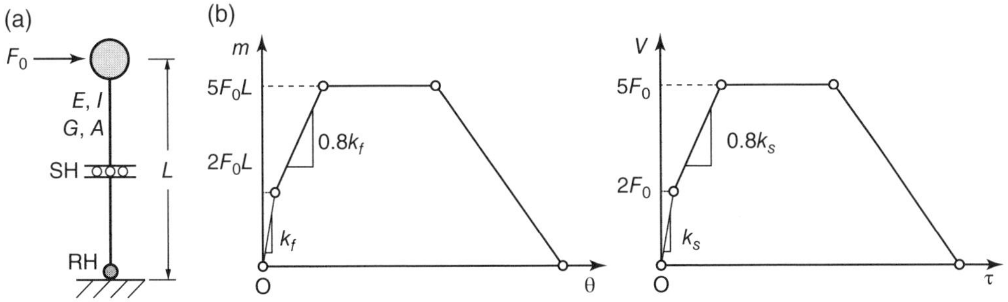

Figure 4.8 The SDOF system: (a) numerical model; (b) backbone curves.

The total displacement x solved by the first equation in Eq. (4.23) gives:

$$x = \frac{F_0}{k_0} \tag{4.24}$$

Now, substituting Eq. (4.24) back into Eq. (4.23) and solving for the moment M and shear force V yields:

$$\begin{cases} m = (k_0 L)\left(\dfrac{F_0}{k_0}\right) = F_0 L \\ V = (k_0)\left(\dfrac{F_0}{k_0}\right) = F_0 \end{cases} \tag{4.25}$$

Since $m = F_0$L and $V = F_0$ are less than the corresponding capacities of RH and SH, the hypothesis of the SDOF system being elastic is correct, and therefore the total displacement x given in Eq. (4.24) and the moment m and shear force V given in Eq. (4.25) represent the final answers.

Example 4.2 SDOF System with Nonlinear Response
Consider the SDOF column shown in Figure 4.8(a) which now is subjected to an applied force of $3F_0$. According to the load – deformation relationships shown in Figure 4.8(b), the SDOF column will behave nonlinearly because the moment and shear force demands at the RH and SH exceed the corresponding capacities.

Similar to Example 4.1, first assume that the structure remains elastic and the plastic rotation θ'' at RH and plastic deformation τ'' at SH are zero, then Eq. (4.20) becomes

$$\begin{bmatrix} k_0 & k_0 L & k_0 \\ k_0 L & k_0 L^2 & k_0 L \\ k_0 & k_0 L & k_0 \end{bmatrix} \begin{Bmatrix} x \\ 0 \\ 0 \end{Bmatrix} = \begin{Bmatrix} 3F_0 \\ m \\ V \end{Bmatrix} \tag{4.26}$$

Using the first equation of Eq. (4.26) and solving for the total displacement x gives:

$$x = \frac{3F_0}{k_0} \tag{4.27}$$

Now substituting Eq. (4.27) back in Eq. (4.26) to solve for the moment m and shear force V gives:

$$\begin{cases} m = (k_0 L)\left(\dfrac{3F_0}{k_0}\right) = 3F_0 L \\ V = (k_0)\left(\dfrac{3F_0}{k_0}\right) = 3F_0 \end{cases} \tag{4.28}$$

Since both the moment and shear force exceed the corresponding cracking capacities $m_{cr} = 2F_0L$ and $V_{cr} = 2\,F_0$, the SDOF column is in the plastic domain and the assumption that θ'' and τ'' are zero is incorrect.

Now assume that the behaviors of RH and SH are governed by the hardening branches. Then the moment m and shear force V can be expressed in terms of the plastic rotation θ'' and the plastic deformation τ'' as expressed in Eq. (4.5) as

$$\begin{cases} m = 2F_0 L + 4k_f \theta'' \\ V = 2F_0 + 4k_s \tau'' \end{cases} \tag{4.29}$$

Substituting Eq. (4.29) into Eq. (4.26) gives:

$$\left[\begin{array}{c|cc} k_0 & k_0 L & k_0 \\ \hline k_0 L & k_0 L^2 + 4k_f & k_0 L \\ k_0 & k_0 L & k_0 + 4k_s \end{array}\right] \left\{\begin{array}{c} x \\ \hline -\theta'' \\ -\tau'' \end{array}\right\} = \left\{\begin{array}{c} 3F_0 \\ \hline 2F_0 L \\ 2F_0 \end{array}\right\} \tag{4.30}$$

Now solving for Eq. (4.30) gives:

$$\left\{\begin{array}{c} x \\ \hline -\theta'' \\ -\tau'' \end{array}\right\} = \begin{bmatrix} L^2/4k_f + 1/4k_s + 1/k_0 & -L/4k_f & -1/4k_s \\ -L/4k_f & 1/4k_f & 0 \\ -1/4k_s & 0 & 1/4k_s \end{bmatrix} \left\{\begin{array}{c} 3F_0 \\ \hline 2F_0 L \\ 2F_0 \end{array}\right\} = \left\{\begin{array}{c} 13F_0/4k_0 \\ \hline -F_0 L/4k_f \\ -F_0/4k_s \end{array}\right\} \tag{4.31}$$

Solving for Eq. (4.31) yields

$$x = \frac{13F_0}{4k_0}, \quad \theta'' = \frac{F_0 L}{4k_f}, \quad \tau'' = \frac{F_0}{4k_s} \tag{4.32}$$

and substituting Eq. (4.32) into Eq. (4.29) gives:

$$\begin{cases} m = 2F_0 L + 4k_f \left(\dfrac{F_0 L}{4k_f}\right) = 3F_0 L \\ V = 2F_0 + 4k_s \left(\dfrac{F_0}{4k_s}\right) = 3F_0 \end{cases} \tag{4.33}$$

in which m is exactly equal to the applied force $3F_0$ multiplied by the length of the column L and V is exactly equal to the applied force $3F_0$.

Example 4.3 Static Analysis of SDOF Column with Input Displacement (1)

Consider that the SDOF column with the length $L = 3$ m extracted from a RC frame with two RHs and one SH shown in Figure 4.9(a), the sectional parameters are shown in Figure 4.9(b). The SDOF column is subjected by the input lateral displacement x and the axial load $P = 500$ kN. The concrete strength is 30 MPa, and the yield strengths of the longitudinal reinforcement and transverse reinforcement are 335 MPa and 235 MPa, respectively. The symmetrical parameters of the bending and shear primary curves are listed in Table 4.1 and Table 4.2. The input displacement x is specified as 2 mm.

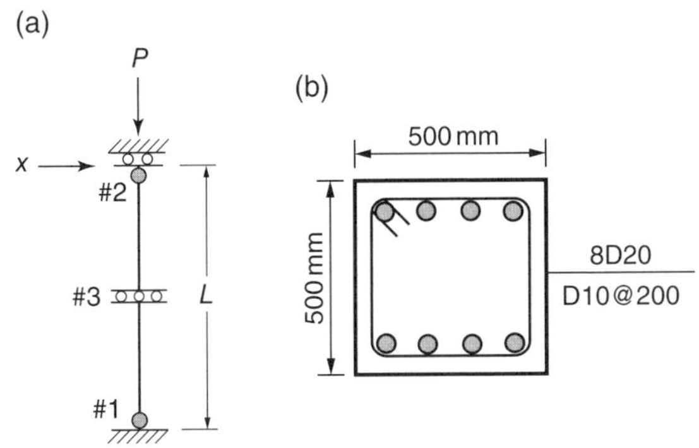

Figure 4.9 The SDOF column with two RHs and one SH: (a) Numerical model; (b) Sectional parameters.

Table 4.1 The parameters of the bending primary curve.

m_{cr} (kN-m)	θ_{cr} (rad)	m_y (kN-m)	θ_y (rad)	θ_u (rad)	θ_f (rad)
91.95	0.00032	229.5	0.0024	0.0165	0.036

Table 4.2 The parameters of the shear primary curve.

V_{cr} (kN)	τ_{cr} (m)	V_y (kN)	τ_y (m)	τ_u (m)	τ_f (m)
61.3	0.00008	153	0.0025	0.0075	0.012

The global stiffness matrices of the SDOF column can be written as:

$$\mathbf{K} = \left[\frac{12}{1+\chi}\frac{EI}{L^3}\right]$$

$$\mathbf{K}' = \left[-\frac{6}{1+\chi}\frac{EI}{L^2} \quad -\frac{6}{1+\chi}\frac{EI}{L^2} \quad \frac{12}{1+\chi}\frac{EI}{L^3}\right]$$

$$\mathbf{K}'' = \begin{bmatrix} \frac{4+\chi}{1+\chi}\frac{EI}{L} & \frac{2-\chi}{1+\chi}\frac{EI}{L} & -\frac{6}{1+\chi}\frac{EI}{L^2} \\ \frac{2-\chi}{1+\chi}\frac{EI}{L} & \frac{4+\chi}{1+\chi}\frac{EI}{L} & -\frac{6}{1+\chi}\frac{EI}{L^2} \\ -\frac{6}{1+\chi}\frac{EI}{L^2} & -\frac{6}{1+\chi}\frac{EI}{L^2} & \frac{12}{1+\chi}\frac{EI}{L^3} \end{bmatrix} \tag{4.34}$$

where $\chi = 12EIb/GAL^2$. The shear effect is introduced to the stiffness matrices $\mathbf{K}$, $\mathbf{K}'$ and $\mathbf{K}''$ in Eq. (4.34) by the parameter χ, and the procedure of obtaining the stiffness matrices can be found as below. By specifying $E = 30$ GPa and $G = 11.5$ GPa, the governing equation can be written as:

$$\left[\begin{array}{c|ccc} 0.6391 & -0.9586 & -0.9586 & 0.6391 \\ \hline -0.9586 & 1.9587 & 0.9171 & -0.9586 \\ -0.9586 & 0.9171 & 1.9587 & -0.9586 \\ 0.6391 & -0.9586 & -0.9586 & 0.6391 \end{array}\right] \times 10^5 \left\{\begin{array}{c} x \\ \hline -\theta_1'' \\ -\theta_2'' \\ -\tau'' \end{array}\right\} = \left\{\begin{array}{c} F \\ \hline m_1 \\ m_2 \\ V \end{array}\right\} \tag{4.35}$$

First, assume that the SDOF column is in the elastic domain, this gives $\theta_1'' = \theta_2'' = 0$ and $\tau'' = 0$. Then the restoring force F, shear force V, moment m_1 and m_2 can be calculated using Eq. (4.35):

$$\begin{Bmatrix} F \\ m_1 \\ m_2 \\ V \end{Bmatrix} = \begin{Bmatrix} 127.8119\,\text{kN} \\ -191.7178\,\text{kN-m} \\ -191.7178\,\text{kN-m} \\ 127.8119\,\text{kN} \end{Bmatrix} \tag{4.36}$$

Comparing these moment and shear demands in Eq. (4.36) with the cracking and yielding capacities in Tables 4.1 and 4.2, the SDOF column is in the plastic domain, which means the assumption that the SDOF column is in the elastic domain is incorrect. Therefore, assume that the structure is in the plastic domain and is located in the hardening branches, which means that the RHs and SH begin to deform in the hardening branches. The $k_f = 2.9\times10^5$ kN m/rad, $\alpha_f = 6.6\times10^4$ kN m/rad, $k_s = 7.7\times10^5$ kN/m and $\alpha_s = 3.8\times10^4$ kN/m can be computed according to the parameters illustrated in Tables 4.1 and 4.2. The moments m_1 and m_2 and shear force V can be expressed in the terms of θ_1'', θ_2'' and τ'' as:

$$\begin{cases} m_1 = -91.95 + 8.6 \times 10^4\, \theta_1'' \\ m_2 = -91.95 + 8.6 \times 10^4\, \theta_2'' \\ V = 61.3 + 4.0 \times 10^4\, \tau'' \end{cases} \tag{4.37}$$

where the minus sign in Eq. (4.37) denotes that the negative cracking capacities has been exceeded. Therefore, Eq. (4.35) becomes:

$$\left[\begin{array}{c|cccc} 0.6391 & -0.9586 & -0.9586 & 0.6391 \\ \hline -0.9586 & 1.9587+0.86 & 0.9171 & -0.9586 \\ -0.9586 & 0.9171 & 1.9587+0.86 & -0.9586 \\ 0.6391 & -0.9586 & -0.9586 & 0.6391+0.40 \end{array}\right] \times 10^5 \left\{\begin{array}{c} 0.002 \\ \hline -\theta_1'' \\ -\theta_2'' \\ -\tau'' \end{array}\right\} = \left\{\begin{array}{c} F \\ \hline -91.95 \\ -91.95 \\ 61.3 \end{array}\right\} \tag{4.38}$$

Solving for the plastic deformations θ_1'', θ_2'' and τ'' in Eq. (4.38) gives:

$$\begin{Bmatrix} \theta_1'' \\ \theta_2'' \\ \tau'' \end{Bmatrix} = \begin{Bmatrix} -0.1952\,\text{rad} \\ -0.1952\,\text{rad} \\ 0.2804\,\text{m} \end{Bmatrix} \times 10^{-3} \tag{4.39}$$

Substituting Eq. (4.39) back into Eq. (4.35) and solving for the restoring force F, the moments m_1 and m_2 at the RHs and the shear force V gives:

$$\begin{Bmatrix} F \\ m_1 \\ m_2 \\ V \end{Bmatrix} = \begin{Bmatrix} 72.4767\,\text{kN} \\ -108.7150\,\text{kN-m} \\ -108.7150\,\text{kN-m} \\ 72.4767\,\text{kN} \end{Bmatrix} \tag{4.40}$$

Comparing these demands in Eq. (4.40) with the corresponding capacities in Tables 4.1 and 4.2, the structure is in the plastic domain and the RHs and SH behave on the hardening branches, which means that the latest assumption is made correctly and the solution in Eq. (4.40) are the final results.

Example 4.4 Static Analysis of SDOF Column with Input Displacement (2)
Consider again the SDOF column as shown in Figure 4.9 with the same parameters in Tables 4.1 and 4.2. Now let the input displacement be equal to 20 mm.

According to results obtained in Example 4.3, first assume that the SDOF column is in the plastic domain and the two RHs and one SH behave in the hardening branches. Then Eq. (4.38) can be rewritten here by specifying the total displacement x as 0.02:

$$\left[\begin{array}{c|ccc} 0.6391 & -0.9586 & -0.9586 & 0.6391 \\ \hline -0.9586 & 1.9587+0.86 & 0.9171 & -0.9586 \\ -0.9586 & 0.9171 & 1.9587+0.86 & -0.9586 \\ 0.6391 & -0.9586 & -0.9586 & 0.6391+0.40 \end{array}\right] \times 10^5 \begin{Bmatrix} 0.02 \\ \hline -\theta_1'' \\ -\theta_2'' \\ -\tau'' \end{Bmatrix} = \begin{Bmatrix} F \\ \hline -91.95 \\ -91.95 \\ 61.3 \end{Bmatrix} \tag{4.41}$$

Solving for the plastic deformations θ_1'', θ_2'' and τ'' in Eq. (4.41) gives:

$$\begin{Bmatrix} \theta_1'' \\ \theta_2'' \\ \tau'' \end{Bmatrix} = \begin{Bmatrix} -0.0036\,\text{rad} \\ -0.0036\,\text{rad} \\ 0.0051\,\text{m} \end{Bmatrix} \tag{4.42}$$

Substituting Eq. (4.42) back into Eq. (4.35) and solving for the restoring force F, the moments m_1 and m_2 at the RHs and the shear force V gives:

$$\begin{Bmatrix} F \\ m_1 \\ m_2 \\ V \end{Bmatrix} = \begin{Bmatrix} 265.7743\,\text{kN} \\ -398.6614\,\text{kN-m} \\ -398.6614\,\text{kN-m} \\ 265.7743\,\text{kN} \end{Bmatrix} \tag{4.43}$$

The rotations θ_1 and θ_2 and the shear deformation τ of the structure as illustrated in Figure 4.7 can be calculated according to the displacement decomposition expressed in Section 4.1.2:

$$\theta_1 = \theta_2 = -0.0050\,\text{rad}, \quad \tau = 0.0054\,\text{m} \tag{4.44}$$

Comparing the demands in Eqs. (4.43) and (4.44) with the corresponding capacities in Tables 4.1 and 4.2, the structure should be represent by the yielding platform.

The α_f and α_s here are specified by 0.1 that close to zero to simulate the yielding platform branches. Then the moments m_1 and m_2 and shear force V can be expressed in the terms of θ_1'', θ_2'' and τ'' according to Eq. (4.5):

$$\begin{cases} m_1 = -229.4998 + 0.1\,\theta_1'' \\ m_2 = -229.4998 + 0.1\,\theta_2'' \\ V = 152.9998 + 0.1\,\tau'' \end{cases} \tag{4.45}$$

After adding the tangent stiffnesses expressed in Eq. (4.45) to the stiffness matrix $\mathbf{K}''$, Eq. (4.41) becomes:

$$\left[\begin{array}{c|ccc} 0.6391 & -0.9586 & -0.9586 & 0.6391 \\ \hline -0.9586 & 1.958701 & 0.9171 & -0.9586 \\ -0.9586 & 0.9171 & 1.958701 & -0.9586 \\ 0.6391 & -0.9586 & -0.9586 & 0.639101 \end{array}\right] \times 10^5 \begin{Bmatrix} 0.02 \\ \hline -\theta_1'' \\ -\theta_2'' \\ -\tau'' \end{Bmatrix} = \begin{Bmatrix} F \\ \hline -229.4998 \\ -229.4998 \\ 152.9998 \end{Bmatrix} \tag{4.46}$$

Solving for the plastic deformations θ_1'', θ_2'' and τ'' in Eq. (4.46) gives:

$$\begin{Bmatrix} \theta_1'' \\ \theta_2'' \\ \tau'' \end{Bmatrix} = \begin{Bmatrix} -0.0045\,\text{rad} \\ -0.0045\,\text{rad} \\ 0.0042\,\text{m} \end{Bmatrix} \tag{4.47}$$

Substituting Eq. (4.47) back into Eq. (4.35) and solving for the restoring force F, the bending moments m_1 and m_2 at the RHs and the shear force V gives:

$$\begin{Bmatrix} F \\ m_1 \\ m_2 \\ V \end{Bmatrix} = \begin{Bmatrix} 153.0002\,\text{kN} \\ -229.5003\,\text{kN-m} \\ -229.5003\,\text{kN-m} \\ 153.0002\,\text{kN} \end{Bmatrix} \tag{4.48}$$

in which the results are quite close to the yielding capacities m_y and V_y listed in Tables 4.1 and 4.2, which means that the yielding platform behaviors can be well simulated. The rotations θ_1 and θ_2 and the shear deformation τ of the structure can be obtained as:

$$\theta_1 = \theta_2 = -0.0053\text{ rad},\quad \tau = 0.0044\text{ m} \tag{4.49}$$

Comparing the demands in Eq. (4.48) and Eq. (4.49) with the corresponding capacities in Tables 4.1 and 4.2, the structure is in the plastic domain and the RHs and SH behave on the yielding platforms, which means that the latest assumption is made correctly and the solution in Eq. (4.48) are the final results.

Example 4.5 Static Analysis of SDOF Column with Input Displacement (3)
Consider again the SDOF column as shown in Figure 4.9 with the same parameters in Tables 4.1 and 4.2. Now let the input displacement be equal to 60 mm.

According to the results obtained in Example 4.4, first assume that the SDOF column is in the plastic domain and the two RHs and one SH behave in the yielding platform branches. Then Eq. (4.46) can be rewritten here by specifying the total displacement x as 0.06:

$$\left[\begin{array}{c|ccc} 0.6391 & -0.9586 & -0.9586 & 0.6391 \\ \hline -0.9586 & 1.958701 & 0.9171 & -0.9586 \\ -0.9586 & 0.9171 & 1.958701 & -0.9586 \\ 0.6391 & -0.9586 & -0.9586 & 0.639101 \end{array}\right] \times 10^5 \left\{\begin{array}{c} 0.06 \\ \hline -\theta_1'' \\ -\theta_2'' \\ -\tau'' \end{array}\right\} = \left\{\begin{array}{c} F \\ \hline -229.4998 \\ -229.4998 \\ 152.9998 \end{array}\right\} \tag{4.50}$$

Solving for the plastic deformations θ_1'', θ_2'' and τ'' in Eq. (4.50) gives:

$$\begin{Bmatrix} \theta_1'' \\ \theta_2'' \\ \tau'' \end{Bmatrix} = \begin{Bmatrix} -0.0154\,\text{rad} \\ -0.0154\,\text{rad} \\ 0.0115\,\text{m} \end{Bmatrix} \tag{4.51}$$

Substituting Eq. (4.51) back into Eq. (4.50) and solving for the restoring force F, the moments m_1 and m_2 at the RHs and the shear force V gives:

$$\begin{Bmatrix} F \\ m_1 \\ m_2 \\ V \end{Bmatrix} = \begin{Bmatrix} 153.0009\ \text{kN} \\ -229.5014\ \text{kN-m} \\ -229.5014\ \text{kN-m} \\ 153.0009\ \text{kN} \end{Bmatrix} \tag{4.52}$$

Then, the rotations θ_1 and θ_2 at the column ends and the shear deformation τ of the structure can be obtained:

$$\theta_1 = \theta_2 = -0.0162\ \text{rad}, \quad \tau = 0.0117\ \text{m} \tag{4.53}$$

Comparing the demands in Eqs. (4.52) and (4.53) with the corresponding capacities in Tables 4.1 and 4.2, we can find that the strength of SH starts to deteriorate. However, due to the same primary curve parameters of the two RHs shown in Table 4.1 and the assumption of the procedure of dividing the total primary curve to the bending and shear ones for the SDOF system in previous section, the RHs and SH should behave on the same type of branches at the same force level. Therefore, assume that the RHs and the SH behave on the corresponding softening branches and the moments m_1 and m_2 and shear force V can be expressed in the terms of θ_1'', θ_2'' and τ'' according to Eq. (4.5):

$$\begin{cases} m_1 = -407.02 - 1.1 \times 10^4\ \theta_1'' \\ m_2 = -407.02 - 1.1 \times 10^4\ \theta'' \\ V = 390.67 - 3.3 \times 10^4\ \tau'' \end{cases} \tag{4.54}$$

Therefore, Eq. (4.50) becomes:

$$\left[\begin{array}{c|ccc} 0.6391 & -0.9586 & -0.9586 & 0.6391 \\ \hline -0.9586 & 1.9587-0.11 & 0.9171 & -0.9586 \\ -0.9586 & 0.9171 & 1.9587-0.11 & -0.9586 \\ 0.6391 & -0.9586 & -0.9586 & 0.6391-0.33 \end{array}\right] \times 10^5 \left\{\begin{array}{c} 0.06 \\ \hline -\theta_1'' \\ -\theta_2'' \\ -\tau'' \end{array}\right\} = \left\{\begin{array}{c} F \\ \hline -407.02 \\ -407.02 \\ 390.67 \end{array}\right\} \tag{4.55}$$

Solving for the plastic deformations θ_1'', θ_2'' and τ'' in Eq. (4.55) gives:

$$\begin{Bmatrix} \theta_1'' \\ \theta_2'' \\ \tau'' \end{Bmatrix} = \begin{Bmatrix} -0.0167\ \text{rad} \\ -0.0167\ \text{rad} \\ 0.0075\ \text{m} \end{Bmatrix} \tag{4.56}$$

Substituting Eq. (4.56) back into Eq. (4.35) and solving for the restoring force F, the moments m_1 and m_2 at the RHs and the shear force V gives:

$$\begin{Bmatrix} F \\ m_1 \\ m_2 \\ V \end{Bmatrix} = \begin{Bmatrix} 145.25\ \text{kN} \\ -217.88\ \text{kN-m} \\ -217.88\ \text{kN-m} \\ 145.25\ \text{kN} \end{Bmatrix} \tag{4.57}$$

Then, the rotation deformations θ_1 and θ_2 at the column ends and the shear deformation τ of the structure can be obtained:

$$\theta_1 = \theta_2 = -0.0175\,\text{rad}, \quad \tau = 0.0077\,\text{m} \tag{4.58}$$

Comparing the demands in Eqs. (4.57) and (4.58) with the corresponding capacities in Tables 4.1 and 4.2, the structure is in the plastic domain and the RHs and SH behave on the softening branches, which means that the latest assumption is made correctly and the solution in Eq. (4.57) are the final results.

4.3.2 Force Analogy Method for Static Multi-Degree-of-Freedom Systems

For a multi-degree-of-freedom (MDOF) system with n degrees of freedom (DOFs), the displacement can be written in vector form as:

$$\mathbf{x} = \mathbf{x}' + \mathbf{x}'' = \begin{Bmatrix} x_1' \\ x_2' \\ \vdots \\ x_n' \end{Bmatrix} + \begin{Bmatrix} x_1'' \\ x_2'' \\ \vdots \\ x_n'' \end{Bmatrix} \tag{4.59}$$

where $\mathbf{x}$ represents the total displacement vector, $\mathbf{x}'$ is the elastic displacement vector, and $\mathbf{x}''$ is the inelastic displacement vector.

Consider a RC frame constructed by flexural members, the q_f RHs are located at two ends of the members to represent the plastic bending behavior and q_s SHs are assigned to the columns to represent the plastic shear behavior. Then the moment vector $\mathbf{m}$ at the RHs and shear force vector $\mathbf{V}$ according to the SHs can be described by the force vector $\mathbf{f}$ as:

$$\mathbf{f} = \begin{Bmatrix} \mathbf{m} \\ \hdashline \mathbf{V} \end{Bmatrix} = \begin{Bmatrix} \mathbf{m}' \\ \hdashline \mathbf{V}' \end{Bmatrix} + \begin{Bmatrix} \mathbf{m}'' \\ \hdashline \mathbf{V}'' \end{Bmatrix} = \begin{Bmatrix} m_1' \\ \vdots \\ m_{q_f}' \\ \hdashline V_{q_f+1}' \\ \vdots \\ V_{q_f+q_s}' \end{Bmatrix} + \begin{Bmatrix} m_1'' \\ \vdots \\ m_{q_f}'' \\ \hdashline V_{q_f+1}'' \\ \vdots \\ V_{q_f+q_s}'' \end{Bmatrix} \tag{4.60}$$

where $\mathbf{m}'$ and $\mathbf{V}'$ are the vectors of elastic moments and elastic shear forces respectively due to elastic displacement $\mathbf{x}'$, and $\mathbf{m}''$ and $\mathbf{V}''$ are the vectors of inelastic moments and shear forces respectively due to inelastic displacement $\mathbf{x}''$.

Corresponding to the inelastic moment and shear force vectors $\mathbf{m}''$ and $\mathbf{V}''$, the inelastic deformations θ'' and τ'' can be written in vector form as:

$$\boldsymbol{\Lambda}'' = \begin{Bmatrix} \boldsymbol{\Theta}'' \\ \hdashline \boldsymbol{\Delta}'' \end{Bmatrix} = \begin{Bmatrix} \theta''_1 \\ \vdots \\ \theta''_{q_f} \\ \hdashline \tau''_{q_f+1} \\ \vdots \\ \tau''_{q_f+q_s} \end{Bmatrix} \tag{4.61}$$

Similar to the derivation of the FAM for the MDOF systems illustrated in Chapter 2, the governing equation for RC frame structures with flexural members can be obtained as:

$$\begin{bmatrix} \mathbf{K} & \mathbf{K}' \\ \mathbf{K}'^T & \mathbf{K}'' \end{bmatrix} \begin{Bmatrix} \mathbf{x} \\ -\boldsymbol{\Lambda}'' \end{Bmatrix} = \begin{Bmatrix} \mathbf{F} \\ \mathbf{f} \end{Bmatrix} \tag{4.62}$$

Equation (4.62) shows that both the total displacement vector $\mathbf{x}$ (n unknowns) and the plastic deformation vector $\boldsymbol{\Lambda}''$ ($q_f + q_s$ unknowns) make up the unknown vector that is typically required to be solved in the nonlinear structural analysis problem. For any applied static force pattern $\mathbf{F}$, the $2(q_f + q_s) + n$ unknowns are $\mathbf{x}$, $\boldsymbol{\Lambda}''$, and $\mathbf{f}$. However, Eq. (4.62) by itself only provides $q_f + q_s + n$ independent equations, and $q_f + q_s$ relationships of load plastic mechanisms of RHs and SHs should be incorporated to solve the $2(q_f + q_s) + n$ unknowns uniquely. As the shear behavior for columns is considered, the stiffness matrices $\mathbf{K}$, $\mathbf{K}'$, and $\mathbf{K}''$ in Eq. (4.62) should be modified from the original version expressed in Chapter 2, and the procedure of obtaining these stiffness matrices is expressed in Example 4.6.

Example 4.6 Stiffness Matrices Assembling of One-Story Frame

Consider a one-story one-bay RC frame with three flexural members as shown in Figure 4.10. Six RHs #1 to #6 are located at the members' ends, and two SHs #7 and #8 are assigned to the two columns. Assume that the members are axially rigid, this results in a three degrees of freedom system ($n = 3$) with six plastic rotation hinge locations ($q_f = 6$) and two slide hinge locations ($q_s = 2$). Assume also that the beam is of different length and size from the two columns.

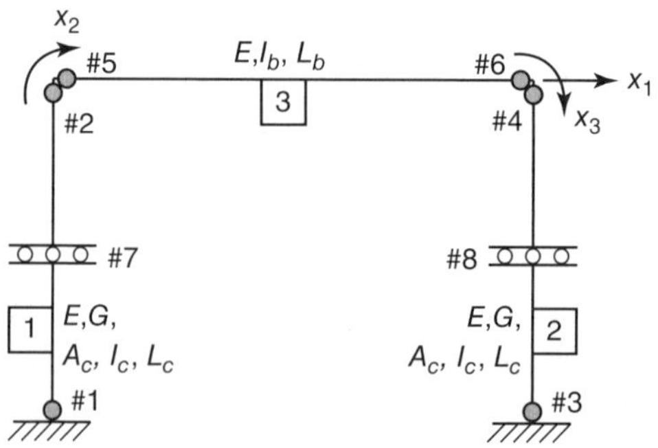

Figure 4.10 The one-story one-bay RC frame.

Due to ignoring the shear effect for the beam, the element stiffness matrices at the local coordinate is expressed by

$$\mathbf{K}_3 = \begin{bmatrix} \frac{12EI_b}{L_b^3} & \frac{6EI_b}{L_b^2} & -\frac{12EI_b}{L_b^3} & \frac{6EI_b}{L_b^2} \\ \frac{6EI_b}{L_b^2} & \frac{4EI_b}{L_b} & -\frac{6EI_b}{L_b^2} & \frac{2EI_b}{L_b} \\ -\frac{12EI_b}{L_b^3} & -\frac{6EI_b}{L_b^2} & \frac{12EI_b}{L_b^3} & -\frac{6EI_b}{L_b^2} \\ \frac{6EI_b}{L_b^2} & \frac{2EI_b}{L_b} & -\frac{6EI_b}{L_b^2} & \frac{4EI_b}{L_b} \end{bmatrix} \tag{4.63}$$

$$\mathbf{K}_3' = \begin{bmatrix} \frac{6EI_b}{L_b^2} & \frac{6EI_b}{L_b^2} \\ \frac{4EI_b}{L_b} & \frac{2EI_b}{L_b} \\ -\frac{6EI_b}{L_b^2} & -\frac{6EI_b}{L_b^2} \\ \frac{2EI_b}{L_b} & \frac{4EI_b}{L_b} \end{bmatrix} \tag{4.64}$$

$$\mathbf{K}_3'' = \begin{bmatrix} \frac{4EI_b}{L_b} & \frac{2EI_b}{L_b} \\ \frac{2EI_b}{L_b} & \frac{4EI_b}{L_b} \end{bmatrix} \tag{4.65}$$

For the two columns, the element stiffness matrices must be modified in order to incorporate the shear effect. The shear behavior is represented as additional shear displacement to the individual member, and this gives the element stiffness matrices at the local coordinate of the form:

$$\mathbf{K}_1 = \mathbf{K}_2 = \begin{bmatrix} \frac{12}{1+\chi}\frac{EI_c}{L_c^3} & \frac{6}{1+\chi}\frac{EI_c}{L_c^2} & -\frac{12}{1+\chi}\frac{EI_c}{L_c^3} & \frac{6}{1+\chi}\frac{EI_c}{L_c^2} \\ \frac{6}{1+\chi}\frac{EI_c}{L_c^2} & \frac{4+\chi}{1+\chi}\frac{EI_c}{L_c} & -\frac{6}{1+\chi}\frac{EI_c}{L_c^2} & \frac{2-\chi}{1+\chi}\frac{EI_c}{L_c} \\ -\frac{12}{1+\chi}\frac{EI_c}{L_c^3} & -\frac{6}{1+\chi}\frac{EI_c}{L_c^2} & \frac{12}{1+\chi}\frac{EI_c}{L_c^3} & -\frac{6}{1+\chi}\frac{EI_c}{L_c^2} \\ \frac{6}{1+\chi}\frac{EI_c}{L_c^2} & \frac{2-\chi}{1+\chi}\frac{EI_c}{L_c} & -\frac{6}{1+\chi}\frac{EI_c}{L_c^2} & \frac{4+\chi}{1+\chi}\frac{EI_c}{L_c} \end{bmatrix} \tag{4.66}$$

$$\mathbf{K}_1' = \mathbf{K}_2' = \begin{bmatrix} \frac{6}{1+\chi}\frac{EI_c}{L_c^2} & \frac{6}{1+\chi}\frac{EI_c}{L_c^2} & \frac{12}{1+\chi}\frac{EI_c}{L_c^3} \\ \frac{4+\chi}{1+\chi}\frac{EI_c}{L_c} & \frac{2-\chi}{1+\chi}\frac{EI_c}{L_c} & \frac{6}{1+\chi}\frac{EI_c}{L_c^2} \\ -\frac{6}{1+\chi}\frac{EI_c}{L_c^2} & -\frac{6}{1+\chi}\frac{EI_c}{L_c^2} & -\frac{12}{1+\chi}\frac{EI_c}{L_c^3} \\ \frac{2-\chi}{1+\chi}\frac{EI_c}{L_c} & \frac{4+\chi}{1+\chi}\frac{EI_c}{L_c} & \frac{6}{1+\chi}\frac{EI_c}{L_c^2} \end{bmatrix} \tag{4.67}$$

$$\mathbf{K}_1'' = \mathbf{K}_2'' = \begin{bmatrix} \frac{4+\chi}{1+\chi}\frac{EI_c}{L_c} & \frac{2-\chi}{1+\chi}\frac{EI_c}{L_c} & \frac{6}{1+\chi}\frac{EI_c}{L_c^2} \\ \frac{2-\chi}{1+\chi}\frac{EI_c}{L_c} & \frac{4+\chi}{1+\chi}\frac{EI_c}{L_c} & \frac{6}{1+\chi}\frac{EI_c}{L_c^2} \\ \frac{6}{1+\chi}\frac{EI_c}{L_c^2} & \frac{6}{1+\chi}\frac{EI_c}{L_c^2} & \frac{12}{1+\chi}\frac{EI_c}{L_c^3} \end{bmatrix} \tag{4.68}$$

where χ in Eqs. (4.66) to (4.68) is the parameter that introduces the shear effect to the stiffness matrices and can be computed by

$$\chi = \frac{12EI_c b}{GA_c L_c^2} \tag{4.69}$$

where the parameter b is the shear correction factor and equals to 1.2 and 1.1 for the rectangular sections and circular sections, respectively. For more details on the derivation of introducing the shear effect to the element stiffness matrix, the readers can refer to the presentation by *Ghali* and *Nevilli* (1989).

Now assembling each of the element stiffness matrices into the global stiffness matrices gives:

$$\mathbf{K} = \begin{bmatrix} \frac{12}{1+\chi}\frac{EI_c}{L_c^3} + \frac{12}{1+\chi}\frac{EI_c}{L_c^3} & -\frac{6}{1+\chi}\frac{EI_c}{L_c^2} & -\frac{6}{1+\chi}\frac{EI_c}{L_c^2} \\ -\frac{6}{1+\chi}\frac{EI_c}{L_c^2} & \frac{4+\chi}{1+\chi}\frac{EI_c}{L_c} + \frac{4EI_b}{L_b} & \frac{2EI_b}{L_b} \\ -\frac{6}{1+\chi}\frac{EI_c}{L_c^2} & \frac{2EI_b}{L_b} & \frac{4+\chi}{1+\chi}\frac{EI_c}{L_c} + \frac{4EI_b}{L_b} \end{bmatrix} \begin{matrix} \leftarrow x_1 \\ \leftarrow x_2 \\ \leftarrow x_3 \end{matrix} \tag{4.70}$$

$$\mathbf{K}' = \begin{bmatrix} -\frac{6}{1+\chi}\frac{EI_c}{L_c^2} & -\frac{6}{1+\chi}\frac{EI_c}{L_c^2} & -\frac{6}{1+\chi}\frac{EI_c}{L_c^2} & -\frac{6}{1+\chi}\frac{EI_c}{L_c^2} & 0 & 0 & \frac{12}{1+\chi}\frac{EI_c}{L_c^3} & \frac{12}{1+\chi}\frac{EI_c}{L_c^3} \\ \frac{2-\chi}{1+\chi}\frac{EI_c}{L_c} & \frac{4+\chi}{1+\chi}\frac{EI_c}{L_c} & 0 & 0 & \frac{4EI_b}{L_b} & \frac{2EI_b}{L_b} & -\frac{6}{1+\chi}\frac{EI_c}{L_c^2} & 0 \\ 0 & 0 & \frac{2-\chi}{1+\chi}\frac{EI_c}{L_c} & \frac{4+\chi}{1+\chi}\frac{EI_c}{L_c} & \frac{2EI_b}{L_b} & \frac{4EI_b}{L_b} & 0 & -\frac{6}{1+\chi}\frac{EI_c}{L_c^2} \end{bmatrix} \begin{matrix} \leftarrow x_1 \\ \leftarrow x_2 \\ \leftarrow x_3 \end{matrix} \tag{4.71}$$

$$
\mathbf{K}'' = \begin{bmatrix}
\frac{4+\chi}{1+\chi}\frac{EI_c}{L_c} & \frac{2-\chi}{1+\chi}\frac{EI_c}{L_c} & 0 & 0 & 0 & 0 & -\frac{6}{1+\chi}\frac{EI_c}{L_c^2} & 0 \\
\frac{2-\chi}{1+\chi}\frac{EI_c}{L_c} & \frac{4+\chi}{1+\chi}\frac{EI_c}{L_c} & 0 & 0 & 0 & 0 & -\frac{6}{1+\chi}\frac{EI_c}{L_c^2} & 0 \\
0 & 0 & \frac{4+\chi}{1+\chi}\frac{EI_c}{L_c} & \frac{2-\chi}{1+\chi}\frac{EI_c}{L_c} & 0 & 0 & 0 & -\frac{6}{1+\chi}\frac{EI_c}{L_c^2} \\
0 & 0 & \frac{2-\chi}{1+\chi}\frac{EI_c}{L_c} & \frac{4+\chi}{1+\chi}\frac{EI_c}{L_c} & 0 & 0 & 0 & -\frac{6}{1+\chi}\frac{EI_c}{L_c^2} \\
0 & 0 & 0 & 0 & \frac{4EI_b}{L_b} & \frac{2EI_b}{L_b} & 0 & 0 \\
0 & 0 & 0 & 0 & \frac{2EI_b}{L_b} & \frac{4EI_b}{L_b} & 0 & 0 \\
-\frac{6}{1+\chi}\frac{EI_c}{L_c^2} & -\frac{6}{1+\chi}\frac{EI_c}{L_c^2} & 0 & 0 & 0 & 0 & \frac{12}{1+\chi}\frac{EI_c}{L_c^3} & 0 \\
0 & 0 & -\frac{6}{1+\chi}\frac{EI_c}{L_c^2} & -\frac{6}{1+\chi}\frac{EI_c}{L_c^2} & 0 & 0 & 0 & \frac{12}{1+\chi}\frac{EI_c}{L_c^3}
\end{bmatrix}
\begin{matrix} \leftarrow \theta''_1 \\ \leftarrow \theta''_2 \\ \leftarrow \theta''_3 \\ \leftarrow \theta''_4 \\ \leftarrow \theta''_5 \\ \leftarrow \theta''_6 \\ \leftarrow \tau''_7 \\ \leftarrow \tau''_8 \end{matrix}
\tag{4.72}
$$

Substituting the above stiffness matrices $\mathbf{K}$, $\mathbf{K}'$, and $\mathbf{K}''$ into the governing equation shown in Eq. (4.62) gives:

$$
\left[\begin{array}{ccc|cccccccc}
\frac{12}{1+\chi}\frac{EI_c}{L_c^3}+\frac{12}{1+\chi}\frac{EI_c}{L_c^3} & -\frac{6}{1+\chi}\frac{EI_c}{L_c^2} & -\frac{6}{1+\chi}\frac{EI_c}{L_c^2} & -\frac{6}{1+\chi}\frac{EI_c}{L_c^2} & -\frac{6}{1+\chi}\frac{EI_c}{L_c^2} & -\frac{6}{1+\chi}\frac{EI_c}{L_c^2} & -\frac{6}{1+\chi}\frac{EI_c}{L_c^2} & 0 & 0 & \frac{12}{1+\chi}\frac{EI_c}{L_c^3} & \frac{12}{1+\chi}\frac{EI_c}{L_c^3} \\
-\frac{6}{1+\chi}\frac{EI_c}{L_c^2} & \frac{4+\chi}{1+\chi}\frac{EI_c}{L_c}+\frac{4EI_b}{L_b} & \frac{2EI_b}{L_b} & \frac{2-\chi}{1+\chi}\frac{EI_c}{L_c} & \frac{4+\chi}{1+\chi}\frac{EI_c}{L_c} & 0 & 0 & \frac{4EI_b}{L_b} & \frac{2EI_b}{L_b} & -\frac{6}{1+\chi}\frac{EI_c}{L_c^2} & 0 \\
-\frac{6}{1+\chi}\frac{EI_c}{L_c^2} & \frac{2EI_b}{L_b} & \frac{4+\chi}{1+\chi}\frac{EI_c}{L_c}+\frac{4EI_b}{L_b} & 0 & 0 & \frac{2-\chi}{1+\chi}\frac{EI_c}{L_c} & \frac{4+\chi}{1+\chi}\frac{EI_c}{L_c} & \frac{2EI_b}{L_b} & \frac{4EI_b}{L_b} & 0 & -\frac{6}{1+\chi}\frac{EI_c}{L_c^2} \\
\hline
-\frac{6}{1+\chi}\frac{EI_c}{L_c^2} & \frac{2-\chi}{1+\chi}\frac{EI_c}{L_c} & 0 & \frac{4+\chi}{1+\chi}\frac{EI_c}{L_c} & \frac{2-\chi}{1+\chi}\frac{EI_c}{L_c} & 0 & 0 & 0 & 0 & -\frac{6}{1+\chi}\frac{EI_c}{L_c^2} & 0 \\
-\frac{6}{1+\chi}\frac{EI_c}{L_c^2} & \frac{4+\chi}{1+\chi}\frac{EI_c}{L_c} & 0 & \frac{2-\chi}{1+\chi}\frac{EI_c}{L_c} & \frac{4+\chi}{1+\chi}\frac{EI_c}{L_c} & 0 & 0 & 0 & 0 & -\frac{6}{1+\chi}\frac{EI_c}{L_c^2} & 0 \\
-\frac{6}{1+\chi}\frac{EI_c}{L_c^2} & 0 & \frac{2-\chi}{1+\chi}\frac{EI_c}{L_c} & 0 & 0 & \frac{4+\chi}{1+\chi}\frac{EI_c}{L_c} & \frac{2-\chi}{1+\chi}\frac{EI_c}{L_c} & 0 & 0 & 0 & -\frac{6}{1+\chi}\frac{EI_c}{L_c^2} \\
-\frac{6}{1+\chi}\frac{EI_c}{L_c^2} & 0 & \frac{4+\chi}{1+\chi}\frac{EI_c}{L_c} & 0 & 0 & \frac{2-\chi}{1+\chi}\frac{EI_c}{L_c} & \frac{4+\chi}{1+\chi}\frac{EI_c}{L_c} & 0 & 0 & 0 & -\frac{6}{1+\chi}\frac{EI_c}{L_c^2} \\
0 & \frac{4EI_b}{L_b} & \frac{2EI_b}{L_b} & 0 & 0 & 0 & 0 & \frac{4EI_b}{L_b} & \frac{2EI_b}{L_b} & 0 & 0 \\
0 & \frac{2EI_b}{L_b} & \frac{4EI_b}{L_b} & 0 & 0 & 0 & 0 & \frac{2EI_b}{L_b} & \frac{4EI_b}{L_b} & 0 & 0 \\
\frac{12}{1+\chi}\frac{EI_c}{L_c^3} & -\frac{6}{1+\chi}\frac{EI_c}{L_c^2} & 0 & -\frac{6}{1+\chi}\frac{EI_c}{L_c^2} & -\frac{6}{1+\chi}\frac{EI_c}{L_c^2} & 0 & 0 & 0 & 0 & \frac{12}{1+\chi}\frac{EI_c}{L_c^3} & 0 \\
\frac{12}{1+\chi}\frac{EI_c}{L_c^3} & 0 & -\frac{6}{1+\chi}\frac{EI_c}{L_c^2} & 0 & 0 & -\frac{6}{1+\chi}\frac{EI_c}{L_c^2} & -\frac{6}{1+\chi}\frac{EI_c}{L_c^2} & 0 & 0 & 0 & \frac{12}{1+\chi}\frac{EI_c}{L_c^3}
\end{array}\right]
\left\{\begin{array}{c} x_1 \\ x_2 \\ x_3 \\ \hline \theta''_1 \\ \theta''_2 \\ \theta''_3 \\ \theta''_4 \\ \theta''_5 \\ \theta''_6 \\ \tau''_7 \\ \tau''_8 \end{array}\right\}
=
\left\{\begin{array}{c} F_1 \\ F_2 \\ F_3 \\ \hline m_1 \\ m_2 \\ m_3 \\ m_4 \\ m_5 \\ m_6 \\ V_7 \\ V_8 \end{array}\right\}
\tag{4.73}
$$

Then, Eq. (4.73) can be used for the nonlinear static analysis of the one-story one-bay RC frame. In order to reduce the size of the problem, the procedure of static condensation expressed in Section 2.6 can also be used here.

Example 4.7 Nonlinear Static Analysis of the One-Story One-Bay RC Frame
Consider a one-story one-bay RC frame with flexural members as shown in Figure 4.11. The columns of the RC frame are constructed with the SDOF flexural member shown in Example 4.3, and the sectional parameters are illustrated in Figure 4.9(b). Therefore, the behaviors of hinges #1, #2, #3, #4, #7, and #8 can be obtained from Tables 4.1 and 4.2. With the same materials and strengths of the concrete and reinforcement, the beam with the sectional parameters are shown in Figure 4.11, and the parameters of the bending primary curve of the beam are listed in Table 4.3. Let the length of the beam $L_b = L_c = 3$ m. Assume that only a lateral force $F_1 = 150$ kN is applied at the horizontal degree of freedom x_1, and external moments $F_2 = F_3 = 0$.

First, the global stiffness matrices $\mathbf{K}$, $\mathbf{K}'$, and $\mathbf{K}''$ should be assembled according to the procedure presented in Example 4.6 and are written as follows:

$$\mathbf{K} = \begin{bmatrix} 127812 & -95859 & -95859 \\ -95859 & 238544 & 21333 \\ -95859 & 21333 & 238544 \end{bmatrix} \begin{matrix} \leftarrow x_1 \\ \leftarrow x_2 \\ \leftarrow x_3 \end{matrix} \ (\text{kN/m}) \tag{4.74}$$

$$\mathbf{K}' = \begin{bmatrix} -95859 & -95859 & -95859 & -95859 & 0 & 0 & 63906 & 63906 \\ 91705 & 195872 & 0 & 0 & 42667 & 21333 & -95859 & 0 \\ 0 & 0 & 91705 & 195872 & 21333 & 42667 & 0 & -95859 \end{bmatrix} \begin{matrix} \leftarrow x_1 \\ \leftarrow x_2 \\ \leftarrow x_3 \end{matrix} \ (\text{kN/rad}) \tag{4.75}$$

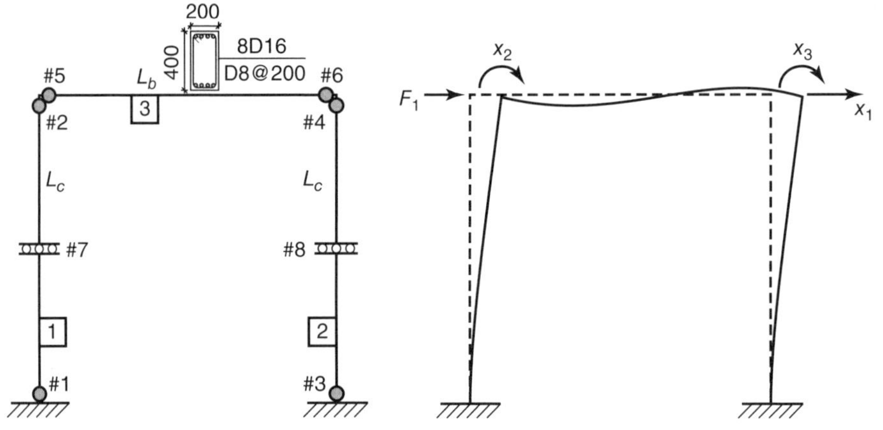

Figure 4.11 The one-story one-bay RC frame.

Table 4.3 The parameters of the bending primary curve.

m_{cr} (kN-m)	θ_{cr} (rad)	m_y (kN-m)	θ_y (rad)	θ_u (rad)	θ_f (rad)
21	0.000166	100	0.00129	0.00335	0.011

$$\mathbf{K}'' = \begin{bmatrix} 195872 & 91705 & 0 & 0 & 0 & 0 & -95859 & 0 \\ 91705 & 195872 & 0 & 0 & 0 & 0 & -95859 & 0 \\ 0 & 0 & 195872 & 91705 & 0 & 0 & 0 & -95859 \\ 0 & 0 & 91705 & 195872 & 0 & 0 & 0 & -95859 \\ 0 & 0 & 0 & 0 & 42667 & 21333 & 0 & 0 \\ 0 & 0 & 0 & 0 & 21333 & 42667 & 0 & 0 \\ -95859 & -95859 & 0 & 0 & 0 & 0 & 63906 & \\ 0 & 0 & -95859 & -95859 & 0 & 0 & & 63906 \end{bmatrix} \begin{matrix} \leftarrow \theta_1'' \\ \leftarrow \theta_2'' \\ \leftarrow \theta_3'' \\ \leftarrow \theta_4'' \\ \leftarrow \theta_5'' \\ \leftarrow \theta_6'' \\ \leftarrow \tau_7'' \\ \leftarrow \tau_8'' \end{matrix} \text{(kN-m/rad)} \tag{4.76}$$

To analyze the structural behavior subjected to the applied later load F_1, the first step is to assume that the response is linear, i.e. $\theta_1'' = \theta_2'' = \theta_3'' = \theta_4'' = \theta_5'' = \theta_6'' = 0$ and $\tau_7'' = \tau_8'' = 0$. As a result, the governing equation expressed in Eq. (4.62) becomes:

$$\left[\begin{array}{ccc|cccccccc} 127812 & -95859 & -95859 & -95859 & -95859 & -95859 & -95859 & 0 & 0 & 63906 & 63906 \\ -95859 & 238544 & 21333 & 91705 & 195872 & 0 & 0 & 42667 & 21333 & -95859 & 0 \\ -95859 & 21333 & 238544 & 0 & 0 & 91705 & 195872 & 21333 & 42667 & 0 & -95859 \\ \hline -95859 & 91705 & 0 & 195872 & 91705 & 0 & 0 & 0 & 0 & -95859 & 0 \\ -95859 & 195872 & 0 & 91705 & 195872 & 0 & 0 & 0 & 0 & -95859 & 0 \\ -95859 & 0 & 91705 & 0 & 0 & 195872 & 91705 & 0 & 0 & 0 & -95859 \\ -95859 & 0 & 195872 & 0 & 0 & 91705 & 195872 & 0 & 0 & 0 & -95859 \\ 0 & 42667 & 21333 & 0 & 0 & 0 & 0 & 42667 & 21333 & 0 & 0 \\ 0 & 21333 & 42667 & 0 & 0 & 0 & 0 & 21333 & 42667 & 0 & 0 \\ 63906 & -95859 & 0 & -95859 & -95859 & 0 & 0 & 0 & 0 & 63906 & 0 \\ 63906 & 0 & -95859 & 0 & 0 & -95859 & -95859 & 0 & 0 & 0 & 63906 \end{array}\right] \begin{Bmatrix} x_1 \\ x_2 \\ x_3 \\ \hline 0 \\ 0 \\ 0 \\ 0 \\ 0 \\ 0 \\ 0 \\ 0 \end{Bmatrix} = \begin{Bmatrix} 150 \\ 0 \\ 0 \\ \hline m_1 \\ m_2 \\ m_3 \\ m_4 \\ m_5 \\ m_6 \\ V_7 \\ V_8 \end{Bmatrix} \tag{4.77}$$

Solving for the displacements gives:

$$\begin{Bmatrix} x_1 \\ x_2 \\ x_3 \end{Bmatrix} = \begin{Bmatrix} 0.0026\,\text{m} \\ 0.0010\,\text{m} \\ 0.0010\,\text{m} \end{Bmatrix} \tag{4.78}$$

Substituting Eq. (4.78) back in Eq. (4.77) and solving for the moments at RHs and shear forces due to SHs gives:

$$\begin{Bmatrix} m_1 \\ m_2 \\ m_3 \\ m_4 \\ m_5 \\ m_6 \\ V_7 \\ V_8 \end{Bmatrix} = \begin{Bmatrix} -162.98\,\text{kN-m} \\ -62.02\,\text{kN-m} \\ -162.98\,\text{kN-m} \\ -62.02\,\text{kN-m} \\ 62.02\,\text{kN-m} \\ 62.02\,\text{kN-m} \\ 75\,\text{kN} \\ 75\,\text{kN} \end{Bmatrix} \tag{4.79}$$

Comparing the demands in Eq. (4.79) with the corresponding capacities listed in Tables 4.1, 4.2, and 4.3 shows that the RHs #1, #3, #5, and #6 begin to rotate and the SHs #7 and #8 begin to slide. Therefore, the second step is to assume that plastic hinges, RHs and SHs, are formed at these locations and deformed on the corresponding hardening branches. The $k_f = 1.3 \times 10^5$ kN m/rad, $\alpha_f = 7.0 \times 10^4$ kN m/rad for the beam can be computed according to the parameters listed in Table 4.3. Then the moments m_1, m_3, m_5, and m_6 and shear forces V_7 and V_8 can be expressed in the terms of θ_1'', θ_3'', θ_5'', θ_6'', τ_7'' and τ_8'' according to Eq. (4.5):

$$\begin{cases} m_1 = -91.95 + 8.6 \times 10^4\, \theta_1'' \\ m_3 = -91.95 + 8.6 \times 10^4\, \theta_3'' \\ m_5 = 21 + 1.5 \times 10^5\, \theta_5'' \\ m_6 = 21 + 1.5 \times 10^5\, \theta_6'' \\ V_7 = 61.3 + 4.0 \times 10^4\, \tau_7'' \\ V_8 = 61.3 + 4.0 \times 10^4\, \tau_8'' \end{cases} \tag{4.80}$$

Substituting Eq. (4.80) into Eq. (4.77) gives:

$$\left[\begin{array}{ccc|cccccc} 127812 & -95859 & -95859 & -95859 & -95859 & 0 & 0 & 63906 & 63906 \\ -95859 & 238544 & 21333 & 91705 & 0 & 42667 & 21333 & -95859 & 0 \\ -95859 & 21333 & 238544 & 0 & 91705 & 21333 & 42667 & 0 & -95859 \\ \hline -95859 & 91705 & 0 & 281872 & 0 & 0 & 0 & -95859 & 0 \\ -95859 & 0 & 91705 & 0 & 281872 & 0 & 0 & 0 & -95859 \\ 0 & 42667 & 21333 & 0 & 0 & 192667 & 21333 & 0 & 0 \\ 0 & 21333 & 42667 & 0 & 0 & 21333 & 192667 & 0 & 0 \\ 63906 & -95859 & 0 & -95859 & 0 & 0 & 0 & 103906 & 0 \\ 63906 & 0 & -95859 & 0 & -95859 & 0 & 0 & 0 & 103906 \end{array}\right] \left\{\begin{array}{c} x_1 \\ x_2 \\ x_3 \\ \hline -\theta_1'' \\ -\theta_3'' \\ -\theta_5'' \\ -\theta_6'' \\ -\tau_7'' \\ -\tau_8'' \end{array}\right\} = \left\{\begin{array}{c} 150 \\ 0 \\ 0 \\ \hline -91.95 \\ -91.95 \\ 21 \\ 21 \\ 61.3 \\ 61.3 \end{array}\right\} \tag{4.81}$$

Note that the tangent stiffnesses for RHs #1, #3, #5, #6 and SHs #7, #8 have been incorporated into Eq. (4.81) through the addition of stiffness in the corresponding diagonal terms of the stiffness matrix. Now, solving for the displacements and plastic deformations in Eq. (4.81) gives:

$$\left\{\begin{array}{c} x_1 \\ x_2 \\ x_3 \\ \hline -\theta_1'' \\ -\theta_3'' \\ -\theta_5'' \\ -\theta_6'' \\ -\tau_7'' \\ -\tau_8'' \end{array}\right\} = \left\{\begin{array}{l} 0.0048\,\text{m} \\ 0.0015\,\text{m} \\ 0.0015\,\text{m} \\ \hline 0.0007\,\text{rad} \\ 0.0007\,\text{rad} \\ -0.0003\,\text{rad} \\ -0.0003\,\text{rad} \\ -0.0003\,\text{m} \\ -0.0003\,\text{m} \end{array}\right\} \tag{4.82}$$

Substituting Eq. (4.82) back in Eq. (4.77) and solving for the moments at the RHs and shear forces due to the SHs gives:

$$\begin{Bmatrix} m_1 \\ m_2 \\ m_3 \\ m_4 \\ m_5 \\ m_6 \\ V_7 \\ V_8 \end{Bmatrix} = \begin{Bmatrix} -152.58\,\text{kN-m} \\ -72.42\,\text{kN-m} \\ -152.58\,\text{kN-m} \\ -72.42\,\text{kN-m} \\ 72.42\,\text{kN-m} \\ 72.42\,\text{kN-m} \\ 75\,\text{kN} \\ 75\,\text{kN} \end{Bmatrix} \quad (4.83)$$

Comparing again the demands in Eq. (4.83) with the corresponding capacities listed in Tables 4.1, 4.2, and 4.3, the assumption that the locations of RHs #1, #3, #5, #6 and SHs #7, #8 are in the plastic domain while the locations of RHs #2 and #4 remain elastic is correct. In summary, there are

$$\begin{Bmatrix} x_1 \\ x_2 \\ x_3 \end{Bmatrix} = \begin{Bmatrix} 0.0048\,\text{m} \\ 0.0015\,\text{m} \\ 0.0015\,\text{m} \end{Bmatrix}, \quad \begin{Bmatrix} \theta_1'' \\ \theta_2'' \\ \theta_3'' \\ \theta_4'' \\ \theta_5'' \\ \theta_6'' \\ \tau_7'' \\ \tau_8'' \end{Bmatrix} = \begin{Bmatrix} -0.0007\,\text{rad} \\ 0\,\text{rad} \\ -0.0007\,\text{rad} \\ 0\,\text{rad} \\ 0.0003\,\text{rad} \\ 0.0003\,\text{rad} \\ 0.0003\,\text{m} \\ 0.0003\,\text{m} \end{Bmatrix}, \quad \begin{Bmatrix} m_1 \\ m_2 \\ m_3 \\ m_4 \\ m_5 \\ m_6 \\ V_7 \\ V_8 \end{Bmatrix} = \begin{Bmatrix} -152.58\,\text{kN-m} \\ -72.42\,\text{kN-m} \\ -152.58\,\text{kN-m} \\ -72.42\,\text{kN-m} \\ 72.42\,\text{kN-m} \\ 72.42\,\text{kN-m} \\ 75\,\text{kN} \\ 75\,\text{kN} \end{Bmatrix} \quad (4.84)$$

Note that the joint equilibriums are satisfied when $m_2 = -m_5$ and $m_4 = -m_6$ and the force equilibriums of the columns are satisfied when $V_7 = -(m_1 + m_2)/L_c$ and $V_8 = -(m_3 + m_4)/L_c$. Figure 4.12 summarizes these results.

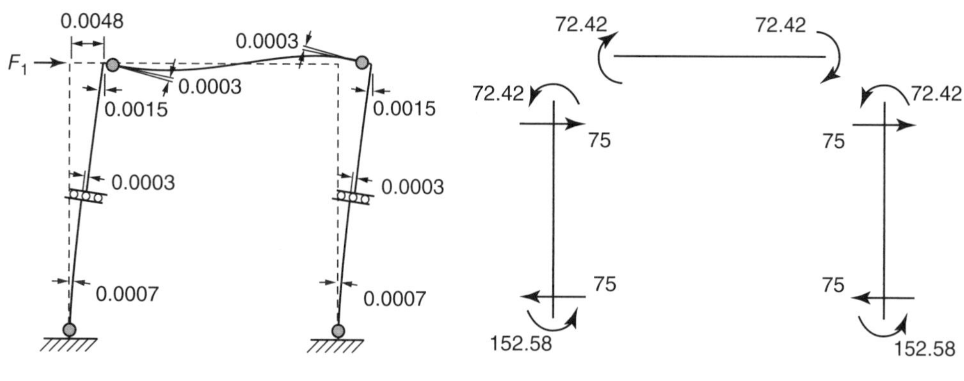

Figure 4.12 Graphical illustration of the response of nonlinear one-story one-bay RC frame.

Example 4.8 Nonlinear Static Analysis of the Two-Story One-Bay RC Frame
Consider the two-story one-bay RC frame as shown in Figure 4.13. Assume that all the members are axially rigid, which therefore gives a total of 6 DOFs (2 translations and 4 rotations), 12 RHs, and 4 SHs. The columns in this example are the same as that used in Example 4.7, while the beams here are of the same sectional parameters comparing that used in Example 4.7 but with the length $L_b = 2L_c = 6\,\text{m}$. Therefore, the bending and shear behaviors of the columns and beams can be obtained in Tables 4.1, 4.2, and 4.3. The RC frame is subjected to the lateral forces $F_1 = 80\,\text{kN}$ and $F_2 = 60\,\text{kN}$ on the first and second floors, which means the lateral forces F_1 and F_2 are applied at the horizontal degree of freedoms x_1 and x_2 while $F_3 = F_4 = F_5 = F_6 = 0$. In order to reduce the nonlinear problem, static condensation procedure is used in this example.

For this two-story RC frame with $n = 6$, $q_f = 12$, and $q_s = 4$, the 6×6 stiffness matrix $\mathbf{K}$, the 6×16 stiffness matrix $\mathbf{K}'$, and the 16×16 stiffness matrix $\mathbf{K}''$ are:

$$\begin{Bmatrix} x_1 \\ x_2 \\ x_3 \\ \hdashline -\theta''_1 \\ -\theta''_3 \\ -\theta''_5 \\ -\theta''_6 \\ -\tau''_7 \\ -\tau''_8 \end{Bmatrix} = \begin{Bmatrix} 0.0048\,\text{m} \\ 0.0015\,\text{m} \\ 0.0015\,\text{m} \\ \hdashline 0.0007\,\text{rad} \\ 0.0007\,\text{rad} \\ -0.0003\,\text{rad} \\ -0.0003\,\text{rad} \\ -0.0003\,\text{m} \\ -0.0003\,\text{m} \end{Bmatrix} \tag{4.85}$$

$$\mathbf{K}' = \begin{bmatrix} \mathbf{K}'_d \\ \hdashline \mathbf{K}'_r \end{bmatrix} = \begin{bmatrix} -9.59 & -9.59 & 9.59 & 9.59 & -9.59 & -9.59 & 9.59 & 9.59 & 0 & 0 & 0 & 0 & 6.39 & -6.39 & 6.39 & -6.39 \\ 0 & 0 & -9.59 & -9.59 & 0 & 0 & -9.59 & -9.59 & 0 & 0 & 0 & 0 & 0 & 6.39 & 0 & 6.39 \\ \hdashline 9.17 & 19.59 & 19.59 & 9.17 & 0 & 0 & 0 & 0 & 2.13 & 1.07 & 0 & 0 & -9.59 & -9.59 & 0 & 0 \\ 0 & 0 & 0 & 0 & 9.17 & 19.59 & 19.59 & 9.17 & 1.07 & 2.13 & 0 & 0 & 0 & 0 & -9.59 & -9.59 \\ 0 & 0 & 9.17 & 19.59 & 0 & 0 & 0 & 0 & 0 & 0 & 2.13 & 1.07 & 0 & -9.59 & 0 & 0 \\ 0 & 0 & 0 & 0 & 0 & 0 & 9.17 & 19.59 & 0 & 0 & 1.07 & 2.13 & 0 & 0 & 0 & -9.59 \end{bmatrix} \times 10^4\ (\text{kN/rad}) \tag{4.86}$$

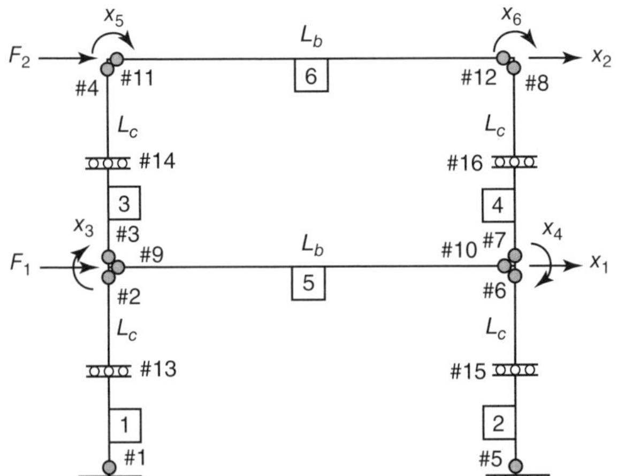

Figure 4.13 The two-story one-bay RC frame.

$$\mathbf{K}'' = \begin{bmatrix} 19.59 & 9.17 & 0 & 0 & 0 & 0 & 0 & 0 & 0 & 0 & 0 & 0 & -9.59 & 0 & 0 & 0 \\ 9.17 & 19.59 & 0 & 0 & 0 & 0 & 0 & 0 & 0 & 0 & 0 & 0 & -9.59 & 0 & 0 & 0 \\ 0 & 0 & 19.59 & 9.17 & 0 & 0 & 0 & 0 & 0 & 0 & 0 & 0 & 0 & -9.59 & 0 & 0 \\ 0 & 0 & 9.17 & 19.59 & 0 & 0 & 0 & 0 & 0 & 0 & 0 & 0 & 0 & -9.59 & 0 & 0 \\ 0 & 0 & 0 & 0 & 19.59 & 9.17 & 0 & 0 & 0 & 0 & 0 & 0 & 0 & 0 & -9.59 & 0 \\ 0 & 0 & 0 & 0 & 9.17 & 19.59 & 0 & 0 & 0 & 0 & 0 & 0 & 0 & 0 & -9.59 & 0 \\ 0 & 0 & 0 & 0 & 0 & 0 & 19.59 & 9.17 & 0 & 0 & 0 & 0 & 0 & 0 & 0 & -9.59 \\ 0 & 0 & 0 & 0 & 0 & 0 & 9.17 & 19.59 & 0 & 0 & 0 & 0 & 0 & 0 & 0 & -9.59 \\ 0 & 0 & 0 & 0 & 0 & 0 & 0 & 0 & 2.13 & 1.07 & 0 & 0 & 0 & 0 & 0 & 0 \\ 0 & 0 & 0 & 0 & 0 & 0 & 0 & 0 & 1.07 & 2.13 & 0 & 0 & 0 & 0 & 0 & 0 \\ 0 & 0 & 0 & 0 & 0 & 0 & 0 & 0 & 0 & 0 & 2.13 & 1.07 & 0 & 0 & 0 & 0 \\ 0 & 0 & 0 & 0 & 0 & 0 & 0 & 0 & 0 & 0 & 1.07 & 2.13 & 0 & 0 & 0 & 0 \\ -9.59 & -9.59 & 0 & 0 & 0 & 0 & 0 & 0 & 0 & 0 & 0 & 0 & 6.39 & 0 & 0 & 0 \\ 0 & 0 & -9.59 & -9.59 & 0 & 0 & 0 & 0 & 0 & 0 & 0 & 0 & 0 & 6.39 & 0 & 0 \\ 0 & 0 & 0 & 0 & -9.59 & -9.59 & 0 & 0 & 0 & 0 & 0 & 0 & 0 & 0 & 6.39 & 0 \\ 0 & 0 & 0 & 0 & 0 & 0 & -9.59 & -9.59 & 0 & 0 & 0 & 0 & 0 & 0 & 0 & 6.39 \end{bmatrix} \times 10^4 \text{ (kN-m/rad)} \tag{4.87}$$

and finally the governing equation of the FAM can be written as:

$$\left[\begin{array}{c|c} \mathbf{K}_{6\times6} & \mathbf{K}'_{6\times16} \\ \hline \mathbf{K}'^{T}_{16\times6} & \mathbf{K}''_{16\times16} \end{array}\right] \left\{\begin{array}{c} \mathbf{x}_{6\times1} \\ \hline -\boldsymbol{\Lambda}''_{16\times1} \end{array}\right\} = \left\{\begin{array}{c} \mathbf{F}_{6\times1} \\ \hline \mathbf{f}_{16\times1} \end{array}\right\} \tag{4.88}$$

Recognizing $F_3 = F_4 = F_5 = F_6 = 0$, static condensation can be performed here according to the expression in Chapter 2. Therefore, the condensed stiffness matrices $\bar{\mathrm{K}}$ $\bar{\mathrm{K}}'$, and $\bar{\mathrm{K}}''$ can be obtained by:

$$\bar{\mathbf{K}} = \mathbf{K}_{dd} - \mathbf{K}_{dr}\mathbf{K}_{rr}^{-1}\mathbf{K}_{rd} = \begin{bmatrix} 16.73 & -5.86 \\ -5.86 & 3.02 \end{bmatrix} \times 10^4 \text{ (kN/m)} \tag{4.89}$$

$$\bar{\mathbf{K}}' = \mathbf{K}'_d - \mathbf{K}_{dr}\mathbf{K}_{rr}^{-1}\mathbf{K}'_r$$

$$= \begin{bmatrix} -8.67 & -7.63 & 7.31 & 1.47 & -8.67 & -7.63 & 7.31 & 1.47 & 0.32 & 0.32 & -1.47 & -1.47 & 5.43 & -2.93 & 5.43 & -2.93 \\ 1.36 & 2.90 & -3.37 & -1.16 & 1.36 & 2.90 & -3.37 & -1.16 & 0.47 & 0.47 & 1.16 & 1.16 & -1.42 & 1.51 & -1.42 & 1.51 \end{bmatrix} \times 10^4 \text{ (kN/rad)} \tag{4.90}$$

$$\bar{\mathbf{K}}'' = \mathbf{K}'' - \mathbf{K}'^{T}_{r}\mathbf{K}_{rr}^{-1}\mathbf{K}'_r$$

$$= \begin{bmatrix} 173.38 & 43.66 & -38.51 & -2.13 & 0.76 & 1.61 & 0.83 & -0.93 & -5.15 & -2.44 & 2.13 & 0.93 & -72.34 & 13.55 & -0.79 & 0.03 \\ 43.66 & 93.25 & -82.26 & -4.54 & 1.61 & 3.45 & 1.77 & -1.98 & -10.99 & -5.21 & 4.54 & 1.98 & -45.63 & 28.93 & -1.69 & 0.07 \\ -38.51 & -82.26 & 91.12 & 5.15 & 0.83 & 1.77 & 2.52 & 2.44 & -8.86 & -4.29 & -5.15 & -2.44 & 40.26 & -32.09 & -0.86 & -1.65 \\ -2.13 & -4.54 & 5.15 & 18.57 & -0.93 & -1.98 & 2.44 & 8.51 & -0.60 & -0.46 & -18.57 & -8.51 & 2.22 & -7.90 & 0.97 & -3.65 \\ 0.76 & 1.61 & 0.83 & -0.93 & 173.38 & 43.66 & -38.51 & -2.13 & -2.44 & -5.15 & 0.93 & 2.13 & -0.79 & 0.03 & -72.34 & 13.55 \\ 1.61 & 3.45 & 1.77 & -1.98 & 43.66 & 93.25 & -82.26 & -4.54 & -5.21 & -10.99 & 1.98 & 4.54 & -1.69 & 0.07 & -45.63 & 28.93 \\ 0.83 & 1.77 & 2.52 & 2.44 & -38.51 & -82.26 & 91.12 & 5.15 & -4.29 & -8.86 & -2.44 & -5.15 & -0.86 & -1.65 & 40.26 & -32.09 \\ -0.93 & -1.98 & 2.44 & 8.51 & -2.13 & -4.54 & 5.15 & 18.57 & -0.46 & -0.60 & -8.51 & -18.57 & 0.97 & -3.65 & 2.22 & -7.90 \\ -5.15 & -10.99 & -8.86 & -0.60 & -2.44 & -5.21 & -4.29 & -0.46 & 19.85 & 9.50 & 0.60 & 0.46 & 5.38 & 3.16 & 2.55 & 1.58 \\ -2.44 & -5.21 & -4.29 & -0.46 & -5.15 & -10.99 & -8.86 & -0.60 & 9.50 & 19.85 & 0.46 & 0.60 & 2.55 & 1.58 & 5.38 & 3.16 \\ 2.13 & 4.54 & -5.15 & -18.57 & 0.93 & 1.98 & -2.44 & -8.51 & 0.60 & 0.46 & 18.57 & 8.51 & -2.22 & 7.90 & -0.97 & 3.65 \\ 0.93 & 1.98 & -2.44 & -8.51 & 2.13 & 4.54 & -5.15 & -18.57 & 0.46 & 0.60 & 8.51 & 18.57 & -0.97 & 3.65 & -2.22 & 7.90 \\ -72.34 & -45.63 & 40.26 & 2.22 & -0.79 & -1.69 & -0.86 & 0.97 & 5.38 & 2.55 & -2.22 & -0.97 & 39.33 & -14.16 & 0.83 & -0.03 \\ 13.55 & 28.93 & -32.09 & -7.90 & 0.03 & 0.07 & -1.65 & -3.65 & 3.16 & 1.58 & 7.90 & 3.65 & -14.16 & 13.33 & -0.03 & 1.77 \\ -0.79 & -1.69 & -0.86 & 0.97 & -72.34 & -45.63 & 40.26 & 2.22 & 2.55 & 5.38 & -0.97 & -2.22 & 0.83 & -0.03 & 39.33 & -14.16 \\ 0.03 & 0.07 & -1.65 & -3.65 & 13.55 & 28.93 & -32.09 & -7.90 & 1.58 & 3.16 & 3.65 & 7.90 & -0.03 & 1.77 & -14.16 & 13.33 \end{bmatrix} \times 10^3 \text{ (kN-m/rad)} \tag{4.91}$$

and the governing equation of the FAM written in Eq. (4.88) can be rewritten as:

$$\begin{bmatrix} \bar{\mathbf{K}}_{2\times2} & \bar{\mathbf{K}}'_{2\times16} \\ \bar{\mathbf{K}}'^{T}_{16\times2} & \bar{\mathbf{K}}''_{16\times16} \end{bmatrix} \begin{Bmatrix} \mathbf{x}_{d2\times1} \\ -\mathbf{\Lambda}''_{16\times1} \end{Bmatrix} = \begin{Bmatrix} \mathbf{F}_{ad2\times1} \\ \mathbf{f}_{16\times1} \end{Bmatrix} \tag{4.92}$$

in which,

$$\mathbf{x}_d = \begin{Bmatrix} x_1 \\ x_2 \end{Bmatrix}, \quad \mathbf{F}_{ad} = \begin{Bmatrix} F_1 \\ F_2 \end{Bmatrix} \tag{4.93}$$

The structure is first assumed to be linear, this gives $\mathbf{\Lambda}'' = 0$. Using the first matrix equation of Eq. (4.92) gives:

$$\begin{bmatrix} 16.73 & -5.86 \\ -5.86 & 3.02 \end{bmatrix} \times 10^4 \begin{Bmatrix} x_1 \\ x_2 \end{Bmatrix} = \begin{Bmatrix} 80 \\ 60 \end{Bmatrix} \tag{4.94}$$

Solving for the displacements in Eq. (4.94) gives:

$$\begin{Bmatrix} x_1 \\ x_2 \end{Bmatrix} = \begin{Bmatrix} 0.0037\,\text{m} \\ 0.0091\,\text{m} \end{Bmatrix} \tag{4.95}$$

Then, substituting Eq. (4.95) into the second matrix equation of Eq. (4.92), the moments at the RHs and shear forces according to the SHs are calculated as:

$$\begin{Bmatrix} m_1 \\ m_2 \\ m_3 \\ m_4 \\ m_5 \\ m_6 \\ m_7 \\ m_8 \\ m_9 \\ m_{10} \\ m_{11} \\ m_{12} \\ V_{13} \\ V_{14} \\ V_{15} \\ V_{16} \end{Bmatrix} = \begin{Bmatrix} -194.16\,\text{kN-m} \\ -15.84\,\text{kN-m} \\ -38.94\,\text{kN-m} \\ -51.06\,\text{kN-m} \\ -194.16\,\text{kN-m} \\ -15.84\,\text{kN-m} \\ -38.94\,\text{kN-m} \\ -51.06\,\text{kN-m} \\ 54.78\,\text{kN-m} \\ 54.78\,\text{kN-m} \\ 51.06\,\text{kN-m} \\ 51.06\,\text{kN-m} \\ 70.00\,\text{kN} \\ 30.00\,\text{kN} \\ 70.00\,\text{kN} \\ 30.00\,\text{kN} \end{Bmatrix} \tag{4.96}$$

Comparing the demands in Eq. (4.96) with the corresponding capacities listed in Tables 4.1, 4.2, and 4.3 shows that the RHs #1, #5, #9, #10, #11, and #12 begin to rotate and the SHs #13, and #15 begin to slide. Therefore, assume that the plastic hinges, RHs and SHs, located at these locations behave on the corresponding hardening branches. Then, the moments m_1, m_5, m_9, m_{10}, m_{11}, m_{12} and shear forces V_{13}, V_{15} can be expressed in the terms of θ_1'', θ_5'', θ_9'', θ_{10}'', θ_{11}'', θ_{12}'', τ_{13}'' and τ_{15}'' according to Eq. (4.5):

$$\begin{cases} m_1 = -91.95 + 8.6 \times 10^4 \theta_1'' \\ m_5 = -91.95 + 8.6 \times 10^4 \theta_5'' \\ m_9 = 21 + 1.5 \times 10^5 \theta_9'' \\ m_{10} = 21 + 1.5 \times 10^5 \theta_{10}'' \\ m_{11} = 21 + 1.5 \times 10^5 \theta_{11}'' \\ m_{12} = 21 + 1.5 \times 10^5 \theta_{12}'' \\ V_{13} = 61.3 + 4.0 \times 10^4 \tau_{13}'' \\ V_{15} = 61.3 + 4.0 \times 10^4 \tau_{15}'' \end{cases} \tag{4.97}$$

From Eq. (4.92), it follows that:

$$\begin{bmatrix} 167.3 & -58.6 & -86.7 & -86.7 & 3.2 & 3.2 & -14.7 & -14.7 & 54.3 & 54.3 \\ -58.6 & 30.2 & 13.6 & 13.6 & 4.7 & 4.7 & 11.6 & 11.6 & -14.2 & -14.2 \\ -86.7 & 13.6 & 173.38+86 & 0.76 & -5.15 & -2.44 & 2.13 & 0.93 & -72.34 & -0.79 \\ -86.7 & 13.6 & 0.76 & 173.38+86 & -2.44 & -5.15 & 0.93 & 2.13 & -0.79 & -72.34 \\ 3.2 & 4.7 & -5.15 & -2.44 & 19.85+150 & 9.50 & 0.60 & 0.46 & 5.38 & 2.55 \\ 3.2 & 4.7 & -2.44 & -5.15 & 9.50 & 19.85+150 & 0.46 & 0.60 & 2.55 & 5.38 \\ -14.7 & 11.6 & 2.13 & 0.93 & 0.60 & 0.46 & 18.57+150 & 8.51 & -2.22 & -0.97 \\ -14.7 & 11.6 & 0.93 & 2.13 & 0.46 & 0.60 & 8.51 & 18.57+150 & -0.97 & -2.22 \\ 54.3 & -14.2 & -72.34 & -0.79 & 5.38 & 2.55 & -2.22 & -0.97 & 39.33+40 & 0.83 \\ 54.3 & -14.2 & -0.79 & -72.34 & 2.55 & 5.38 & -0.97 & -2.22 & 0.83 & 39.33+40 \end{bmatrix} \times 10^3 \begin{Bmatrix} x_1 \\ x_2 \\ -\theta_1'' \\ -\theta_5'' \\ -\theta_9'' \\ -\theta_{10}'' \\ -\theta_{11}'' \\ -\theta_{12}'' \\ -\tau_{13}'' \\ -\tau_{15}'' \end{Bmatrix} = \begin{Bmatrix} 80 \\ 60 \\ -91.95 \\ -91.95 \\ 21 \\ 21 \\ 21 \\ 21 \\ 61.3 \\ 61.3 \end{Bmatrix} \tag{4.98}$$

Now, solving for the displacements and plastic deformations in Eq. (4.98) gives:

$$\begin{Bmatrix} x_1 \\ x_2 \\ -\theta_1'' \\ -\theta_5'' \\ -\theta_9'' \\ -\theta_{10}'' \\ -\theta_{11}'' \\ -\theta_{12}'' \\ -\tau_{13}'' \\ -\tau_{15}'' \end{Bmatrix} = \begin{Bmatrix} 0.0063\,\text{m} \\ 0.0134\,\text{m} \\ 0.00098\,\text{rad} \\ 0.00098\,\text{rad} \\ -0.00030\,\text{rad} \\ -0.00030\,\text{rad} \\ -0.00025\,\text{rad} \\ -0.00025\,\text{rad} \\ -0.00022\,\text{m} \\ -0.00022\,\text{m} \end{Bmatrix} \tag{4.99}$$

Substituting Eq. (4.99) back into Eq. (4.92) and solving for the moments at the RHs and shear forces due to the SHs gives:

$$\begin{Bmatrix} m_1 \\ m_2 \\ m_3 \\ m_4 \\ m_5 \\ m_6 \\ m_7 \\ m_8 \\ m_9 \\ m_{10} \\ m_{11} \\ m_{12} \\ V_{13} \\ V_{14} \\ V_{15} \\ V_{16} \end{Bmatrix} = \begin{Bmatrix} -176.01\,\text{kN-m} \\ -33.99\,\text{kN-m} \\ -31.44\,\text{kN-m} \\ -58.56\,\text{kN-m} \\ -176.01\,\text{kN-m} \\ -33.99\,\text{kN-m} \\ -31.44\,\text{kN-m} \\ -58.56\,\text{kN-m} \\ 65.43\,\text{kN-m} \\ 65.43\,\text{kN-m} \\ 58.56\,\text{kN-m} \\ 58.56\,\text{kN-m} \\ 70.00\,\text{kN} \\ 30.00\,\text{kN} \\ 70.00\,\text{kN} \\ 30.00\,\text{kN} \end{Bmatrix} \tag{4.100}$$

Comparing again the demands in Eq. (4.100) with the corresponding capacities listed in Tables 4.1, 4.2, and 4.3, the assumption that the locations of RHs #1, #5, #9, #10, #11, #12 and SHs #13, #15 are in the plastic domain and behave on the corresponding hardening branches while the other locations of RHs and SHs remain elastic is correct.

Finally, the DOFs that were statically condensed, $\mathbf{x}_r = \{x_3\ x_4\ x_5\ x_6\}^T$, can be calculated as:

$$\mathbf{x}_r = \begin{Bmatrix} x_3 \\ x_4 \\ x_5 \\ x_6 \end{Bmatrix} = \mathbf{K}_{rr}^{-1}\mathbf{K}_r'\boldsymbol{\Lambda}'' - \mathbf{K}_{rr}^{-1}\mathbf{K}_{rd}\mathbf{x}_d = \begin{Bmatrix} 0.0029\,\text{m} \\ 0.0029\,\text{m} \\ 0.0018\,\text{m} \\ 0.0018\,\text{m} \end{Bmatrix} \tag{4.101}$$

Therefore, in summary,

$$\begin{Bmatrix} x_1 \\ x_2 \\ x_3 \\ x_4 \\ x_5 \\ x_6 \end{Bmatrix} = \begin{Bmatrix} 0.0063\,\text{m} \\ 0.0134\,\text{m} \\ 0.0029\,\text{m} \\ 0.0029\,\text{m} \\ 0.0018\,\text{m} \\ 0.0018\,\text{m} \end{Bmatrix}, \begin{Bmatrix} \theta_1'' \\ \theta_2'' \\ \theta_3'' \\ \theta_4'' \\ \theta_5'' \\ \theta_6'' \\ \theta_7'' \\ \theta_8'' \\ \theta_9'' \\ \theta_{10}'' \\ \theta_{11}'' \\ \theta_{12}'' \\ \tau_{13}'' \\ \tau_{14}'' \\ \tau_{15}'' \\ \tau_{16}'' \end{Bmatrix} = \begin{Bmatrix} -0.00098\,\text{rad} \\ 0\,\text{rad} \\ 0\,\text{rad} \\ 0\,\text{rad} \\ -0.00098\,\text{rad} \\ 0\,\text{rad} \\ 0\,\text{rad} \\ 0\,\text{rad} \\ 0.00030\,\text{rad} \\ 0.00030\,\text{rad} \\ 0.00025\,\text{rad} \\ 0.00025\,\text{rad} \\ 0.00022\,\text{m} \\ 0\,\text{m} \\ 0.00022\,\text{m} \\ 0\,\text{m} \end{Bmatrix}, \begin{Bmatrix} m_1 \\ m_2 \\ m_3 \\ m_4 \\ m_5 \\ m_6 \\ m_7 \\ m_8 \\ m_9 \\ m_{10} \\ m_{11} \\ m_{12} \\ V_{13} \\ V_{14} \\ V_{15} \\ V_{16} \end{Bmatrix} = \begin{Bmatrix} -176.01\,\text{kN-m} \\ -33.99\,\text{kN-m} \\ -31.44\,\text{kN-m} \\ -58.56\,\text{kN-m} \\ -176.01\,\text{kN-m} \\ -33.99\,\text{kN-m} \\ -31.44\,\text{kN-m} \\ -58.56\,\text{kN-m} \\ 65.43\,\text{kN-m} \\ 65.43\,\text{kN-m} \\ 58.56\,\text{kN-m} \\ 58.56\,\text{kN-m} \\ 70.00\,\text{kN} \\ 30.00\,\text{kN} \\ 70.00\,\text{kN} \\ 30.00\,\text{kN} \end{Bmatrix} \tag{4.102}$$

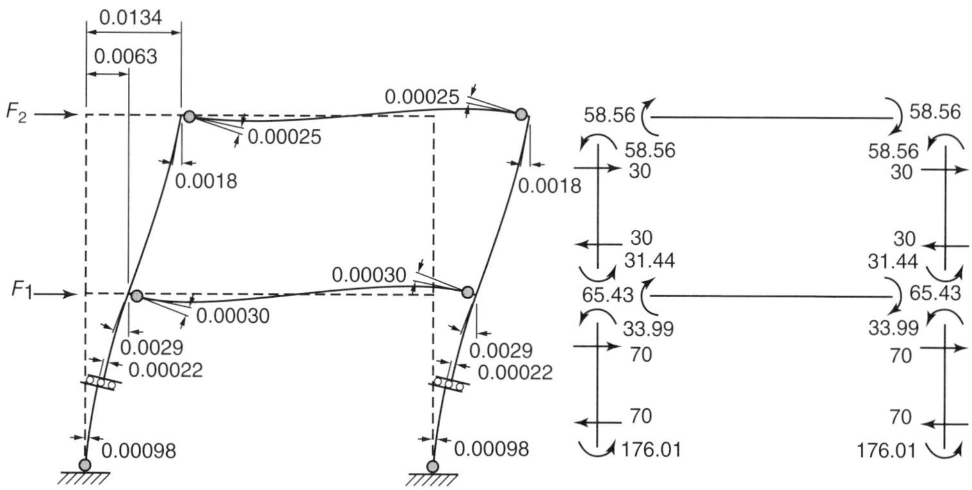

Figure 4.14 Graphical illustration of the responses of the two-story one-bay RC frame.

Figure 4.14 summarizes of these results in graphical form.

4.4 Nonlinear Dynamic Analysis of Structures with Flexural Members

4.4.1 *Hysteretic Behaviors of the Flexural Members*

As previous illustration in Figure 4.2, the hysteretic loops of the bending and shear behaviors mainly contain three components: the loading branches, unloading branches, and reloading branches. The loading branches that can be directly represented by the primary curves that are envelops of the bending and shear hysteretic loops are discussed in Section 4.3. Therefore, the main properties of the unloading and reloading branches are detailed here.

Consider the loading branches, regions A_1–B_1, B_1–C_1, C_1–D_1 for the bending behavior and regions A_2–B_2, B_2–C_2, C_2–D_2 for the shear behavior in Figure 4.15, their tangent stiffnesses α_f and α_s can be directly obtained by the bending and shear primary curves, respectively. In addition, the corresponding tangent stiffnesses α_f'' and α_s'' of the RH and SH behaviors can be calculated according to Eq. (4.5).

Consider the unloading branches, regions E_1–F_1, G_1–H_1 for the bending behavior and regions E_2–F_2, G_2–H_2 for the shear behavior, shown in Figure 4.16. The tangent stiffness for the bending behavior can be computed by

$$\alpha_f^{E_1F_1} = \alpha_f^{G_1H_1} = 1.7k_{f1}\left(\frac{\theta_y}{\theta}\right)^{0.4} \tag{4.103}$$

where k_{f1} is the slope of the line connecting the cracking point and yield point; θ is the maximum deformation attained in the loading direction and is also where unloading takes

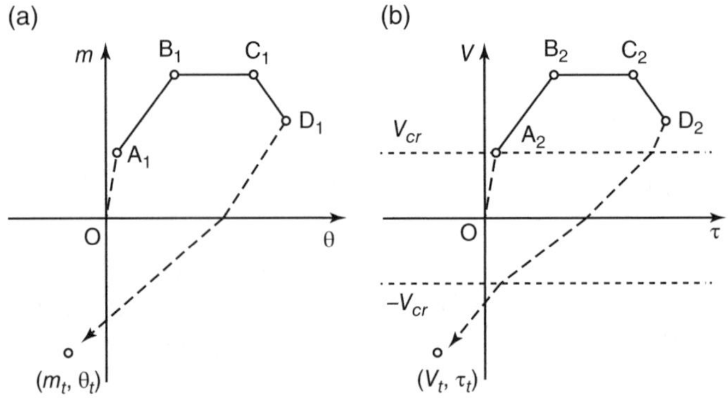

Figure 4.15 Loading branches: (a) Bending behavior; (b) Shear behavior.

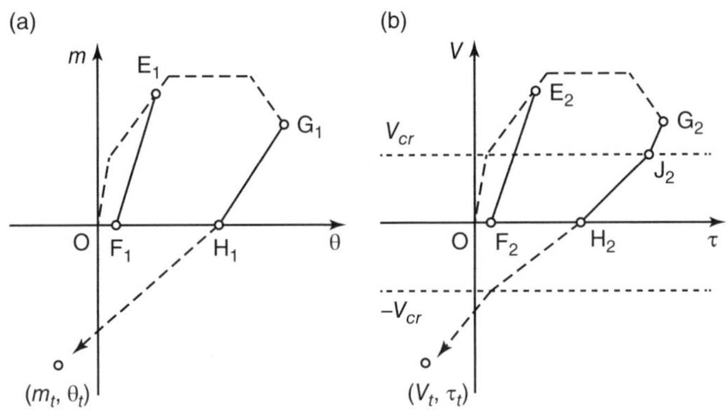

Figure 4.16 Unloading branches: (a) Bending behavior; (b) Shear behavior.

place. While for the shear behavior, the tangent stiffness for pre-yield cracked unloading branches E_2F_2 can be calculated by

$$\alpha_s^{E_2F_2} = k_s - \frac{\tau - \tau_{cr}}{\tau_y - \tau_{cr}}(k_s - k_{s2}) \tag{4.104}$$

where k_s is the slope connecting origin to the crack point as shown in Figure 4.16(b); k_{s2} is the slope connecting the yield point to the crack point in the opposite side; τ is the maximum deformation attained in the loading direction and is also where unloading takes place. However, the tangent stiffness α_s for the post-yield unloading branch G_2H_2 should be computed according to components above and below the cracking shear capacity V_{cr}. Then, the tangent stiffness α_s above the cracking shear level is given by Eq. (4.105), and below the cracking shear level by Eq. (4.106). The tangent stiffnesses α_f'' and α_s'' of the RH and SH behaviors can be calculated according to Eq. (4.5).

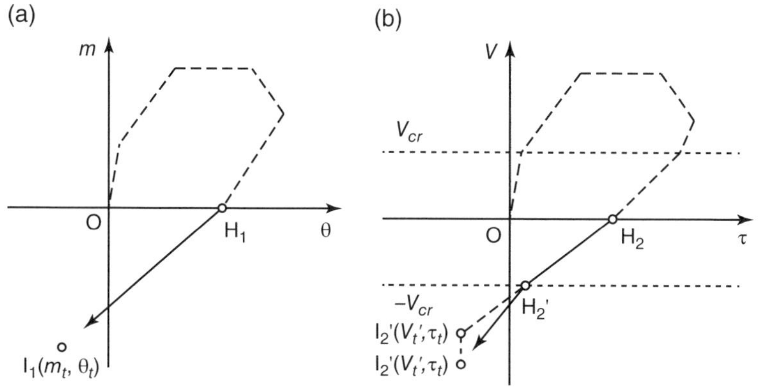

Figure 4.17 Reloading branches: (a) Bending behavior; (b) Shear behavior.

$$\alpha_s^{G_2J_2} = 1.4k_{s2}e^{-0.35(\tau/\tau_y)^{0.01}}\left(1-0.02\tau/\tau_y\right)^{3.5} \quad (4.105)$$

$$\alpha_s^{J_2H_2} = 0.78k_{s2}e^{-0.35(\tau/\tau_y)^{0.01}}\left(1-0.02\tau/\tau_y\right)^{5.5} \quad (4.106)$$

Consider the reloading branches, region H_1–I_1 for the bending behavior and region H_2–I_2 for the shear behavior, as shown in Figure 4.17. The tangent stiffnesses α_f and α_s can be obtained according to the reloading points H_1, H_2 and the target point I_1, I_2. For the bending behavior, the target point I_1 is the maximum deformation that attained in the previous cycle, and can be calculated as:

$$\alpha_f = \frac{m_t - m_r}{\theta_t - \theta_r} \quad (4.107)$$

where m_r and θ_r are the moment and rotation at the reloading point H_1. If the target point I_1 is the unloading point at the previous reloading branch, after reaching point H_1 at the slope computed using Eq. (4.107), additional branch that is on target to the peak point attained on the primary curve should be connected with. However, for the shear behavior, the reloading branch below the cracking capacity V_{cr} is on target to point I_2':

$$\alpha_s = \frac{V_{t'} - V_r}{\tau_t - \tau_r} \quad (4.108)$$

in which

$$V_{t'} = V_t e^{[0.82(N/N_0)-0.14](\tau_t/\tau_y)} \quad (4.109)$$

where V_r and τ_r are the shear force and deformation at the reloading point H_2; N/N_0 is the ratio of applied load to nominal compressive capacity of column and $0.82(N/N_0)–0.14 < 0$. As for the reloading branch $H_2'I_2$ above the cracking capacity V_{cr}, here assume that be targeted at the peak deformation τ_t and corresponding shear force V_t reached during previous cycle when the point I_2 is on the primary curve, and the α_s can be calculated as the slope of this branch:

$$\alpha_s = \frac{V_t - V_g}{\tau_t - \tau_g} \tag{4.110}$$

where V_g and τ_g are the shear force and deformation at point H_2'. Then the tangent stiffnesses α_f'' and α_s'' of the RH and SH behaviors can be calculated according to Eq. (4.5).

For a SDOF system, the governing equation (4.20) can be rewritten here in incremental form:

$$\left[\begin{array}{c|cc} k_0 & k_0 L & k_0 \\ \hline k_0 L & k_0 L^2 & k_0 L \\ k_0 & k_0 L & k_0 \end{array}\right] \left\{\begin{array}{c} x_{k+1} \\ \hline -(\theta_k'' + d\theta'') \\ -(\delta_k'' + d\tau'') \end{array}\right\} = \left\{\begin{array}{c} F_{k+1} \\ m_{k+1} \\ V_{k+1} \end{array}\right\} \tag{4.111}$$

where the subscript k denotes the kth time step and the parameters with subscript k, θ_k'' and τ_k'', are known for the time step $k+1$; $d\theta''$ and $d\tau''$ are the incremental plastic deformations of RH and SH respectively at time step $k+1$. For a given displacement at time step $k+1$, x_{k+1}, there are five unknowns, namely F_{k+1}, m_{k+1}, V_{k+1}, $d\theta''$, and $d\tau''$, in Eq. (4.111) with three equations provided. Therefore, together with two additional local plastic mechanisms, moment m versus plastic rotation θ'' and shear force V versus plastic lateral deformation τ'' as described in previous section, the five unknowns can be uniquely solved.

Example 4.9 Hysteretic response analysis of a RC column

Consider a RC column with length $L = 1.4732$ m as shown in Figure 4.18(a), Figure 4.18(b) illustrates the sectional parameters of the column. The strength of the concrete is 33.1 MPa, while the yield strengths of the longitudinal and transverse reinforcement are 331 MPa and 399.9 MPa, respectively. An axial load $P = 503$ kN is applied to the column. The bending and shear behaviors of the flexural member are summarized in Tables 4.4 and 4.5. Now the column is subjected to a series of cyclic input displacements $\mathbf{x}$ as shown in Figure 4.19.

Using the governing equation expressed in Eq. (4.111), the hysteretic response of the RC column can be computed step by step with the set of input displacements $\mathbf{x}$ shown in Figure 4.19. The obtained hysteretic loops are shown in Figure 4.20, and the detailed results of the first hysteretic response of the RC column are listed in Table 4.6.

4.4.2 *Solution Procedure of the FAM*

Consider the second governing equation in Eq. (4.62) with input displacement $\mathbf{x}_{k+1}$ at time step $k+1$ coming from the state space method of analysis expressed in Chapter 3. To take care of the loading and unloading issues of the flexural members in a dynamic analysis, let the plastic deformation vector $\boldsymbol{\Lambda}_{k+1}''$ at time step $k+1$ be written in incremental form as:

$$\boldsymbol{\Lambda}_{k+1}'' = \boldsymbol{\Lambda}_k'' + d\boldsymbol{\Lambda}'' \tag{4.112}$$

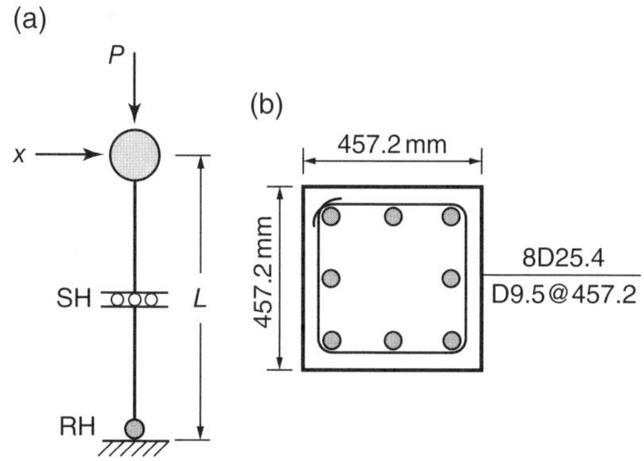

Figure 4.18 The RC column: (a) Numerical model; (b) Sectional parameters.

Table 4.4 The parameters of the bending primary curve.

m_{cr} (kN-m)	θ_{cr} (rad)	m_y (kN-m)	θ_y (rad)	θ_u (rad)	θ_f (rad)
88.392	0.00012	338.836	0.0045	0.012	0.018

Table 4.5 The parameters of the shear primary curve.

V_{cr} (kN)	τ_{cr} (m)	V_y (kN)	τ_y (m)	τ_u (m)	τ_f (m)
60	0.00012	230	0.0044	0.012	0.018

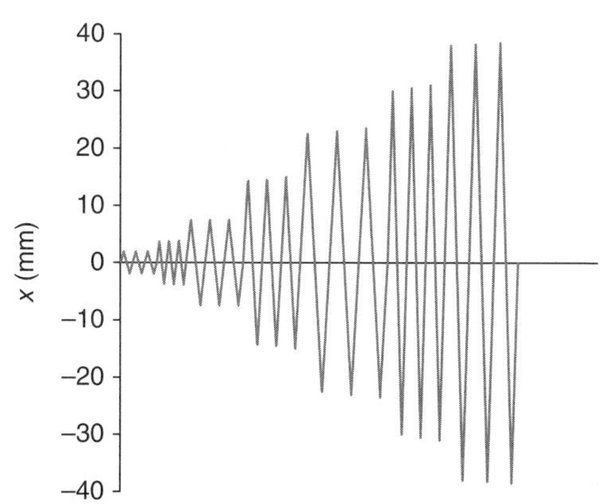

Figure 4.19 Cyclic loading history.

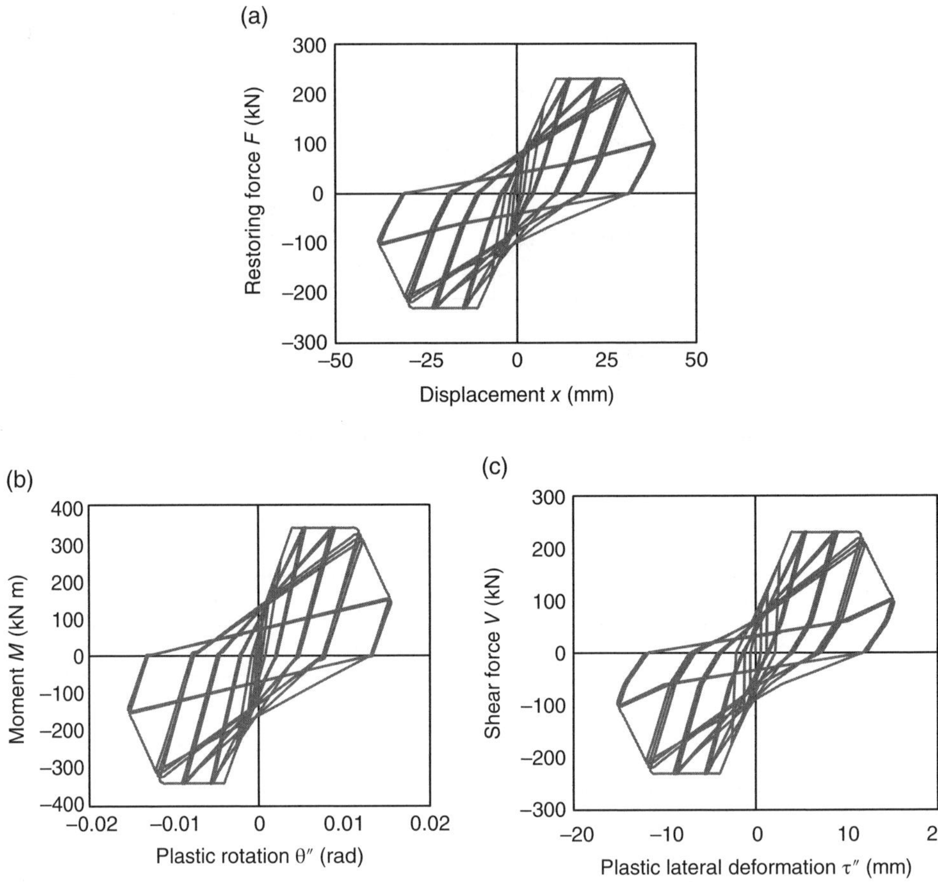

Figure 4.20 Hysteretic loops of the RC column: (a) Restoring force F versus displacement x hysteretic behavior; (b) Moment M versus plastic rotation θ'' hysteretic behavior at RH; (c) Shear force V versus plastic deformation τ'' hysteretic behavior at SH.

where $d\boldsymbol{\Lambda}''$ is the change in plastic deformations at RHs and SHs from time step k to time step $k+1$, and $\boldsymbol{\Lambda}_k''$ is the discretized form of $\boldsymbol{\Lambda}''(t)$. Note that the plastic rotation $\boldsymbol{\Theta}''(t)$ and plastic deformation $\boldsymbol{\Delta}''(t)$ are assembled in the vector $\boldsymbol{\Lambda}''(t)$ as expressed in Eq. (4.61). Then referring to the Eq. (3.59), the second governing equation in Eq. (4.62) can be rewritten here in incremental form as:

$$\mathbf{f}_{k+1} + \mathbf{K}'' d\boldsymbol{\Lambda}'' = \mathbf{K}'^T \mathbf{x}_{k+1} - \mathbf{K}''\boldsymbol{\Lambda}_k'' \tag{4.113}$$

where $\mathbf{f}_{k+1}$ is the discretized form of $\mathbf{f}(t)$. Each item on the right side of Eq. (4.113) is known, the objective is to calculate $\mathbf{f}_{k+1}$ and $d\boldsymbol{\Lambda}''$, a total of $2(q_f + q_s)$ unknowns. However, Eq. (4.113) provides $q_f + q_s$ equations, and m_f moment versus plastic rotation relationships and m_s shear force

Table 4.6 Numerical results of the RC column subjected to cyclic loadings

Branches	Input displacements x (mm)	Elastic displacements x' (mm)	Plastic deformations θ'' (rad)	τ'' (mm)	Forces m (kN-m)	V (kN)
Elastic	0	0	0	0	0	0
	0.10	0.10	0	0	29.46	20.00
	0.20	0.20	0	0	58.93	40.00
	0.30	0.30	0	0	88.39	60.00
Hardening	0.40	0.31	0.000037	0.000037	90.73	61.59
	0.50	0.32	0.000075	0.000074	93.08	63.18
	⋮	⋮	⋮	⋮	⋮	⋮
	1.80	0.42	0.00056	0.00055	123.50	83.83
	1.90	0.43	0.00060	0.00059	125.84	85.42
Unloading	1.80	0.39	0.00056	0.00059	114.76	77.90
	1.70	0.35	0.00052	0.00058	103.70	70.39
	⋮	⋮	⋮	⋮	⋮	⋮
	0.90	0.051	0.00019	0.00057	15.12	10.26
	0.80	0.014	0.00015	0.00056	4.04	2.74
Reloading	0.70	−0.018	0.00013	0.00053	−5.27	−3.58
	0.60	−0.046	0.00011	0.00048	−13.60	−9.23
	⋮	⋮	⋮	⋮	⋮	⋮
	−0.20	−0.27	0.000013	0.000053	−80.08	−54.36
	−0.30	−0.30	0	0.00	−88.39	−60.00
Hardening	−0.40	−0.31	−0.000037	−0.000037	−90.73	−61.59
	−0.50	−0.32	−0.000075	−0.000074	−93.08	−63.18
	⋮	⋮	⋮	⋮	⋮	⋮
	−1.80	−0.42	−0.00056	−0.00055	−123.50	−83.83
	−1.90	−0.43	−0.00060	−0.00059	−125.84	−85.42
Unloading	−0.18	−0.39	−0.00056	−0.00059	−114.76	−77.90
	−0.17	−0.35	−0.00052	−0.00058	−103.70	−70.39
	⋮	⋮	⋮	⋮	⋮	⋮
	−0.90	−0.051	−0.00019	−0.00057	−15.12	−10.26
	−0.80	−0.014	−0.00015	−0.00056	−4.04	−2.74
⋮	⋮	⋮	⋮	⋮	⋮	⋮

versus plastic deformation relationships for RHs and SHs, respectively, should be involved for solving the $2(q_f + q_s)$ unknowns uniquely. Once $\mathbf{f}_{k+1}$ and $d\boldsymbol{\Lambda}''$ are solved, the plastic deformation at time step $k+1$ $\boldsymbol{\Lambda}_{k+1}''$ can be obtained using Eq. (4.112). Then the plastic displacement $\mathbf{x}_{k+1}''$ can be calculated as:

$$\mathbf{x}_{k+1}'' = \mathbf{K}^{-1}\mathbf{K}'\boldsymbol{\Lambda}_{k+1}'' \tag{4.114}$$

This completes the portion of using the FAM in computing the internal forces of the RC structures with flexural members, and the inelastic displacement $\mathbf{x}''_{k+1}$ is passed back to the state space method to compute the response at the next step.

In addition, because of the introduction of the more complicated local plastic mechanisms (i.e. the RH and SH behaviors), there are a few differences of the computational procedure comparing that expressed in Chapter 3. For a known input displacement at time step $k+1$ $\mathbf{x}_{k+1}$, in any event, first assume that the structure is unloading. Therefore, for bending and shear behaviors of the flexural members, the points with internal forces m_k, V_k and equivalent SDOF deformation θ_k, τ_k at time step k serve as the unloading points. Substituting these values into the corresponding equations from Eqs. (4.103) (4.106) to calculate the unloading tangent stiffnesses α_f and α_s at time step $k+1$, and then the unloading tangent stiffnesses α_f'' and α_s'' of the local plastic mechanisms, RHs and SHs, at time step $k+1$ can be obtained using the Eq. (4.5). Once the tangent stiffnesses α_f'' and α_s'' of RHs and SHs at time step $k+1$ are known, the additional q_f+q_s equations are ready for solving the Eq. (4.113), and $\mathbf{f}_{k+1}$ and $d\mathbf{\Lambda}''$ can be calculated. Finally, comparing the values between m_{k+1}, V_{k+1} and m_k, V_k for each RH and SH, if the unloading conditions $|m_{k+1}|<|m_k|$ and $|V_{k+1}|<|V_k|$ are satisfied, the assumption that the RH and SH are unloading at time step $k+1$ is correct, otherwise, the RH and SH are loading or reloading at time step $k+1$.

Example 4.10 Nonlinear dynamic analysis of a RC frame subjected to cyclic increasing excitation

Consider a 6-story RC frame with flexural members as shown in Figure 4.21. Assuming the members are axially rigid, this gives total of 30 DOFs (i.e. $n=30$), 84 RHs and 24 SHs (i.e. $q_f=84$, $q_s=24$). No mass moment of inertia is assumed at the rotational joints, and therefore static condensation is used to eliminate the 24 rotational DOFs (i.e. $r=24$), resulting in only 6 translational DOFs for the RC frame (i.e. $d=6$). Let the masses be 4.5×10^4 kg for the first to fifth floor, and 4.4×10^4 kg for the top floor. The strength of concrete is 30 MPa, while the yield strengths of the longitudinal and transverse reinforcement are 335 MPa and 235 MPa, respectively. The sectional parameters of the members are illustrated in Figure 4.21, and the parameters of the bending and shear behaviors are listed in Tables 4.7 and 4.8. Now the RC frame is subjected to the cyclic increasing excitation with peak acceleration 1.14 m/s^2 and time step 0.01 s as shown in Figure 4.22.

After static condensation, the global stiffness matrix is written as:

$$\bar{\mathbf{K}}=\begin{bmatrix} 218104 & -136033 & 36765 & -6471 & 1138 & -159 \\ -136033 & 206525 & -134120 & 36386 & -6284 & 868 \\ 36765 & -134120 & 206187 & -133941 & 35743 & -4883 \\ -6471 & 36386 & -133941 & 205544 & -130517 & 27969 \\ 1138 & -6284 & 35743 & -130517 & 186625 & -86519 \\ -159 & 868 & -4883 & 27969 & -86519 & 62698 \end{bmatrix} \text{(kN/m)} \quad (4.115)$$

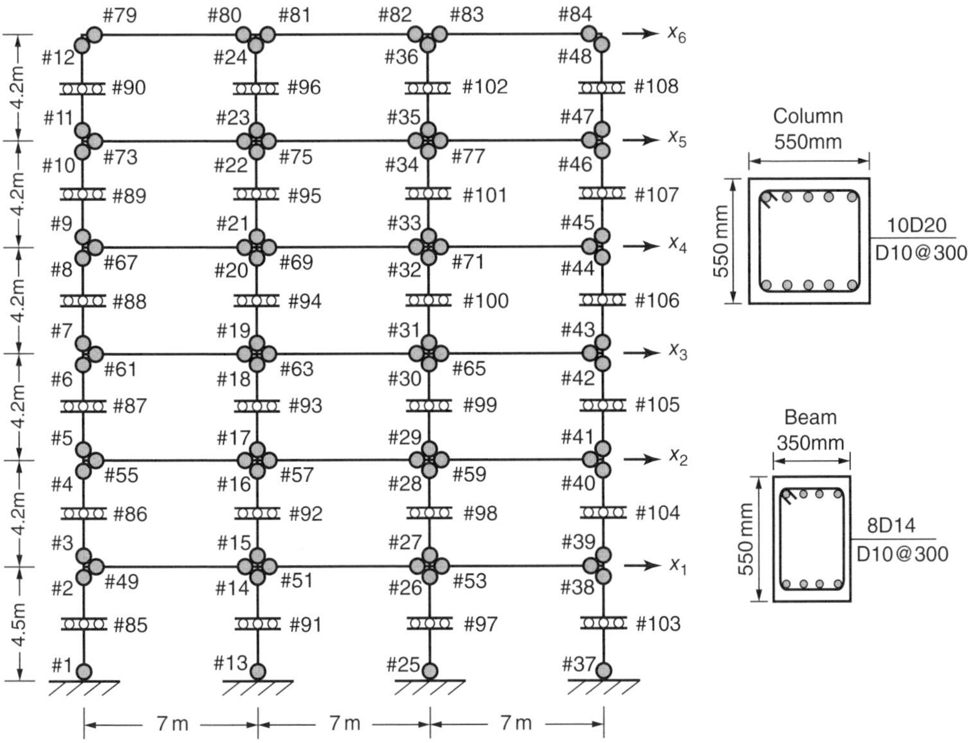

Figure 4.21 Six-story RC frame with flexural members.

Table 4.7 The parameters of the bending primary curves for the columns and beams.

Members	m_{cr} (kN-m)	θ_{cr} (rad)	m_y (kN-m)	θ_y (rad)	θ_u (rad)	θ_f (rad)
Columns	88	0.0003	252	0.0026	0.012	0.040
Beams	33.3	0.0005	100	0.0025	0.0056	0.027

Table 4.8 The parameters of the shear primary curves for the columns.

Members	V_{cr} (kN)	τ_{cr} (m)	V_y (kN)	τ_y (m)	τ_u (m)	τ_f (m)
Columns	41.9	0.00016	120	0.0011	0.0056	0.0168

and the $\bar{\mathrm{K}}'$ and $\bar{\mathrm{K}}''$ matrices can be similarly be calculated according to the expression of the static condensation procedure in Chapter 2. The mass matrix $\mathbf{M}_{dd}$ can be written as:

$$\mathbf{M}_{dd} = \begin{bmatrix} 45000 & 0 & 0 & 0 & 0 & 0 \\ 0 & 45000 & 0 & 0 & 0 & 0 \\ 0 & 0 & 45000 & 0 & 0 & 0 \\ 0 & 0 & 0 & 45000 & 0 & 0 \\ 0 & 0 & 0 & 0 & 45000 & 0 \\ 0 & 0 & 0 & 0 & 0 & 44000 \end{bmatrix} (\mathrm{kg}) \tag{4.116}$$

With the damping ratio 5%, the damping matrix $\mathbf{C}_{dd}$ can be written as:

$$\mathbf{C}_{dd} = \begin{bmatrix} 771 & -466 & 126 & -22 & 4 & -0.5 \\ -466 & 732 & -460 & 125 & -22 & 3 \\ 126 & -460 & 730 & -460 & 122 & -17 \\ -22 & 125 & -460 & 728 & -447 & 96 \\ 4 & -22 & 122 & -447 & 663 & -296 \\ -0.5 & 3 & -17 & 96 & -296 & 238 \end{bmatrix} (\text{kN-s/m}) \tag{4.117}$$

and the equations of motion for the structure can be written as:

$$\mathbf{M}_{dd}\ddot{\mathbf{X}}_d(t) + \mathbf{C}_{dd}\dot{\mathbf{X}}_d(t) + \bar{\mathbf{K}}\mathbf{x}'_d(t) = -\mathbf{M}_{dd}\ddot{\mathbf{g}}(t) \tag{4.118}$$

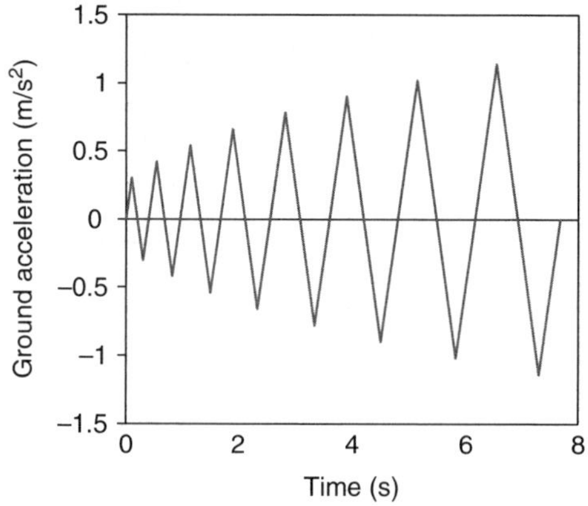

Figure 4.22 The cyclic increasing ground acceleration.

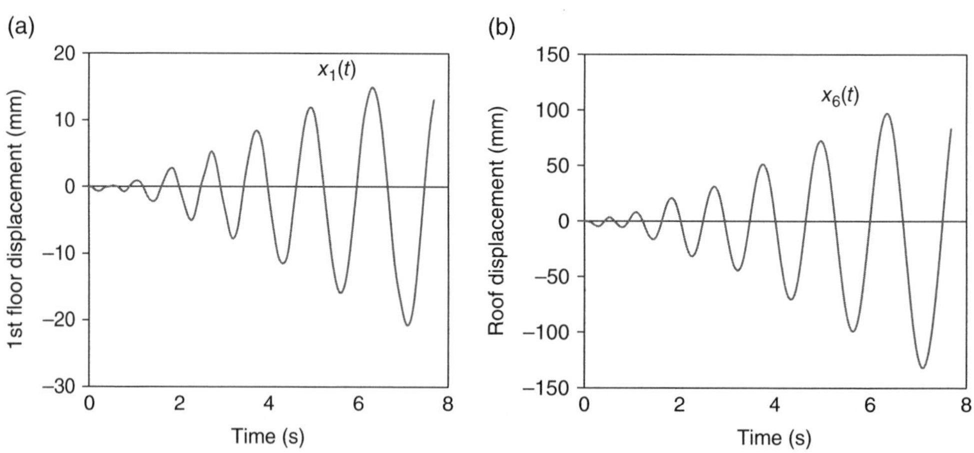

Figure 4.23 Global displacement responses: (a) displacement response of the first floor; (b) displacement response of the roof.

The transition matrices **A**, **H**, and **G** can be calculated using Eq. (3.139), and then the discrete form $\mathbf{F}_d$, $\mathbf{H}_d$, and $\mathbf{G}_d$ can be obtained using Eq. (3.138).

By subjecting the 6-story frame to the excitation shown in Figure 4.22, the displacement and velocity responses of selected floors are shown in Figure 4.23 and Figure 4.24. The hysteretic loops of restoring force versus lateral displacement of selected floors are plotted in Figure 4.25. Figure 4.26 and Figure 4.27 show the responses and hysteretic loops of selected hinges.

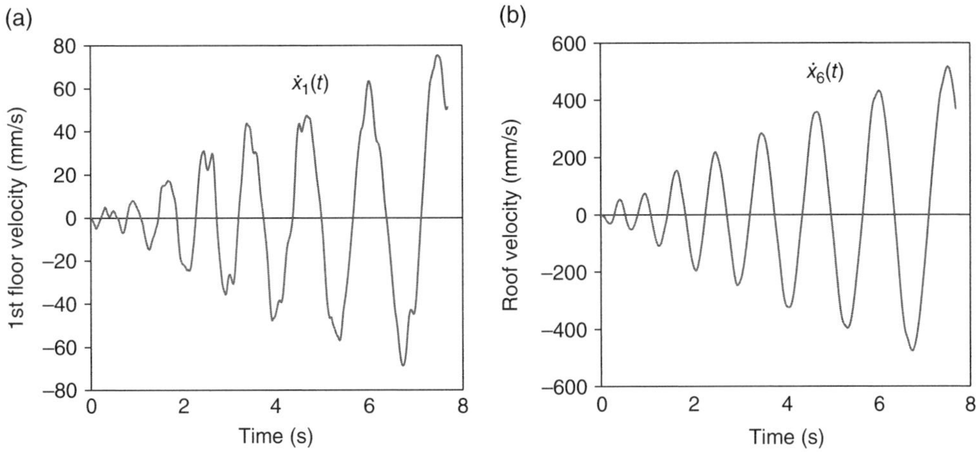

Figure 4.24 Global velocity responses: (a) velocity response of the first floor; (b) velocity response of the roof.

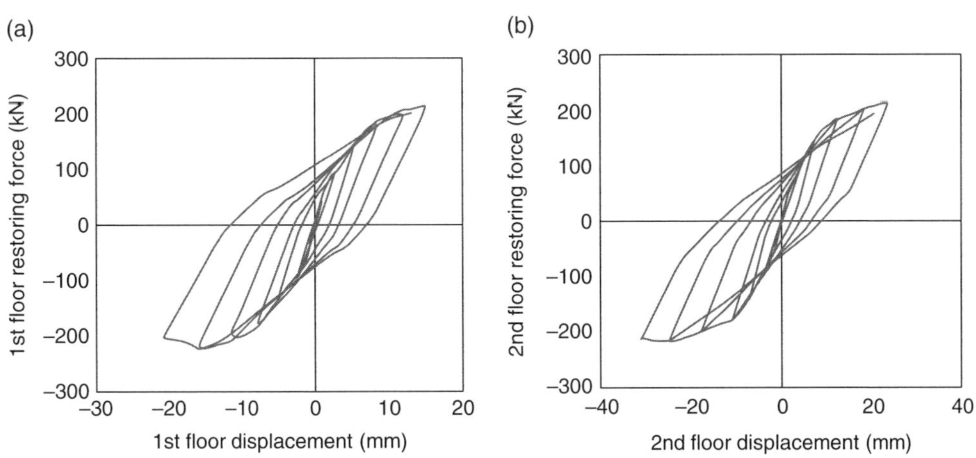

Figure 4.25 Hysteretic loops of restoring forces versus displacements: (a) hysteretic loop of the 1st floor; (b) hysteretic loop of the 2nd floor.

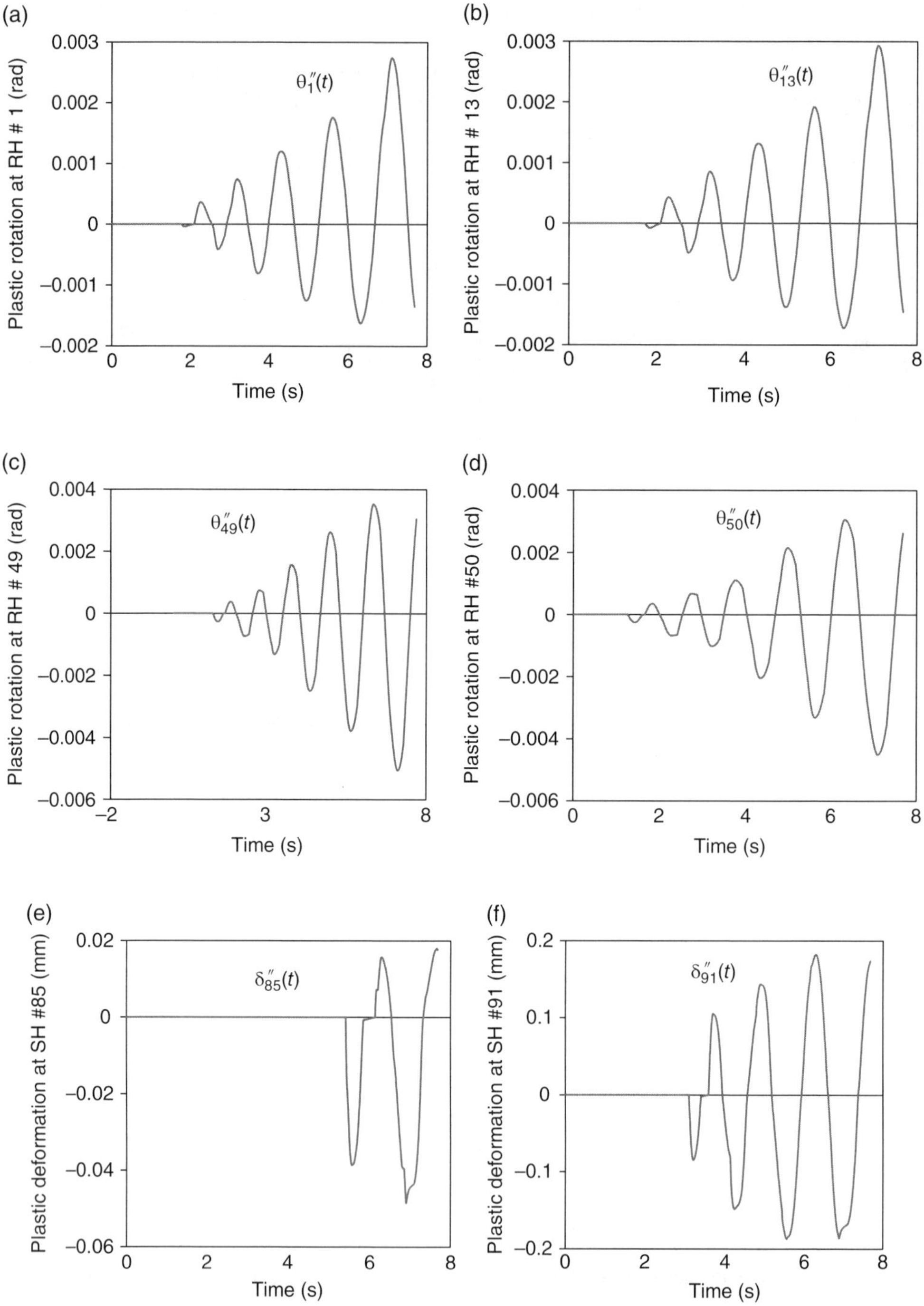

Figure 4.26 Local plastic responses: local plastic response at RH #1; (b) local plastic response at RH #13; (c) local plastic response at RH #49; (d) local plastic response at RH #50; (e) local plastic response at SH #85; (f) local plastic response at SH #91.

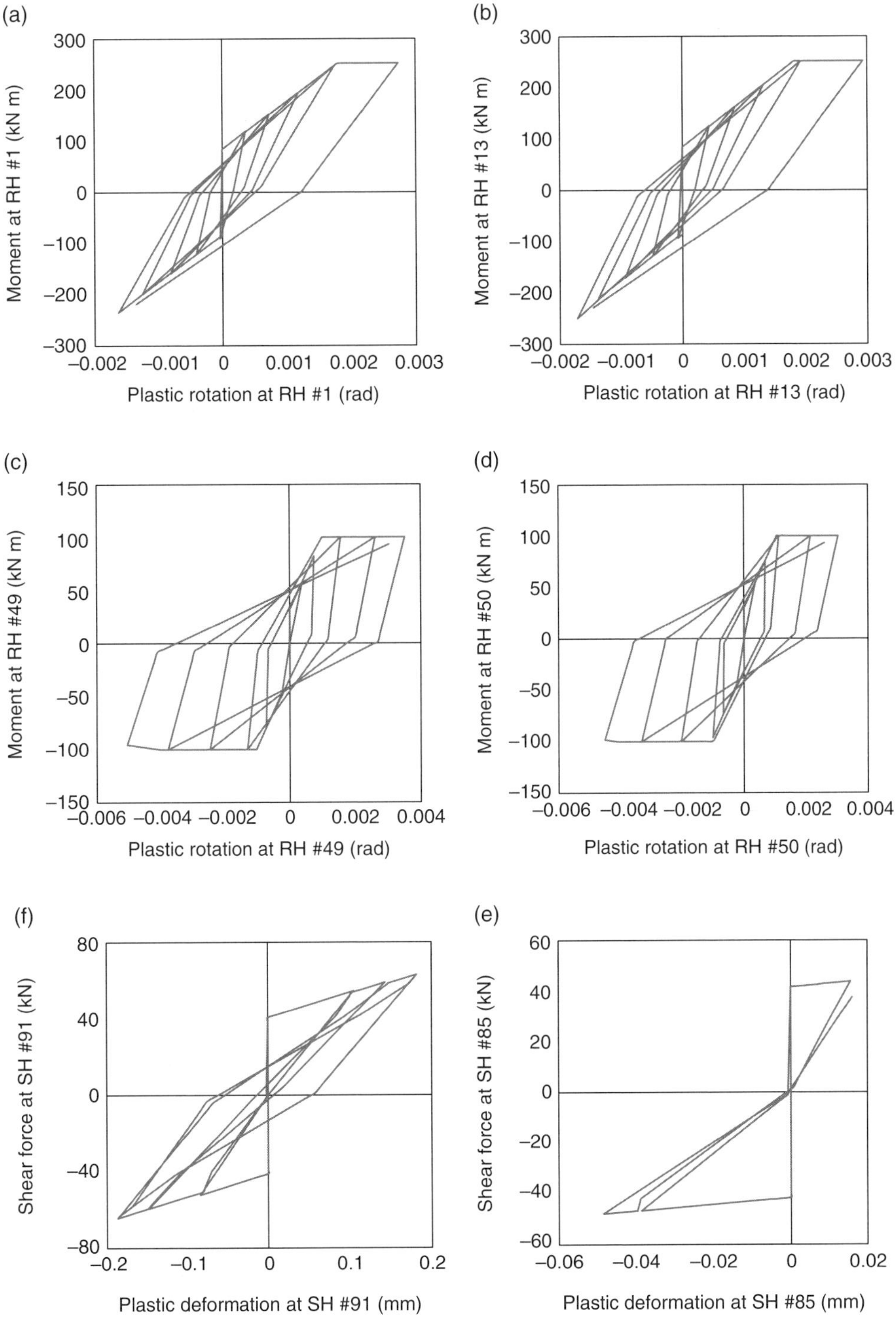

Figure 4.27 Hysteretic loops of selected RHs and SHs: (a) hysteretic loop of RH #1; (b) hysteretic loop of RH #13; (c) hysteretic loop of RH #49; (d) hysteretic loop of RH #50; (e) hysteretic loop of SH #85; (f) hysteretic loop of SH #91.

Example 4.11 Nonlinear dynamic analysis of a RC frame subjected to earthquake excitation

Consider again the 6-story RC frame with flexural members as shown in Figure 4.21. Now the structure is subjected to the 1994 Northridge earthquake ground motion as shown in Figure 3.3 with peak acceleration scaled to 2.5 m/s^2. The displacement and velocity responses of selected floors are shown in Figure 4.28 and Figure 4.29. The hysteretic loops of restoring force versus lateral displacement of selected floors are plotted in Figure 4.30. Figure 4.31 and Figure 4.32 show the responses and hysteretic loops of selected hinges.

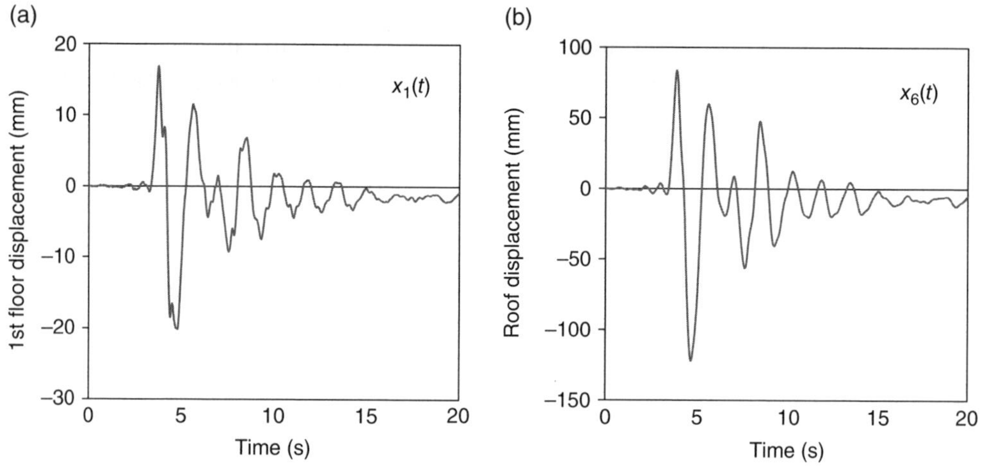

Figure 4.28 Global displacement responses: (a) displacement response of the first floor; (b) displacement response of the second floor.

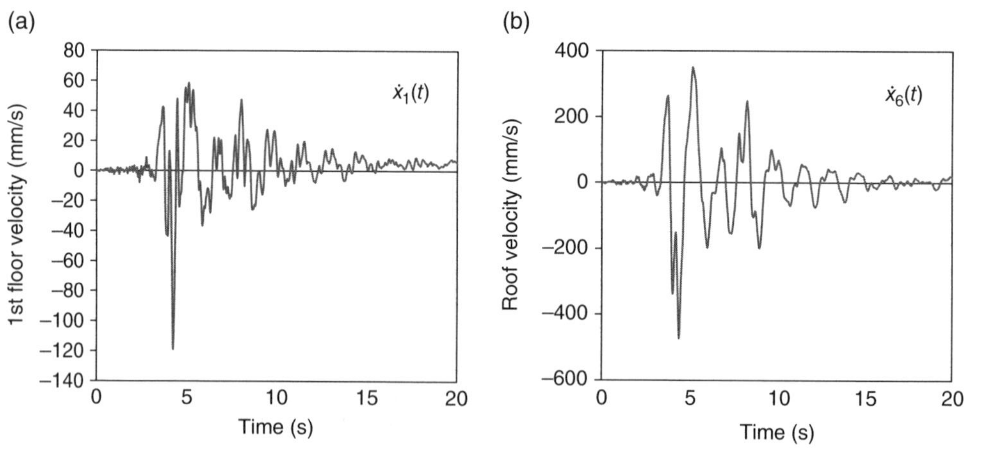

Figure 4.29 Global velocity responses: (a) velocity response of the first floor; (b) velocity response of the second floor.

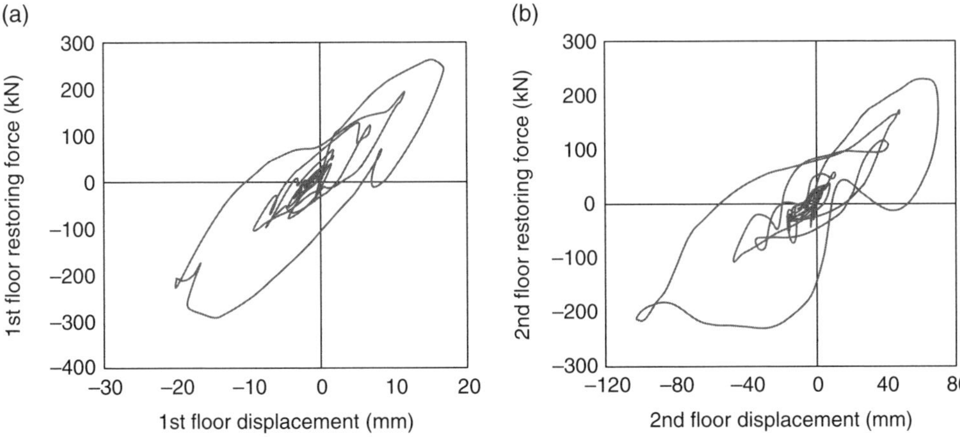

Figure 4.30 Hysteretic loops of restoring forces versus displacements: (a) hysteretic loop of the first floor; (b) hysteretic loop of the second floor.

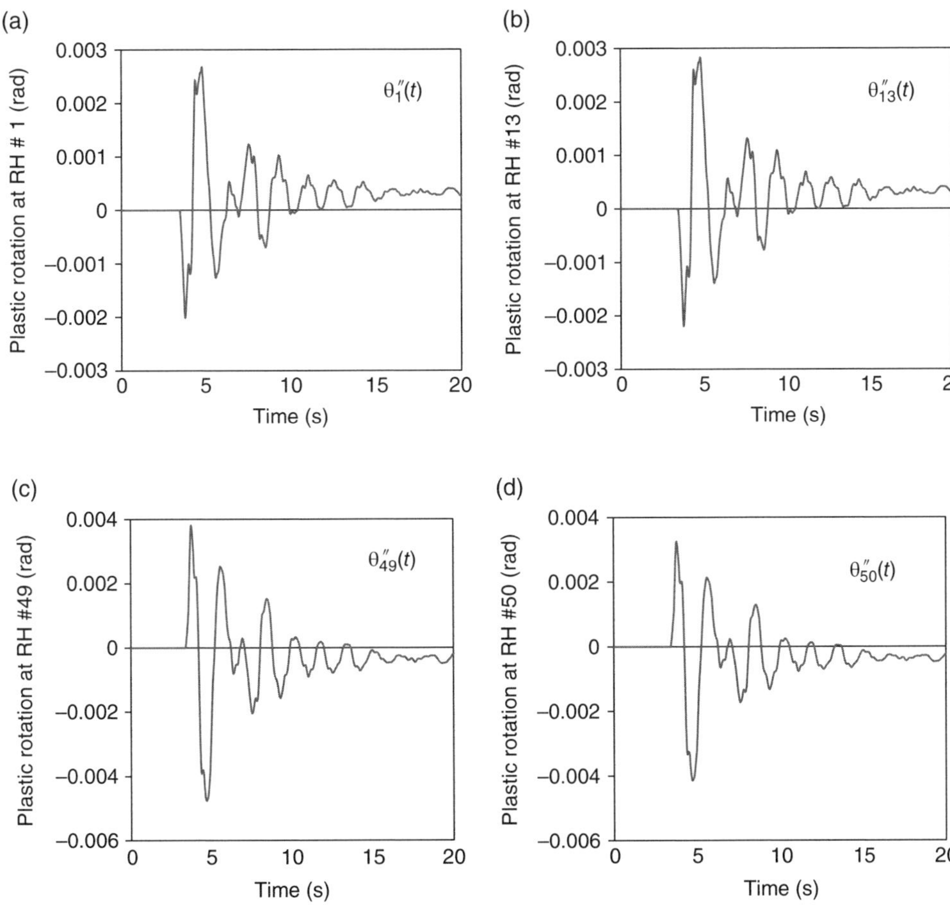

Figure 4.31 Local plastic responses: (a) local plastic response at RH #1; (b) local plastic response at RH #13; (c) local plastic response at RH #49; (d) local plastic response at RH #50;

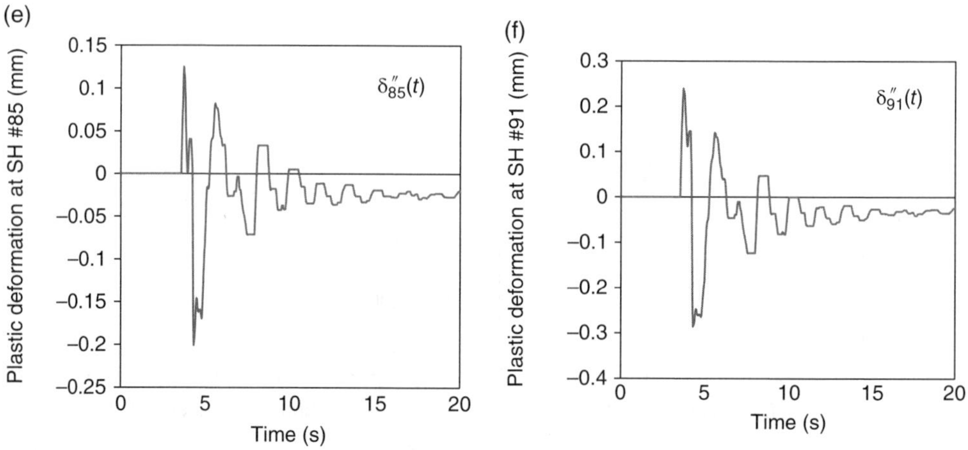

Figure 4.31 (*continued*) (e) local plastic response at SH #85; (f) local plastic response at SH #91.

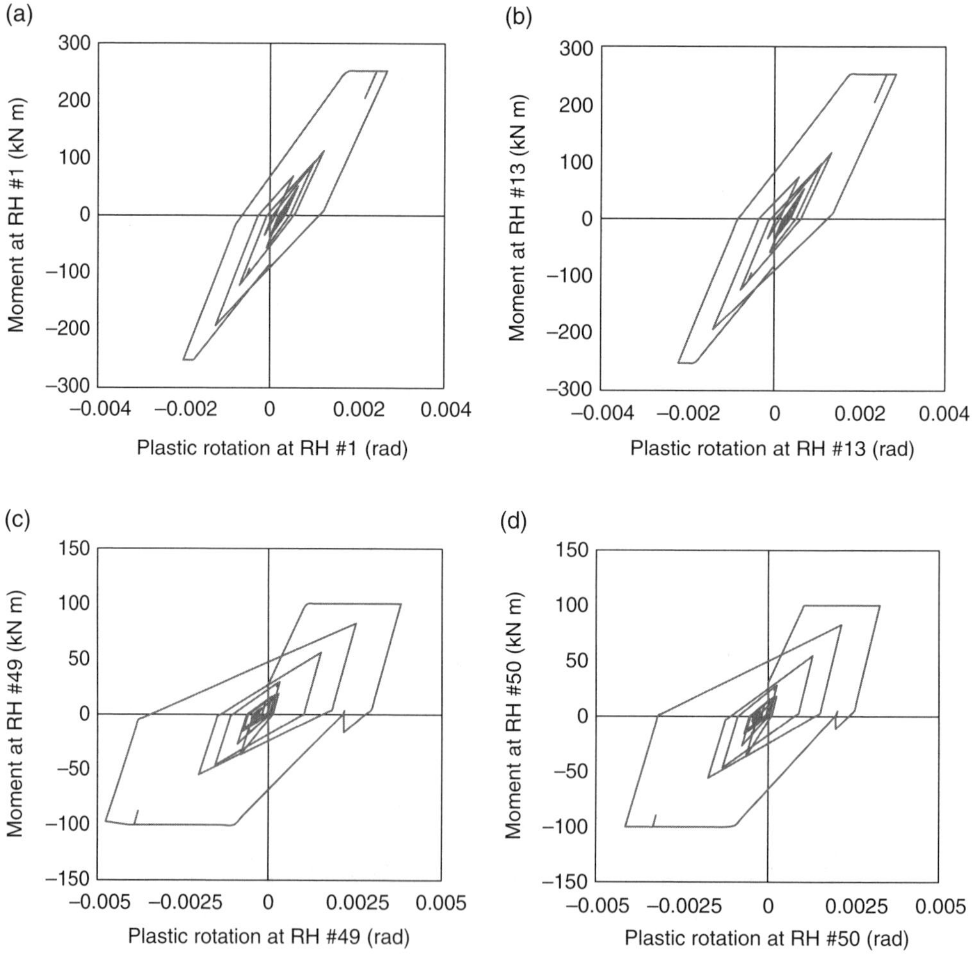

Figure 4.32 Hysteretic loops of selected RHs and SHs: (a) hysteretic loop of RH #1; (b) hysteretic loop of RH #13; (c) hysteretic loop of RH #49; (d) hysteretic loop of RH #50;

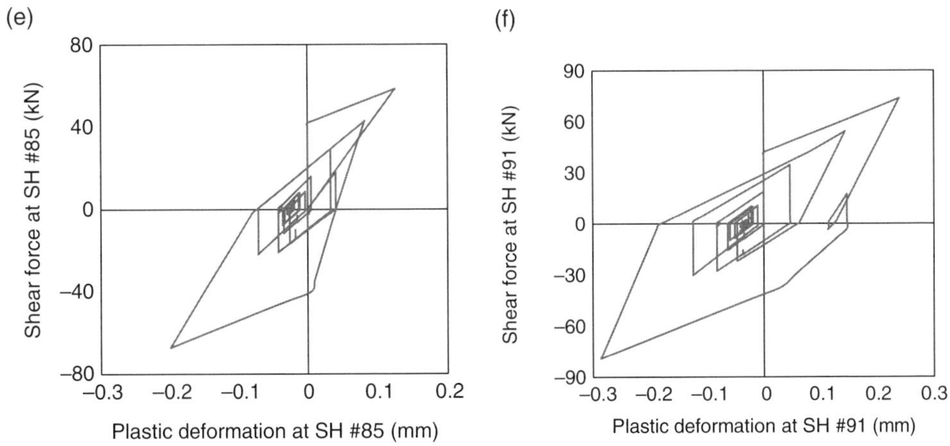

Figure 4.32 (*continued*) (e) hysteretic loop of SH #85; (f) hysteretic loop of SH #91.

References

Park R. and Paulay T. (1975). *Reinforced Concrete Structures*, John Wiley & Sons, New York.

Takeda T., Sozen M.A. and Nielson N.N. (1970). "Reinforced concrete response to simulated earthquakes." *Journal of the Structural Division*, 96(12): 2557–2573.

Ozcebe G. and Saatcioglu M. (1989). "Hysteretic shear model for reinforced concrete members." *Journal of Structural Engineering*, 115(1): 132–148.

Sezen H. and Chowdhury T. (2009). "Hysteretic model for reinforced concrete columns including the effect of shear and axial load failure." *Journal of Structural Engineering*, 135(2): 139–146.

Xu S.Y. and Zhang J. (2011) "Hysteretic shear-flexure interaction model of reinforced concrete columns for seismic response assessment of bridges." *Earthquake Engineering and Structural Dynamics*, 40(3): 315–337.

Vecchio F.J. and Collins M.P. (1988) "Predicting the response of reinforced concrete beams subjected to shear using modified compression field theory." *ACI Structural Journal*, 85(3): 258–268.

Ghali A. and Nevilli A.M. (1989). *Structural Analysis: a Unified Classical and Matrix Approach*, New York: Chapman and Hall, London.

5

Axial Deformation Member

The axial deformation member is used widely in structural engineering and its buckling behavior often attracts more attention in the engineering design and analysis when subjected to compressive loads. In this chapter, the physical theory model previously presented by **Dicleli** and **Calik** (2008) is selected as a baseline model to develop a modified version for incorporating it in the force analogy method (FAM). How to use the force analogy method to simulate the inelastic behavior of axial members under static and dynamical loads is interpreted through steel tube acted to compressive and tensile loads examples. In addition, the brace as one type of axial member that is an important structural component in the concentrically braced framed (CBF) system, so application of the FAM in the CBF is presented to verify the procedure for solving nonlinear static and dynamic problems.

5.1 Physical Theory Models for Axial Members

Three broad models: finite element models, phenomenological models and physical theory models are often used to calculate the response of axial deformation members. Physical theory models, which are based on fundamental structural behaviors, provide an effective and simple method and typically represent a brace using a beam-column element that can develop a plastic hinge at mid-length, so they can simulate overall force–deformation response reasonably well, and a variety of models in this category have been developed.

In this chapter, the physical theory model presented by **Dicleli** and **Calik** (2008) is selected as a baseline model to develop a modified version for incorporating it in the FAM. This model

Theory of Nonlinear Structural Analysis: The Force Analogy Method for Earthquake Engineering,
First Edition. Gang Li and Kevin K.F. Wong.

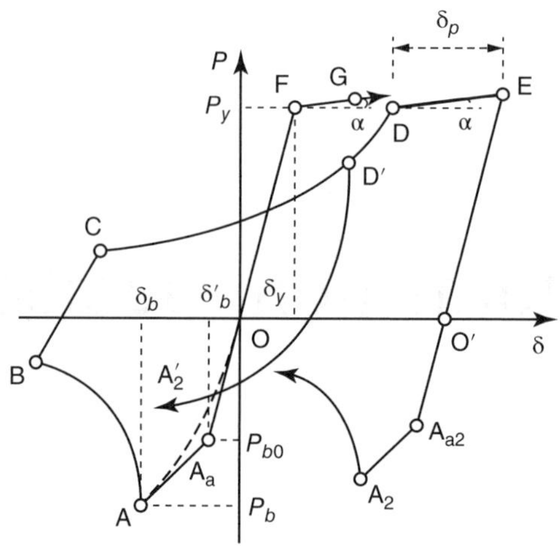

Figure 5.1 Hysteretic loop of the axial member.

appropriately depicts changing stiffness in compression. However, since the coincident initial stiffness is required for the FAM, minor modifications are made as follows: line OA, as shown in Figure 5.1, was divided into two parts: (1) line OA_a, which is the opposite extension of line OF such that the initial stiffness is the same in tension and compression, and (2) line A_aA, where the axial buckling load P_b occurs at Point A, and the intermediate axial load P_{b0} at Point A_a is set equal to half of the buckling load. In addition, the axial members under axial cycle loads have relatively complex nonlinear displacements responses so that the model of the axial member response is not a continuous function. In order to investigate the inelastic cyclic behavior of axial members, the axial force-displacement hysteretic loop can be broken into ten regions, O–A_a, A_a–A, A–B, B–C, C–D, D–E, D′–A_2', E–O′, O′–A_{a2}, and A_{a2}–A_2 in first cycle and will be discussed respectively below.

5.1.1 General Parameters

The nonlinear hysteretic behavior of axial members is usually associated with following parameters: the axial total load P, transverse displacement Δ, at the middle of axial members and axial displacement δ. In addition, several basic parameters representing the axial member characteristics: the section area A, the moment of inertia for the section I, the length l_b, and elastic module E of the member are necessary in this model. The buckling load P_b and reduced plastic moment capacity m_{pr} are governing parameters to depict buckling behaviors, and the initial axial load eccentricity, e is the main cause of buckling behaviors, can be expressed as:

$$e = -\frac{m_{pr}}{P_b\left(1 - \dfrac{l_b^2 P_b}{8EI\left(1 + \dfrac{P_b l b^2}{\pi^2 EI}\right)}\right)} \tag{5.1}$$

5.1.2 Displacement Decomposition

In the FAM, the key is to decompose the total displacement in response into elastic and inelastic components. Take the line A_aA, shown in Figure 5.2(a), as an illustration. Point M, which is located on line A_aA, represents the axial displacement δ of the axial member under the axial load *P*. Extend line OA_a until it reaches the load *P*, at point N, the axial displacement at point N is the elastic displacement δ′. This elastic displacement is equal to the difference between the total displacement δ and the inelastic displacement δ″:

$$\delta' = \delta - \delta'' \tag{5.2}$$

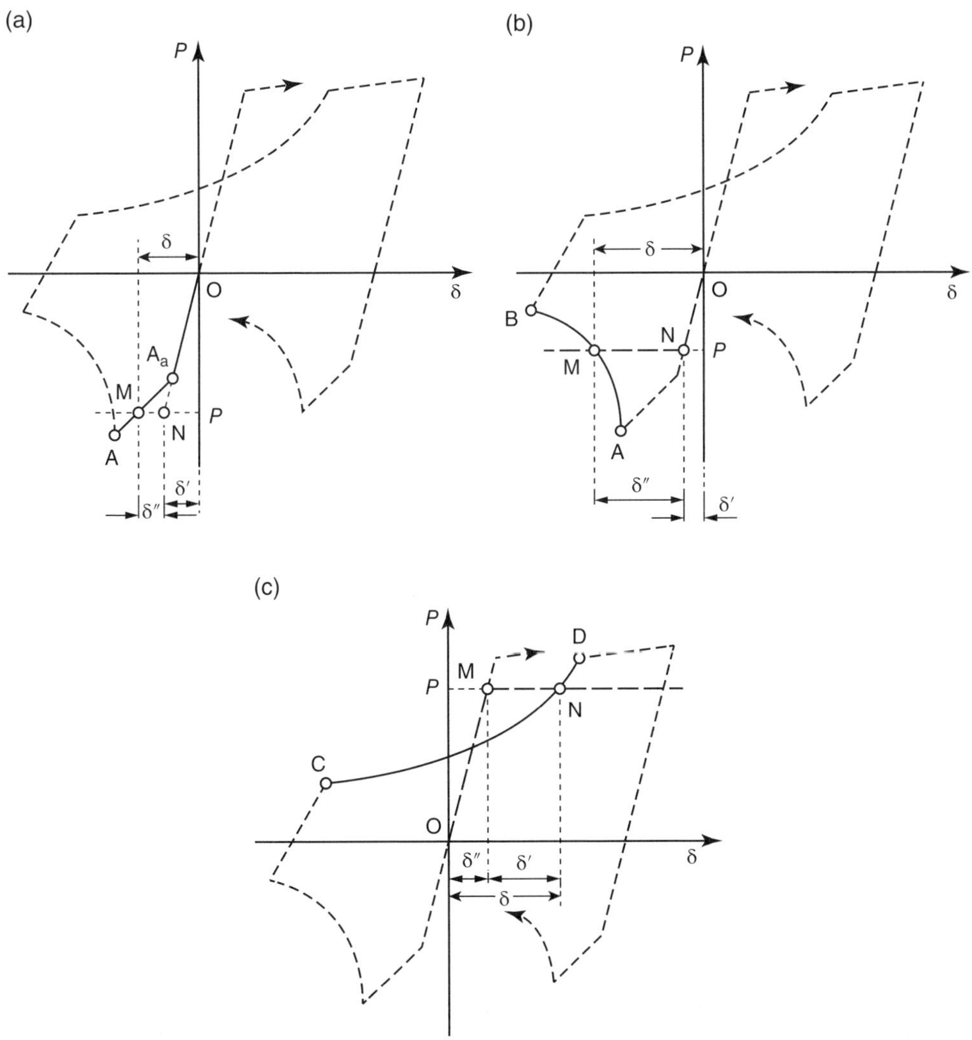

Figure 5.2 Decomposition of brace axial displacement: (a) Region A_a–A; (b) Region A–B; (c) Region C–D.

The force at point M is satisfied by:

$$P = P_N \tag{5.3}$$

where P_N is satisfied by:

$$P_N = k_b \delta' \tag{5.4}$$

in which, the axial member elastic stiffness is

$$k_b = EA/l_b \tag{5.5}$$

Substituting Eqs. (5.2) and (5.5) into Eq. (5.4) yields

$$P = k_b(\delta - \delta'') \tag{5.6}$$

In addition, the total displacement in other regions can be also divided into elastic and inelastic components, such as region A–B and region C–D shown in Figure 5.2(b) to 5.2(c).

5.2 Sliding Hinge Mechanisms

In the physical theory brace model, the axial inelastic displacement δ'' is assumed to be equal to the sum of two displacements produced by two sliding hinge (SH) plastic mechanisms, called SH1and SH2 and written as (Li et al. 2013):

$$\delta'' = \delta_1 + \delta_2 \tag{5.7}$$

Where δ_1 represents the displacement due to transverse bending that occurs in the presence of eccentricity when the axial member is in compression, as shown in Figure 5.3(a); δ_2 is the sum of two displacements: (1) the growth effect, which is interpreted as the displacement at which the axial member yields in tension is larger than the prior cycle to this force level due to the axial member has buckled in compression, as shown in Figure 5.4(a) (**Dicleli** and **Calik**, 2008) and (2) plastic uniaxial elongation.

To acquire the governing behavior of the sliding hinge SH1, SH2 is locked and axial displacement is only generated from SH1, as shown Figure 5.3(b). Based on the physical theory brace model, the axial inelastic displacement δ_1 is related to the transverse deformation Δ at any time. According to the geometric deformation relation, the inelastic axial displacement of the member δ_1 is expressed as:

$$\delta_1 = \sqrt{l_b{}^2 - 4\Delta^2} - l_b \tag{5.8}$$

where Δ illustrated in Figure 5.3 is the transverse displacement of the axial member.

For the sliding hinge SH2, the axial plastic displacement in tension and the growth effect illustrated in Figure 5.4(b) are involved. In order to define the relationship between the axial force P and the inelastic axial displacement δ_2 associated with SH2, SH1 is locked and axial displacement is only generated from SH2.

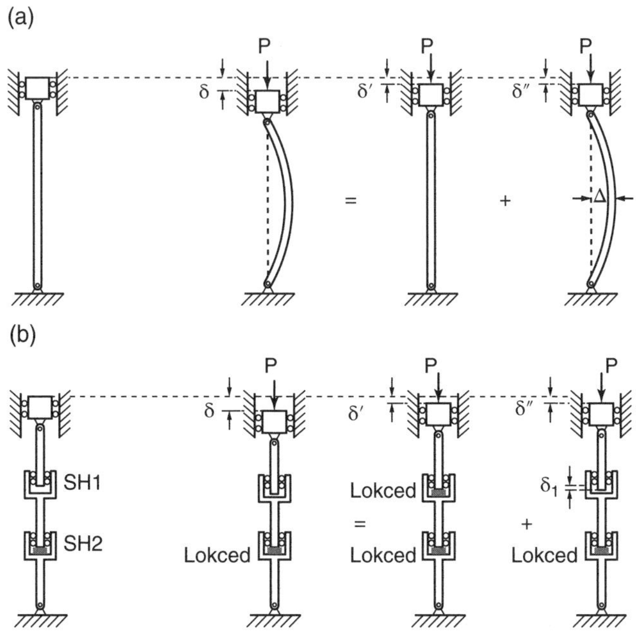

Figure 5.3 Axial member in compression: (a) Displacement decomposition; (b) Two sliding hinges plastic mechanisms.

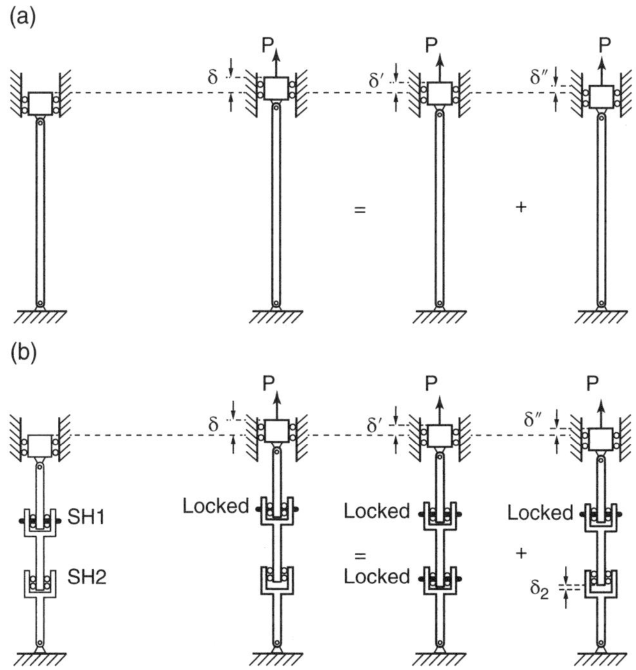

Figure 5.4 Axial member in tension: (a) Displacement decomposition; (b) Two sliding hinges plastic mechanisms.

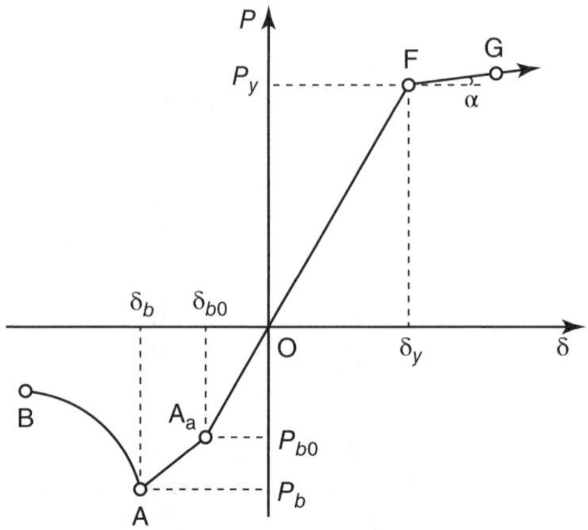

Figure 5.5 Backbone curve of axial member.

5.3 Force Analogy Method for Static Axial Members

Figure 5.5 shows the relation of the axial load and displacement of the axial member. It can be seen from this figure that there are totally three categories: (1) elastic regions O–F and O–A_a; (2) post-yield tensile region F–G, which indicates that the axial member is beyond tensile yielding strength and is nonlinear; (3) post-yield compressive regions, A_a–A and A–B, which mean the axial member steps into inelasticity after the buckling behavior in compression.

5.3.1 Regions O–A_a and O–F

When the axial member is subjected to the static axial load *P* in elastic domain, the axial load *P* is satisfied by

$$\textit{if} \quad \delta_{b0} < \delta < \delta_y, \qquad \textit{then} \quad P = k_b \delta, \quad \delta'' = 0 \tag{5.9}$$

in which

$$P_y = \sigma_y A \tag{5.10}$$

$$\delta_y = \sigma_y A / k_b \tag{5.11}$$

$$P_{b0} = 0.5 P_b \tag{5.12}$$

where σ_y is the material yielding stress of the axial member.

5.3.2 Region F–G

When the axial member is acted to the tensile load *P* and keeps in plastic domain, the behavior in this region is associated with the material nonlinearity and is expressed by $\delta > \delta_y$. The inelastic displacement of the SH1 and SH2 are satisfied by

$$\text{If} \quad \delta > \delta_y, \quad \text{then} \quad \begin{cases} \delta_1 = 0 \\ \delta_2 = (1-\alpha)\left[\delta - \delta_y\right] \end{cases} \tag{5.13}$$

Substituting Eq. (5.13) into Eq. (5.7) gives

$$\delta'' = (1-\alpha)\left(\delta - \delta_y\right) \tag{5.14}$$

Then, substitute Eq. (5.14) back into Eq. (5.6), and the axial resistant force P can be expressed by:

$$P = k_b(\delta - \delta'') = \alpha k_b\left(\delta - \delta_y\right) + P_y \tag{5.15}$$

where α is the second hardening factor of axial stiffness, as shown in Figure 5.5.

5.3.3 Regions A_a–A and A–B

The inelastic deformation is dominated by the bending behavior of the axial member due to the rotation of the plastic hinge formed at the middle of the deformed member at buckling. The transverse displacement is satisfied by

$$\text{If} \quad \begin{cases} \delta_b < \delta < \delta_{b0} \\ \delta < \delta_b \end{cases}, \quad \text{then} \quad \begin{cases} \Delta = -l_b{}^2 P_b e\left(1 + P_b l_b{}^2/\pi^2 EI\right)[P/P_b - 1/2]/4EI \\ \Delta = -m_{pr}/P - e \end{cases} \tag{5.16}$$

in which Point A is the intersection point of Line A_aA and Line AB, so the buckling load P_b is obtained through solving the following equation as

$$-l_b{}^2 P_b e\left(1 + P_b l_b{}^2/\pi^2 EI\right)/8EI = -m_{pr}/P_b - e \tag{5.17}$$

Substituting Eq. (5.16) into Eq. (5.8), and the axial inelastic displacement of the SH1 and SH2 are satisfied by

$$\text{if} \quad \begin{cases} \delta_b < \delta < \delta_{b0} \\ \delta < \delta_b \end{cases}, \quad \text{then} \quad \begin{cases} \delta_1 = \left[l_b{}^2 - 4\left(l_b{}^2 P_b e\left(1 + P_b l_b{}^2/\pi^2 EI\right)[P/P_b - 1/2]/4EI\right)^2\right]^{1/2} - l_b \\ \delta_1 = \left[l_b{}^2 - 4\left(-m_{pr}/P - e\right)^2\right]^{1/2} - l_b \end{cases} \tag{5.18}$$

and

$$\delta_2 = 0 \tag{5.19}$$

Substitute Eqs. (5.18) and (5.19) into Eq. (5.7), and the inelastic displacement is expressed by

$$\text{if} \quad \begin{cases} \delta_b < \delta < \delta_{b0} \\ \delta < \delta_b \end{cases}, \quad \text{then} \quad \begin{cases} \delta'' = \left[l_b{}^2 - 4\left(l_b{}^2 P_b e\left(1 + P_b l_b{}^2/\pi^2 EI\right)[P/P_b - 1/2]/4EI\right)^2\right]^{1/2} - l_b \\ \delta'' = \left[l_b{}^2 - 4\left(-m_{pr}/P - e\right)^2\right]^{1/2} - l_b \end{cases} \tag{5.20}$$

For convenient expression, Eq. (5.20) is simplified as:

$$if \quad \begin{cases} \delta_b < \delta < \delta_{b0} \\ \delta < \delta_b \end{cases}, \quad then \quad \begin{cases} P = f_{A_aA}(\delta'') \\ P = f_{AB}(\delta'') \end{cases} \tag{5.21}$$

where $f_{AaA}(\delta'')$ is the function of the equation as:

$$\delta'' = \left[l_b{}^2 - 4\left(l_b{}^2 P_b e\left(1 + P_b l_b{}^2/\pi^2 EI\right)[P/P_b - 1/2]/4EI\right)^2\right]^{1/2} - l_b \tag{5.22}$$

and f_{AB} (δ'') is the function of the equation as

$$\delta'' = \left[l_b{}^2 - 4\left(-m_{pr}/P - e\right)^2\right]^{1/2} - l_b \tag{5.23}$$

By defining that the right item in Eqs. (5.22) equals that in Eq. (5.23), the buckling load P_b can be solved. Then, the buckling displacement can be expressed by:

$$\delta_b = 1 - \left[l_b{}^2 - 4\left(-m_{pr}/P_b - e\right)^2\right]^{1/2} - l_b \tag{5.24}$$

Example 5.1 A steel tube in tension

Consider a 4.9-m long pin-ended square steel tube with cross-sectional dimensions $100 \times 10 \times 5$ mm, as shown in Figure 5.6. The yielding moment at the middle of the tube is thought of a given value of 4680 N-m. Material properties for the steel tube are $E = 2.1$ GPa, $\alpha = 0.1$ and $\sigma_y = 205$ MPa. Let us make sense how to solve the axial force when the tube subjected to (*a*) 0.003 m axial displacement, and (*b*) 0.006 m axial displacement input at the support.

Solution

(1) 3 mm displacement input

By using Eqs. (5.5), (5.10) and (5.11), there are

$$k_b = EA/l_b = 2.1 \times 10^{11} \times \left(0.1^2 - 0.09^2\right)/4.9 = 8.14 \times 10^7\,\text{N/m} \tag{5.25}$$

$$P_y = \sigma_y A = 205000000 \times \left(0.1^2 - 0.09^2\right) = 389500\,\text{N} \tag{5.26}$$

$$\delta_y = P_y/k_b = 389500/\left(8.14 \times 10^7\right) = 0.0048\,\text{m} \tag{5.27}$$

Since the axial displacement input $\delta = 0.003$ m is less than the yielding displacement $\delta_y = 0.0048$ m, the steel tube is linear and the axial force at the support should be applied by

$$P = k_b \delta = 8.14 \times 10^7 \times 3 \times 10^{-3} = 2.44 \times 10^5\,\text{N} \tag{5.28}$$

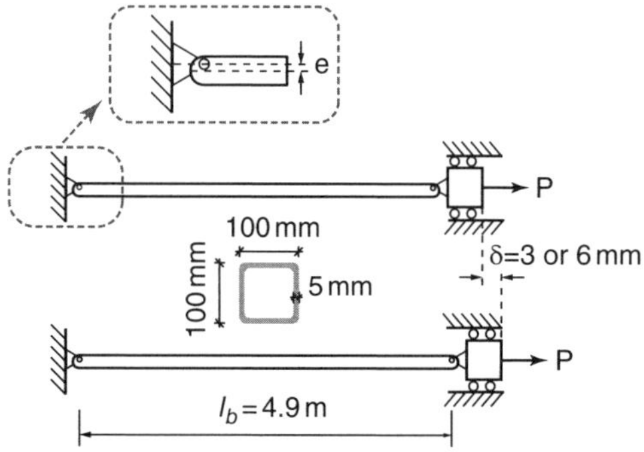

Figure 5.6 Model of the steel tube in tension.

(2) 6 mm displacement input

Since the axial displacement $\delta = 6$ mm is larger than the yielding displacement, $\delta_y = 0.0048$ m, the steel tube has stepped into plasticity. The axial inelastic displacement and axial force can be solved through Eqs. (5.14) and (5.15) as

$$\delta'' = (1-\alpha)(\delta-\delta_y) = (1-0.1)(0.006-0.0048) = 0.0011 \text{ m} \tag{5.29}$$

$$P = \alpha k_b(\delta-\delta_y) + P_y = 0.1 \times 8.14 \times 10^7 \times 0.0011 + 389500 = 398454 \text{ N} \tag{5.30}$$

Example 5.2 A steel tube in compression

Consider the steel tube is now subjected to –0.005 m axial displacement at the support as shown in Figure 5.7. Let us investigate how to solve the axial force should be applied on the steel tube.

Solution

According to Eqs. (5.17) and (5.24), the buckling load and corresponding displacement are obtained as

$$P_b = -180210 \text{ N} \tag{5.31}$$

$$\delta_b = -0.00265 \text{ m} \tag{5.32}$$

and the axial force and displacement at intermediate Point A_a based on Point A are calculated as:

$$P_{b0} = -90105 \text{ N} \tag{5.33}$$

$$\delta_{b0} = -0.0013 \text{ m} \tag{5.34}$$

Since the given displacement $\delta = -0.005$ m at the support is less than the intermediate displacement $\delta_b = -0.0013$ m, the steel tube has stepped into the nonlinear compressive region A–B. By substituting Eq. (5.22) into Eq. (5.6), the axial load is obtained as:

$$P = -138400 \text{ N} \tag{5.35}$$

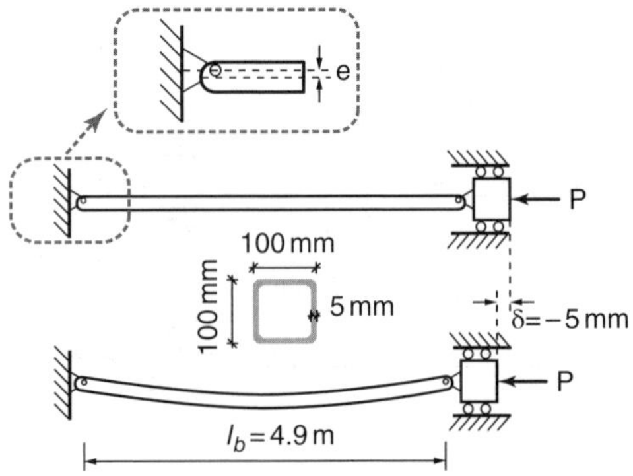

Figure 5.7 Model of steel tube in compression.

Substituting Eq. (5.35) back into Eq. (5.22) yields

$$\delta'' = -0.0041 \text{ m} \tag{5.36}$$

5.4 Force Analogy Method for Cycling Response Analysis of Axial Members

The behavior of the axial member under cycling loads is associated with the loading or input displacement path. The hysteresis loop for expressing the relation of the load and displacement has been broken into ten regions. In Section 5.3, the usage of the FAM in the regions O–A_a, A_a–A, O–D and D–E has been addressed, so the remainders will be formulated as below.

5.4.1 Region B–C

The behavior of the axial member is related with the elastic unloading in this region, as shown in Figure 5.8. The deformed shape of the axial member is assumed by two straight lines, because the transverse plastic displacement that is produced from the plastic rotations at the middle of the axial member within this region is far larger than that produced from the elastic deformation.

The transverse displacement of the axial member is expressed by

$$\Delta = -\frac{m_{pr}}{P_B} - e - \frac{0.5e - \dfrac{m_{pr}}{P_B}}{\dfrac{12EI}{(P-P_B)l_b^2} + 1} \tag{5.37}$$

where P_B is the load at Point B.

Substitute Eq. (5.37) into Eq. (5.8), and the displacement of SH1 equals

$$\delta_1 = \sqrt{l_b^2 - 4\left[-m_{pr}/P_B - e - (0.5e - m_{pr}/P_B)/\left(12EI/(P-P_B)/l_b^2 + 1\right)\right]^2} - l_b \tag{5.38}$$

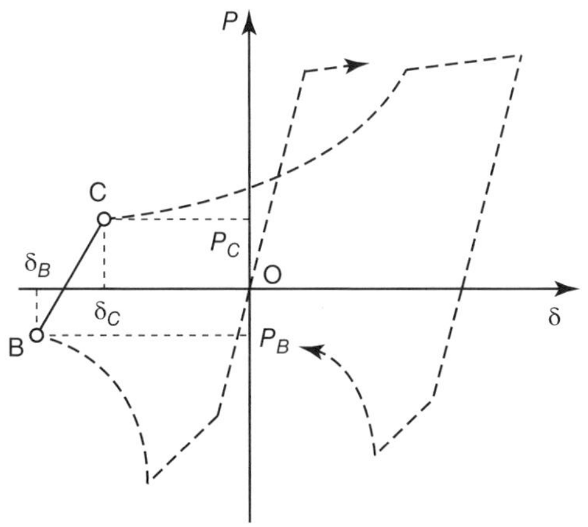

Figure 5.8 Relation of the load and displacement in Region B–C.

In addition, the growth effect is involved in this region. Since the axial member is assumed to be a function of the cumulative transverse plastic displacement in compression plus the cumulative transverse plastic displacement in tension, the axial inelastic displacement of SH2 is defined as

$$\delta_2 = \delta_{Gn}(P - P_B)l_b/EA \tag{5.39}$$

The parameter δ_{Gn} defined as normalized member growth by **Dicleli** and **Calik** (2008) is determined as a function of normalized cumulative plastic deformation. The inelastic behavior in tension depends on δ_{Gn} and the plastic axial deformation δ_p as shown in Figure 5.1.

Substitute Eqs. (5.38) and (5.39) into Eq. (5.7), and the total inelastic displacement yields

$$\begin{aligned}\delta'' = &\sqrt{{l_b}^2 - 4\left[-m_{pr}/P_B - e - (0.5e - m_{pr}/P_B)/\left(12EI/(P-P_B)/{l_b}^2 + 1\right)\right]^2} - l_b \\ &+ \delta_{Gn}(P - P_B)l_b/EA\end{aligned} \tag{5.40}$$

For a give load or input displacement, the corresponding deformation or axial force can be solved through the combination of Eq. (5.40) and Eq. (5.6).

5.4.2 Region C–D

In this region, the tensile load P and the existed transverse displacement Δ generated a second-order moment to satisfy static moment equilibrium. Thus, the plastic rotation behavior of the axial member is involved. The moment of the plastic hinge formed at the middle of the member increases in the reverse direction following the increasing of the axial load P in tension as shown in Figure 5.9.

The transverse displacement of the axial member is satisfied by

$$\Delta = m_{pr}/P - e\left[(P_y - P)/(P_y - P_C)\right] \tag{5.41}$$

where P_C is the load of Point C

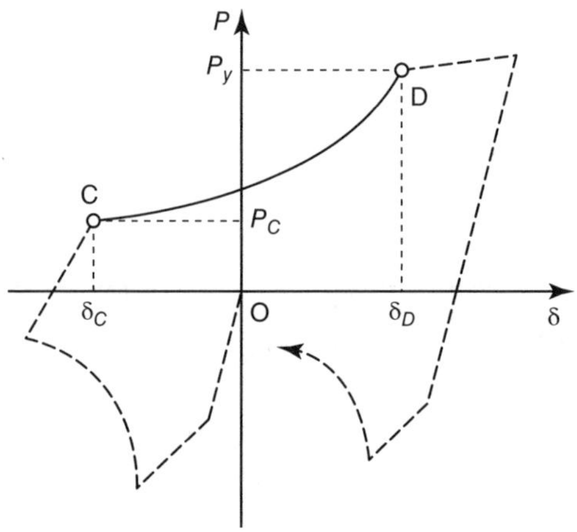

Figure 5.9 Relation of the load and displacement in Region C–D.

Substitute Eq. (5.41) in Eq. (5.8), and the axial displacement of the SH1 can be expressed by:

$$\delta_1 = \sqrt{l_b^2 - 4\left[m_{pr}/P - e\left[(P_y - P)/(P_y - P_C)\right]\right]^2} - l_b \tag{5.42}$$

The axial displacement of the SH2 in this region is also satisfied by Eq. (5.18).

Substituting Eqs. (5.42) and (5.39) into Eq. (5.7) gives

$$\delta'' = \sqrt{l_b^2 - 4\left[m_{pr}/P - e\left[(P_y - P)/(P_y - P_C)\right]\right]^2} - l_b + \delta_{Gn}(P - P_B)l_b/EA \tag{5.43}$$

The load P is the solution of the equation set (5.43) and (5.6).

5.4.3 Region D′–A$_2$

This region describes unloading and reverse loading behaviors of the axial member, but unlike other regions, the starting point in this region begins on the point within the line CD rather than the endpoint D, as shown in Figure 5.10. Thus, the maximum load P in this region is less than the tensile yield strength, P_y.

In this region, the transverse displacement is satisfied by

$$\Delta = \left[\frac{m_{pr}}{P_{D'}} - e\left(\frac{P_y - P_{D'}}{P_y - P_C}\right)\right]\left[1 + \left(\frac{P_{D'}}{l_b^2 P_{D'}(12-\pi^2) + 12\pi^2 EI} - \frac{P}{l_b^2 P(12-\pi^2) + 12\pi^2 EI}\right)\pi^2 l_b^2\right] \tag{5.44}$$

where $P_{D'}$ is the load at Point D'.

Since the axial displacement of the SH2 equals zero, the total axial inelastic displacement is

$$\delta'' = \sqrt{l_b^2 - 4\left[\left[\frac{m_{pr}}{P_{D'}} - e\left(\frac{P_y - P_{D'}}{P_y - P_C}\right)\right]\left[1 + \left(\frac{P_{D'}}{L^2 P_{D'}(12-\pi^2) + 12\pi^2 EI} - \frac{P}{l_b^2 P(12-\pi^2) + 12\pi^2 EI}\right)\pi^2 L^2\right]\right]^2} - l_b \tag{5.45}$$

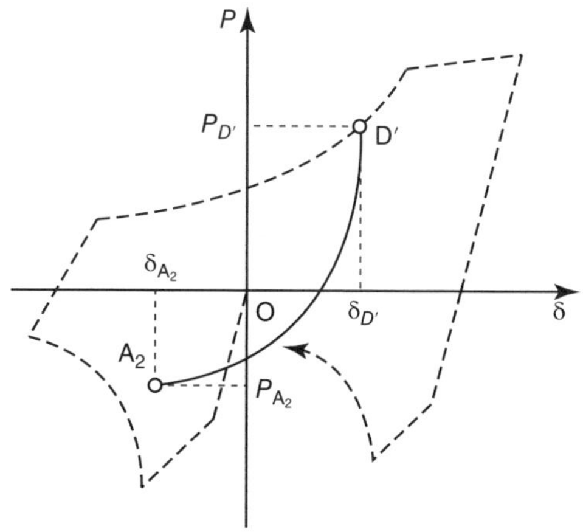

Figure 5.10 Relation of the load and displacement in Region D′–A₂.

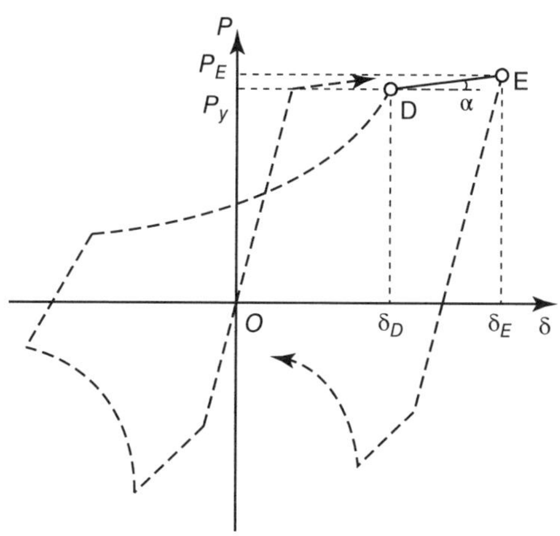

Figure 5.11 Relation of the load and displacement in Region D–E.

The axial load P is the solution of the equation set (5.45) and (5.6).

5.4.4 Region D–E

When the axial load P is beyond to the yielding force, P_y, the axial member is located in plasticity in tensile and remains straight, as shown in Figure 5.11. Thus, there are

$$\Delta = 0 \tag{5.46}$$

$$\delta'' = 0 \tag{5.47}$$

$$P = P_y + \alpha k_b(\delta - \delta_D) \tag{5.48}$$

5.4.5 Region E–F

The load–displacement relation in this region is again associated with the elastic behavior as shown in Figure 5.12, therefore, the transverse displacement and plastic displacement like region O–F are

$$\Delta = 0 \tag{5.49}$$

$$\delta'' = 0 \tag{5.50}$$

$$P = \alpha k_b(\delta - \delta_E) + P_E \tag{5.51}$$

in which, δ_E and P_E are the axial total displacement and load at Point E.

5.4.6 Region A_{a2}–A_2

The behavior of the axial member in this region is similar to that in the region A_a–A, see Figure 5.13. The transverse displacement is expressed as

$$\Delta = -2\left(\frac{P}{P_b'} - \frac{1}{2}\right)\frac{l_b^2 P_b' e}{8E_t I\left(1 + \dfrac{P_b' l_b^2}{\pi^2 E_t I}\right)} \tag{5.52}$$

Substituting Eq. (5.52) into Eq. (5.8) yields

$$-l_b^2 P_b' e\left(1 + P_b' l_b^2 / \pi^2 E_t I\right) / 8E_t I = -m_{pr}/P_b' - e \tag{5.53}$$

where E_t is defined as elastic modulus considering Baushinger effect degradation by **Dicleli** and **Calik** (2008) is determined as a function of normalized cumulative plastic deformation. The buckling load P'_b in the second cyclic loop is the solution of Eq. (5.53).

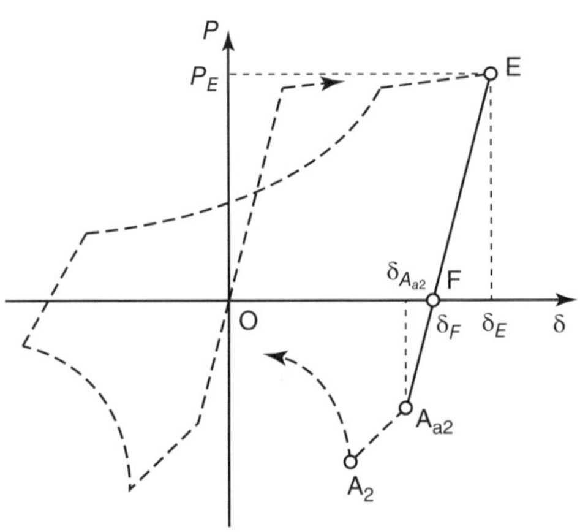

Figure 5.12 Relation of the load and displacement in Region E-F.

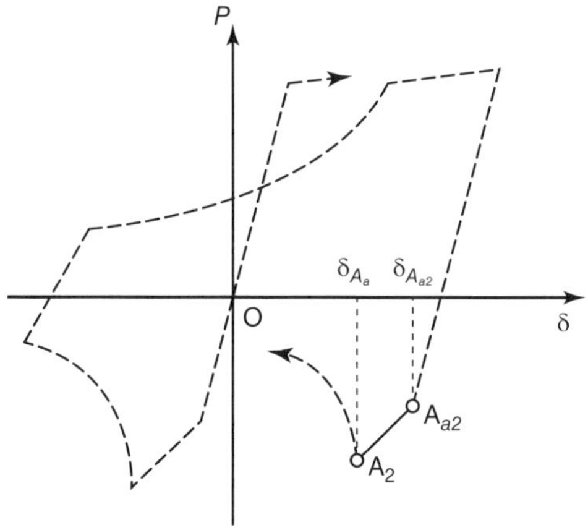

Figure 5.13 Relation of the load and displacement in Region A_2–A_{a2}.

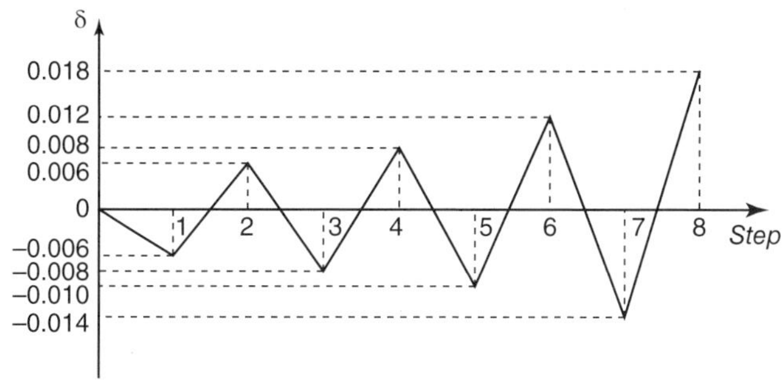

Figure 5.14 Cycle loading history.

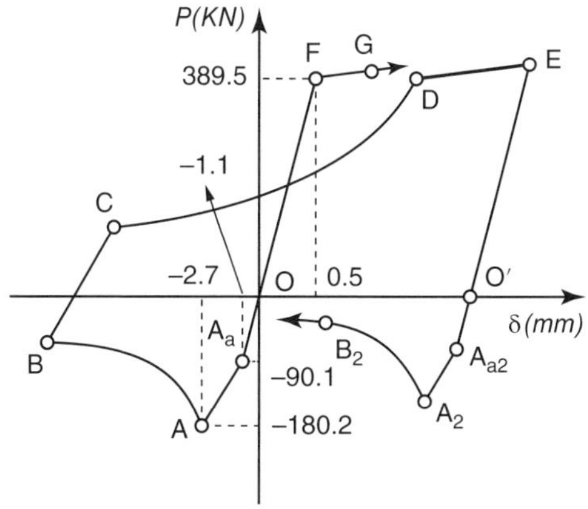

Figure 5.15 General parameters summary of the steel tube.

Example 5.3 Hysteretic response analysis of a steel tube under cycling loads set
Consider the same 2-m long pin-ended square steel tube used in Example 5.1 with only one difference that the tube is now subjected to a set of cycling support displacements. The displacement track with the step of 0.1 mm is shown in Figure 5.14.

The axial displacement and corresponding load at points A, A_a and E have been solved in Example 5.1 and are summarized in Figure 5.15. Figure 5.16 shows the hysteretic loop of the steel tube under the whole cyclic support displacement. Detailed results of the tube subjected the first cycle support displacements at each step are listed in Table 5.1.

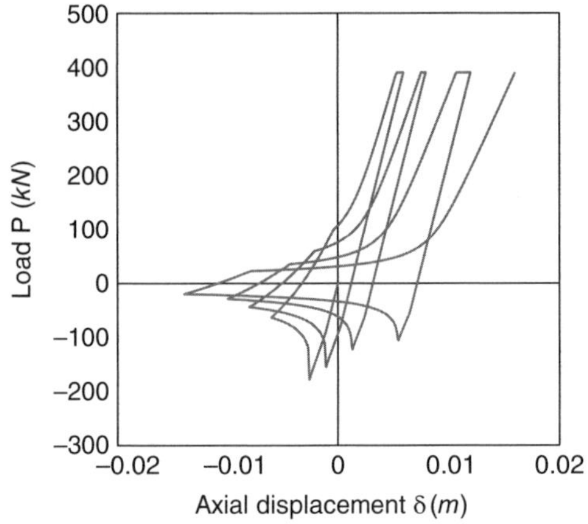

Figure 5.16 Hysteretic loop of the steel tube.

Table 5.1 Numerical results of the steel tube under cycling loads.

Lines	Support displacement δ (mm)	Transverse displacement Δ (mm)	Elastic displacement δ′(mm)	Plastic displacement δ″ (mm)			Axial Force P (kN)	Elastic modulus E (GPa)
				δ_1	δ_2	δ''		
	0	0	0	0	0	0	0	
	−0.1	0	−0.1	0	0	0	−8.1	
	−0.2	0	−0.2	0	0	0	−1.6	
O–A_a	⋮	⋮	⋮		⋮		⋮	2.1
	−1.0	0	−1.0	0	0	0	−81.4	
	−1.1	0	−1.1	0	0	0	−89.6	
	−1.2	0	−1.2	0	0	0	−95.5	
	−1.3	0	−1.3	0	0	0	−101.3	
A_a–A	⋮	⋮	⋮		⋮		⋮	2.1
	−2.5	0	−2.5	0	0	0	−170.8	
	−2.6	0	−2.6	0	0	0	−176.6	

Table 5.1 (*continued*)

Lines	Support displacement δ (mm)	Transverse displacement Δ (mm)	Elastic displacement δ′(mm)	Plastic displacement δ″ (mm)			Axial Force *P* (kN)	Elastic modulus *E* (GPa)
				δ_1	δ_2	δ″		
	−2.7	54.8	−1.5	−1.2	0	−1.2	−120.2	
	−2.8	58.8	−1.4	−1.4	0	−1.4	−113.1	
A–B	⋮	⋮	⋮		⋮		⋮	2.1
	−5.9	111.9	−0.8	−5.1	0	−5.1	−63.8	
	−6.0	113.1	−0.7	−5.2	0	−5.2	−63.2	
	−5.9	112.3	−0.7	−5.1	0	−5.1	−61.3	
	−5.8	111.5	−0.7	−5.1	0	−5.1	−59.4	
B–C	⋮	⋮	⋮		⋮		⋮	2.1
	−0.5	68.9	1.2	−2.0	0.3	−1.7	95.4	
	−0.4	68.1	1.2	−1.9	0.3	−1.6	99.5	
	−0.3	66.8	1.2	−1.8	0.3	−1.5	101.4	
	−0.2	65.4	1.3	−1.7	0.3	−1.5	103.4	
C–D	⋮	⋮	⋮		⋮		⋮	2.1
	5.2	20.2	4.6	−0.2	0.7	0.6	377.1	
	5.3	20.1	4.7	−0.2	0.7	0.6	384.1	
	5.4	0	4.7	0	0.7	0.7	389.5	
	5.5	0	4.7		0.8	0.8	389.5	
D–E	⋮	⋮	⋮		⋮		⋮	2.1
	5.9	0	4.7		1.2	1.2	389.5	
	6.0	0	4.7	0	1.3	1.3	389.5	
	5.9	0	4.6	0	1.3	1.3	381.4	
	5.8	0	4.5	0	1.3	1.3	373.2	
	⋮	⋮	⋮	⋮	⋮	⋮	⋮	
E–A_{a2}	1.3	0	0	0	1.3	1.3	6.8	1.84
	1.2	0	0	0	1.3	1.3	−1.2	
	⋮	⋮	⋮	⋮	⋮	⋮	⋮	
	0.4	0	−0.9	0	1.3	1.3	−66.5	
	0.3	0	−1.0	0	1.3	1.3	−74.6	
	0.2	0	−1.1	0	1.3	1.3	−81.5	
A_{a2}–A_2	0.1	0	−1.2	0	1.3	1.3		1.84
	⋮	⋮	⋮	⋮	⋮	⋮	⋮	
	−1.0	0	−2.3	0	1.3	1.3	−148.3	
	−1.1	0	−2.4	0	1.3	1.3	−153.8	
	−1.2	50.1	−1.4	−1.1	1.3	0.2	−133.3	
A_2–B_2	−1.3	54.7	−1.3	−1.3	1.3	0	−105.1	1.84
	−1.4	58.4	−1.2	−1.5	1.3	−0.2	−99.5	
	⋮	⋮	⋮	⋮	⋮	⋮	⋮	

5.5 Application of the Force Analogy Method in Concentrically Braced Frames

5.5.1 Force Analogy Method for Static SDOF CBFs

The brace component in CBFs is a typical axial member and has the capacity of providing a large lateral stiffness and strength. Thus, the brace has a significant influence on overall CBF performance improvement. Since the axial member model in the FAM presented above and the moment frame model have been mentioned previously, the overall CBF model can be implemented by a simple superposition of these two models.

Take the one-bay one-story CBF shown in Figure 5.17 as an illustration, in which the framed members are selected from Example 2.3. The CBF is subjected to a horizontal applied force F_0. As the growth effect is neglected in the static analysis, it results in a totally three-degrees-of-freedom system with seven plastic mechanisms including six plastic hinges for the frame and one sliding hinge for the brace. The brace element stiffness matrix of the form in the local coordinate system is

$$\mathbf{k}_e = \begin{bmatrix} EA_b/l_b & 0 & 0 & -EA_b/l_b & 0 & 0 \\ 0 & 12EI/l_b^3 & 6EI/l_b^2 & 0 & -12EI/l_b^3 & 6EI/l_b^2 \\ 0 & 6EI/l_b^2 & 4EI/l & 0 & -6EI/l_b^2 & 4EI/l \\ -EA_b/l_b & 0 & 0 & EA_b/l_b & 0 & 0 \\ 0 & -12EI/l_b^3 & -6EI/l_b^2 & 0 & 12EI/l_b^3 & -6EI/l_b^2 \\ 0 & 6EI/l_b^2 & 4EI/l_b & 0 & -6EI/l_b^2 & 4EI/l_b \end{bmatrix} \tag{5.54}$$

In addition, the angle translate matrix associated with the brace angle β is written as:

$$\mathbf{T} = \begin{bmatrix} \cos\beta & \sin\beta & 0 & 0 & 0 & 0 \\ -\sin\beta & \cos\beta & 0 & 0 & 0 & 0 \\ 0 & 0 & 1 & 0 & 0 & 0 \\ 0 & 0 & 0 & \cos\beta & \sin\beta & 0 \\ 0 & 0 & 0 & -\sin\beta & \cos\beta & 0 \\ 0 & 0 & 0 & 0 & 0 & 1 \end{bmatrix} \tag{5.55}$$

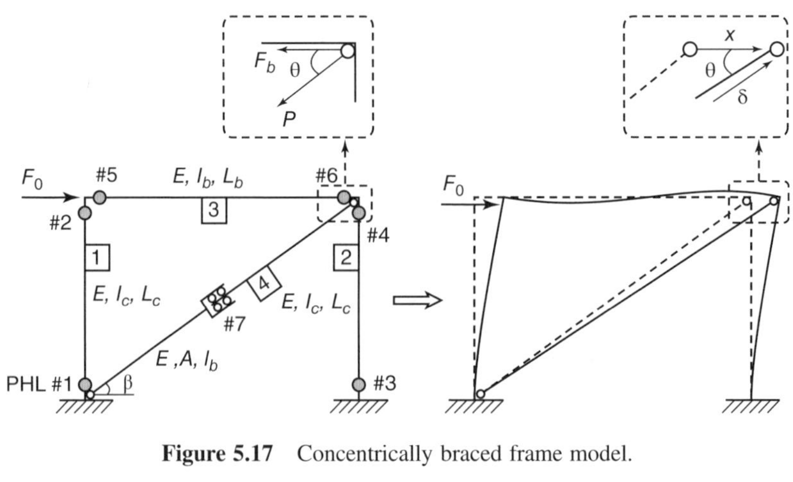

Figure 5.17 Concentrically braced frame model.

Thus, the brace element stiffness matrices of the form in the global coordinate system can be expressed by

$$\mathbf{k_b} = \mathbf{T}^{\mathrm{T}}\mathbf{k}_e\mathbf{T} \tag{5.56}$$

Since the brace is usually used as an axial member in structures, the axial stiffness is mainly considered and the shearing and bending stiffness are neglected here. Now, assembling the frame and brace element matrices into the global stiffness matrix gives:

$$\mathbf{K} = \begin{bmatrix} 24EI_c/L_c^3 + EA_b\cos^2\beta/l_b & 6EI_c/L_c^2 & 6EI_c/L_c^2 \\ 6EI_c/L_c^2 & 4EI_c/L_c + 4EI_b/L_b & 2EI_b/L_b \\ 6EI_c/L_c^2 & 2EI_b/L_b & 4EI_c/L_c + 4EI_b/L_b \end{bmatrix} \begin{matrix} \leftarrow x_1 \\ \leftarrow x_2 \\ \leftarrow x_3 \end{matrix} \tag{5.57}$$

$$\mathbf{K}' = \left[\begin{array}{cc:cc:cc:c} 6EI_c/L_c^2 & 6EI_c/L_c^2 & 6EI_c/L_c^2 & 6EI_c/L_c^2 & 0 & 0 & EA\cos^2\beta/l_b \\ 2EI_c/L_c & 4EI_c/L_c & 0 & 0 & 4EI_b/L_b & 2EI_b/L_b & 0 \\ 0 & 0 & 2EI_c/L_c & 4EI_c/L_c & 2EI_b/L_b & 4EI_b/L_b & 0 \end{array}\right] \begin{matrix} \leftarrow x_1 \\ \leftarrow x_2 \\ \leftarrow x_3 \end{matrix} \tag{5.58}$$

$$\mathbf{K}'' = \left[\begin{array}{cc:cc:cc:c} 4EI_c/L_c & 2EI_c/L_c & 0 & 0 & 0 & 0 & 0 \\ 2EI_c/L_c & 4EI_c/L_c & 0 & 0 & 0 & 0 & 0 \\ \hdashline 0 & 0 & 4EI_c/L_c & 2EI_c/L_c & 0 & 0 & 0 \\ 0 & 0 & 2EI_c/L_c & 4EI_c/L_c & 0 & 0 & 0 \\ \hdashline 0 & 0 & 0 & 0 & 4EI_b/L_b & 2EI_b/L_b & 0 \\ 0 & 0 & 0 & 0 & 2EI_b/L_b & 4EI_b/L_b & 0 \\ \hdashline 0 & 0 & 0 & 0 & 0 & 0 & EA_b/l_b \end{array}\right] \begin{matrix} \leftarrow \theta_1'' \\ \leftarrow \theta_2'' \\ \leftarrow \theta_3'' \\ \leftarrow \theta_4'' \\ \leftarrow \theta_5'' \\ \leftarrow \theta_6'' \\ \leftarrow \delta'' \end{matrix} \tag{5.59}$$

Based on the above the global matrices are presented from Eqs. (5.57) to (5.59), and the governing equation of the FAM for this one-bay one-story CBF becomes:

$$\left[\begin{array}{ccc:ccccccc} 24EI_c/L_c^3+EA_b\cos^2\beta/l_b & 6EI_c/L_c^2 & 6EI_c/L_c^2 & 6EI_c/L_c^2 & 6EI_c/L_c^2 & 6EI_c/L_c^2 & 6EI_c/L_c^2 & 0 & 0 & EA_b\cos^2\beta/l_b \\ 6EI_c/L_c^2 & 4EI_c/L_c+4EI_b/L & EI_b/L_b & 2EI_c/L_c & 4EI_c/L_c & 0 & 0 & 4EI_b/L_b & 2EI_b/L_b & 0 \\ 6EI_c/L_c^2 & 2EI_b/L_b & 4EI_c/L_c+4EI_b/L_b & 0 & 0 & 2EI_c/L_c & 4EI_c/L_c & 2EI_b/L_b & 4EI_b/L_b & 0 \\ \hdashline 6EI_c/L_c^2 & 2EI_c/L_c & 0 & 4EI_c/L_c & 2EI_c/L_c & 0 & 0 & 0 & 0 & 0 \\ 6EI_c/L_c^2 & 4EI_c/L_c & 0 & 2EI_c/L_c & 4EI_c/L_c & 0 & 0 & 0 & 0 & 0 \\ 6EI_c/L_c^2 & 0 & 2EI_c/L_c & 0 & 0 & 4EI_c/L_c & 2EI_c/L_c & 0 & 0 & 0 \\ 6EI_c/L_c^2 & 0 & 4EI_c/L_c & 0 & 0 & 2EI_c/L_c & 4EI_c/L_c & 0 & 0 & 0 \\ 0 & 4EI_b/L_b & 2EI_b/L_b & 0 & 0 & 0 & 0 & 4EI_b/L_b & 2EI_b/L_b & 0 \\ 0 & 2EI_b/L_b & 4EI_b/L_b & 0 & 0 & 0 & 0 & 2EI_b/L_b & 4EI_b/L_b & 0 \\ EA_b\cos^2\beta/l_b & 0 & 0 & 0 & 0 & 0 & 0 & 0 & 0 & EA_b/l_b \end{array}\right] \begin{Bmatrix} x_1 \\ x_2 \\ x_3 \\ \hdashline -\theta_1'' \\ -\theta_2'' \\ -\theta_3'' \\ -\theta_4'' \\ -\theta_5'' \\ -\theta_6'' \\ -\delta \end{Bmatrix} = \begin{Bmatrix} F_1 \\ F_2 \\ F_3 \\ \hdashline m_1 \\ m_2 \\ m_3 \\ m_4 \\ m_5 \\ m_6 \\ P \end{Bmatrix} \tag{5.60}$$

The rotation-moment relation of the plastic hinges and axial force–deformation relation of the brace sliding hinges have already mentioned in Sections 2.2.1 and 2.3.3.

Example 5.4 Nonlinear static analysis for a one-bay one-story CBF

A one-bay one-story CBF as shown in Figure 5.18 is selected to illustrate the application of the FAM in CBF models. The columns in the frame are square steel tubes with dimensions of $250 \times 250 \times 6$ mm and length of $L_c = 3.0$ m, and the beams in the frame are square steel tubes with dimensions of $200 \times 150 \times 4$ mm and length of $L_b = 3.9$ m. The brace is selected to the same brace used in Example 5.1. The brace is made of square tube sections with dimensions $100 \times 100 \times 5$ mm. The material properties for both braces and columns are $E = 2.1$ GPa and $\sigma_y = 275$ MPa. The CBF is subjected to a horizontal displacement of −3 mm on the roof, now solve how much of load should be applied on the CBF.

Solution

First, stiffness matrices in the FAM are

$$\mathbf{K} = \begin{bmatrix} 62967286 & 8140000 & 8140000 \\ 8140000 & 19779000 & 1749500 \\ 8140000 & 1749500 & 19779000 \end{bmatrix} \begin{matrix} \leftarrow x_1 \\ \leftarrow x_2 \\ \leftarrow x_3 \end{matrix} \ \text{N/m} \tag{5.61}$$

$$\mathbf{K}' = \left[\begin{array}{cc|cc|cc|c} 8140000 & 8140000 & 8140000 & 8140000 & 0 & 0 & 52114286 \\ 8140000 & 16280000 & 0 & 0 & 3498872 & 1749500 & 0 \\ 0 & 0 & 8140000 & 16280000 & 1749500 & 3498872 & 0 \end{array}\right] \begin{matrix} \leftarrow x_1 \\ \leftarrow x_2 \\ \leftarrow x_3 \end{matrix} \tag{5.62}$$

$$\mathbf{K}'' = \left[\begin{array}{cc|cc|cc|c} 16280000 & 8140000 & 0 & 0 & 0 & 0 & 0 \\ 8140000 & 16280000 & 0 & 0 & 0 & 0 & 0 \\ \hline 0 & 0 & 16280000 & 8140000 & 0 & 0 & 0 \\ 0 & 0 & 8140000 & 16280000 & 0 & 0 & 0 \\ \hline 0 & 0 & 0 & 0 & 3498872 & 1749500 & 0 \\ 0 & 0 & 0 & 0 & 1749500 & 498872 & 0 \\ \hline 0 & 0 & 0 & 0 & 0 & 0 & 814285714 \end{array}\right] \begin{matrix} \leftarrow \theta_1'' \\ \leftarrow \theta_2'' \\ \leftarrow \theta_3'' \\ \leftarrow \theta_4'' \\ \leftarrow \theta_5'' \\ \leftarrow \theta_6'' \\ \leftarrow \delta'' \end{matrix} \tag{5.63}$$

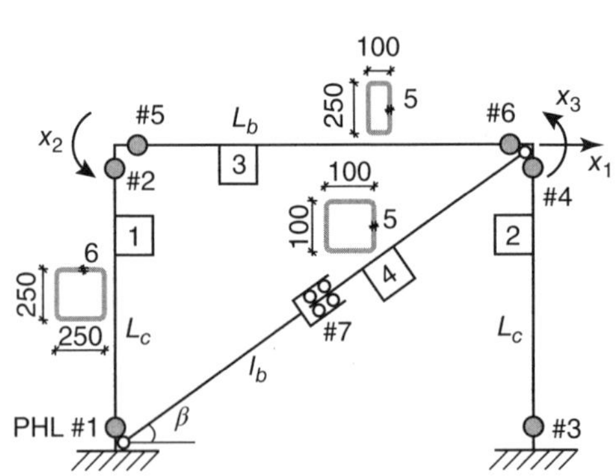

Figure 5.18 one-bay one-story CBF model.

Secondly, assume that the CBF response is linear, i.e. $\theta''_1=\theta''_2=\theta''_3=\theta''_4=\theta''_5=\theta''_6=\delta''=0$. Then, Eq. (5.60) becomes

$$\begin{bmatrix} 24EI_c/L_c^3+EA_b\cos^2\beta/l_b & 6EI_c/L_c^2 & 6EI_c/L_c^2 \\ 6EI_c/L_c^2 & 4EI_c/L_c+4EI_b/L_b & 2EI_b/L_b \\ 6EI_c/L_c^2 & 2EI_b/L_b & 4EI_c/L_c+4EI_b/L_b \end{bmatrix} \begin{Bmatrix} x_1 \\ x_2 \\ x_3 \end{Bmatrix} = \begin{Bmatrix} F_0 \\ 0 \\ 0 \end{Bmatrix} \tag{5.64}$$

Substituting materials properties in Eq. (5.64) gives

$$\begin{bmatrix} 62967286 & 8140000 & 8140000 \\ 8140000 & 19779000 & 1749500 \\ 8140000 & 1749500 & 19779000 \end{bmatrix} \begin{Bmatrix} -0.03 \\ x_2 \\ x_3 \end{Bmatrix} = \begin{Bmatrix} F_0 \\ 0 \\ 0 \end{Bmatrix} \tag{5.65}$$

The solution of Eq. (5.65) is

$$\begin{Bmatrix} x_1 \\ x_2 \\ x_3 \end{Bmatrix} = \begin{Bmatrix} -0.03\text{ m} \\ 0.0113\text{ rad} \\ 0.0113\text{ rad} \end{Bmatrix} \quad \text{and} \quad F_0 = -1704400\text{ N} \tag{5.66}$$

Substitute Eq. (5.66) back in Eq. (5.60) and solving for the moments and axial force gives

$$\begin{Bmatrix} m_1 \\ m_2 \\ m_3 \\ m_4 \\ m_5 \\ m_6 \\ P \end{Bmatrix} = \begin{bmatrix} 8140000 & 8140000 & 0 \\ 8140000 & 16280000 & 0 \\ 8140000 & 0 & 8140000 \\ 8140000 & 0 & 16280000 \\ 0 & 3498872 & 1749500 \\ 0 & 1749500 & 3498872 \\ 52114286 & 0 & 0 \end{bmatrix} \begin{Bmatrix} -0.03 \\ 0.0113 \\ 0.0113 \end{Bmatrix} = \begin{Bmatrix} -151870\text{ N-m} \\ -59533\text{ N-m} \\ -151870\text{ N-m} \\ -59533\text{ N-m} \\ 59533\text{ N-m} \\ 59533\text{ N-m} \\ -1563400\text{ N} \end{Bmatrix} \tag{5.67}$$

Comparing the moments demands in Eq. (5.67) with the corresponding yield moments show that moments at the PHL #5 and #6 have exceeded their corresponding yield moment. For PHLs #1, #2, #3 and #4, their moment demands are less than the corresponding yield moment, so the assumption that these four PHLs remain elastic is correct. In addition, the axial force of the brace also has exceeded their maximum capacity in compression, so the SH#7 has stepped in buckling stage, region A–B. Therefore, there are the following relations:

$$\begin{Bmatrix} m_1 \\ m_2 \\ m_3 \\ m_4 \\ m_5 \\ m_6 \\ P \end{Bmatrix} = \begin{Bmatrix} -151870\text{ N-m} \\ -59533\text{ N-m} \\ -151870\text{ N-m} \\ -59533\text{ N-m} \\ 50000\text{ N-m} \\ 50000\text{ N-m} \\ -1563400\text{ N} \end{Bmatrix} \tag{5.68}$$

Substituting Eq. (5.68) in Eq. (5.60) gives

$$\left[\begin{array}{ccc|ccccccc} 62967286 & 8140000 & 8140000 & 8140000 & 8140000 & 8140000 & 8140000 & 0 & 0 & 52114286 \\ 8140000 & 19779000 & 1749500 & 8140000 & 16280000 & 0 & 0 & 3498872 & 1749500 & 0 \\ 8140000 & 1749500 & 19779000 & 0 & 0 & 8140000 & 16280000 & 1749500 & 3498872 & 0 \\ \hline 8140000 & 8140000 & 0 & 16280000 & 8140000 & 0 & 0 & 0 & 0 & 0 \\ 8140000 & 16280000 & 0 & 8140000 & 16280000 & 0 & 0 & 0 & 0 & 0 \\ 8140000 & 0 & 8140000 & 0 & 0 & 16280000 & 8140000 & 0 & 0 & 0 \\ 8140000 & 0 & 16280000 & 0 & 0 & 8140000 & 16280000 & 0 & 0 & 0 \\ 0 & 3498900 & 1749500 & 0 & 0 & 0 & 0 & 3498872 & 1749500 & 0 \\ 0 & 1749500 & 3498900 & 0 & 0 & 0 & 0 & 1749500 & 3498872 & 0 \\ 52114286 & 0 & 0 & 0 & 0 & 0 & 0 & 0 & 0 & 81428571 \end{array}\right] \left\{\begin{array}{c} -0.03 \\ x_2 \\ x_3 \\ \hline -\theta''_1 \\ -\theta''_2 \\ -\theta''_3 \\ -\theta''_4 \\ -\theta''_5 \\ -\theta''_6 \\ -\delta'' \end{array}\right\} = \left\{\begin{array}{c} F_0 \\ 0 \\ 0 \\ \hline -151870 \\ -59500 \\ -151870 \\ -59500 \\ 50000 \\ 50000 \\ f_{A-B}(\delta'') \end{array}\right\} \tag{5.69}$$

The solution of Eq. (5.69) is

$$\left\{\begin{array}{c} x_1 \\ x_2 \\ x_3 \end{array}\right\} = \left\{\begin{array}{c} -0.03\text{ m} \\ 0.0119\text{ rad} \\ 0.0119\text{ rad} \end{array}\right\}, \left\{\begin{array}{c} \theta''_1 \\ \theta''_2 \\ \theta''_3 \\ \theta''_4 \\ \theta''_5 \\ \theta''_6 \\ \delta'' \end{array}\right\} = \left\{\begin{array}{c} 0\text{ rad} \\ 0\text{ rad} \\ 0\text{ rad} \\ 0\text{ rad} \\ -0.0024\text{ rad} \\ -0.0024\text{ rad} \\ -0.0188\text{ m} \end{array}\right\}, \left\{\begin{array}{c} m_1 \\ m_2 \\ m_3 \\ m_4 \\ m_5 \\ m_6 \\ P \end{array}\right\} = \left\{\begin{array}{c} -147100\text{ N-m} \\ -50000\text{ N-m} \\ -147100\text{ N-m} \\ -50000\text{ N-m} \\ 5000\text{ N-m} \\ 5000\text{ N-m} \\ -30943\text{ N} \end{array}\right\} \tag{5.70}$$

Figure 5.19 summarizes of these results in graphical form.

5.5.2 *Force Analogy Method for Static MDOF CBFs*

Take a one-bay two-story CBF shown in Figure 5.20 as an illustration for exhibiting matrices of the CBFs and governing equations in the FAM. The CBF is subjected a horizontal applied load vector $\mathbf{F}_s = [F_0, F_0]^T$. There is a totally six-degrees-of-freedom system with fourteen plastic mechanisms including twelve plastic hinges for frame and two sliding hinges for brace in this example.

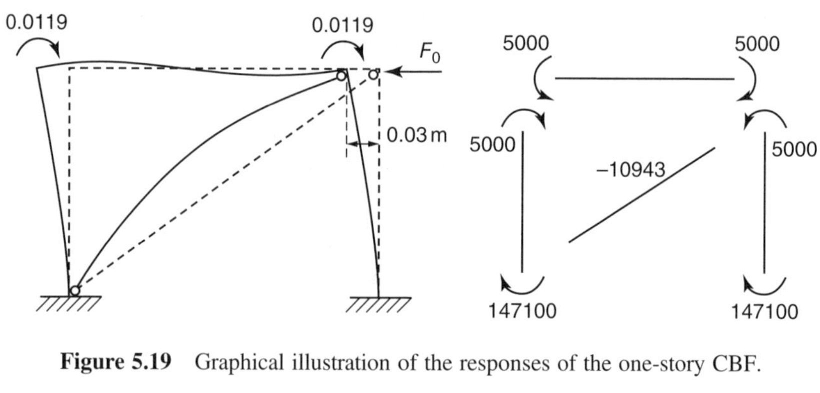

Figure 5.19 Graphical illustration of the responses of the one-story CBF.

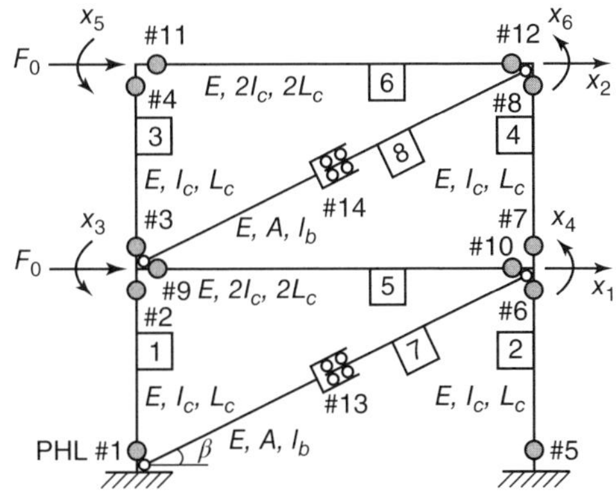

Figure 5.20 Two-story CBF model.

The same element stiffness matrix and angles translate matrix of the form in the local coordinate system expressed by Eqs. (5.54) and (5.55) are used here. Finally, the global stiffness matrixes are written as:

$$
\mathbf{K}=\begin{bmatrix}
\dfrac{48EI_c}{L_c^3}+\dfrac{2EA\cos^2\beta}{l_b} & -\dfrac{24EI_c}{L_c^2}-\dfrac{EA\cos^2\beta}{l_b} & 0 & 0 & \dfrac{-6EI_c}{L_c^2} & \dfrac{-6EI_c}{L_c^2} \\
-\dfrac{24EI_c}{L_c^2}-\dfrac{EA\cos^2\beta}{l_b} & \dfrac{24EI_c}{L_c^2}+\dfrac{EA\cos^2\beta}{l_b} & \dfrac{6EI_c}{L_c^2} & \dfrac{6EI_c}{L_c^2} & \dfrac{6EI_c}{L_c^2} & \dfrac{6EI_c}{L_c^2} \\
0 & 6EI_c/L_c^2 & \dfrac{8EI_c}{L_c}+\dfrac{4EI_b}{L_b} & 2EI_b/L_b & 2EI_c/L_c & 0 \\
0 & 6EI_c/L_c^2 & 2EI_b/L_b & \dfrac{8EI_c}{L_c}+\dfrac{4EI_b}{L_b} & 0 & 2EI_c/L_c \\
-6EI_c/L_c^2 & 6EI_c/L_c^2 & 2EI_c/L_c & 0 & \dfrac{4EI_c}{L_c}+\dfrac{4EI_b}{L_b} & 2EI_b/L_b \\
-6EI_c/L_c^2 & 6EI_c/L_c^2 & 0 & 2EI_c/L_c & 2EI_b/L_b & \dfrac{4EI_c}{L_c}+\dfrac{4EI_b}{L_b}
\end{bmatrix}
\tag{5.71}
$$

$$
\mathbf{K}'=\left[\begin{array}{cc:cc:cc:cc:cc:cc:cc}
\dfrac{6EI_c}{L_c^2} & \dfrac{6EI_c}{L_c^2} & -\dfrac{6EI_c}{L_c^2} & -\dfrac{6EI_c}{L_c^2} & \dfrac{6EI_c}{L_c^2} & \dfrac{6EI_c}{L_c^2} & -\dfrac{6EI_c}{L_c^2} & -\dfrac{6EI_c}{L_c^2} & 0 & 0 & 0 & 0 & \dfrac{EA_b\cos^2\beta}{l_b} & -\dfrac{EA_b\cos^2\beta}{l_b} \\
0 & 0 & \dfrac{6EI_c}{L_c^2} & \dfrac{6EI_c}{L_c^2} & 0 & 0 & \dfrac{6EI_c}{L_c^2} & \dfrac{6EI_c}{L_c^2} & 0 & 0 & 0 & 0 & -\dfrac{EA_b\cos^2\beta}{l_b} & 0 \\
\dfrac{2EI_c}{L_c} & \dfrac{4EI_c}{L_c} & \dfrac{4EI_c}{L_c} & \dfrac{2EI_c}{L_c} & 0 & 0 & 0 & 0 & \dfrac{4EI_b}{L_b} & \dfrac{2EI_b}{L_b} & 0 & 0 & 0 & 0 \\
0 & 0 & 0 & 0 & \dfrac{2EI_c}{L_c} & \dfrac{4EI_c}{L_c} & \dfrac{4EI_c}{L_c} & \dfrac{2EI_c}{L_c} & \dfrac{2EI_b}{L_b} & \dfrac{4EI_b}{L_b} & 0 & 0 & 0 & 0 \\
0 & 0 & \dfrac{2EI_c}{L_c} & \dfrac{4EI_c}{L_c} & 0 & 0 & 0 & 0 & 0 & 0 & \dfrac{4EI_b}{L_b} & \dfrac{2EI_b}{L_b} & 0 & 0 \\
0 & 0 & 0 & 0 & 0 & 0 & \dfrac{2EI_c}{L_c} & \dfrac{4EI_c}{L_c} & 0 & 0 & \dfrac{2EI_b}{L_b} & \dfrac{4EI_b}{L_b} & 0 & 0
\end{array}\right]
\tag{5.72}
$$

$$\mathbf{K}''=\begin{bmatrix}
\frac{4EI_c}{L_c} & \frac{2EI_c}{L_c} & 0 & 0 & 0 & 0 & 0 & 0 & 0 & 0 & 0 & 0 & 0 & 0\\
\frac{2EI_c}{L_c} & \frac{4EI_c}{L_c} & 0 & 0 & 0 & 0 & 0 & 0 & 0 & 0 & 0 & 0 & 0 & 0\\
0 & 0 & \frac{4EI_c}{L_c} & \frac{2EI_c}{L_c} & 0 & 0 & 0 & 0 & 0 & 0 & 0 & 0 & 0 & 0\\
0 & 0 & \frac{2EI_c}{L_c} & \frac{4EI_c}{L_c} & 0 & 0 & 0 & 0 & 0 & 0 & 0 & 0 & 0 & 0\\
0 & 0 & 0 & 0 & \frac{4EI_c}{L_c} & \frac{2EI_c}{L_c} & 0 & 0 & 0 & 0 & 0 & 0 & 0 & 0\\
0 & 0 & 0 & 0 & \frac{2EI_c}{L_c} & \frac{4EI_c}{L_c} & 0 & 0 & 0 & 0 & 0 & 0 & 0 & 0\\
0 & 0 & 0 & 0 & 0 & 0 & \frac{4EI_c}{L_c} & \frac{2EI_c}{L_c} & 0 & 0 & 0 & 0 & 0 & 0\\
0 & 0 & 0 & 0 & 0 & 0 & \frac{2EI_c}{L_c} & \frac{4EI_c}{L_c} & 0 & 0 & 0 & 0 & 0 & 0\\
0 & 0 & 0 & 0 & 0 & 0 & 0 & 0 & \frac{4EI_b}{L_b} & \frac{2EI_b}{L_b} & 0 & 0 & 0 & 0\\
0 & 0 & 0 & 0 & 0 & 0 & 0 & 0 & \frac{2EI_b}{L_b} & \frac{4EI_b}{L_b} & 0 & 0 & 0 & 0\\
0 & 0 & 0 & 0 & 0 & 0 & 0 & 0 & 0 & 0 & \frac{4EI_b}{L_b} & \frac{2EI_b}{L_b} & 0 & 0\\
0 & 0 & 0 & 0 & 0 & 0 & 0 & 0 & 0 & 0 & \frac{2EI_b}{L_b} & \frac{4EI_b}{L_b} & 0 & 0\\
0 & 0 & 0 & 0 & 0 & 0 & 0 & 0 & 0 & 0 & 0 & 0 & -\frac{EA_b}{l_b} & 0\\
0 & 0 & 0 & 0 & 0 & 0 & 0 & 0 & 0 & 0 & 0 & 0 & 0 & -\frac{EA_b}{l_b}
\end{bmatrix} \tag{5.73}$$

The governing equation the FAM for this one-bay two-story CBF is

$$\begin{Bmatrix}\mathbf{F}_s\\ \mathbf{m}\end{Bmatrix}=\begin{bmatrix}\mathbf{K} & \mathbf{K}'\\ \mathbf{K}'^{\mathrm{T}} & \mathbf{K}''\end{bmatrix}\begin{Bmatrix}\mathbf{x}\\ \boldsymbol{\theta}''\end{Bmatrix} \tag{5.74}$$

Example 5.5 Nonlinear static analysis of a one-bay two-story CBF

A one-bay two-story CBF is selected again to illustrate the application of the CBF model as shown in Figure 5.21. The section dimensions and lengths of the columns, beams and braces are the same to that in Example 5.4. The material properties for both braces and columns are $E = 2.1$ GPa and $\sigma_y = 275$ MPa. The CBF is subjected to –0.03 m and –0.06 m displacement on the first and second floors. Now we show how to solve the resistant force at the supports.

Solution

Now, assembling the frame and brace element matrices based on Eqs. (5.71), (5.72) and (5.73) into the global stiffness matrix gives:

$$\mathbf{K}=\begin{bmatrix}
125.93 & -62.97 & 0 & 0 & -8.14 & -8.14\\
-62.97 & 62.97 & 8.14 & 8.14 & 8.14 & 8.14\\
0 & 8.14 & 36.06 & 1.75 & 8.14 & 0\\
0 & 8.14 & 1.75 & 36.06 & 0 & 8.14\\
-8.14 & 8.14 & 8.14 & 0 & 19.78 & 1.75\\
-8.14 & 8.14 & 0 & 8.14 & 1.75 & 19.78
\end{bmatrix}\times 10^6\ \mathrm{N/m} \tag{5.75}$$

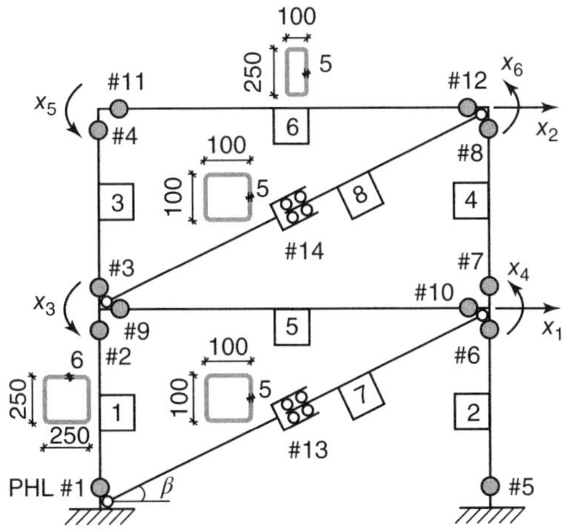

Figure 5.21 An actual two-story CBF.

$$
\mathbf{K}' = \left[\begin{array}{cc:cc:cc:cc:cc:cc:cc}
8.14 & 8.14 & -8.14 & -8.14 & 8.14 & 8.14 & -8.14 & -8.14 & 0 & 0 & 0 & 0 & -52.11 & 52.11 \\
0 & 0 & 8.14 & 8.14 & 0 & 0 & 8.14 & 8.14 & 0 & 0 & 0 & 0 & 52.11 & 0 \\
8.14 & 16.28 & 16.28 & 8.14 & 0 & 0 & 0 & 0 & 3.50 & 1.75 & 0 & 0 & 0 & 0 \\
0 & 0 & 0 & 0 & 8.14 & 16.28 & 16.28 & 8.14 & 1.75 & 3.50 & 0 & 0 & 0 & 0 \\
0 & 0 & 8.14 & 16.28 & 0 & 0 & 0 & 0 & 0 & 0 & 3.50 & 1.75 & 0 & 0 \\
0 & 0 & 0 & 0 & 0 & 0 & 8.14 & 16.28 & 0 & 0 & 1.75 & 3.50 & 0 & 0
\end{array}\right] \times 10^6 \text{ N/m} \tag{5.76}
$$

$$
\mathbf{K}'' = \left[\begin{array}{cc:cc:cc:cc:cc:cc:cc}
16.28 & 8.14 & 0 & 0 & 0 & 0 & 0 & 0 & 0 & 0 & 0 & 0 & 0 & 0 \\
8.14 & 16.28 & 0 & 0 & 0 & 0 & 0 & 0 & 0 & 0 & 0 & 0 & 0 & 0 \\ \hdashline
0 & 0 & 16.28 & 8.14 & 0 & 0 & 0 & 0 & 0 & 0 & 0 & 0 & 0 & 0 \\
0 & 0 & 8.14 & 16.28 & 0 & 0 & 0 & 0 & 0 & 0 & 0 & 0 & 0 & 0 \\ \hdashline
0 & 0 & 0 & 0 & 16.28 & 8.14 & 0 & 0 & 0 & 0 & 0 & 0 & 0 & 0 \\
0 & 0 & 0 & 0 & 8.14 & 16.28 & 0 & 0 & 0 & 0 & 0 & 0 & 0 & 0 \\ \hdashline
0 & 0 & 0 & 0 & 0 & 0 & 16.28 & 8.14 & 0 & 0 & 0 & 0 & 0 & 0 \\
0 & 0 & 0 & 0 & 0 & 0 & 8.14 & 16.28 & 0 & 0 & 0 & 0 & 0 & 0 \\ \hdashline
0 & 0 & 0 & 0 & 0 & 0 & 0 & 0 & 3.50 & 1.75 & 0 & 0 & 0 & 0 \\
0 & 0 & 0 & 0 & 0 & 0 & 0 & 0 & 1.75 & 3.50 & 0 & 0 & 0 & 0 \\ \hdashline
0 & 0 & 0 & 0 & 0 & 0 & 0 & 0 & 0 & 0 & 3.50 & 1.75 & 0 & 0 \\
0 & 0 & 0 & 0 & 0 & 0 & 0 & 0 & 0 & 0 & 1.75 & 3.50 & 0 & 0 \\ \hdashline
0 & 0 & 0 & 0 & 0 & 0 & 0 & 0 & 0 & 0 & 0 & 0 & -81.42 & 0 \\
0 & 0 & 0 & 0 & 0 & 0 & 0 & 0 & 0 & 0 & 0 & 0 & 0 & -81.42
\end{array}\right] \times 10^6 \text{ (N-m/rad)} \tag{5.77}
$$

Secondly, assume that the CBF response is linear, i.e.
$\theta''_1=\theta''_2=\theta''_3=\theta''_4=\theta''_5=\theta''_6=\theta''_7=\theta''_8=\theta''_9=\theta''_{10}=\theta''_{11}=\theta''_{12}=\delta''_1=\delta''_2=0$
and there is

$$\begin{bmatrix} 125.93 & -62.97 & 0 & 0 & -8.14 & -8.14 \\ -62.97 & 62.97 & 8.14 & 8.14 & 8.14 & 8.14 \\ 0 & 8.14 & 36.06 & 1.75 & 8.14 & 0 \\ 0 & 8.14 & 1.75 & 36.06 & 0 & 8.14 \\ -8.14 & 8.14 & 8.14 & 0 & 19.78 & 1.75 \\ -8.14 & 8.14 & 0 & 8.14 & 1.75 & 19.78 \end{bmatrix} \times 10^6 \begin{Bmatrix} -0.03 \\ -0.06 \\ x_3 \\ x_4 \\ x_5 \\ x_6 \end{Bmatrix} = \begin{Bmatrix} F_1 \\ F_2 \\ 0 \\ 0 \\ 0 \\ 0 \end{Bmatrix} \tag{5.78}$$

The solution of Eq. (5.78) is

$$\begin{Bmatrix} x_1 \\ x_2 \\ x_3 \\ x_4 \\ x_5 \\ x_6 \end{Bmatrix} = \begin{Bmatrix} -0.03\text{ m} \\ -0.06\text{ m} \\ 0.0114\text{ rad} \\ 0.0114\text{ rad} \\ 0.0070\text{ rad} \\ 0.0070\text{ rad} \end{Bmatrix} \quad \text{and} \quad \begin{Bmatrix} F_1 \\ F_2 \end{Bmatrix} = \begin{Bmatrix} -114470 \\ -1588900 \end{Bmatrix} \text{N} \tag{5.79}$$

Then, the moments and axial resistant force are expressed by

$$\begin{Bmatrix} m_1 \\ m_2 \\ m_3 \\ m_4 \\ m_5 \\ m_6 \\ m_7 \\ m_8 \\ m_9 \\ m_{10} \\ m_{11} \\ m_{12} \\ P_1 \\ P_2 \end{Bmatrix} = \mathbf{K}'^{\mathrm{T}} \begin{Bmatrix} x_1 \\ x_2 \\ x_3 \\ x_4 \\ x_5 \\ x_6 \end{Bmatrix} = \begin{bmatrix} 8.14 & 0 & 8.14 & 0 & 0 & 0 \\ 8.14 & 0 & 16.28 & 0 & 0 & 0 \\ -8.14 & 8.14 & 16.28 & 0 & 8.14 & 0 \\ -8.14 & 8.14 & 8.14 & 0 & 16.28 & 0 \\ 8.14 & 0 & 0 & 8.14 & 0 & 0 \\ 8.14 & 0 & 0 & 16.28 & 0 & 0 \\ -8.14 & 8.14 & 0 & 16.28 & 0 & 8.14 \\ -8.14 & 8.14 & 0 & 8.14 & 0 & 16.28 \\ 0 & 0 & 3.50 & 1.75 & 0 & \\ 0 & 0 & 1.75 & 3.50 & 0 & 0 \\ 0 & 0 & 0 & 0 & 3.50 & 1.75 \\ 0 & 0 & 0 & 0 & 1.75 & 3.50 \\ 52.11 & 52.11 & 0 & 0 & 0 & 0 \\ -52.11 & 0 & 0 & 0 & 0 & 0 \end{bmatrix} \begin{Bmatrix} -0.03 \\ -0.06 \\ 0.0114 \\ 0.0114 \\ 0.0070 \\ 0.0070 \end{Bmatrix} \times 10^6 = \begin{Bmatrix} -151370\text{ N-m} \\ -58543\text{ N-m} \\ -1309\text{ N-m} \\ -36903\text{ N-m} \\ -151370\text{ N-m} \\ -58543\text{ N-m} \\ -1309\text{ N-m} \\ -36903\text{ N-m} \\ 59852\text{ N-m} \\ 59852\text{ N-m} \\ 36903\text{ N-m} \\ 36903\text{ N-m} \\ -1563400\text{ N} \\ -1563400\text{ N} \end{Bmatrix} \tag{5.80}$$

Comparing moments demands in Eq. (5.80) with the corresponding yield moments shows that the moments at PHL #1, #5, #9, and #10 have exceeded their corresponding yield moment. For other PHLs, their moment demands are less than the corresponding yield moment, so the assumption that these four PHLs remain elastic is correct. In addition, the axial force of the brace has also exceeded their maximum capacity in compression, so the SH#7 has stepped into its buckling stage, region A–B. Therefore, there are the following relations: $\theta''_2=\theta''_3=\theta''_4=\theta''_6=\theta''_7=\theta''_8=\theta''_{11}=\theta''_{12}=\delta''_2=0$

The moments at the PHs and axial resistant force at the SHs are expressed by

$$\mathbf{m} = [\mathbf{m_c}\ \mathbf{m_c}\ \mathbf{m_{b1}}\ \mathbf{m_{b2}}\ F_1\ F_2] \tag{5.81}$$

in which, $\mathbf{m_c} = [-130000\ -130000\ 1110208111]$N-m, $\mathbf{m}_{b1} = [50000\ 50000]$N-m and $\mathbf{m}_{b2} = [-8111\ -8111]$N-m

Finally, the governing equation set is satisfied by

$$
\left[\begin{array}{cccccc|cccccc}
125.93 & -62.97 & 0 & 0 & -8.14 & -8.14 & 8.14 & 8.14 & 0 & 0 & -52.11 & 52.11 \\
-62.97 & 62.97 & 8.14 & 8.14 & 8.14 & 8.14 & 0 & 0 & 0 & 0 & 52.11 & 0 \\
0 & 8.14 & 36.06 & 1.75 & 8.14 & 0 & 8.14 & 0 & 3.50 & 1.75 & 0 & 0 \\
0 & 8.14 & 1.75 & 36.06 & 0 & 8.14 & 0 & 8.14 & 1.75 & 3.50 & 0 & 0 \\
-8.14 & 8.14 & 8.14 & 0 & 19.78 & 1.75 & 0 & 0 & 0 & 0 & 0 & 0 \\
-8.14 & 8.14 & 0 & 8.14 & 1.75 & 19.78 & 0 & 0 & 0 & 0 & 0 & 0 \\
\hline
8.14 & 0 & 8.14 & 0 & 0 & 0 & 16.28 & 0 & 0 & 0 & 0 & 0 \\
8.14 & 0 & 0 & 8.14 & 0 & 0 & 0 & 16.28 & 0 & 0 & 0 & 0 \\
0 & 0 & 3.50 & 1.75 & 0 & 0 & 0 & 0 & 3.50 & 1.75 & 0 & 0 \\
0 & 0 & 1.75 & 3.50 & 0 & 0 & 0 & 0 & 1.75 & 3.50 & 0 & 0 \\
-52.11 & 52.11 & 0 & 0 & 0 & 0 & 0 & 0 & 0 & 0 & -81.42 & 0 \\
52.11 & 0 & 0 & 0 & 0 & 0 & 0 & 0 & 0 & 0 & 0 & -81.42
\end{array}\right]
\left\{\begin{array}{c}
-0.03 \\ -0.06 \\ x_3 \\ x_4 \\ x_5 \\ x_6 \\ \hline \theta_1'' \\ \theta_5'' \\ \theta_9'' \\ \theta_{10}'' \\ \delta_1'' \\ \delta_2''
\end{array}\right\} \times 10^6 =
\left\{\begin{array}{c}
F_1 \\ F_2 \\ 0 \\ 0 \\ 0 \\ 0 \\ \hline -150000 \\ -150000 \\ 50000 \\ 50000 \\ f_{A\text{-}B}(\delta_1'') \\ f_{A\text{-}B}(\delta_2'')
\end{array}\right\}
\tag{5.82}
$$

The solution of Eq. (5.82) is

$$
\begin{Bmatrix} x_1 \\ x_2 \\ x_3 \\ x_4 \\ x_5 \\ x_6 \end{Bmatrix} =
\begin{Bmatrix} -0.03\ \text{m} \\ -0.06\ \text{m} \\ 0.0118\ \text{rad} \\ 0.0118\ \text{rad} \\ 0.0069\ \text{rad} \\ 0.0069\ \text{rad} \end{Bmatrix},
\begin{Bmatrix} \theta_1'' \\ \theta_2'' \\ \theta_3'' \\ \theta_4'' \\ \theta_5'' \\ \theta_6'' \\ \theta_7'' \\ \theta_8'' \\ \theta_9'' \\ \theta_{10}'' \\ \theta_{11}'' \\ \theta_{12}'' \\ \delta_1'' \\ \delta_2'' \end{Bmatrix} =
\begin{Bmatrix} -0.0001\ \text{rad} \\ 0\ \text{rad} \\ 0\ \text{rad} \\ 0\ \text{rad} \\ -0.0001\ \text{rad} \\ 0\ \text{rad} \\ 0\ \text{rad} \\ 0\ \text{rad} \\ -0.0022\ \text{rad} \\ -0.0022\ \text{rad} \\ 0\ \text{rad} \\ 0\ \text{rad} \\ -0.0188\ \text{m} \\ -0.0188\ \text{m} \end{Bmatrix},
\begin{Bmatrix} m_1 \\ m_2 \\ m_3 \\ m_4 \\ m_5 \\ m_6 \\ m_7 \\ m_8 \\ m_9 \\ m_{10} \\ m_{11} \\ m_{12} \\ P_1 \\ P_2 \end{Bmatrix} =
\begin{Bmatrix} -150000\ \text{N-m} \\ -53453\ \text{N-m} \\ 3453\ \text{N-m} \\ -36187\ \text{N-m} \\ -150000\ \text{N-m} \\ -53453\ \text{N-m} \\ 3453\ \text{N-m} \\ -36187\ \text{N-m} \\ 50000\ \text{N-m} \\ 50000\ \text{N-m} \\ 36187\ \text{N-m} \\ 36187\ \text{N-m} \\ -32571\ \text{N} \\ -32571\ \text{N} \end{Bmatrix}
\tag{5.83}
$$

Figure 5.22 summarizes of these results in graphical form.

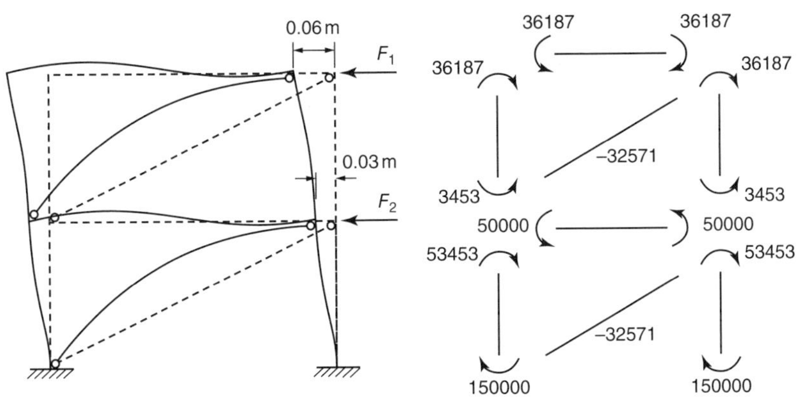

Figure 5.22 Graphical illustration of the responses of the actual CBF.

5.5.3 Force Analogy Method for Dynamical CBFs under Earthquake Loads

Example 5.6 Seismic dynamical analysis for a one-bay one-story CBF

Consider the same illustration presented in Section 5.5.1, but the only difference is that the SDOF CBF is subjected to ground acceleration $\ddot{x}_g(t)$, as shown in Figure 5.23 and the governing equation of motion:

$$M\ddot{x}(t) + C\dot{x}(t) + F(t) = -M\ddot{x}_g(t) \tag{5.84}$$

where $\dot{x}(t)$ and $\ddot{x}(t)$ represent the velocity and acceleration of the system, respectively, m is the system mass, and c is the viscous damping coefficient. Let's discuss how to solve the response of the system under the ground motion using the FAM.

Solution

By performing static condensation using Eqs. (5.57), (5.58) and (5.59), there are

$$\left[\begin{array}{c:c} \mathbf{K}_{dd} & \mathbf{K}_{dr} \\ \hdashline \mathbf{K}_{rd} & \mathbf{K}_{rr} \end{array}\right] = \left[\begin{array}{c:cc} 24EI_c/L_c^3 + EA_b\cos^2\beta/l_b & 6EI_c/L_c^2 & 6EI_c/L_c^2 \\ \hdashline 6EI_c/L_c^2 & 4EI_c/L_c^2 + 4EI_b/L_b & 2EI_b/L_b \\ 6EI_c/L_c^2 & 2EI_b/L_b & 4EI_c/L_c^2 + 4EI_b/L_b \end{array}\right] \tag{5.85}$$

$$\left[\begin{array}{c} \mathbf{K}'_d \\ \hdashline \mathbf{K}'_r \end{array}\right] = \left[\begin{array}{ccccccc} 6EI_c/L_c^2 & 6EI_c/L_c^2 & 6EI_c/L_c^2 & 6EI_c/L_c^2 & 0 & 0 & EA_b\cos^2\beta/l_b \\ \hdashline 2EI_c/L_c^2 & 4EI_c/L_c^2 & 0 & 0 & 4EI_b/L_b^2 & 2EI_b/L_b^2 & 0 \\ 0 & 0 & 2EI_c/L_c^2 & 4EI_c/L_c^2 & 2EI_b/L_b^2 & 4EI_b/L_b^2 & 0 \end{array}\right] \tag{5.86}$$

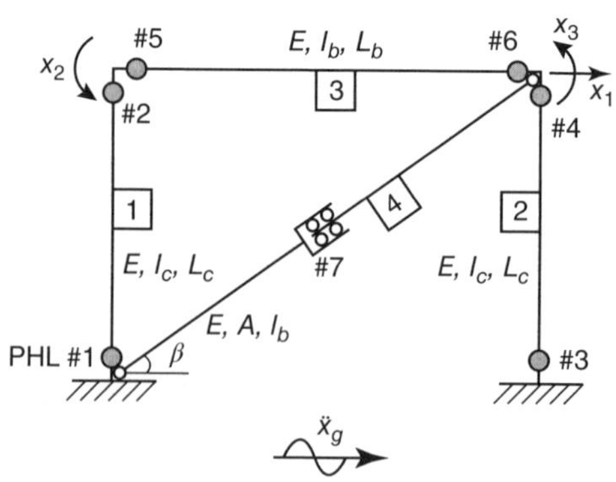

Figure 5.23 One-bay one-story model under earthquakes.

$$
\mathbf{K}'' = \begin{bmatrix} 4EI_c/L_c & 2EI_c/L_c & 0 & 0 & 0 & 0 & 0 \\ 2EI_c/L_c & 4EI_c/L_c & 0 & 0 & 0 & 0 & 0 \\ 0 & 0 & 4EI_c/L_c & 2EI_c/L_c & 0 & 0 & 0 \\ 0 & 0 & 2EI_c/L_c & 4EI_c/L_c & 0 & 0 & 0 \\ 0 & 0 & 0 & 0 & 4EI_b/L_b & 2EI_b/L_b & 0 \\ 0 & 0 & 0 & 0 & 2EI_b/L_b & 4EI_b/L_b & 0 \\ 0 & 0 & 0 & 0 & 0 & 0 & EA_b/l_b \end{bmatrix} \tag{5.87}
$$

According to the static condensation by Eq. (2.149), the condensed stiffness matrix is written as

$$
\begin{aligned} \bar{\mathbf{K}} = \mathbf{K}_{dd} - \mathbf{K}_{dr}\mathbf{K}_{rr}^{-1}\mathbf{K}_{rd} &= \left(24EI_c/L_c^3 + EA_b\cos^2\beta/l_b\right) \\ &- \left[6EI_c/L_c^2 \;\; 6EI_c/L_c^2\right] \begin{bmatrix} 4EI_c/L_c^2 + 4EI_b/L_b & 2EI_b/L_b \\ 2EI_b/L_b & 4EI_c/L_c^2 + 4EI_b/L_b \end{bmatrix}^{-1} \begin{bmatrix} 6EI_c/L_c^2 \\ 6EI_c/L_c^2 \end{bmatrix} \end{aligned} \tag{5.88}
$$

Thus, the governing equation is written as:

$$
\begin{Bmatrix} F(t) \\ \hdashline m_1(t) \\ m_2(t) \\ m_3(t) \\ m_4(t) \\ m_5(t) \\ m_6(t) \\ P(t) \end{Bmatrix} = \left[\begin{array}{c:c} \bar{\mathbf{K}} & \bar{\mathbf{K}}' \\ \hdashline \bar{\mathbf{K}}'^T & \bar{\mathbf{K}}'' \end{array}\right] \begin{bmatrix} x(t) \\ \hdashline -\theta_1''(t) \\ -\theta_2''(t) \\ -\theta_3''(t) \\ -\theta_4''(t) \\ -\theta_5''(t) \\ -\theta_6''(t) \\ -\delta''(t) \end{bmatrix} \tag{5.89}
$$

Substituting the first equation in Eq. (5.89) in (5.84) gives

$$
m\ddot{x}(t) + c\dot{x}(t) + \bar{\mathbf{K}}x(t) = -m\ddot{x}_g(t) + \bar{\mathbf{K}}' \begin{bmatrix} \theta_1''(t) \\ \theta_2''(t) \\ \theta_3''(t) \\ \theta_4''(t) \\ \theta_5''(t) \\ \theta_6''(t) \\ \delta''(t) \end{bmatrix} \tag{5.90}
$$

Using the state space numerical integration method of analysis, Eq. (5.84) is expressed as

$$
\mathbf{z}(t) = \mathbf{A}\mathbf{z}(t) + \mathbf{H}a_g(t) + \mathbf{F}_\mathbf{s}(t) \tag{5.91}
$$

where

$$
\mathbf{z}(t) = \begin{Bmatrix} x(t) \\ x'(t) \end{Bmatrix}, \quad \mathbf{A} = \begin{bmatrix} 0 & 1 \\ -\bar{\mathbf{K}}/m & -c/m \end{bmatrix}, \quad \mathbf{H} = \begin{bmatrix} 0 \\ 1 \end{bmatrix}
$$

$$\mathbf{F}_s(t) = \begin{bmatrix} \frac{1}{m}\bar{\mathbf{K}}' \left[\theta_1''(t) \ \ \theta_2''(t) \ \ \theta_3''(t) \ \ \theta_4''(t) \ \ \theta_5''(t) \ \ \theta_6''(t) \ \ \delta''(t) \right]^T \\ 0 \end{bmatrix}$$

The solution to Eq. (5.91) is

$$\mathbf{Z}(t) = \mathbf{e}^{\mathbf{A}(t-t_0)} z(t_0) + \mathbf{e}^{\mathbf{A}t} \int_{t_o}^{t} \mathbf{e}^{\mathbf{A}s} \left[\mathbf{H}\mathbf{a}_g(s) + \mathbf{F}_s \right] ds \tag{5.92}$$

Let $t_{k+1} = t$, $t_k = t_0$, and $\Delta t = t - t_0$, and represent the ground acceleration vector and the inelastic displacement vector using pulses within the small time step Δt. Performing integration gives

$$\mathbf{Z}(t_{k+!}) = \mathbf{e}^{\mathbf{A}\Delta t} + \mathbf{e}^{\mathbf{A}\Delta t} \mathbf{H} \Delta t \mathbf{a}(t_k) + \mathbf{e}^{\mathbf{A}\Delta t} \mathbf{F}_s \Delta t \tag{5.93}$$

In addition, the moments at the locations where plastic hinges may form is solved through the way similar to the static solution method. Assume that the response at time step t_{k+1} is the same as that at previous step of t_k. Thus, there is

$$\begin{Bmatrix} \widetilde{m}_1(t_{k+1}) \\ \widetilde{m}_2(t_{k+1}) \\ \widetilde{m}_3(t_{k+1}) \\ \widetilde{m}_4(t_{k+1}) \\ \widetilde{m}_5(t_{k+1}) \\ \widetilde{m}_6(t_{k+1}) \\ \widetilde{P}(t_{k+1}) \end{Bmatrix} = \bar{\mathbf{K}}'^T x_1(t_k) - \bar{\mathbf{K}}'' \begin{bmatrix} \theta_1''(t_k) \\ \theta_2''(t_k) \\ \theta_3''(t_k) \\ \theta_4''(t_k) \\ \theta_5''(t_k) \\ \theta_6''(t_k) \\ \delta''(t_k) \end{bmatrix} \tag{5.94}$$

Comparing the temporary moments and axial force at time step of t_{k+1} with the corresponding yield moments and changing axial force of the CBF by tracking loading path, and define the modified moments and axial force as

$$\left[m_1(t_{k+1}) \, m_2(t_{k+1}) \, m_3(t_{k+1}) \, m_4(t_{k+1}) \, m_5(t_{k+1}) \, m_6(t_{k+1}) \, P(t_{k+1}) \right] \tag{5.95}$$

to replace the originals. Substituting Eq. (5.95) back into Eq. (5.89) gives

$$\begin{Bmatrix} F(t_{k+1}) \\ \hline m_1(t_{k+1}) \\ m_2(t_{k+1}) \\ m_3(t_{k+1}) \\ m_4(t_{k+1}) \\ m_5(t_{k+1}) \\ m_6(t_{k+1}) \\ P(t_{k+1}) \end{Bmatrix} = \left[\begin{array}{c|c} \bar{\mathbf{K}} & \bar{\mathbf{K}}' \\ \hline \bar{\mathbf{K}}'^{T} & \bar{\mathbf{K}}'' \end{array} \right] \begin{Bmatrix} x(t_{k+1}) \\ \hline -\theta_1''(t_{k+1}) \\ -\theta_2''(t_{k+1}) \\ -\theta_3''(t_{k+1}) \\ -\theta_4''(t_{k+1}) \\ -\theta_5''(t_{k+1}) \\ -\theta_6''(t_{k+1}) \\ -\delta''(t_{k+1}) \end{Bmatrix} \tag{5.96}$$

The final moments and axial force at time step of t_{k+1} are determined by Eq. (5.96).

Example 5.7 Seismic dynamical analysis for a one-bay two-story CBF

In order to clarify the application of the FAM in multi-story CBFs, now take the one-bay two-story CBF in Section 5.5.2 again as an illustration but the applied force is replaced by the earthquake input as shown in Figure 5.24.

The equations of motion for the two-degree-of-freedom CBF subjected to ground acceleration $\ddot{x}_g(t)$:

$$\mathbf{M}\ddot{\mathbf{X}}(\mathbf{t})+\mathbf{C}\dot{\mathbf{X}}(\mathbf{t})+\mathbf{F}(\mathbf{t})=-\mathbf{M}\boldsymbol{\iota}\ddot{x}_g(t) \tag{5.97}$$

where $\mathbf{M}$ is the mass matrix, $\mathbf{C}$ is the structural damping matrix constructed based on Rayleigh damping, $\boldsymbol{\iota}$ is the influence coefficient vector, $\mathbf{X(t)}$, $\dot{\mathbf{X}}\mathbf{(t)}$ and $\ddot{\mathbf{X}}\mathbf{(t)}$ are the displacement, velocity and acceleration vectors, respectively. The restoring force vector $\mathbf{F(t)}$ is equal to the sum of the restoring force vectors related to the frame and to the brace.

Solution

First, performing the static condensation yields

$$\left[\begin{array}{c:c}\mathbf{K}_{dd} & \mathbf{K}_{dr}\\ \hdashline \mathbf{K}_{rd} & \mathbf{K}_{rr}\end{array}\right]=\left[\begin{array}{cc:cccc}
\frac{48EI_c}{L_c^3}+\frac{2EA\cos^2\beta}{l_b} & \frac{24EI_c}{L_c^2}-\frac{EA_b\cos^2\beta}{l_b} & 0 & 0 & \frac{-6EI_c}{L_c^2} & \frac{-6EI_c}{L_c^2}\\
\frac{24EI_c}{L_c^2}-\frac{EA\cos^2\beta}{l_b} & \frac{24EI_c}{L_c^2}+\frac{EA_b\cos^2\beta}{l_b} & \frac{6EI_c}{L_c^2} & \frac{6EI_c}{L_c^2} & \frac{6EI_c}{L_c^2} & \frac{6EI_c}{L_c^2}\\ \hdashline
0 & 6EI_c/L_c^2 & \frac{8EI_c}{L_c}+\frac{4EI_b}{L_b} & 2EI_b/L_b & 2EI_c/L_c & 0\\
0 & 6EI_c/L_c^2 & 2EI_b/L_b & \frac{8EI_c}{L_c}+\frac{4EI_b}{L_b} & 0 & 2EI_c/L_c\\
-6EI_c/L_c^2 & 6EI_c/L_c^2 & 2EI_c/L_c & 0 & \frac{4EI_c}{L_c}+\frac{4EI_b}{L_b} & 2EI_b/L_b\\
-6EI_c/L_c^2 & 6EI_c/L_c^2 & 0 & 2EI_c/L_c & 2EI_b/L_b & \frac{4EI_c}{L_c}+\frac{4EI_b}{L_b}
\end{array}\right] \tag{5.98}$$

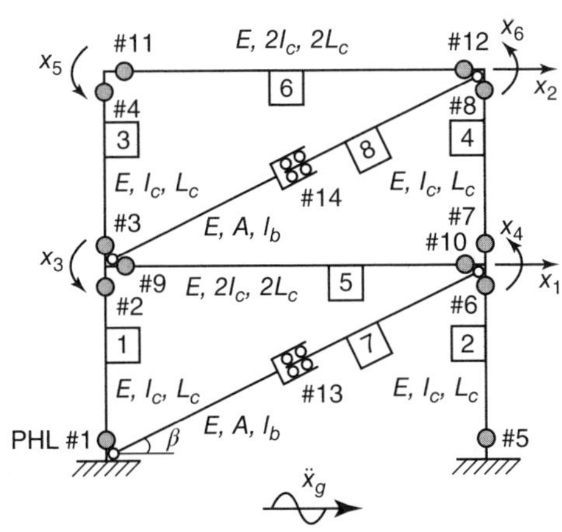

Figure 5.24 A one-bay two-story model under earthquakes.

$$
\begin{bmatrix} \mathbf{K}_d' \\ \hdashline \mathbf{K}_r' \end{bmatrix} = \left[\begin{array}{cccccccccccccc}
\frac{6EI_c}{L_c^2} & \frac{6EI_c}{L_c^2} & -\frac{6EI_c}{L_c^2} & -\frac{6EI_c}{L_c^2} & \frac{6EI_c}{L_c^2} & \frac{6EI_c}{L_c^2} & -\frac{6EI_c}{L_c^2} & -\frac{6EI_c}{L_c^2} & 0 & 0 & 0 & 0 & \frac{EA_b\cos^2\beta}{l_b} & -\frac{EA_b\cos^2\beta}{l_b} \\
0 & 0 & \frac{6EI_c}{L_c^2} & \frac{6EI_c}{L_c^2} & 0 & 0 & \frac{6EI_c}{L_c^2} & \frac{6EI_c}{L_c^2} & 0 & 0 & 0 & 0 & -\frac{EA_b\cos^2\beta}{l_b} & 0 \\
\hdashline
\frac{2EI_c}{L_c} & \frac{4EI_c}{L_c} & \frac{4EI_c}{L_c} & \frac{2EI_c}{L_c} & 0 & 0 & 0 & 0 & \frac{4EI_b}{L_b} & \frac{2EI_b}{L_b} & 0 & 0 & 0 & 0 \\
0 & 0 & 0 & 0 & \frac{2EI_c}{L_c} & \frac{4EI_c}{L_c} & \frac{4EI_c}{L_c} & \frac{2EI_c}{L_c} & \frac{2EI_b}{L_b} & \frac{4EI_b}{L_b} & 0 & 0 & 0 & 0 \\
0 & 0 & \frac{2EI_c}{L_c} & \frac{4EI_c}{L_c} & 0 & 0 & 0 & 0 & 0 & 0 & \frac{4EI_b}{L_b} & \frac{2EI_b}{L_b} & 0 & 0 \\
0 & 0 & 0 & 0 & 0 & 0 & \frac{2EI_c}{L_c} & \frac{4EI_c}{L_c} & 0 & 0 & \frac{2EI_b}{L_b} & \frac{4EI_b}{L_b} & 0 & 0
\end{array}\right] \tag{5.99}
$$

$$
\bar{\mathbf{K}} = \mathbf{K}_{dd} - \mathbf{K}_{dr}\mathbf{K}_{rr}^{-1}\mathbf{K}_{rd} = \begin{bmatrix} \frac{48EI_c}{L_c^3} + \frac{2EA_b\cos^2\beta}{l_b} & -\frac{24EI_c}{L_c^2} - \frac{EA_b\cos^2\beta}{l_b} \\ -\frac{24EI_c}{L_c^2} - \frac{EA_b\cos^2\beta}{l_b} & \frac{24EI_c}{L_c^2} + \frac{EA_b\cos^2\beta}{l_b} \end{bmatrix}
$$

$$
- \begin{bmatrix} 0 & 0 & \frac{-6EI_c}{L_c^2} & \frac{-6EI_c}{L_c^2} \\ \frac{6EI_c}{L_c^2} & \frac{6EI_c}{L_c^2} & \frac{6EI_c}{L_c^2} & \frac{6EI_c}{L_c^2} \end{bmatrix} \begin{bmatrix} \frac{8EI_c}{L_c} + \frac{4EI_b}{L_b} & 2EI_b/L_b & 2EI_c/L_c & 0 \\ 2EI_b/L_b & \frac{8EI_c}{L_c} + \frac{4EI_b}{L_b} & 0 & 2EI_c/L_c \\ 2EI_c/L & 0 & \frac{4EI_c}{L_c} + \frac{4EI_b}{L_b} & 2EI_b/L_b \\ 0 & 2EI_c/L_c & 2EI_b/L_b & \frac{4EI_c}{L_c} + \frac{4EI_b}{L_b} \end{bmatrix}^{-1} \begin{bmatrix} 0 & \frac{6EI_c}{L_c^2} \\ 0 & \frac{6EI_c}{L_c^2} \\ \frac{-6EI_c}{L_c^2} & \frac{6EI_c}{L_c^2} \\ \frac{-6EI_c}{L_c^2} & \frac{6EI_c}{L_c^2} \end{bmatrix} \tag{5.100}
$$

Thus, the governing equation is written as:

$$
\begin{aligned}
\mathbf{F}(\mathbf{t}) &= \begin{Bmatrix} F_1(t) \\ F_2(t) \end{Bmatrix} = \bar{\mathbf{K}} \begin{Bmatrix} x_1(t) \\ x_2(t) \end{Bmatrix} - \bar{\mathbf{K}}' \begin{Bmatrix} \theta_1''(t) \\ \vdots \\ \theta_{12}''(t) \\ \delta_1''(t) \\ \delta_2''(t) \end{Bmatrix}; \\
\mathbf{m}(\mathbf{t}) &= \begin{Bmatrix} m_1 \\ \vdots \\ m_{12} \\ P_1 \\ P_2 \end{Bmatrix} = \bar{\mathbf{K}}'^{\mathrm{T}} \begin{Bmatrix} x_1(t) \\ x_2(t) \end{Bmatrix} - \bar{\mathbf{K}}'' \begin{Bmatrix} \theta_1''(t) \\ \vdots \\ \theta_{12}''(t) \\ \delta_1''(t) \\ \delta_2''(t) \end{Bmatrix}
\end{aligned} \tag{5.101}
$$

Substituting Eq. (5.101) into Eq. (5.97) yields

$$\mathbf{M}\ddot{\mathbf{X}}(\mathbf{t})+\mathbf{C}\dot{\mathbf{X}}(\mathbf{t})+\bar{\mathbf{K}}\mathbf{X}(\mathbf{t})=-\mathbf{M}\iota\ddot{x}_g(t)+\bar{\mathbf{K}}'\begin{Bmatrix}\theta_1''(t)\\ \vdots \\ \theta_{12}''(t)\\ \delta_1''(t)\\ \delta_2''(t)\end{Bmatrix} \tag{5.102}$$

Using the state space numerical integration method of analysis, Eq. (5.102) is expressed as (5.103)

$$\mathbf{z}(t)=\mathbf{A}\mathbf{z}(t)+\mathbf{H}a_g(t)+\mathbf{F}_s(t) \tag{5.103}$$

where

$$\mathbf{z}(t)=\begin{Bmatrix}\mathbf{X}(\mathbf{t})\\ \mathbf{X}'(\mathbf{t})\end{Bmatrix},\quad \mathbf{A}=\begin{bmatrix}0 & 1\\ -\bar{\mathbf{K}}/\mathbf{M} & -\mathbf{C}/\mathbf{M}\end{bmatrix},\quad \mathbf{H}=\begin{bmatrix}0\\ 1\end{bmatrix}$$

$$\mathbf{F}_s(t)=\begin{bmatrix}\bar{\mathbf{K}}'\left[\theta_1''(\mathrm{t})\ \ \theta_2''(\mathrm{t})\ \ \theta_3''(\mathrm{t})\ \ \theta_4''(\mathrm{t})\ \ \theta_5''(\mathrm{t})\ \ \theta_6''(\mathrm{t})\ \ \delta''(\mathrm{t})\right]^{\mathrm{T}}/\mathbf{M}\\ 0\end{bmatrix}$$

The solution to Eq. (5.91) is

$$\mathrm{Z}(t)=e^{A(t-t_0)}z(t0)+e^{At}\int_{t_0}^{t}e^{\mathrm{A}s}\left[\mathrm{Ha_g}(s)+\mathrm{F}_s\right]ds \tag{5.104}$$

Let $t_{k+1}=t$, $t_k=t_0$, and $\Delta t=t\text{–}t_0$, and represent the ground acceleration vector and the inelastic displacement vector using pulses within the small time step Δt. Performing integration gives

$$\mathrm{Z}(t_{k+1})=e^{\mathrm{A}\Delta t}+e^{\mathrm{A}\Delta t}\mathrm{H}\Delta t\mathrm{a}(t_k)+e^{\mathrm{A}\Delta t}F_s\Delta t \tag{5.105}$$

In addition, the moments at the locations where plastic hinges may form is solved through a similar way to the static solution method.

Assume that the response at time step of t_{k+1} is the same as that at the previous step of t_k. Thus, there is

$$\begin{Bmatrix}\widetilde{m}_1(t_{k+1})\\ \vdots \\ \widetilde{m}_{12}(t_{k+1})\\ \widetilde{P}_1(t_{k+1})\\ \widetilde{P}_2(t_{k+1})\end{Bmatrix}=\bar{\mathbf{K}}'^{\mathrm{T}}\begin{Bmatrix}x_1(t_k)\\ x_2(t_k)\end{Bmatrix}-\bar{\mathbf{K}}''\begin{Bmatrix}\theta_1''(t_k)\\ \vdots \\ \theta_{12}''(t_k)\\ \delta_1''(t_k)\\ \delta_2''(t_k)\end{Bmatrix} \tag{5.106}$$

Comparing the temporary moments and axial force at time step t_{k+1} with the corresponding yield moments and changing axial force of the CBF by tracking loading path, and define the modified moments and axial force as

$$\left[m_1(t_{k+1}) \quad \cdots \quad m_{12}(t_{k+1}) \quad P_1(t_{k+1}) \quad P_2(t_{k+1}) \right] \tag{5.107}$$

replace the originals. Substituting Eq. (5.95) back into Eq. (5.89) gives

$$\begin{Bmatrix} F_1(t_{k+1}) \\ F_2(t_{k+1}) \\ \hdashline m_1(t_{k+1}) \\ \vdots \\ m_{12}(t_{k+1}) \\ P_1(t_{k+1}) \\ P_2(t_{k+1}) \end{Bmatrix} = \begin{bmatrix} \bar{\mathbf{K}} & \bar{\mathbf{K}}' \\ \bar{\mathbf{K}}'^{\mathrm{T}} & \bar{\mathbf{K}}'' \end{bmatrix} \begin{Bmatrix} x_1(t_k) \\ x_2(t_k) \\ \hdashline \theta_1''(t_k) \\ \vdots \\ \theta_{12}''(t_k) \\ \delta_1''(t_k) \\ \delta_2''(t_k) \end{Bmatrix} \tag{5.108}$$

The final moments and resistant force at time step of t_{k+1} are determined by Eq. (5.96).

References

Dicleli M. and Calik E.E. (2008). "Physical theory hysteretic model for steel braces." *Journal of Structural Engineering*, 134(7): 1215–1228.

Li H.N., Li G. and Fahnestock L.A. (2013). "Simulation of steel brace hysteretic response using the force analogy method." *Journal of Structural Engineering ASCE*, 139(4): 526–536.

6

Shear Member

The main role of the reinforced concrete (RC) shear wall members is to enhance the global stiffness of building structures, and this type of structural member is widely used especially in tall buildings. In this chapter, the macro model presented by *Linde* and *Bachmann* (1994) is selected as a baseline model to establish an improved shear wall numerical model that can be incorporated in the force analogy method (FAM). The nonlinear behavior of the RC shear wall member under static and dynamic loads is interpreted. Besides, the multi-layer RC frame-shear wall structures are selected to address the procedures of nonlinear static and dynamic analyses with the FAM.

6.1 Physical Theory Models of Shear Members

A macro model shown in Figure 6.1 for simulating the hysteretic response of shear wall members presented by *Linde* and *Bachmann* (1994) is selected here as a baseline model to develop an improved version for incorporating it in the force analogy method (FAM). The model shown in Figure 6.1 consists of four springs that are connected with rigid beams. The model properties are derived based on the elastic theory for cantilever shear wall members and the nonlinear physical properties that are based on simple nonlinear physical cross-sectional behavior and empirical data.

6.1.1 Flexural Behavior

The flexural behavior of the shear wall member is governed by the two outer vertical springs with stiffness K^v as shown in Figure 6.1. In addition, several general parameters representing

Theory of Nonlinear Structural Analysis: The Force Analogy Method for Earthquake Engineering,
First Edition. Gang Li and Kevin K.F. Wong.

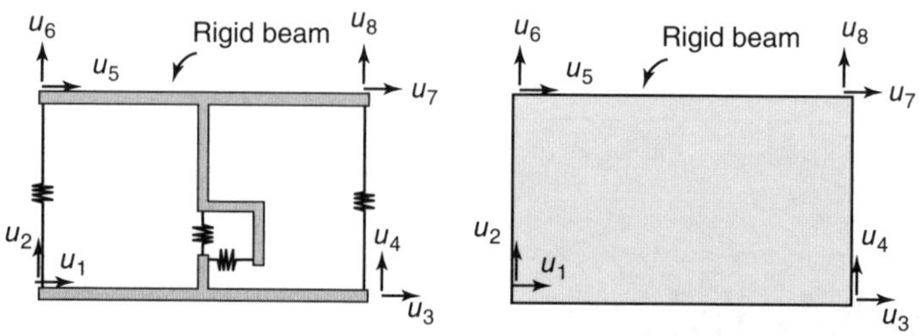

Figure 6.1 The macro model proposed by *Linde* and *Bachmann* (1994).

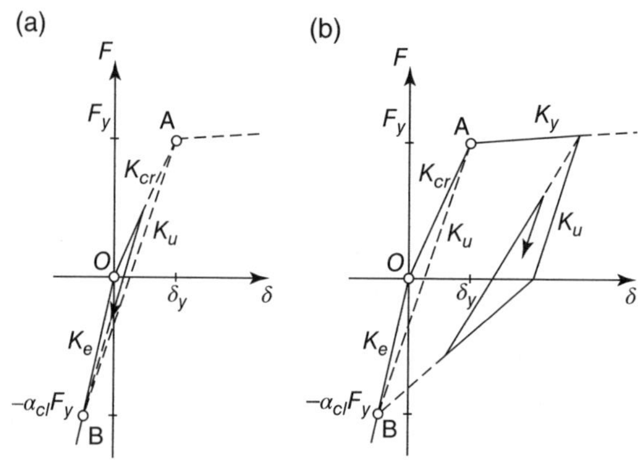

Figure 6.2 The spring force versus displacement relationship of the outer vertical springs: (a) Small cycle; (b) Big cycle.

the shear wall members' characteristics are: the sectional area A, the moment of inertia for the section I, the length L_w, and the height h_w. The elastic modulus E of the member is used for calculating the stiffnesses of the springs. The relationship of the spring force versus displacement is illustrated in Figure 6.2, in which the initial stiffness in tension is written by

$$K_{cr} = \alpha_w \frac{A_s E}{h_w} \tag{6.1}$$

in which, α_w represents the stiffness proportion taken by the outer vertical springs, and is given by

$$\alpha_w = \frac{2I}{AL_w^2} \tag{6.2}$$

The initial stiffness K_{cr} is the stiffness of the RC shear wall member after cracking.

The compressive stiffness is satisfied by

$$K_e = \frac{A_s E}{h_w} \tag{6.3}$$

where A_s is the spring area and can be obtained as

$$A_s = \frac{2I}{L_w^2} \tag{6.4}$$

In addition, the stiffness of the post-yield range equals

$$K_y = 0.01K_e \tag{6.5}$$

As shown in Figure 6.2, the stiffness in the unloading range equals to the slope of the line AB, in which Point A represents the yielding status in tension with yielding strength F_y and displacement δ_y, and Point B denotes the status of flexural cracks closed by multiplying a factor a_{cl} with F_y. Moreover, the stiffness during unloading changes at the zero-force level and becomes smaller than K_u in Figure 6.2 and the response is target at the point B.

6.1.2 Axial Behavior

The stiffness of the central vertical spring of the model can be expressed by

$$K^c = EA\frac{(1-2\alpha_w)}{h_w} \tag{6.6}$$

6.1.3 Shear Behavior

The horizontal spring for simulating the shear behavior is often assumed to be located at round $h_c = h_w/3$ from the bottom of the model. The spring force versus displacement relationship is shown in Figure 6.3 and its stiffness can be expressed as

$$K^s = \frac{GA}{h_w\kappa} \tag{6.7}$$

where κ is a form factor.

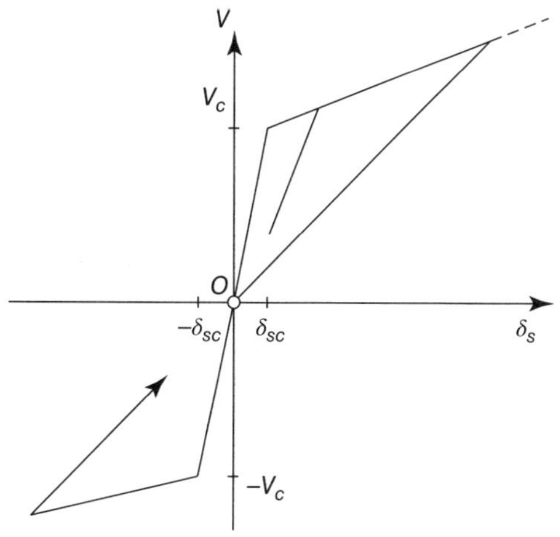

Figure 6.3 Spring force versus displacement relationship of the horizontal spring.

6.2 Local Plastic Mechanisms in the FAM

In order to simulate the inelastic response of shear wall members in the FAM, an improved model including three local plastic mechanisms is developed here based on the above theory. In this model, assume that the two rigid beams and the central vertical spring shown in Figure 6.1 are axially rigid and the axial deformations of the two outer vertical springs are equivalent to a rotation degree of freedom in the RC shear wall member. Therefore, totally four degrees of freedom, including two horizontal and two rotational degrees of freedom, x_1, x_2, x_3, and x_4, are retained, as shown in Figure 6.4. Then, the element stiffness matrices can be written as

$$\mathbf{K}^{\mathbf{e}} = \begin{bmatrix} K^s & -K^s(h_w-h_c) & -K^s & -K^sh_c \\ -K^s(h_w-h_c) & K^s(h_w-h_c)^2+K_e\dfrac{l_w^2}{2} & K^s(h_w-h_c) & K^s(h_w-h_c)h_c-K_e\dfrac{l_w^2}{2} \\ -K^s & K^s(h_w-h_c) & K^s & K^sh_c \\ -K^sh_w & K^s(h_w-h_c)h_c-K_e\dfrac{l_w^2}{2} & K^sh_c & K^sh_c^2+K_e\dfrac{l_w^2}{2} \end{bmatrix} \tag{6.8}$$

in which l_w is the distance between the two outer vertical springs.

The improved numerical model is illustrated in Figure 6.4, in which there are three local plastic mechanisms, including two vertical sliding hinges (VSHs) and one horizontal sliding hinge (HSH).

6.2.1 Displacement Decomposition

Based on the FAM, the displacements of each spring in Figure 6.2 and Figure 6.3 are expressed as the sum of the elastic components and plastic components. Thus, there are

$$\delta_1 = \delta_1' + \delta_1'' \tag{6.9}$$

$$\delta_2 = \delta_2' + \delta_2'' \tag{6.10}$$

$$\delta_3 = \delta_3' + \delta_3'' \tag{6.11}$$

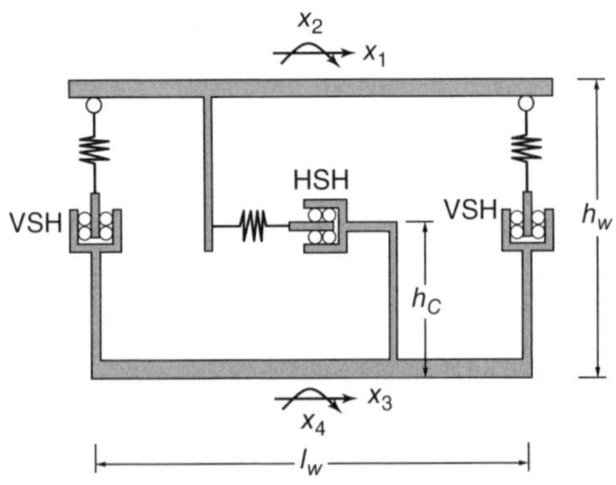

Figure 6.4 Shear wall member model in the FAM.

in which, δ_s, δ_s' and δ_s'' represent the total displacement, elastic displacement and plastic displacement of the sth spring, respectively. Thus, each spring force can be written as:

$$F_1 = K_e\delta_1' = K_e(\delta_1 - \delta_1'') \quad (6.12)$$

$$F_2 = K_e\delta_2' = K_e(\delta_2 - \delta_2'') \quad (6.13)$$

$$F_3 = K^s\delta_3' = K^s(\delta_3 - \delta_3'') \quad (6.14)$$

Although it can be observed from Figure 6.2 that the initial stiffness in the tension branch, K_{cr}, is different from that in compression branch, K_e, the stiffness in tension branch is assumed as K_e because a constant initial elastic stiffness is necessary in the FAM. Thus, the line OA is thought of as the nonlinear state in the improved model. In addition, one difference in the improved model from the origin model (*Linde* and *Bachmann*) is that the elastic and plastic displacements are generated from the corresponding spring and sliding hinges (VSHs and HSH), respectively.

6.2.2 Behavior of VSH

The outer vertical springs are subjected to the same magnitude of force F_1 but with opposite directions, $F_1 = -F_2$. Based on the relationship of the total displacement δ_1 versus the spring force F_1 in Figure 6.2, the relationship of spring forces F_1 and F_2 versus plastic displacement δ_1'' and δ_2'', respectively, can be derived as illustrated in Figure 6.5. With the known information of K_e, K_{cr}, K_y, K_u, F_y, and α_{cl}, the following rules can be obtained.

(1) Region O–C
$$F_1 = \frac{K_eK_{cr}}{K_e - K_{cr}}\delta_1'' \quad (6.15)$$

(2) Region A–B
$$F_1 = \frac{F_A + \alpha_{cl}F_y}{\delta_{1A}''}(\delta_1'' - \delta_{1A}'') + F_A \quad (6.16)$$

(3) Region C–D
$$F_1 = \frac{K_eK_y}{K_e - K_y}\left(\delta_1'' - F_y\frac{K_e - K_{cr}}{K_eK_{cr}}\right) + F_y \quad (6.17)$$

(4) Region D–E
$$F_1 = \frac{K_eK_u}{K_e - K_u}(\delta_1'' - \delta_{1D}'') + F_D \quad (6.18)$$

(5) Region E–B
$$F_1 = \frac{\alpha_{cl}F_yK_eK_u}{K_eK_u\delta_{1D}'' - F_D(K_e - K_u)}\left(\delta_1'' + \frac{F_D(K_e - K_u)}{K_eK_u} - \delta_{1D}''\right) \quad (6.19)$$

(6) Region F–D
$$F_1 = \frac{F_F - F_D}{\delta_F'' - \delta_D''}(\delta_1'' - \delta_{1F}'') + F_F \quad (6.20)$$

(7) Region G–H
$$F_1 = \frac{K_eK_u}{K_e - K_u}(\delta_1'' - \delta_{1G}'') + F_G \quad (6.21)$$

(8) Region B–D
$$F_1 = \frac{\alpha_{cl}F_y - F_D}{-\delta_D''}\delta_1'' + F_B \quad (6.22)$$

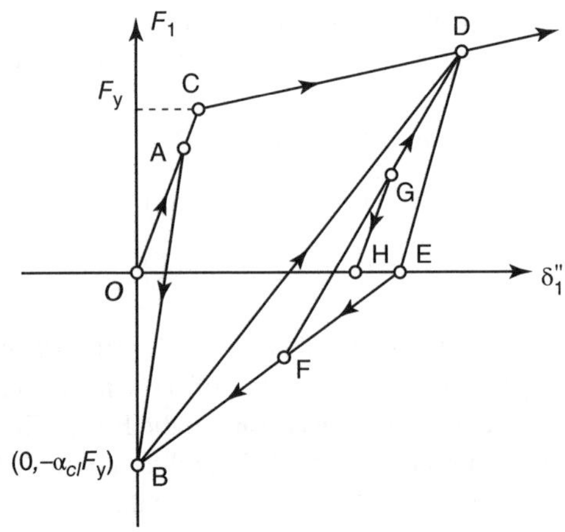

Figure 6.5 The VSH hysteretic behavior.

in which the subscripts of F, such as A and B, etc. represent the spring force at the corresponding points as shown in Figure 6.5.

6.2.3 Behavior of HSH

Due to the horizontal spring being connected with the HSH in series, they are acted on by the same applied force F_3. Based on the relationship of the total displacement δ_3 and spring force F_3 shown in Figure 6.3, the relationship of spring force F_3 versus plastic displacement δ_3'' can be derived as shown in Figure 6.6 and satisfied by the following rules. With the known information of K^s and K_{scr}, there are:

(1) Region A–B

$$F_3 = \frac{K^s K_{scr}}{K^s - K_{scr}} \delta_3'' \tag{6.23}$$

(2) Region B–O

$$F_3 = \frac{K^s K_{BO}}{K^s - K_{BO}} (\delta_3'' - \delta_{3B}'') + F_B \tag{6.24}$$

(3) Region C–D

$$F_3 = \frac{K^s K_{scr}}{K^s - K_{scr}} \delta_3'' \tag{6.25}$$

(4) Region D–O

$$F_3 = \frac{K^s K_{DO}}{K^s - K_{DO}} (\delta_3'' - \delta_{3D}'') + F_D \tag{6.26}$$

in which the subscripts of F, such as A and D represent the spring force at the corresponding points; the subscripts of K, such as DO and BO represent the stiffnesses of that branches as shown in Figure 6.6.

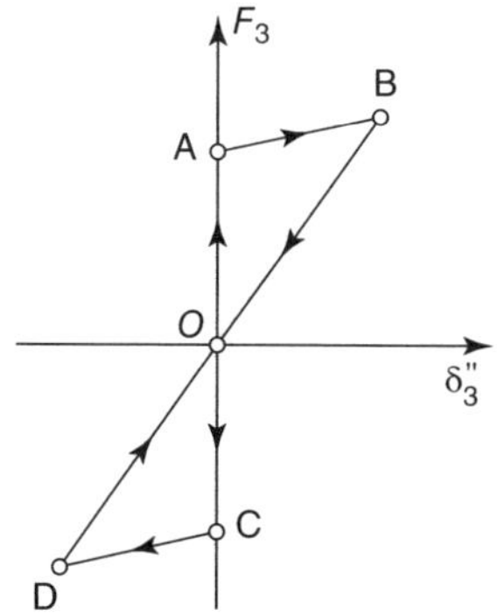

Figure 6.6 The HSH hysteretic behavior.

6.3 Nonlinear Static Analysis of the Shear Wall Structures

A RC shear wall member subjected to a lateral applied force F as shown in Figure 6.7 is used to demonstrate how to use the concept of the FAM to perform the static analysis for shear wall members. The shear wall member has two degrees of freedom labeled as x and θ, and two VSHs and one HSH. According to Eq. (6.8), the global stiffness matrices are given as:

$$\mathbf{K} = \begin{bmatrix} K^s & -K^s(h_w - h_c) \\ -K^s(h_w - h_c) & K^s(h_w - h_c)^2 + K_e \dfrac{l_w^2}{2} \end{bmatrix} \begin{matrix} \leftarrow x \\ \leftarrow \theta \end{matrix} \tag{6.27}$$

$$\mathbf{K}' = \begin{bmatrix} 0 & 0 & K^s \\ K_e \dfrac{l_w}{2} & -K_e \dfrac{l_w}{2} & -K^s(h_w - h_c) \end{bmatrix} \begin{matrix} \leftarrow x_1 \\ \leftarrow x_2 \end{matrix} \tag{6.28}$$

$$\mathbf{K}'' = \begin{bmatrix} \dfrac{K_e}{2} & -\dfrac{K_e}{2} & 0 \\ -\dfrac{K_e}{2} & \dfrac{K_e}{2} & 0 \\ 0 & 0 & K^s \end{bmatrix} \begin{matrix} \leftarrow \delta_1'' \\ \leftarrow \delta_2'' \\ \leftarrow \delta_3'' \end{matrix} \tag{6.29}$$

Thus, the governing equation of the FAM for this RC shear wall member becomes:

$$\left[\begin{array}{cc|ccc} K^s & -K^s(h_w - h_c) & 0 & 0 & K^s \\ -K^s(h_w - h_c) & K^s(h_w - h_c)^2 + K_e \dfrac{l_w^2}{2} & K_e \dfrac{l_w}{2} & -K_e \dfrac{l_w}{2} & -K^s(h_w - h_c) \\ \hline 0 & K_e \dfrac{l_w}{2} & \dfrac{K_e}{2} & -\dfrac{K_e}{2} & 0 \\ 0 & -K_e \dfrac{l_w}{2} & -\dfrac{K_e}{2} & \dfrac{K_e}{2} & 0 \\ K^s & -K^s(h_w - h_c) & 0 & 0 & K^s \end{array}\right] \left\{\begin{array}{c} x \\ \theta \\ \hline -\delta_1'' \\ -\delta_2'' \\ -\delta_3'' \end{array}\right\} = \left\{\begin{array}{c} F \\ M \\ \hline f_1 \\ f_2 \\ f_3 \end{array}\right\} \tag{6.30}$$

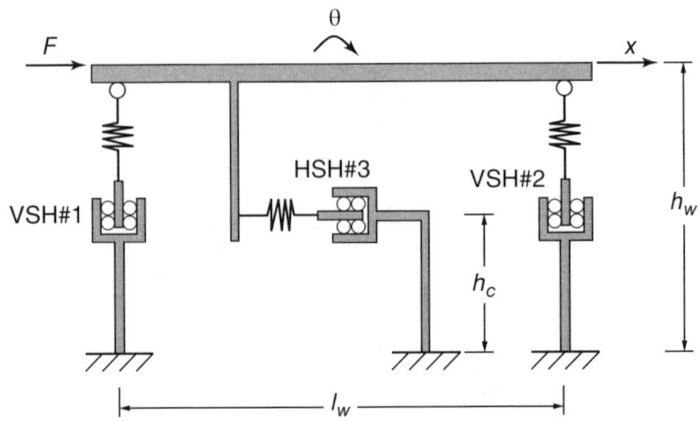

Figure 6.7 The numerical model of RC shear wall.

or simply

$$\begin{bmatrix} \mathbf{K} & \mathbf{K}' \\ \mathbf{K}'^T & \mathbf{K}'' \end{bmatrix} \begin{Bmatrix} \mathbf{x} \\ -\boldsymbol{\Delta}'' \end{Bmatrix} = \begin{Bmatrix} \mathbf{F} \\ \mathbf{f} \end{Bmatrix} \tag{6.31}$$

If the shear wall member illustrated in Figure 6.7 remains in elasticity when subjecting to the lateral force F, the response is linear, i.e. $\delta_1'' = \delta_2'' = \delta_3'' = 0$. Therefore, Eq. (6.30) becomes

$$\begin{bmatrix} K^s & -K^s(h_w-h_c) & 0 & 0 & K^s \\ -K^s(h_w-h_c) & K^s(h_w-h_c)^2 + K_e\dfrac{l^2}{2} & K_e\dfrac{l_w}{2} & -K_e\dfrac{l_w}{2} & -K^s(h_w-h_c) \\ 0 & K_e\dfrac{l_w}{2} & \dfrac{K_e}{2} & -\dfrac{K_e}{2} & 0 \\ 0 & -K_e\dfrac{l_w}{2} & -\dfrac{K_e}{2} & \dfrac{K_e}{2} & 0 \\ K^s & -K^s(h_w-h_c) & 0 & 0 & K^s \end{bmatrix} \begin{Bmatrix} x \\ \theta \\ 0 \\ 0 \\ 0 \end{Bmatrix} = \begin{Bmatrix} F \\ M \\ f_1 \\ f_2 \\ f_3 \end{Bmatrix} \tag{6.32}$$

and the elastic response of the RC shear wall member can be computed by Eq. (6.32).

Example 6.1 Nonlinear Static Analysis of the Shear Wall Structure with Input Force

Consider a shear wall structure as shown in Figure 6.8(a) with $l_w = 6$ m and $h_w = 4$ m. The sectional parameters are shown in Figure 6.8(b) and the numerical parameters are provided in Table 6.1. Let us attempt to solve the response of the shear wall structure subjected to the lateral force $F = 5$ MN in the FAM.

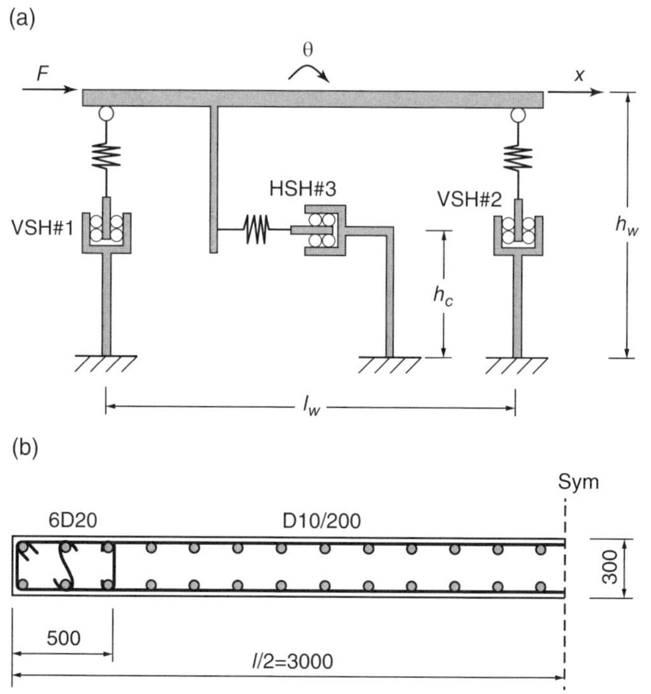

Figure 6.8 Numerical model: (a) Basic information; (b) Sectional parameters.

Solution

The global stiffness matrices, **K**, **K′**, and **K″**, of the shear wall structure can be written as

$$\mathbf{K} = \begin{bmatrix} K^s & -K^s(h_w - h_c) \\ -K^s(h_w - h_c) & K^s(h_w - h_c)^2 + K_e \dfrac{l_w^{\,2}}{2} \end{bmatrix} \tag{6.33}$$

$$\mathbf{K}' = \begin{bmatrix} 0 & 0 & K^s \\ K_e \dfrac{l_w}{2} & -K_e \dfrac{l_w}{2} & -K^s(h_w - h_c) \end{bmatrix} \tag{6.34}$$

$$\mathbf{K}'' = \begin{bmatrix} \dfrac{K_e}{2} & -\dfrac{K_e}{2} & 0 \\ -\dfrac{K_e}{2} & \dfrac{K_e}{2} & 0 \\ 0 & 0 & K^s \end{bmatrix} \tag{6.35}$$

Then, the governing equation can be written as:

$$\left[\begin{array}{cc:ccc} 2500 & -6500 & 0 & 0 & 2500 \\ -6500 & 43900 & 4500 & -4500 & -6500 \\ \hdashline 0 & 4500 & 750 & -750 & 0 \\ 0 & -4500 & -750 & 750 & 0 \\ 2500 & -6500 & 0 & 0 & 2500 \end{array}\right]\left\{\begin{array}{c} x \\ \theta \\ \hdashline -\delta_1'' \\ -\delta_2'' \\ -\delta_3'' \end{array}\right\}=\left\{\begin{array}{c} 5 \\ 0 \\ \hdashline f_1 \\ f_2 \\ f_3 \end{array}\right\} \tag{6.36}$$

First, assume that the shear wall structure is in the elastic domain, this gives $\delta_1''=\delta_2''=0$ and $\delta_3''=0$. Then, the displacement x, rotation θ, spring force f_1, f_2 and f_3 can be calculated using Eq. (6.36):

$$\left\{\begin{array}{c} x \\ \theta \end{array}\right\}=\left\{\begin{array}{l} 0.0033\ \text{m} \\ 0.0005\ \text{rad} \end{array}\right\},\quad \left\{\begin{array}{c} f_1 \\ f_2 \\ f_3 \end{array}\right\}=\left\{\begin{array}{c} 2.1667 \\ -2.1667 \\ 5.0000 \end{array}\right\}\text{MN} \tag{6.37}$$

Comparing these demands in Eq. (6.37) with the cracking and yielding capacities provided in Table 6.1, the shear wall structure is in the plastic domain, which means the assumption that the shear wall structure is in the elastic domain is incorrect. Therefore, assume that the structure is in the plastic domain, which means that the VSH#1 and HSH#3 begin to deform in their hardening branches. For local plastic mechanisms VSH#1 and HSH#3, the stiffness $K_e = 1500$ MN/m, $K_y = 15$ MN/m, $K_{cr} = 1050$ MN/m, $F_y = 1.32$ MN/m, $K^s = 2500$ MN/m and $K_{scr} = 400$ MN/m can be obtained or computed according to the parameters listed in Table 6.1. Then, the spring forces f_1 and f_3 can be expressed in terms of plastic displacements δ_1'', δ_3'' according to Eq. (6.17) and Eq. (6.23):

$$\begin{cases} f_1 = 15.15\,\delta_1'' + 1.314 \\ f_3 = 476.19\,\delta_3'' + 3.00 \end{cases} \tag{6.38}$$

Therefore, Eq. (6.36) becomes:

Table 6.1 The Parameters of the shear wall structure.

Parameters	Values
Cross sectional area, A	1.8 m^2
Cross sectional moment of inertia, I	5.4 m^4
Young's modulus, E	20 GPa
Yield moment, M_y	7.9 MN-m
Yield force of vertical spring, F_y	1.32 MN
Shear force at shear cracking, V_c	3.0 MN
Height of plastic hinge, h_c	1.4 m
stiffness proportion, α_{cr}	0.7
stiffness proportion, α_y	0.01
stiffness proportion, α_s	0.16
Uncracked stiffness of vertical spring, K_e	1500 MN/m
Uncracked stiffness of shear spring, K^s	2500 MN/m

$$\left[\begin{array}{cc:cc} 2500 & -6500 & 0 & 2500 \\ -6500 & 43900 & 4500 & -6500 \\ \hdashline 0 & 4500 & 750+15.15 & 0 \\ 2500 & -6500 & 0 & 2500+476.19 \end{array}\right]\left\{\begin{array}{c} x \\ \theta \\ \hdashline -\delta_1'' \\ -\delta_3'' \end{array}\right\}=\left\{\begin{array}{c} 5 \\ 0 \\ \hdashline 1.314 \\ 3 \end{array}\right\} \tag{6.39}$$

Solving for Eq. (6.39) gives:

$$\begin{Bmatrix} x \\ \theta \end{Bmatrix}=\begin{Bmatrix} 0.0318\ \text{m} \\ 0.0099\ \text{rad} \end{Bmatrix}, \begin{Bmatrix} \delta_1'' \\ \delta_3'' \end{Bmatrix}=\begin{Bmatrix} 0.0563\ \text{m} \\ 0.0042\ \text{m} \end{Bmatrix} \tag{6.40}$$

Substituting Eq. (6.40) back into Eq. (6.36) and solving for the spring forces gives

$$\begin{Bmatrix} x \\ \theta \end{Bmatrix}=\begin{Bmatrix} 0.0318\ \text{m} \\ 0.0099\ \text{rad} \end{Bmatrix}, \begin{Bmatrix} \delta_1'' \\ \delta_3'' \end{Bmatrix}=\begin{Bmatrix} 0.0563\ \text{m} \\ 0.0042\ \text{m} \end{Bmatrix}, \begin{Bmatrix} f_1 \\ f_2 \\ f_3 \end{Bmatrix}=\begin{Bmatrix} 2.1667\ \text{MN} \\ -2.1667\ \text{MN} \\ 5.0000\ \text{MN} \end{Bmatrix} \tag{6.41}$$

Example 6.2 Nonlinear Static Analysis of the One-Story RC Frame-Shear Wall Structure

Consider the one-story RC frame-shear wall structure as shown in Figure 6.9. The frame and the shear wall connected by the linking beam are satisfied by the compatibility condition. There are 6 DOFs (5 translations and 1 rotation), 14 RHs, 2 VSHs and 1 HSH in the structure. The numerical properties of the frame are provided in Table 6.2, while the shear wall member expressed in Example 6.1 is used in this RC frame-shear wall structure. The RC frame-shear wall structure is subjected to the input displacement $X_1 = 0.03$ m at the horizontal degree of freedom x_1.

Solution

First, for the frame part, the global stiffness matrices, $\mathbf{K_f}$, $\mathbf{K_f'}$, and $\mathbf{K_f''}$, can be assembled as

$$\mathbf{K_f}=\begin{bmatrix} 129 & -90 & -39 & -39 & -90 \\ -90 & 312 & 36 & 0 & 0 \\ -39 & 36 & 211 & 17 & 0 \\ -39 & 0 & 17 & 211 & 36 \\ -90 & 0 & 0 & 36 & 312 \end{bmatrix}\begin{matrix} \leftarrow x_1 \\ \leftarrow x_2 \\ \leftarrow x_3 \\ \leftarrow x_4 \\ \leftarrow x_5 \end{matrix}\ \text{MN/m} \tag{6.42}$$

$$\mathbf{K_f'}=\begin{bmatrix} -90 & -90 & -39 & -39 & -39 & -39 & -90 & -90 & 0 & 0 & 0 & 0 & 0 & 0 \\ 240 & 120 & 0 & 0 & 0 & 0 & 0 & 0 & 72 & 36 & 0 & 0 & 0 & 0 \\ 0 & 0 & 104 & 52 & 0 & 0 & 0 & 0 & 36 & 72 & 34.67 & 17.33 & 0 & 0 \\ 0 & 0 & 0 & 0 & 104 & 52 & 0 & 0 & 0 & 0 & 17.33 & 34.67 & 72 & 36 \\ 0 & 0 & 0 & 0 & 0 & 0 & 240 & 120 & 0 & 0 & 0 & 0 & 36 & 72 \end{bmatrix}\begin{matrix} \leftarrow x_1 \\ \leftarrow x_2 \\ \leftarrow x_3 \\ \leftarrow x_4 \\ \leftarrow x_5 \end{matrix}\ \text{MN/m} \tag{6.43}$$

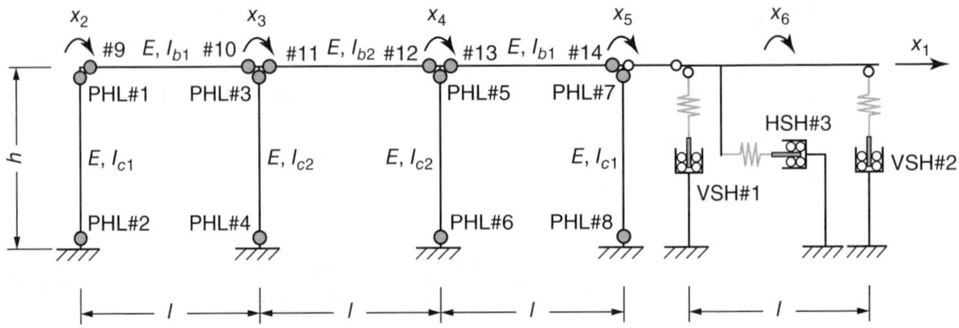

Figure 6.9 Numerical model.

Table 6.2 The numerical Parameters of the RC frame-shear wall structure.

Parameters	Values
Column c1 yield moment, M_{pc1}	1.25 MN-m
Column c2 yield moment, M_{pc2}	0.95 MN-m
Beam b1 yield moment, M_{pb1}	0.75 MN-m
Beam b2 yield moment, M_{pb2}	0.45 MN-m
Cross sectional moment of inertia, I_{c1}	0.012 m^4
Cross sectional moment of inertia, I_{c2}	0.0052 m^4
Cross sectional moment of inertia, I_{b1}	0.0054 m^4
Cross sectional moment of inertia, I_{b2}	0.0026 m^4
Young's modulus, E	20 GPa

$$\mathbf{K}_\mathrm{f}'' = \begin{bmatrix}
240 & 120 & 0 & 0 & 0 & 0 & 0 & 0 & 0 & 0 & 0 & 0 & 0 & 0 \\
120 & 240 & 0 & 0 & 0 & 0 & 0 & 0 & 0 & 0 & 0 & 0 & 0 & 0 \\
0 & 0 & 104 & 52 & 0 & 0 & 0 & 0 & 0 & 0 & 0 & 0 & 0 & 0 \\
0 & 0 & 52 & 104 & 0 & 0 & 0 & 0 & 0 & 0 & 0 & 0 & 0 & 0 \\
0 & 0 & 0 & 0 & 104 & 52 & 0 & 0 & 0 & 0 & 0 & 0 & 0 & 0 \\
0 & 0 & 0 & 0 & 52 & 104 & 0 & 0 & 0 & 0 & 0 & 0 & 0 & 0 \\
0 & 0 & 0 & 0 & 0 & 0 & 240 & 120 & 0 & 0 & 0 & 0 & 0 & 0 \\
0 & 0 & 0 & 0 & 0 & 0 & 120 & 240 & 0 & 0 & 0 & 0 & 0 & 0 \\
0 & 0 & 0 & 0 & 0 & 0 & 0 & 0 & 72 & 36 & 0 & 0 & 0 & 0 \\
0 & 0 & 0 & 0 & 0 & 0 & 0 & 0 & 36 & 72 & 0 & 0 & 0 & 0 \\
0 & 0 & 0 & 0 & 0 & 0 & 0 & 0 & 0 & 0 & 34.67 & 17.33 & 0 & 0 \\
0 & 0 & 0 & 0 & 0 & 0 & 0 & 0 & 0 & 0 & 17.33 & 34.67 & 0 & 0 \\
0 & 0 & 0 & 0 & 0 & 0 & 0 & 0 & 0 & 0 & 0 & 0 & 72 & 36 \\
0 & 0 & 0 & 0 & 0 & 0 & 0 & 0 & 0 & 0 & 0 & 0 & 36 & 72
\end{bmatrix}
\begin{matrix}
\leftarrow \theta_1'' \\ \leftarrow \theta_2'' \\ \leftarrow \theta_3'' \\ \leftarrow \theta_4'' \\ \leftarrow \theta_5'' \\ \leftarrow \theta_6'' \\ \leftarrow \theta_7'' \\ \leftarrow \theta_8'' \\ \leftarrow \theta_9'' \\ \leftarrow \theta_{10}'' \\ \leftarrow \theta_{11}'' \\ \leftarrow \theta_{12}'' \\ \leftarrow \theta_{13}'' \\ \leftarrow \theta_{14}''
\end{matrix}
\ \mathrm{MN/m} \tag{6.44}$$

The governing equation of the FAM for the frame part can be written as:

$$\left[\begin{array}{c:c}\mathbf{K}_{\mathbf{f}5\times5} & \mathbf{K}'_{\mathbf{f}5\times14}\\ \hdashline \mathbf{K}'^{T}_{\mathbf{f}14\times5} & \mathbf{K}''_{\mathbf{f}14\times14}\end{array}\right]\left\{\begin{array}{c}\mathbf{x}_{5\times1}\\ \hdashline -\boldsymbol{\theta}''_{14\times1}\end{array}\right\}=\left\{\begin{array}{c}\mathbf{F}_{5\times1}\\ \hdashline \mathbf{m}_{14\times1}\end{array}\right\} \tag{6.45}$$

in which,

$$\mathbf{x}=\begin{Bmatrix}0.03\\x_2\\x_3\\x_4\\x_5\end{Bmatrix},\qquad \mathbf{F}=\begin{Bmatrix}F_f\\0\\0\\0\\0\end{Bmatrix} \tag{6.46}$$

To analyze the structural behavior subjected to the input displacement X_1, first assume that the structural response is linear, i.e. $\theta''_1=\theta''_2=\theta''_3=\theta''_4=\theta''_5=\theta''_6=\theta''_7=\theta''_8=\theta''_9=\theta''_{10}=\theta''_{11}=\theta''_{12}=\theta''_{13}=\theta''_{14}=0$. Therefore, the governing equation of Eq. (6.45) becomes:

$$\begin{bmatrix}129 & -90 & -39 & -39 & -90\\ -90 & 312 & 36 & 0 & 0\\ -39 & 36 & 211 & 17 & 0\\ -39 & 0 & 17 & 211 & 36\\ -90 & 0 & 0 & 36 & 312\end{bmatrix}\begin{Bmatrix}0.03\\x_2\\x_3\\x_4\\x_5\end{Bmatrix}=\begin{Bmatrix}F_f\\0\\0\\0\\0\end{Bmatrix} \tag{6.47}$$

$$\begin{bmatrix}-90 & 240 & 0 & 0 & 0\\ -90 & 120 & 0 & 0 & 0\\ -39 & 0 & 104 & 0 & 0\\ -39 & 0 & 52 & 0 & 0\\ -39 & 0 & 0 & 104 & 0\\ -39 & 0 & 0 & 52 & 0\\ -90 & 0 & 0 & 0 & 240\\ -90 & 0 & 0 & 0 & 120\\ 0 & 72 & 36 & 0 & 0\\ 0 & 36 & 72 & 0 & 0\\ 0 & 0 & 34.67 & 17.33 & 0\\ 0 & 0 & 17.33 & 34.67 & 0\\ 0 & 0 & 0 & 72 & 36\\ 0 & 0 & 0 & 36 & 72\end{bmatrix}\begin{Bmatrix}0.03\\x_2\\x_3\\x_4\\x_5\end{Bmatrix}=\begin{Bmatrix}m_1\\m_2\\m_3\\m_4\\m_5\\m_6\\m_7\\m_8\\m_9\\m_{10}\\m_{11}\\m_{12}\\m_{13}\\m_{14}\end{Bmatrix} \tag{6.48}$$

Then, the solutions of Eq. (6.47) and Eq. (6.48) are

$$\begin{Bmatrix} x_1 \\ x_2 \\ x_3 \\ x_4 \\ x_5 \end{Bmatrix} = \begin{Bmatrix} 0.03\ \text{m} \\ 0.0082\ \text{rad} \\ 0.0038\ \text{rad} \\ 0.0038\ \text{rad} \\ 0.0082\ \text{rad} \end{Bmatrix}, \begin{Bmatrix} m_1 \\ m_2 \\ m_3 \\ m_4 \\ m_5 \\ m_6 \\ m_7 \\ m_8 \\ m_9 \\ m_{10} \\ m_{11} \\ m_{12} \\ m_{13} \\ m_{14} \end{Bmatrix} = \begin{Bmatrix} -0.7320 \\ -1.7160 \\ -0.7748 \\ -0.9724 \\ -0.7748 \\ -0.9724 \\ -0.7320 \\ -1.7160 \\ 0.7272 \\ 0.5688 \\ 0.1976 \\ 0.1976 \\ 0.5688 \\ 0.7272 \end{Bmatrix} \text{MN-m}, \quad F_f = 2.09\ \text{MN} \tag{6.49}$$

Comparing the moment demands in Eq. (6.49) with the corresponding capacities shows that the moments at PHLs #2, #4, #6, and #8 exceed their corresponding yield moment. While for PHLs #1, #3, #5, #7, #9, #10, #11, #12, #13, and #14, their moment demands are less than the corresponding yield moment, so the assumption that these ten PHLs remain elastic is correct, and this gives:

$$\begin{Bmatrix} m_2 \\ m_4 \\ m_6 \\ m_8 \end{Bmatrix} = \begin{Bmatrix} -1.25 \\ -0.95 \\ -0.95 \\ -1.25 \end{Bmatrix} \text{MN-m} \tag{6.50}$$

Substituting Eq. (6.50) back into Eq.(6.45) gives:

$$\left[\begin{array}{ccccc|cccc} 129 & -90 & -39 & -39 & -90 & -90 & -39 & -39 & -90 \\ -90 & 312 & 36 & 0 & 0 & 120 & 0 & 0 & 0 \\ -39 & 36 & 211 & 17 & 0 & 0 & 52 & 0 & 0 \\ -39 & 0 & 17 & 211 & 36 & 0 & 0 & 52 & 0 \\ -90 & 0 & 0 & 36 & 312 & 0 & 0 & 0 & 120 \\ \hline -90 & 120 & 0 & 0 & 0 & 240 & 0 & 0 & 0 \\ -39 & 0 & 52 & 0 & 0 & 0 & 104 & 0 & 0 \\ -39 & 0 & 0 & 52 & 0 & 0 & 0 & 104 & 0 \\ -90 & 0 & 0 & 0 & 120 & 0 & 0 & 0 & 240 \end{array}\right] \begin{Bmatrix} 0.03 \\ x_2 \\ x_3 \\ x_4 \\ x_5 \\ \hline -\theta_2'' \\ -\theta_4'' \\ -\theta_6'' \\ -\theta_8'' \end{Bmatrix} = \begin{Bmatrix} F_f \\ 0 \\ 0 \\ 0 \\ 0 \\ \hline -1.25 \\ -0.95 \\ -0.95 \\ -1.25 \end{Bmatrix} \tag{6.51}$$

Solving for Eq. (6.51) gives:

$$\begin{Bmatrix} x_1 \\ x_2 \\ x_3 \\ x_4 \\ x_5 \end{Bmatrix} = \begin{Bmatrix} 0.0300 \text{ m} \\ 0.0073 \text{ rad} \\ 0.0040 \text{ rad} \\ 0.0040 \text{ rad} \\ 0.0073 \text{ rad} \end{Bmatrix}, \quad \begin{Bmatrix} F_f \\ \theta_2'' \\ \theta_4'' \\ \theta_6'' \\ \theta_8'' \end{Bmatrix} = \begin{Bmatrix} 1.8088 \text{ MN} \\ -0.0024 \text{ rad} \\ -0.00014 \text{ rad} \\ -0.00014 \text{ rad} \\ -0.0024 \text{ rad} \end{Bmatrix} \tag{6.52}$$

Substituting Eq. (6.52) back into Eq. (6.45) gives

$$\begin{Bmatrix} x_1 \\ x_2 \\ x_3 \\ x_4 \\ x_5 \end{Bmatrix} = \begin{Bmatrix} 0.0300 \text{ m} \\ 0.0073 \text{ rad} \\ 0.0040 \text{ rad} \\ 0.0040 \text{ rad} \\ 0.0073 \text{ rad} \end{Bmatrix}, \quad \begin{Bmatrix} \theta_1'' \\ \theta_2'' \\ \theta_3'' \\ \theta_4'' \\ \theta_5'' \\ \theta_6'' \\ \theta_7'' \\ \theta_8'' \\ \theta_9'' \\ \theta_{10}'' \\ \theta_{11}'' \\ \theta_{12}'' \\ \theta_{13}'' \\ \theta_{14}'' \end{Bmatrix} = \begin{Bmatrix} 0 \\ -0.0024 \\ 0 \\ -0.00014 \\ 0 \\ -0.00014 \\ 0 \\ -0.0024 \\ 0 \\ 0 \\ 0 \\ 0 \\ 0 \\ 0 \end{Bmatrix} \text{rad} \tag{6.53}$$

$$\begin{Bmatrix} m_1 \\ m_2 \\ m_3 \\ m_4 \\ m_5 \\ m_6 \\ m_7 \\ m_8 \\ m_9 \\ m_{10} \\ m_{11} \\ m_{12} \\ m_{13} \\ m_{14} \end{Bmatrix} = \begin{Bmatrix} -0.6589 \\ -1.25 \\ -0.7518 \\ -0.95 \\ -0.7518 \\ -0.95 \\ -0.6589 \\ -1.25 \\ 0.6589 \\ 0.5463 \\ 0.2055 \\ 0.2055 \\ 0.5463 \\ 0.6589 \end{Bmatrix} \text{MN-m}, \quad F_f = 1.8088 \text{ MN} \tag{6.54}$$

Secondly, for the shear wall part, the global stiffness matrices, $\mathbf{K_w}$, $\mathbf{K'_w}$, and $\mathbf{K''_w}$, can be written as

$$\mathbf{K_w} = \begin{bmatrix} 2500 & -6500 \\ -6500 & 43900 \end{bmatrix} \text{MN/m} \tag{6.55}$$

$$\mathbf{K'_w} = \begin{bmatrix} 0 & 0 & 2500 \\ 4500 & -4500 & -6500 \end{bmatrix} \text{MN/m} \tag{6.56}$$

$$\mathbf{K''_w} = \begin{bmatrix} 750 & -750 & 0 \\ -750 & 750 & 0 \\ 0 & 0 & 2500 \end{bmatrix} \text{MN/m} \tag{6.57}$$

To analyze the behavior of the shear wall member, the first step is to assume that the response of the shear wall member is linear, i.e. $\delta''_1 = \delta''_2 = \delta''_3 = 0$. Therefore, the governing equation written in Eq. (6.31) becomes:

$$\left[\begin{array}{cc|ccc} 2500 & -6500 & 0 & 0 & 2500 \\ -6500 & 43900 & 4500 & -4500 & -6500 \\ \hline 0 & 4500 & 750 & -750 & 0 \\ 0 & -4500 & -750 & 750 & 0 \\ 2500 & -6500 & 0 & 0 & 2500 \end{array}\right] \left\{\begin{array}{c} 0.03 \\ x_6 \\ \hline -\delta''_1 \\ -\delta''_2 \\ -\delta''_3 \end{array}\right\} = \left\{\begin{array}{c} F_w \\ 0 \\ \hline f_1 \\ f_2 \\ f_3 \end{array}\right\} \tag{6.58}$$

Solving for Eq. (6.58) gives:

$$\begin{Bmatrix} x_6 \\ F_w \\ f_1 \\ f_2 \\ f_3 \end{Bmatrix} = \begin{Bmatrix} 0.0044 \text{ rad} \\ 46.12 \text{ MN} \\ 19.83 \text{ MN} \\ -19.83 \text{ MN} \\ 46.12 \text{ MN} \end{Bmatrix} \tag{6.59}$$

Comparing these demands in Eq. (6.59) with the cracking and yielding capacities provided in Table 6.1, the shear wall member is in the plastic domain, which means the assumption that the shear wall member is linear is incorrect. Therefore, assume that the shear wall structure is in the plastic domain, which means that the VSH#1 and HSH#3 begin to deform in their hardening branches. The stiffness K_e = 1500 MN/m, K_y = 15 MN/m, K_{cr} = 1050 MN/m, F_y = 1.32 MN/m, K^s = 2500 MN/m and K_{scr} = 400 MN/m can be obtained or computed according to the parameters listed in Table 6.1. Then, the spring force f_1 and f_3 can be expressed in terms of δ''_1, δ''_3 according to Eq. (6.17) and Eq. (6.23):

$$\begin{cases} f_1 = 15.15\,\delta_1'' + 1.314 \\ f_3 = 476.19\,\delta_3'' + 3 \end{cases} \tag{6.60}$$

Therefore, Eq. (6.59) becomes:

$$\begin{bmatrix} 2500 & -6500 & 0 & 2500 \\ -6500 & 43900 & 4500 & -6500 \\ 0 & 4500 & 750+15.15 & 0 \\ 2500 & -6500 & 0 & 2500+476.19 \end{bmatrix} \begin{Bmatrix} 0.03 \\ x_6 \\ -\delta_1'' \\ -\delta_3'' \end{Bmatrix} = \begin{Bmatrix} F_w \\ 0 \\ 1.314 \\ 3 \end{Bmatrix} \tag{6.61}$$

Solving for Eq. (6.39) gives:

$$\begin{Bmatrix} x_6 \\ \delta_1'' \\ \delta_3'' \\ F_w \end{Bmatrix} = \begin{Bmatrix} 0.0093 \text{ rad} \\ 0.0528 \text{ m} \\ 0.0039 \text{ m} \\ 4.8717 \text{ MN} \end{Bmatrix} \tag{6.62}$$

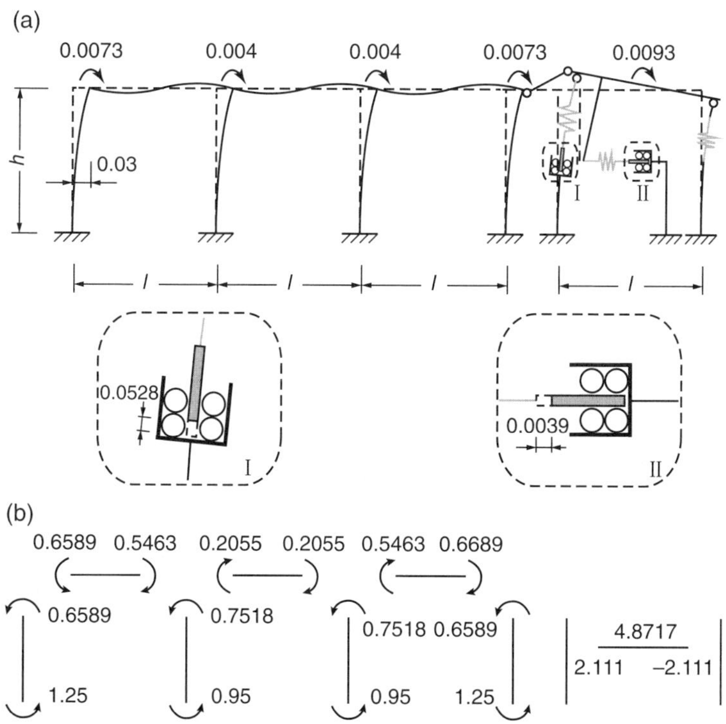

Figure 6.10 Graphical illustration of the response of the RC frame-shear wall structure. (a) Deformations; (b) Internal forces.

Substituting Eq. (6.62) back into Eq. (6.58) and solving for the spring forces gives

$$\begin{Bmatrix} x_1 \\ x_6 \end{Bmatrix} = \begin{Bmatrix} 0.03\ \text{m} \\ 0.0093\ \text{rad} \end{Bmatrix}, \quad \begin{Bmatrix} \delta_1'' \\ \delta_2'' \\ \delta_3'' \end{Bmatrix} = \begin{Bmatrix} 0.0528\ \text{m} \\ 0 \\ 0.0039\ \text{m} \end{Bmatrix}, \quad \begin{Bmatrix} f_1 \\ f_2 \\ f_3 \end{Bmatrix} = \begin{Bmatrix} 2.1111\ \text{MN} \\ -2.1111\ \text{MN} \\ 4.8717\ \text{MN} \end{Bmatrix},$$

$$F_w = 4.8717\ \text{MN} \tag{6.63}$$

Then the restoring force of the RC frame-shear wall structure can be obtained by

$$F = F_f + F_w = 1.8088 + 4.8717 = 6.6805\ \text{MN} \tag{6.64}$$

Example 6.3 Nonlinear Static Analysis of the Two-Story RC Frame-Shear Wall Structure

Consider the two-story RC frame-shear wall structure as shown in Figure 6.11. The behaviors of the frame part and shear wall part are satisfied by the compatibility condition. There are a total of 12 DOFs (2 translations and 10 rotations), 28 RHs, 4 VSHs and 2 HSHs in this RC frame-shear wall structure. The components used in Example 6.2 are still used in this example. The structure is subjected to the input displacements $X_1 = 0.03$ m and $X_2 = 0.08$ m on the first and second floors, which means the input displacements X_1 and X_2 are applied at the horizontal degrees of freedom x_1 and x_2. Recognizing

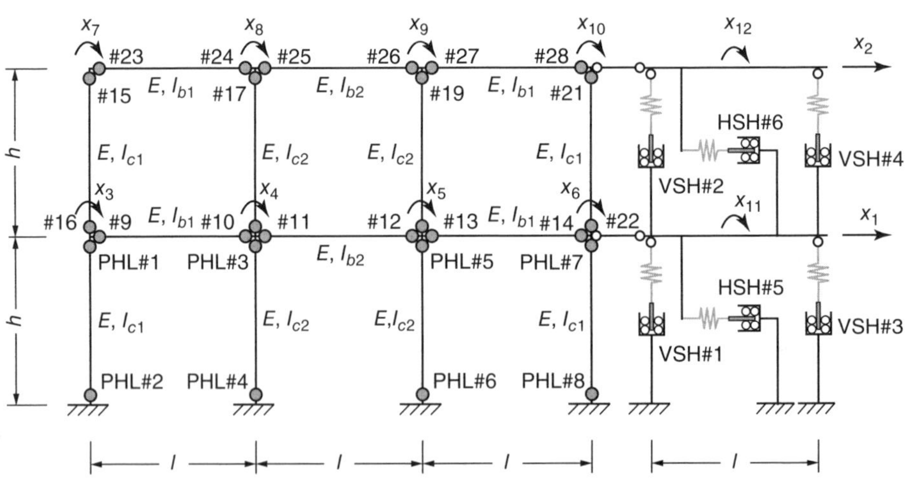

Figure 6.11 The two-story RC frame-shear wall structure.

$F_3 = F_4 = F_5 = F_6 = F_7 = F_8 = F_9 = F_{10} = F_{11} = F_{12} = 0$, the static condensation procedure can be used in this example to reduce the nonlinear problem.

Solution

First, for the RC frame part, the 10×10 stiffness matrix $\mathbf{K_f}$, the 10×28 stiffness matrix $\mathbf{K_f}'$, and the 28×28 stiffness matrix $\mathbf{K_f}''$ can be written as

$$\mathbf{K_f} = \left[\begin{array}{c|c} \mathbf{K}^f_{dd} & \mathbf{K}^f_{dr} \\ \hline \mathbf{K}^f_{rd} & \mathbf{K}^f_{rr} \end{array}\right]$$

$$= \left[\begin{array}{cc|cccccccc}
258 & -129 & 0 & 0 & 0 & 0 & 90 & 39 & 39 & 90 \\
-129 & 129 & -90 & -39 & -39 & -90 & -90 & -39 & -39 & -90 \\ \hline
0 & -90 & 552 & 36 & 0 & 0 & 120 & 0 & 0 & 0 \\
0 & -39 & 36 & 314.67 & 17.33 & 0 & 0 & 52 & 0 & 0 \\
0 & -39 & 0 & 17.33 & 314.67 & 36 & 0 & 0 & 52 & 0 \\
0 & -90 & 0 & 0 & 36 & 552 & 0 & 0 & 0 & 120 \\
90 & -90 & 120 & 0 & 0 & 0 & 312 & 36 & 0 & 0 \\
39 & -39 & 0 & 52 & 0 & 0 & 36 & 210.67 & 17.33 & 0 \\
39 & -39 & 0 & 0 & 52 & 0 & 0 & 17.33 & 210.67 & 36 \\
90 & -90 & 0 & 0 & 0 & 120 & 0 & 0 & 36 & 312
\end{array}\right] \text{MN/m}$$

$$\mathbf{K}'_f = \left[\begin{array}{c} \mathbf{K}^{f'}_d \\ \hline \mathbf{K}^{f'}_r \end{array}\right] \tag{6.65}$$

$$\left[\begin{array}{cccccccccccccccccccccccccccc}
-90 & -90 & -39 & -39 & -39 & -39 & -90 & -90 & 0 & 0 & 0 & 0 & 0 & 0 & 90 & 90 & 39 & 39 & 39 & 39 & 90 & 90 & 0 & 0 & 0 & 0 & 0 & 0 \\
0 & 0 & 0 & 0 & 0 & 0 & 0 & 0 & 0 & 0 & 0 & 0 & 0 & 0 & -90 & -90 & -39 & -39 & -39 & -39 & -90 & -90 & 0 & 0 & 0 & 0 & 0 & 0 \\ \hline
240 & 120 & 0 & 0 & 0 & 0 & 0 & 0 & 72 & 36 & 0 & 0 & 0 & 0 & 120 & 240 & 0 & 0 & 0 & 0 & 0 & 0 & 0 & 0 & 0 & 0 & 0 & 0 \\
0 & 0 & 104 & 52 & 0 & 0 & 0 & 0 & 36 & 72 & 34.67 & 17.33 & 0 & 0 & 0 & 0 & 52 & 104 & 0 & 0 & 0 & 0 & 0 & 0 & 0 & 0 & 0 & 0 \\
0 & 0 & 0 & 0 & 104 & 52 & 0 & 0 & 0 & 0 & 17.33 & 34.67 & 72 & 36 & 0 & 0 & 0 & 0 & 52 & 104 & 0 & 0 & 0 & 0 & 0 & 0 & 0 & 0 \\
0 & 0 & 0 & 0 & 0 & 0 & 240 & 120 & 0 & 0 & 0 & 0 & 36 & 72 & 0 & 0 & 0 & 0 & 0 & 0 & 120 & 240 & 0 & 0 & 0 & 0 & 0 & 0 \\
0 & 0 & 0 & 0 & 0 & 0 & 0 & 0 & 0 & 0 & 0 & 0 & 0 & 0 & 240 & 120 & 0 & 0 & 0 & 0 & 0 & 0 & 72 & 36 & 0 & 0 & 0 & 0 \\
0 & 0 & 0 & 0 & 0 & 0 & 0 & 0 & 0 & 0 & 0 & 0 & 0 & 0 & 0 & 0 & 104 & 52 & 0 & 0 & 0 & 0 & 36 & 72 & 34.67 & 17.33 & 0 & 0 \\
0 & 0 & 0 & 0 & 0 & 0 & 0 & 0 & 0 & 0 & 0 & 0 & 0 & 0 & 0 & 0 & 0 & 0 & 104 & 52 & 0 & 0 & 0 & 0 & 17.33 & 34.67 & 72 & 36 \\
0 & 0 & 0 & 0 & 0 & 0 & 0 & 0 & 0 & 0 & 0 & 0 & 0 & 0 & 0 & 0 & 0 & 0 & 0 & 0 & 240 & 120 & 0 & 0 & 0 & 0 & 36 & 72
\end{array}\right] \text{MN/m} \tag{6.66}$$

$$
\mathbf{K}_{\mathbf{f}}'' = \left[\begin{array}{cccccccccccccccccccccccccccc}
240 & 120 & 0 & 0 & 0 & 0 & 0 & 0 & 0 & 0 & 0 & 0 & 0 & 0 & 0 & 0 & 0 & 0 & 0 & 0 & 0 & 0 & 0 & 0 & 0 & 0 & 0 & 0 \\
120 & 240 & 0 & 0 & 0 & 0 & 0 & 0 & 0 & 0 & 0 & 0 & 0 & 0 & 0 & 0 & 0 & 0 & 0 & 0 & 0 & 0 & 0 & 0 & 0 & 0 & 0 & 0 \\
0 & 0 & 104 & 52 & 0 & 0 & 0 & 0 & 0 & 0 & 0 & 0 & 0 & 0 & 0 & 0 & 0 & 0 & 0 & 0 & 0 & 0 & 0 & 0 & 0 & 0 & 0 & 0 \\
0 & 0 & 52 & 104 & 0 & 0 & 0 & 0 & 0 & 0 & 0 & 0 & 0 & 0 & 0 & 0 & 0 & 0 & 0 & 0 & 0 & 0 & 0 & 0 & 0 & 0 & 0 & 0 \\
0 & 0 & 0 & 0 & 104 & 52 & 0 & 0 & 0 & 0 & 0 & 0 & 0 & 0 & 0 & 0 & 0 & 0 & 0 & 0 & 0 & 0 & 0 & 0 & 0 & 0 & 0 & 0 \\
0 & 0 & 0 & 0 & 52 & 104 & 0 & 0 & 0 & 0 & 0 & 0 & 0 & 0 & 0 & 0 & 0 & 0 & 0 & 0 & 0 & 0 & 0 & 0 & 0 & 0 & 0 & 0 \\
0 & 0 & 0 & 0 & 0 & 0 & 240 & 120 & 0 & 0 & 0 & 0 & 0 & 0 & 0 & 0 & 0 & 0 & 0 & 0 & 0 & 0 & 0 & 0 & 0 & 0 & 0 & 0 \\
0 & 0 & 0 & 0 & 0 & 0 & 120 & 240 & 0 & 0 & 0 & 0 & 0 & 0 & 0 & 0 & 0 & 0 & 0 & 0 & 0 & 0 & 0 & 0 & 0 & 0 & 0 & 0 \\
0 & 0 & 0 & 0 & 0 & 0 & 0 & 0 & 72 & 36 & 0 & 0 & 0 & 0 & 0 & 0 & 0 & 0 & 0 & 0 & 0 & 0 & 0 & 0 & 0 & 0 & 0 & 0 \\
0 & 0 & 0 & 0 & 0 & 0 & 0 & 0 & 36 & 72 & 0 & 0 & 0 & 0 & 0 & 0 & 0 & 0 & 0 & 0 & 0 & 0 & 0 & 0 & 0 & 0 & 0 & 0 \\
0 & 0 & 0 & 0 & 0 & 0 & 0 & 0 & 0 & 0 & 34.67 & 17.33 & 0 & 0 & 0 & 0 & 0 & 0 & 0 & 0 & 0 & 0 & 0 & 0 & 0 & 0 & 0 & 0 \\
0 & 0 & 0 & 0 & 0 & 0 & 0 & 0 & 0 & 0 & 17.33 & 34.67 & 0 & 0 & 0 & 0 & 0 & 0 & 0 & 0 & 0 & 0 & 0 & 0 & 0 & 0 & 0 & 0 \\
0 & 0 & 0 & 0 & 0 & 0 & 0 & 0 & 0 & 0 & 0 & 0 & 72 & 36 & 0 & 0 & 0 & 0 & 0 & 0 & 0 & 0 & 0 & 0 & 0 & 0 & 0 & 0 \\
0 & 0 & 0 & 0 & 0 & 0 & 0 & 0 & 0 & 0 & 0 & 0 & 36 & 72 & 0 & 0 & 0 & 0 & 0 & 0 & 0 & 0 & 0 & 0 & 0 & 0 & 0 & 0 \\
0 & 0 & 0 & 0 & 0 & 0 & 0 & 0 & 0 & 0 & 0 & 0 & 0 & 0 & 240 & 120 & 0 & 0 & 0 & 0 & 0 & 0 & 0 & 0 & 0 & 0 & 0 & 0 \\
0 & 0 & 0 & 0 & 0 & 0 & 0 & 0 & 0 & 0 & 0 & 0 & 0 & 0 & 120 & 240 & 0 & 0 & 0 & 0 & 0 & 0 & 0 & 0 & 0 & 0 & 0 & 0 \\
0 & 0 & 0 & 0 & 0 & 0 & 0 & 0 & 0 & 0 & 0 & 0 & 0 & 0 & 0 & 0 & 104 & 52 & 0 & 0 & 0 & 0 & 0 & 0 & 0 & 0 & 0 & 0 \\
0 & 0 & 0 & 0 & 0 & 0 & 0 & 0 & 0 & 0 & 0 & 0 & 0 & 0 & 0 & 0 & 52 & 104 & 0 & 0 & 0 & 0 & 0 & 0 & 0 & 0 & 0 & 0 \\
0 & 0 & 0 & 0 & 0 & 0 & 0 & 0 & 0 & 0 & 0 & 0 & 0 & 0 & 0 & 0 & 0 & 0 & 104 & 52 & 0 & 0 & 0 & 0 & 0 & 0 & 0 & 0 \\
0 & 0 & 0 & 0 & 0 & 0 & 0 & 0 & 0 & 0 & 0 & 0 & 0 & 0 & 0 & 0 & 0 & 0 & 52 & 104 & 0 & 0 & 0 & 0 & 0 & 0 & 0 & 0 \\
0 & 0 & 0 & 0 & 0 & 0 & 0 & 0 & 0 & 0 & 0 & 0 & 0 & 0 & 0 & 0 & 0 & 0 & 0 & 0 & 240 & 120 & 0 & 0 & 0 & 0 & 0 & 0 \\
0 & 0 & 0 & 0 & 0 & 0 & 0 & 0 & 0 & 0 & 0 & 0 & 0 & 0 & 0 & 0 & 0 & 0 & 0 & 0 & 120 & 240 & 0 & 0 & 0 & 0 & 0 & 0 \\
0 & 0 & 0 & 0 & 0 & 0 & 0 & 0 & 0 & 0 & 0 & 0 & 0 & 0 & 0 & 0 & 0 & 0 & 0 & 0 & 0 & 0 & 72 & 36 & 0 & 0 & 0 & 0 \\
0 & 0 & 0 & 0 & 0 & 0 & 0 & 0 & 0 & 0 & 0 & 0 & 0 & 0 & 0 & 0 & 0 & 0 & 0 & 0 & 0 & 0 & 36 & 72 & 0 & 0 & 0 & 0 \\
0 & 0 & 0 & 0 & 0 & 0 & 0 & 0 & 0 & 0 & 0 & 0 & 0 & 0 & 0 & 0 & 0 & 0 & 0 & 0 & 0 & 0 & 0 & 0 & 34.67 & 17.33 & 0 & 0 \\
0 & 0 & 0 & 0 & 0 & 0 & 0 & 0 & 0 & 0 & 0 & 0 & 0 & 0 & 0 & 0 & 0 & 0 & 0 & 0 & 0 & 0 & 0 & 0 & 34.67 & 17.33 & 0 & 0 \\
0 & 0 & 0 & 0 & 0 & 0 & 0 & 0 & 0 & 0 & 0 & 0 & 0 & 0 & 0 & 0 & 0 & 0 & 0 & 0 & 0 & 0 & 0 & 0 & 0 & 0 & 72 & 36 \\
0 & 0 & 0 & 0 & 0 & 0 & 0 & 0 & 0 & 0 & 0 & 0 & 0 & 0 & 0 & 0 & 0 & 0 & 0 & 0 & 0 & 0 & 0 & 0 & 0 & 0 & 36 & 72
\end{array}\right] \text{MN/m} \tag{6.67}
$$

Then, the governing equation of the FAM can be written as:

$$
\left[\begin{array}{c:c} \mathbf{K}_{\mathbf{f}10\times10} & \mathbf{K}'_{\mathbf{f}10\times28} \\ \hdashline \mathbf{K}'^{T}_{\mathbf{f}28\times10} & \mathbf{K}''_{\mathbf{f}28\times28} \end{array}\right] \left\{\begin{array}{c} \mathbf{x}_{\mathbf{f}10\times1} \\ \hdashline -\boldsymbol{\theta}''_{28\times1} \end{array}\right\} = \left\{\begin{array}{c} \mathbf{F}_{\mathbf{f}10\times1} \\ \hdashline \mathbf{m}_{28\times1} \end{array}\right\} \tag{6.68}
$$

Recognizing $F_3 = F_4 = F_5 = F_6 = F_7 = F_8 = F_9 = F_{10} = 0$, static condensation is performed here according to the expression in Chapter 2. Therefore, the condensed stiffness matrices $\bar{K}_{\mathbf{f}}$, $\bar{K}'_{\mathbf{f}}$, and $\bar{K}_{\mathbf{f}}''$ can be obtained

$$
\bar{\mathbf{K}}_{\mathbf{f}} = \mathbf{K}^{f}_{dd} - \mathbf{K}^{f}_{dr}\left(\mathbf{K}^{f}_{rr}\right)^{-1}\mathbf{K}^{f}_{rd} = \begin{bmatrix} 194.38 & -77.92 \\ -77.92 & 51.91 \end{bmatrix} \text{MN/m} \tag{6.69}
$$

$\bar{\mathbf{K}}'_{\mathbf{f}} = \mathbf{K}^{f'}_{d} - \mathbf{K}^{f}_{dr}\left(\mathbf{K}^{f}_{rr}\right)^{-1}\mathbf{K}^{f'}_{r} =$

Column 1 through 14

$$
\begin{bmatrix} -74.63 & -82.32 & -37.66 & -38.33 & -37.66 & -38.33 & -37.66 & -38.33 & -74.63 & -82.32 & 5.08 & 3.24 & 0.67 & 0.67 \\ 25.52 & 12.76 & 9.16 & 4.58 & 9.16 & 4.58 & 9.16 & 4.58 & 25.52 & 12.76 & 10.83 & 10.18 & 4.58 & 4.58 \end{bmatrix}
$$

Column 14 through 28

$$
\begin{bmatrix} 3.24 & 5.08 & 26.06 & 69.56 & 26.48 & 33.75 & 26.06 & 69.56 & -26.06 & -19.88 & -6.60 & -6.60 & -19.88 & -20.06 \\ 10.17 & 10.83 & -20.98 & -36.35 & -22.57 & -23.91 & -20.98 & -36.35 & 20.98 & 16.64 & 5.93 & 5.93 & 16.64 & 20.98 \end{bmatrix} \text{MN/m} \tag{6.70}
$$

$$\bar{\mathbf{K}}_{\mathbf{f}}'' = \mathbf{K}_{\mathbf{f}}'' - \left(\mathbf{K}_{r}^{f'}\right)^{T} \left(\mathbf{K}_{rr}^{f}\right)^{-1} \mathbf{K}_{r}^{f'} =$$

Column 1 through 14

$$\begin{bmatrix}
124.7327 & 62.3663 & 6.5742 & 3.2871 & -0.4571 & -0.2286 & 0.1093 & 0.0547 & -32.3045 & -12.7387 & 2.1152 & 0.9433 & -0.3001 & -0.1254 \\
62.3663 & 211.1832 & 3.2871 & 1.6436 & -0.2286 & -0.1143 & 0.0547 & 0.0273 & -16.1523 & -6.3694 & 1.0576 & 0.4717 & -0.1500 & -0.0627 \\
6.5742 & 3.2871 & 67.6122 & 33.8061 & 2.2544 & 1.1272 & -0.4571 & -0.2286 & -10.6235 & -24.2054 & -11.7535 & -5.3132 & 1.4922 & 0.6432 \\
3.2871 & 1.6436 & 33.8061 & 94.9031 & 1.1272 & 0.5636 & -0.2286 & -0.1143 & -5.3118 & -12.1027 & -5.8768 & -2.6566 & 0.7461 & 0.3216 \\
-0.4571 & -0.2286 & 2.2544 & 1.1272 & 67.6122 & 33.8061 & 6.5742 & 3.2871 & 0.6432 & 1.4922 & -5.3132 & -11.7535 & -24.2054 & -10.6235 \\
-0.2286 & -0.1143 & 1.1272 & 0.5636 & 33.8061 & 94.9031 & 3.2871 & 1.6436 & 0.3216 & 0.7461 & -2.6566 & -5.8768 & -12.1027 & -5.3118 \\
0.1093 & 0.0547 & -0.4571 & -0.2286 & 6.5742 & 3.2871 & 124.7327 & 62.3663 & -0.1254 & -0.3001 & 0.9433 & 2.1152 & -12.7387 & -32.3045 \\
0.0547 & 0.0273 & -0.2286 & -0.1143 & 3.2871 & 1.6436 & 62.3663 & 211.1832 & -0.0627 & -0.1500 & 0.4717 & 1.0576 & -6.3694 & -16.1523 \\
-32.3045 & -16.1523 & -10.6235 & -5.3118 & 0.6432 & 0.3216 & -0.1254 & -0.0627 & 58.6313 & 23.7996 & -3.4340 & -1.5562 & 0.4265 & 0.1850 \\
-12.7387 & -6.3694 & -24.2054 & -12.1027 & 1.4922 & 0.7461 & -0.3001 & -0.1500 & 23.7996 & 53.3316 & -7.8198 & -3.5368 & 0.9880 & 0.4265 \\
2.1152 & 1.0576 & -11.7535 & -5.8768 & -5.3132 & -2.6566 & 0.9433 & 0.4717 & -3.4340 & -7.8198 & 29.8633 & 13.6034 & -3.5368 & -1.5562 \\
0.9433 & 0.4717 & -5.3132 & -2.6566 & -11.7535 & -5.8768 & 2.1152 & 1.0576 & -1.5562 & -3.5368 & 13.6034 & 29.8633 & -7.8198 & -3.4340 \\
-0.3001 & -0.1500 & 1.4922 & 0.7461 & -24.2054 & -12.1027 & -12.7387 & -6.3694 & 0.4265 & 0.9880 & -3.5368 & -7.8198 & 53.3316 & 23.7996 \\
-0.1254 & -0.0627 & 0.6432 & 0.3216 & -10.6235 & -5.3118 & -32.3045 & -16.1523 & 0.1850 & 0.4265 & -1.5562 & -3.4340 & 23.7996 & 58.6313 \\
-11.9554 & -5.9777 & -1.7627 & -0.8814 & 0.3136 & 0.1568 & -0.1317 & -0.0659 & -4.1968 & -3.0137 & -0.5353 & -0.1893 & 0.1973 & 0.0690 \\
-92.4282 & -46.2141 & 4.0493 & 2.0246 & -0.1861 & -0.0930 & 0.0161 & 0.0081 & -26.3268 & -11.0609 & 1.3187 & 0.6129 & -0.1264 & -0.0596 \\
-1.7627 & -0.8814 & -8.7250 & -4.3625 & -0.2485 & -0.1242 & 0.3136 & 0.1568 & -3.5490 & -6.3048 & -2.9497 & -1.5370 & -0.1250 & 0.0081 \\
4.0493 & 2.0246 & -31.6533 & -15.8267 & 1.5666 & 0.7833 & -0.1861 & -0.0930 & -9.7421 & -21.3064 & -10.2900 & -4.7534 & 1.0566 & 0.4864 \\
0.3136 & 0.1568 & -0.2485 & -0.1242 & -8.7250 & -4.3625 & -1.7627 & -0.8814 & 0.0081 & -0.1250 & -1.5370 & -2.9497 & -6.3048 & -3.5490 \\
-0.1861 & -0.0930 & 1.5666 & 0.7833 & -31.6533 & -15.8267 & 4.0493 & 2.0246 & 0.4864 & 1.0566 & -4.7534 & -10.2900 & -21.3064 & -9.7421 \\
-0.1317 & -0.0659 & 0.3136 & 0.1568 & -1.7627 & -0.8814 & -11.9554 & -5.9777 & 0.0690 & 0.1973 & -0.1893 & -0.5353 & -3.0137 & -4.1968 \\
0.0161 & 0.0081 & -0.1861 & -0.0930 & 4.0493 & 2.0246 & -92.4282 & -46.2141 & -0.0596 & -0.1264 & 0.6129 & 1.3187 & -11.0609 & -26.3268 \\
11.9554 & 5.9777 & 1.7627 & 0.8814 & -0.3136 & -0.1568 & 0.1317 & 0.0659 & 4.1968 & 3.0137 & 0.5353 & 0.1893 & -0.1973 & -0.0690 \\
3.3557 & 1.6778 & 5.7979 & 2.8990 & -0.8711 & -0.4355 & 0.3474 & 0.1737 & 3.0137 & 4.5173 & 1.7875 & 0.6760 & -0.5510 & -0.1973 \\
-1.5929 & -0.7965 & 2.9270 & 1.4635 & 1.1196 & 0.5598 & -0.6609 & -0.3305 & 0.5353 & 1.7875 & 1.1623 & 0.8610 & 0.6760 & 0.1893 \\
-0.6609 & -0.3305 & 1.1196 & 0.5598 & 2.9270 & 1.4635 & -1.5929 & -0.7965 & 0.1893 & 0.6760 & 0.8610 & 1.1623 & 1.7875 & 0.5353 \\
0.3474 & 0.1737 & -0.8711 & -0.4355 & 5.7979 & 2.8990 & 3.3557 & 1.6778 & -0.1973 & -0.5510 & 0.6760 & 1.7875 & 4.5173 & 3.0137 \\
0.1317 & 0.0659 & -0.3136 & -0.1568 & 1.7627 & 0.8814 & 11.9554 & 5.9777 & -0.0690 & -0.1973 & 0.1893 & 0.5353 & 3.0137 & 4.1968
\end{matrix}$$

Column 14 through 28

$$\begin{matrix}
-11.9554 & -92.4282 & -1.7627 & 4.0493 & 0.3136 & -0.1861 & -0.1317 & 0.0161 & 11.9554 & 3.3557 & -1.5929 & -0.6609 & 0.3474 & 0.1317 \\
-5.9777 & -46.2141 & -0.8814 & 2.0246 & 0.1568 & -0.0930 & -0.0659 & 0.0081 & 5.9777 & 1.6778 & -0.7965 & -0.3305 & 0.1737 & 0.0659 \\
-1.7627 & 4.0493 & -8.7250 & -31.6533 & -0.2485 & 1.5666 & 0.3136 & -0.1861 & 1.7627 & 5.7979 & 2.9270 & 1.1196 & -0.8711 & -0.3136 \\
-0.8814 & 2.0246 & -4.3625 & -15.8267 & -0.1242 & 0.7833 & 0.1568 & -0.0930 & 0.8814 & 2.8990 & 1.4635 & 0.5598 & -0.4355 & -0.1568 \\
0.3136 & -0.1861 & -0.2485 & 1.5666 & -8.7250 & -31.6533 & -1.7627 & 4.0493 & -0.3136 & -0.8711 & 1.1196 & 2.9270 & 5.7979 & 1.7627 \\
0.1568 & -0.0930 & -0.1242 & 0.7833 & -4.3625 & -15.8267 & -0.8814 & 2.0246 & -0.1568 & -0.4355 & 0.5598 & 1.4635 & 2.8990 & 0.8814 \\
-0.1317 & 0.0161 & 0.3136 & -0.1861 & -1.7627 & 4.0493 & -11.9554 & -92.4282 & 0.1317 & 0.3474 & -0.6609 & -1.5929 & 3.3557 & 11.9554 \\
-0.0659 & 0.0081 & 0.1568 & -0.0930 & -0.8814 & 2.0246 & -5.9777 & -46.2141 & 0.0659 & 0.1737 & -0.3305 & -0.7965 & 1.6778 & 5.9777 \\
-4.1968 & -26.3268 & -3.5490 & -9.7421 & 0.0081 & 0.4864 & 0.0690 & -0.0596 & 4.1968 & 3.0137 & 0.5353 & 0.1893 & -0.1973 & -0.0690 \\
-3.0137 & -11.0609 & 6.3048 & 21.3064 & 0.1250 & 1.0566 & 0.1973 & -0.1264 & 3.0137 & 4.5173 & 1.7875 & 0.6760 & -0.5510 & -0.1973 \\
-0.5353 & 1.3187 & -2.9497 & -10.2900 & -1.5370 & -4.7534 & -0.1893 & 0.6129 & 0.5353 & 1.7875 & 1.1623 & 0.8610 & 0.6760 & 0.1893 \\
-0.1893 & 0.6129 & -1.5370 & -4.7534 & -2.9497 & -10.2900 & -0.5353 & 1.3187 & 0.1893 & 0.6760 & 0.8610 & 1.1623 & 1.7875 & 0.5353 \\
0.1973 & -0.1264 & -0.1250 & 1.0566 & -6.3048 & -21.3064 & -3.0137 & -11.0609 & -0.1973 & -0.5510 & 0.6760 & 1.7875 & 4.5173 & 3.0137 \\
0.0690 & -0.0596 & 0.0081 & 0.4864 & -3.5490 & -9.7421 & -4.1968 & -26.3268 & -0.0690 & -0.1973 & 0.1893 & 0.5353 & 3.0137 & 4.1968 \\
50.2377 & 16.1523 & 13.2676 & 5.3118 & -1.1136 & -0.3216 & 0.3231 & 0.0627 & -50.2377 & -17.7722 & 4.5046 & 1.9347 & -0.8211 & -0.3231 \\
16.1523 & 118.7550 & 5.3118 & 5.6928 & -0.3216 & -0.3004 & 0.0627 & 0.0435 & -16.1523 & -6.3694 & 1.0576 & 0.4717 & -0.1500 & -0.0627 \\
13.2676 & 5.3118 & 49.0464 & 17.9795 & 4.1937 & 1.9105 & -1.1136 & -0.3216 & -13.2676 & -32.9023 & -16.1441 & -6.9925 & 2.7988 & 1.1136 \\
5.3118 & 5.6928 & 17.9795 & 63.2498 & 1.9105 & 2.1302 & -0.3216 & -0.3004 & -5.3118 & -12.1027 & -5.8768 & -2.6566 & 0.7461 & 0.3216 \\
-1.1136 & -0.3216 & 4.1937 & 1.9105 & 49.0464 & 17.9795 & 13.2676 & 5.3118 & 1.1136 & 2.7988 & -6.9925 & -16.1441 & -32.9023 & -13.2676 \\
-0.3216 & -0.3004 & 1.9105 & 2.1302 & 17.9795 & 63.2498 & 5.3118 & 5.6928 & 0.3216 & 0.7461 & -2.6566 & -5.8768 & -12.1027 & -5.3118 \\
0.3231 & 0.0627 & -1.1136 & -0.3216 & 13.2676 & 5.3118 & 50.2377 & 16.1523 & -0.3231 & -0.8211 & 1.9347 & 4.5046 & -17.7722 & -50.2377 \\
0.0627 & 0.0435 & -0.3216 & -0.3004 & 5.3118 & 5.6928 & 16.1523 & 118.7550 & -0.0627 & -0.1500 & 0.4717 & 1.0576 & -6.3694 & -16.1523 \\
-50.2377 & -16.1523 & -13.2676 & -5.3118 & 1.1136 & 0.3216 & -0.3231 & -0.0627 & 50.2377 & 17.7722 & -4.5046 & -1.9347 & 0.8211 & 0.3231 \\
-17.7722 & -6.3694 & -32.9023 & -12.1027 & 2.7988 & 0.7461 & -0.8211 & -0.1500 & 17.7722 & 44.2970 & -11.3947 & -4.8888 & 2.0899 & 0.8211 \\
4.5046 & 1.0576 & -16.1441 & -5.8768 & -6.9925 & -2.6566 & 1.9347 & 0.4717 & -4.5046 & -11.3947 & 27.5388 & 11.8813 & -4.8888 & -1.9347 \\
1.9347 & 0.4717 & -6.9925 & -2.6566 & -16.1441 & -5.8768 & 4.5046 & 1.0576 & -1.9347 & -4.8888 & 11.8813 & 27.5388 & -11.3947 & -4.5046 \\
-0.8211 & -0.1500 & 2.7988 & 0.7461 & -32.9023 & -12.1027 & -17.7722 & -6.3694 & 0.8211 & 2.0899 & -4.8888 & -11.3947 & 44.2970 & 17.7722 \\
-0.3231 & -0.0627 & 1.1136 & 0.3216 & -13.2676 & -5.3118 & -50.2377 & -16.1523 & 0.3231 & 0.8211 & -1.9347 & -4.5046 & 17.7722 & 50.2377
\end{bmatrix} \text{MN/m} \tag{6.71}$$

and the governing equation of the FAM of Eq. (6.45) can be rewritten as:

$$\begin{bmatrix} \overline{\mathbf{K}}_{\mathrm{f}2\times2} & \overline{\mathbf{K}}'_{\mathrm{f}2\times28} \\ \overline{\mathbf{K}}'^{\mathrm{T}}_{\mathrm{f}28\times2} & \overline{\mathbf{K}}''_{\mathrm{f}28\times28} \end{bmatrix} \begin{Bmatrix} \mathbf{x}_{\mathrm{d}2\times1} \\ -\boldsymbol{\theta}''_{28\times1} \end{Bmatrix} = \begin{Bmatrix} \mathbf{F}_{\mathrm{df}\,2\times1} \\ \mathbf{m}_{28\times1} \end{Bmatrix} \tag{6.72}$$

in which,

$$\mathbf{x}_{\mathrm{d}} = \begin{Bmatrix} x_1 \\ x_2 \end{Bmatrix}, \quad \mathbf{F}_{\mathrm{df}} = \begin{Bmatrix} F_{1f} \\ F_{2f} \end{Bmatrix} \tag{6.73}$$

Now, the frame members are assumed to be linear, this gives $\boldsymbol{\theta}'' = 0$. Then, using the first equation of Eq. (6.72) gives:

$$\begin{bmatrix} 194.38 & -77.92 \\ -77.92 & 51.91 \end{bmatrix} \times 10^6 \begin{Bmatrix} 0.03 \\ 0.08 \end{Bmatrix} = \begin{Bmatrix} F_{1f} \\ F_{2f} \end{Bmatrix} \tag{6.74}$$

Solving for the restoring forces in Eq. (6.74) gives:

$$\begin{Bmatrix} F_{1f} \\ F_{2f} \end{Bmatrix} = \begin{Bmatrix} -0.4019 \\ 1.8149 \end{Bmatrix} \text{MN} \tag{6.75}$$

Substituting Eq. (6.75) back into the second equation of Eq. (6.72), the moments at the RHs are calculated as:

$$\begin{Bmatrix} m_1 \\ m_2 \\ m_3 \\ m_4 \\ m_5 \\ m_6 \\ m_7 \\ m_8 \\ m_9 \\ m_{10} \\ m_{11} \\ m_{12} \\ m_{13} \\ m_{14} \\ m_{15} \\ m_{16} \\ m_{17} \\ m_{18} \\ m_{19} \\ m_{20} \\ m_{21} \\ m_{22} \\ m_{23} \\ m_{24} \\ m_{25} \\ m_{26} \\ m_{27} \\ m_{28} \end{Bmatrix} = \begin{Bmatrix} -0.1972 \\ -1.4486 \\ -0.3967 \\ -0.7834 \\ -0.3967 \\ -0.7834 \\ -0.1972 \\ -1.4486 \\ 1.0185 \\ 0.9108 \\ 0.3866 \\ 0.3866 \\ 0.9108 \\ 1.0185 \\ -0.8967 \\ -0.8213 \\ -1.0112 \\ -0.9006 \\ -1.0112 \\ -0.9006 \\ -0.8967 \\ -0.8213 \\ 0.8967 \\ 0.7351 \\ 0.2761 \\ 0.2761 \\ 0.7351 \\ 0.8967 \end{Bmatrix} \text{MN-m} \tag{6.76}$$

Comparing the moment demands in Eq. (6.76) with the corresponding capacities shows that the moments at the PHLs #2, #8, #9, #10, #13, #14, #17,#19, #23, and #28 have exceeded their corresponding yield moment. While for PHLs #1, #3, #4, #5, #6, #7, #11, #12, #15, #16, #18, #20, #21, #22, #24, #25, #26, and #27, their moment demands are less than the corresponding yield moment, so the assumption that these 18 PHLs remain elastic is correct. So, it gives:

$$\begin{Bmatrix} m_2 \\ m_8 \\ m_9 \\ m_{10} \\ m_{13} \\ m_{14} \\ m_{17} \\ m_{19} \\ m_{23} \\ m_{28} \end{Bmatrix} = \begin{Bmatrix} -1.25 \\ -1.25 \\ 0.75 \\ 0.75 \\ 0.75 \\ 0.75 \\ -0.95 \\ -0.95 \\ 0.75 \\ 0.75 \end{Bmatrix} \text{MN-m} \tag{6.77}$$

Then, Eq. (6.72) can be written as

$$\left[\begin{array}{cc|cccccccccc} 194.38 & -77.92 & -82.3155 & -82.3156 & 5.0761 & 3.2362 & 3.2362 & 5.0761 & 26.477 & 26.477 & -26.0561 & -26.0561 \\ -77.92 & 51.91 & 12.7607 & 12.7607 & 10.8277 & 10.1708 & 10.1708 & 10.8277 & -22.5686 & -22.5686 & 20.980 & 20.98 \\ \hline -82.3155 & 12.7607 & 211.1832 & 0.0273 & -16.1523 & -6.3694 & -0.15 & -0.0627 & -0.8814 & 0.1568 & 5.9777 & 0.0689 \\ -82.3155 & 12.76107 & 0.0273 & 211.1832 & -0.0627 & -0.15 & -6.3694 & -16.1523 & 0.1568 & -0.8814 & 0.0659 & 5.9777 \\ 5.0761 & 10.8277 & -16.1523 & -0.0627 & 58.6313 & 23.7996 & 0.4265 & 0.185 & -3.549 & 0.0081 & 4.1968 & -0.069 \\ 3.2362 & 10.1708 & -6.3694 & -0.15 & 23.7996 & 53.3316 & 0.988 & 0.4265 & -6.3048 & -0.125 & 3.0137 & -0.1973 \\ 3.2362 & 10.1708 & -0.15 & -6.3694 & 0.4265 & 0.988 & 53.3316 & 23.7996 & -0.125 & -0.63048 & -0.1973 & 3.0137 \\ 5.0761 & 10.8277 & -0.0627 & -16.1523 & 0.185 & 0.4265 & 23.7996 & 58.6313 & 0.0081 & -3.549 & -0.069 & 4.1968 \\ 26.477 & -22.5868 & -0.8814 & 0.1568 & -3.549 & -6.3048 & -0.125 & 0.0081 & 49.0464 & 4.1937 & -13.2676 & 1.1136 \\ 26.477 & -22.5868 & 0.1568 & -0.8814 & 0.0081 & -0.125 & -6.3048 & -3.549 & 4.1937 & 49.0464 & 1.1136 & -13.2676 \\ -26.0561 & 20.98 & 5.9777 & 0.0659 & 4.1968 & 3.0137 & -0.1973 & -0.069 & -13.2676 & 1.1136 & 50.2377 & 0.3231 \\ -26.0561 & 20.98 & 0.0659 & 5.9777 & -0.069 & -0.1973 & 3.0137 & 4.1968 & 1.1136 & -13.2676 & 0.3231 & 50.2377 \end{array}\right] \begin{Bmatrix} 0.03 \\ 0.08 \\ \hline -\theta''_2 \\ -\theta''_8 \\ -\theta''_9 \\ -\theta''_{10} \\ -\theta''_{13} \\ -\theta''_{14} \\ -\theta''_{17} \\ -\theta''_{19} \\ -\theta''_{23} \\ -\theta''_{28} \end{Bmatrix} = \begin{Bmatrix} F_{1f} \\ F_{2f} \\ \hline -1.25 \\ -1.25 \\ 0.75 \\ 0.75 \\ 0.75 \\ 0.75 \\ -0.95 \\ -0.95 \\ 0.75 \\ 0.75 \end{Bmatrix} \tag{6.78}$$

Now solving for Eq. (6.78) gives:

$$\begin{Bmatrix} \theta''_2 \\ \theta''_8 \\ \theta''_9 \\ \theta''_{10} \\ \theta''_{13} \\ \theta''_{14} \\ \theta''_{17} \\ \theta''_{19} \\ \theta''_{23} \\ \theta''_{28} \end{Bmatrix} = \begin{Bmatrix} -0.00069 \\ -0.00069 \\ 0.0038 \\ 0.001 \\ 0.001 \\ 0.0038 \\ -0.0002 \\ -0.0002 \\ 0.0026 \\ 0.0026 \end{Bmatrix} \text{rad}, \quad \begin{Bmatrix} F_{1f} \\ F_{2f} \end{Bmatrix} = \begin{Bmatrix} -0.4167 \\ 1.6133 \end{Bmatrix} \text{MN} \tag{6.79}$$

Then, the DOFs of the frame that were statically condensed, $\mathbf{x}_r^f = \{x_3\, x_4\, x_5\, x_6\, x_7\, x_8\, x_9\, x_{10}\}^T$, can be calculated as:

$$\mathbf{x}_r^f = \begin{Bmatrix} x_3 \\ x_4 \\ x_5 \\ x_6 \\ x_7 \\ x_8 \\ x_9 \\ x_{10} \end{Bmatrix} = \left(\mathbf{K}_{rr}^f\right)^{-1}\mathbf{K}_r^f\boldsymbol{\theta}'' - \left(\mathbf{K}_{rr}^f\right)^{-1}\mathbf{K}_{rd}^f\mathbf{x}_\mathrm{d} = \begin{Bmatrix} 0.0107 \\ 0.08 \\ 0.08 \\ 0.0107 \\ 0.0103 \\ 0.0054 \\ 0.0054 \\ 0.0103 \end{Bmatrix} \text{rad} \tag{6.80}$$

Finally, substituting Eq. (6.79) back into Eq. (6.72) gives:

$$\begin{Bmatrix} \theta_1'' \\ \theta_2'' \\ \theta_3'' \\ \theta_4'' \\ \theta_5'' \\ \theta_6'' \\ \theta_7'' \\ \theta_8'' \\ \theta_9'' \\ \theta_{10}'' \\ \theta_{11}'' \\ \theta_{12}'' \\ \theta_{13}'' \\ \theta_{14}'' \\ \theta_{15}'' \\ \theta_{16}'' \\ \theta_{17}'' \\ \theta_{18}'' \\ \theta_{19}'' \\ \theta_{20}'' \\ \theta_{21}'' \\ \theta_{22}'' \\ \theta_{23}'' \\ \theta_{24}'' \\ \theta_{25}'' \\ \theta_{26}'' \\ \theta_{27}'' \\ \theta_{28}'' \end{Bmatrix} = \begin{Bmatrix} 0 \\ -0.00069 \\ 0 \\ 0 \\ 0 \\ 0 \\ 0 \\ -0.00069 \\ 0.0038 \\ 0.001 \\ 0 \\ 0 \\ 0.001 \\ 0.0038 \\ 0 \\ 0 \\ -0.0002 \\ 0 \\ -0.0002 \\ 0 \\ 0 \\ 0 \\ 0.0026 \\ 0 \\ 0 \\ 0 \\ 0 \\ 0.0026 \end{Bmatrix} \text{rad}, \quad \begin{Bmatrix} m_1 \\ m_2 \\ m_3 \\ m_4 \\ m_5 \\ m_6 \\ m_7 \\ m_8 \\ m_9 \\ m_{10} \\ m_{11} \\ m_{12} \\ m_{13} \\ m_{14} \\ m_{15} \\ m_{16} \\ m_{17} \\ m_{18} \\ m_{19} \\ m_{20} \\ m_{21} \\ m_{22} \\ m_{23} \\ m_{24} \\ m_{25} \\ m_{26} \\ m_{27} \\ m_{28} \end{Bmatrix} = \begin{Bmatrix} -0.05 \\ -1.25 \\ -0.3389 \\ -0.7544 \\ -0.3389 \\ -0.7544 \\ -0.5 \\ -1.25 \\ 0.75 \\ 0.75 \\ 0.4156 \\ 0.4156 \\ 0.75 \\ 0.75 \\ -0.75 \\ -0.7 \\ -0.95 \\ -0.8267 \\ -0.95 \\ -0.8267 \\ -0.75 \\ -0.7 \\ 0.75 \\ 0.6679 \\ 0.2821 \\ 0.2821 \\ 0.6679 \\ 0.75 \end{Bmatrix} \text{MN-m}, \quad \begin{Bmatrix} F_{1f} \\ F_{2f} \end{Bmatrix} = \begin{Bmatrix} -0.41667 \\ 1.6133 \end{Bmatrix} \text{MN} \tag{6.81}$$

Secondly, for the shear wall part, the 4×4 stiffness matrix $\mathbf{K}_\mathbf{w}$, the 4×6 stiffness matrix $\mathbf{K}'_\mathbf{w}$, and the 6×6 stiffness matrix $\mathbf{K}''_\mathbf{w}$ can be written as

$$\mathbf{K}_\mathbf{w} = \left[\begin{array}{c|c} \mathbf{K}_{dd}^w & \mathbf{K}_{dr}^w \\ \hline \mathbf{K}_{rd}^w & \mathbf{K}_{rr}^w \end{array}\right] = \left[\begin{array}{cc|cc} 5000 & -2500 & -3000 & 6500 \\ -2500 & 2500 & -3500 & -6500 \\ \hline -3000 & -300 & 75800 & -17900 \\ 6500 & -6500 & -17900 & 43900 \end{array}\right] \text{MN/m} \tag{6.82}$$

$$\mathbf{K}'_{\mathbf{w}} = \begin{bmatrix} \mathbf{K}_d^{w'} \\ \hdashline \mathbf{K}_r^{w'} \end{bmatrix}$$

$$= \begin{bmatrix} 0 & 0 & 0 & 0 & 2500 & -2500 \\ 0 & 0 & 0 & 0 & 0 & 2500 \\ \hdashline 4500 & -4500 & -4500 & 4500 & -6500 & -3500 \\ 0 & 4500 & 0 & -4500 & 0 & -6500 \end{bmatrix} \text{MN/m} \tag{6.83}$$

$$\mathbf{K}''_{\mathbf{w}} = \begin{bmatrix} 750 & 0 & -750 & 0 & 0 & 0 \\ 0 & 750 & 0 & -750 & 0 & 0 \\ -750 & 0 & 750 & 0 & 0 & 0 \\ 0 & -750 & 0 & 750 & 0 & 0 \\ 0 & 0 & 0 & 0 & 2500 & 0 \\ 0 & 0 & 0 & 0 & 0 & 2500 \end{bmatrix} \text{MN/m} \tag{6.84}$$

Then, the governing equation of the shear wall component with the FAM can be written as:

$$\begin{bmatrix} \mathbf{K}_{\mathbf{w}4\times4} & \mathbf{K}'_{\mathbf{w}4\times6} \\ \mathbf{K}'^{T}_{\mathbf{w}6\times4} & \mathbf{K}''_{\mathbf{w}6\times6} \end{bmatrix} \begin{Bmatrix} \mathbf{x}_{\mathbf{w}4\times1} \\ -\boldsymbol{\Delta}''_{6\times1} \end{Bmatrix} = \begin{Bmatrix} \mathbf{F}_{\mathbf{w}4\times1} \\ \mathbf{f}_{6\times1} \end{Bmatrix} \tag{6.85}$$

Similarly, the condensed stiffness matrices, $\bar{\mathbf{K}}_{\mathbf{w}}$, $\bar{\mathbf{K}}'_{\mathbf{w}}$, and $\bar{\mathbf{K}}''_{\mathbf{w}}$, can be obtained:

$$\bar{\mathbf{K}}_{\mathbf{w}} = \mathbf{K}_{dd}^{w} - \mathbf{K}_{dr}^{w}(\mathbf{K}_{rr}^{w})^{-1}\mathbf{K}_{rd}^{w} = \begin{bmatrix} 4035.8 & -1569 \\ -1569 & 985.4 \end{bmatrix} \text{MN/m} \tag{6.86}$$

$$\bar{\mathbf{K}}'_{\mathbf{w}} = \mathbf{K}_d^{w\prime} - \mathbf{K}_{dr}^{w\prime}(\mathbf{K}_{rr}^{w})^{-1}\mathbf{K}_r^{w\prime} = \begin{bmatrix} 23 & -679.9 & -23 & 679.9 & 2466.8 & -1569 \\ 404 & 427 & -404 & -427 & -583.6 & 985.4 \end{bmatrix} \text{MN/m} \tag{6.87}$$

$$\bar{\mathbf{K}}''_{\mathbf{w}} = \mathbf{K}''_{w} - \left(\mathbf{K}_r^{w\prime}\right)^{T}(\mathbf{K}_{rr}^{w})^{-1}\mathbf{K}_r^{w\prime}$$

$$= \begin{bmatrix} 454.4 & 175.1 & -454.4 & -175.1 & 427 & 404 \\ 175.1 & 185 & -175.1 & -185 & -252.9 & 427 \\ -454.4 & -175.1 & 454.4 & 175.1 & -427 & -404 \\ -175.1 & -185 & 175.1 & 185 & 252.9 & -427 \\ 427 & -252.9 & -427 & 252.9 & 1883.2 & -583.6 \\ 404 & 427 & -404 & -427 & -583.6 & 985.4 \end{bmatrix} \text{MN/m} \tag{6.88}$$

and the governing equation expressed in Eq. (6.85) can be written as:

$$\left[\begin{array}{c:c} \bar{\mathbf{K}}_{\mathbf{w}2\times2} & \bar{\mathbf{K}}'_{\mathbf{w}2\times6} \\ \hdashline \bar{\mathbf{K}}'^{T}_{\mathbf{w}6\times2} & \bar{\mathbf{K}}''_{\mathbf{w}6\times6} \end{array}\right]\left\{\begin{array}{c} \mathbf{x}_{\mathbf{d}2\times1} \\ \hdashline -\boldsymbol{\Delta}''_{6\times1} \end{array}\right\}=\left\{\begin{array}{c} \mathbf{F}_{\mathbf{dw}2\times1} \\ \hdashline \mathbf{f}_{6\times1} \end{array}\right\} \tag{6.89}$$

in which,

$$\mathbf{x_d}=\begin{Bmatrix} x_1 \\ x_2 \end{Bmatrix},\quad \mathbf{F_{dw}}=\begin{Bmatrix} F_{1w} \\ F_{2w} \end{Bmatrix} \tag{6.90}$$

Now the shear wall is first assumed to be linear, this gives $\boldsymbol{\Delta}''=0$. Then, using the first equation of Eq. (6.89) gives:

$$\begin{bmatrix} 4035.8 & -1569 \\ -1569 & 985.4 \end{bmatrix}\times 10^6\begin{Bmatrix} 0.03 \\ 0.08 \end{Bmatrix}=\begin{Bmatrix} F_{1w} \\ F_{2w} \end{Bmatrix} \tag{6.91}$$

Solving for the forces in Eq. (6.91) gives:

$$\begin{Bmatrix} F_{1w} \\ F_{2w} \end{Bmatrix}=\begin{Bmatrix} -4.446 \\ 31.762 \end{Bmatrix}\ \text{MN} \tag{6.92}$$

Then substituting Eq. (6.92) back into the second equation of Eq. (6.89) gives:

$$\begin{Bmatrix} f_1 \\ f_2 \\ f_3 \\ f_4 \\ f_5 \\ f_6 \end{Bmatrix}=\begin{Bmatrix} 33.011 \\ 13.763 \\ -33.011 \\ -13.763 \\ 27.317 \\ 31.761 \end{Bmatrix}\ \text{MN} \tag{6.93}$$

Comparing these demands in Eq. (6.93) with the corresponding capacities in Table 6.1, the shear wall members is in the plastic domain, which means the assumption that the shear wall members are in the elastic domain is incorrect. Therefore, assume that the shear wall members are in the plastic domain, which means that the VSH#1, VSH#2, HSH#5 and HSH#6 begin to deform in the hardening branches. The stiffnesses $K_e=1500$ MN/m, $K_y=15$ MN/m, $K_{cr}=1050$ MN/m, $F_y=1.32$ MN/m, $K^s=2500$ MN/m, and $K_{scr}=400$ MN/m can be obtained or computed according to the parameters listed in Tables 6.9. Then, the spring force f_1, f_2, f_5, and f_6 can be expressed in terms of δ_1'', δ_2'', δ_5'', and δ_6'' according to Eq. (6.17) and Eq. (6.23) as:

$$\begin{cases} f_1=15.15\delta_1''+1.314 \\ f_2=15.15\delta_2''+1.314 \\ f_5=476.19\delta_5''+3.00 \\ f_6=476.19\delta_6''+3.00 \end{cases} \tag{6.94}$$

Therefore, Eq. (6.89) becomes:

$$\left[\begin{array}{cc:cccc} 4035.8 & -1569 & 23 & -679.9 & 2466.8 & -1569 \\ -1569 & 985.4 & 404 & 427 & -583.6 & 985.4 \\ \hdashline 23 & 404 & 454.4+15.15 & 175.1 & 427 & 404 \\ -679.9 & 427 & 175.1 & 185+15.15 & -252.9 & 427 \\ 2466.8 & -583.6 & 427 & -25.29 & 188.32+476.19 & -583.6 \\ -1569 & 985.4 & 404 & 427 & -583.6 & 985.4+476.19 \end{array}\right] \left\{\begin{array}{c} 0.03 \\ 0.08 \\ \hdashline -\delta_1'' \\ -\delta_2'' \\ -\delta_5'' \\ -\delta_6'' \end{array}\right\} = \left\{\begin{array}{c} F_{1w} \\ F_{2w} \\ \hdashline 1.314 \\ 1.314 \\ 3 \\ 3 \end{array}\right\} \tag{6.95}$$

Solving for Eq. (6.95) gives:

$$\begin{Bmatrix} f_1 \\ f_2 \\ f_5 \\ f_6 \end{Bmatrix} = \begin{Bmatrix} 3.2922 \\ 1.2574 \\ 3.1332 \\ 2.9017 \end{Bmatrix} \text{MN} \tag{6.96}$$

Comparing these spring force demands in Eq. (6.96) with the cracking and yielding capacities in Table 6.1 shows that the assumption that both the shear wall members are in the plastic domain is incorrect. Therefore, new trial should be assumed according to the status of the spring forces in Eq. (6.96). After two trials, assume that the left outer vertical springs of both the shear wall members and the shear spring of the second floor are in the plastic domain. Then the spring force f_1, f_2, and f_6 can be expressed in terms of δ_1'', δ_2'', and δ_6'' according to Eq. (6.17) and Eq. (6.23) as:

$$\begin{cases} f_1 = 15.15\delta_1'' + 1.314 \\ f_2 = 15.15\delta_2'' + 1.314 \\ f_6 = 476.2\delta_6'' + 3.000 \end{cases} \tag{6.97}$$

Therefore, Eq. (6.89) becomes:

$$\left[\begin{array}{cc:ccc} 4035.8 & -1569 & 23 & -679.9 & -1569 \\ -1569 & 985.4 & 404 & 427 & 985.4 \\ \hdashline 23 & 404 & 454.39+15.15 & 175.08 & 404.03 \\ -679.9 & 427 & 175.08 & 185.03+15.15 & 427 \\ -1569 & 985.4 & 404.03 & 427 & 985.38+476.2 \end{array}\right] \left\{\begin{array}{c} 0.03 \\ 0.08 \\ \hdashline -\delta_1'' \\ -\delta_2'' \\ -\delta_6'' \end{array}\right\} = \left\{\begin{array}{c} F_{1w} \\ F_{2w} \\ \hdashline 1.314 \\ 1.314 \\ 3 \end{array}\right\} \tag{6.98}$$

Solving for Eq. (6.98) gives:

$$\begin{Bmatrix} \delta_1'' \\ \delta_2'' \\ \delta_6'' \end{Bmatrix} = \begin{Bmatrix} 0.0657 \\ 0.0039 \\ 0.0004 \end{Bmatrix} \text{m}, \quad \begin{Bmatrix} F_{1w} \\ F_{2w} \\ f_1 \\ f_2 \\ f_6 \end{Bmatrix} = \begin{Bmatrix} -2.7173 \\ 3.1707 \\ 2.3103 \\ 1.3740 \\ 3.1707 \end{Bmatrix} \text{MN} \tag{6.99}$$

Then, the DOFs of the shear wall that were statically condensed, $\mathbf{x}_r^w = \{x_{11}\ x_{12}\}^T$, can be calculated as:

$$\mathbf{x}_r^w = \begin{Bmatrix} x_{11} \\ x_{12} \end{Bmatrix} = (\mathbf{K}_{rr}^w)^{-1}\mathbf{K}_r^w\boldsymbol{\theta}'' - (\mathbf{K}_{rr}{}^w)^{-1}\mathbf{K}_{rd}^w\mathbf{x}_\mathrm{d} = \begin{Bmatrix} 0.0115 \\ 0.0124 \end{Bmatrix} \text{ rad} \tag{6.100}$$

Then, substituting Eq. (6.99) back into Eq. (6.89) gives:

$$\begin{Bmatrix} \delta_1'' \\ \delta_2'' \\ \delta_3'' \\ \delta_4'' \\ \delta_5'' \\ \delta_6'' \end{Bmatrix} = \begin{Bmatrix} 0.0657 \\ 0.0039 \\ 0 \\ 0 \\ 0 \\ 0.0004 \end{Bmatrix} \text{m}, \quad \begin{Bmatrix} f_1 \\ f_2 \\ f_3 \\ f_4 \\ f_5 \\ f_6 \end{Bmatrix} = \begin{Bmatrix} 2.3103 \\ 1.3740 \\ -2.3103 \\ -1.3740 \\ 0.4535 \\ 3.1707 \end{Bmatrix} \text{MN}, \quad \begin{Bmatrix} F_{1w} \\ F_{2w} \end{Bmatrix} = \begin{Bmatrix} -2.7173 \\ 3.1707 \end{Bmatrix} \text{MN} \tag{6.101}$$

Finally, the restoring force of the RC frame-shear wall structure can be obtained by

$$\mathbf{F} = \mathbf{F_f} + \mathbf{F_w} = \begin{Bmatrix} F_{1f} \\ F_{2f} \end{Bmatrix} + \begin{Bmatrix} F_{1w} \\ F_{2w} \end{Bmatrix} = \begin{Bmatrix} -0.4167 \\ 1.6133 \end{Bmatrix} + \begin{Bmatrix} -2.7173 \\ 3.1707 \end{Bmatrix} = \begin{Bmatrix} -3.134 \\ 4.7840 \end{Bmatrix} \text{MN} \tag{6.102}$$

Figure 6.12 summarizes these results in graphical form.

6.4 Nonlinear Dynamic Analysis of RC Frame-Shear Wall Structures

6.4.1 Hysteretic Behaviors of the RC Shear Wall Members

The behavior of the shear wall member under cyclic loadings is composed by the flexural and shear behaviors. The flexural behavior of the shear wall member is simulated by the two outer vertical springs and the shear behavior is simulated by the horizontal spring. The basic hysteretic rules of above springs are illustrated in Figures 6.5 and 6.6. The detailed behaviors of the above branches of the outer vertical and horizontal springs can be obtained by using the equations Eq. (6.15) to Eq. (6.26).

For a RC shear wall member, the governing equation written in Eq. (6.30) can be rewritten here in incremental form as:

$$\left[\begin{array}{cc|ccc} K^s & -K^s(h-h_c) & 0 & 0 & K^s \\ -K^s(h-h_c) & K^s(h-h_c)^2 + K_e\dfrac{l^2}{2} & K_e\dfrac{l}{2} & -K_e\dfrac{l}{2} & -K^s(h-h_c) \\ \hline 0 & K_e\dfrac{l}{2} & \dfrac{K_e}{2} & -\dfrac{K_e}{2} & 0 \\ 0 & -K_e\dfrac{l}{2} & -\dfrac{K_e}{2} & \dfrac{K_e}{2} & 0 \\ K^s & -K^s(h-h_c) & 0 & 0 & K^s \end{array}\right] \left\{\begin{array}{c} x_{(k+1)} \\ \theta_{(k+1)} \\ \hline -(\delta_{1(k)}'' + d\delta_1'') \\ -(\delta_{2(k)}'' + d\delta_2'') \\ -(\delta_{3(k)}'' + d\delta_3'') \end{array}\right\} = \left\{\begin{array}{c} F_{(k+1)} \\ M_{(k+1)} \\ \hline f_{1(k+1)} \\ f_{2(k+1)} \\ f_{3(k+1)} \end{array}\right\} \tag{6.103}$$

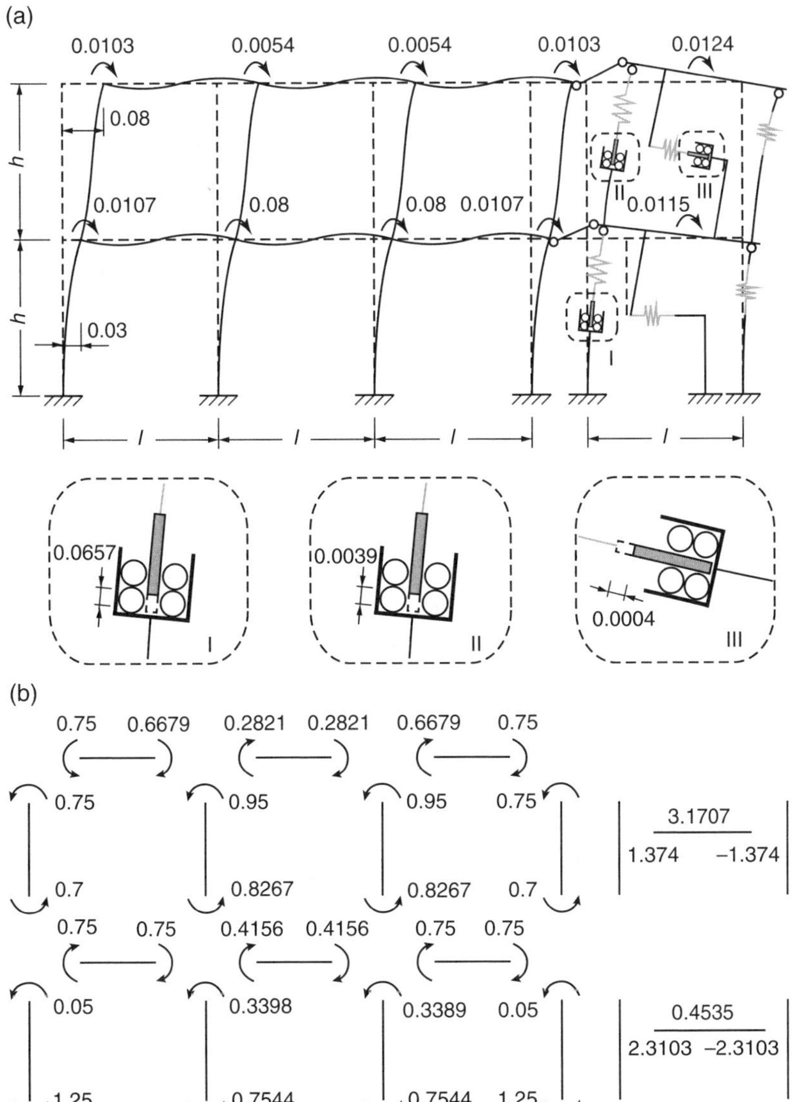

Figure 6.12 Graphical illustration of the response of the RC frame-shear wall structure: (a) Deformations; (b) Internal forces.

where the subscript k denotes the kth time step and the parameters with subscript k, $\delta''_{(k)}$, are known for the time step $k+1$; $d\delta''_1$, $d\delta''_2$and $d\delta''_3$ are the incremental plastic deformations of the VSHs and HSH, respectively, at time step $k+1$. For a given displacement at time step $k+1$, x_{k+1} and θ_{k+1}, there are eight unknowns, namely F_{k+1}, M_{k+1}, $f_{1(k+1)}$, $f_{2(k+1)}$, $f_{3(k+1)}$, $d\delta''_1$, $d\delta''_2$, and $d\delta''_3$, in Eq. (6.103) with five equations provided. Therefore, together with additional three relationships, the left vertical spring force f_1 versus plastic displacement δ''_1, the right vertical spring force f_2 versus plastic displacement δ_2'' and the shear spring f_3 force versus plastic displacement δ_3'' as described in the previous section, the eight unknowns can be uniquely solved.

Example 6.4 Hysteretic Response Analysis of the Shear Wall Member
Consider again the RC shear wall member used in Example 6.1. Now, the shear wall structure is subjected to a set of cyclic input displacements. The input displacements **x** tracking with the step of 0.1 mm are shown in Figure 6.13.

By using Eq. (6.103), the hysteretic response of the shear wall member can be computed step by step with the set of input displacements **x** shown in Figure 6.13. The obtained hysteretic loops are shown in Figures 6.14 and 6.15. The detailed results of the first two hysteretic responses of the shear wall member are listed in Table 6.3.

6.4.2 Solution Procedure of the FAM

For the shear wall member, consider the second equation in Eq. (6.103) with the input displacement $\mathbf{x}_{k+1}$ at time step $k+1$ coming from the state space method of analysis expressed in Chapter 3. To take care of the loading and unloading issues of the springs in the dynamic analysis, let the plastic deformation vector $\mathbf{\Delta}_{k+1}''$ at time step $k+1$ be written in an incremental form as:

$$\mathbf{\Delta}_{\mathrm{k}+1} = \mathbf{\Delta}_{\mathrm{k}} + d\mathbf{\Delta}'' \tag{6.104}$$

where $d\mathbf{\Delta}''$ is the change in plastic deformations at VSHs and HSH from time step k to time step $k+1$, and $\mathbf{\Delta}_k''$ is the discretized form of $\mathbf{\Delta}''(t)$.

Then the second governing equation expressed in Eq. (6.103) can be rewritten here in incremental form as:

$$\mathbf{f}_{\mathrm{k}+1} + \mathbf{K}'' d\mathbf{\Delta}'' = \mathbf{K}'^{\mathrm{T}} \mathbf{x}_{\mathrm{k}+1} - \mathbf{K}'' \mathbf{\Delta}_{\mathrm{k}}'' \tag{6.105}$$

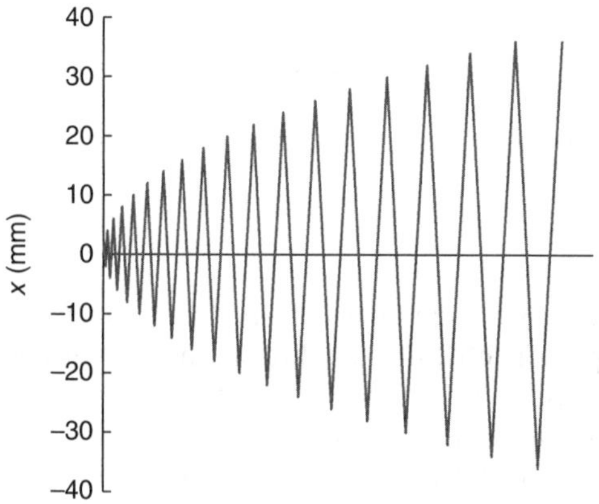

Figure 6.13 Cyclic loading history.

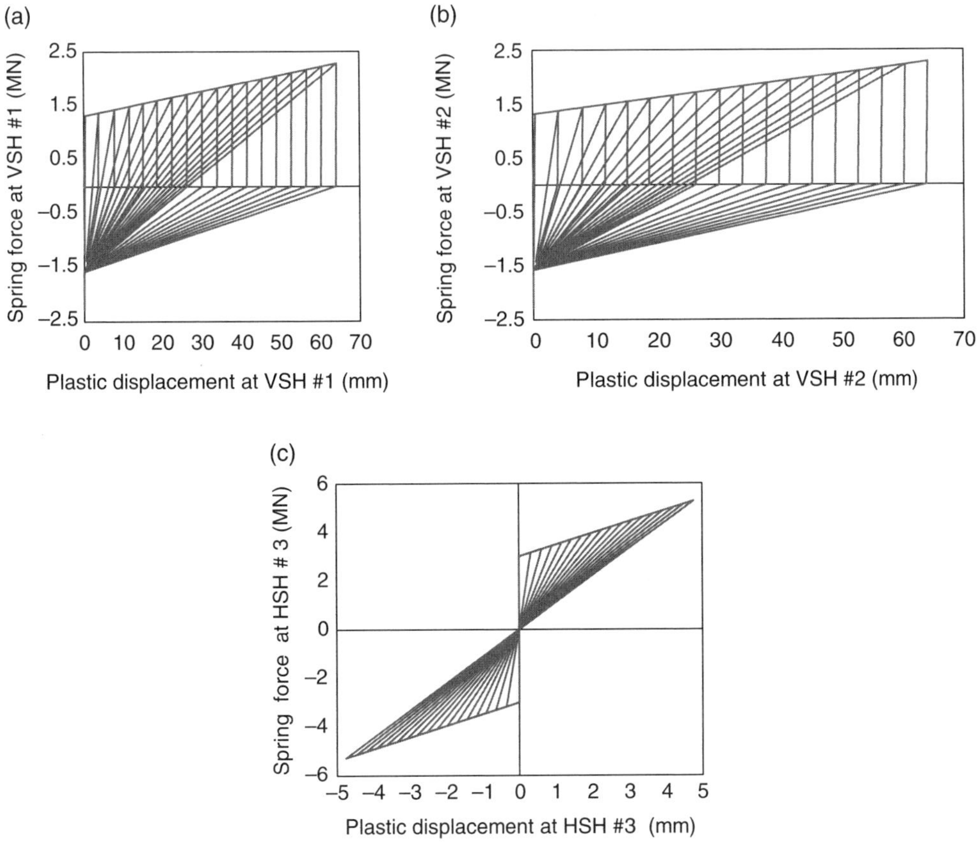

Figure 6.14 Hysteretic behaviors of local plastic mechanisms: (a) hysteretic behavior of VSH #1; (b) hysteretic behavior of VSH #2; (c) hysteretic behavior of HSH #3.

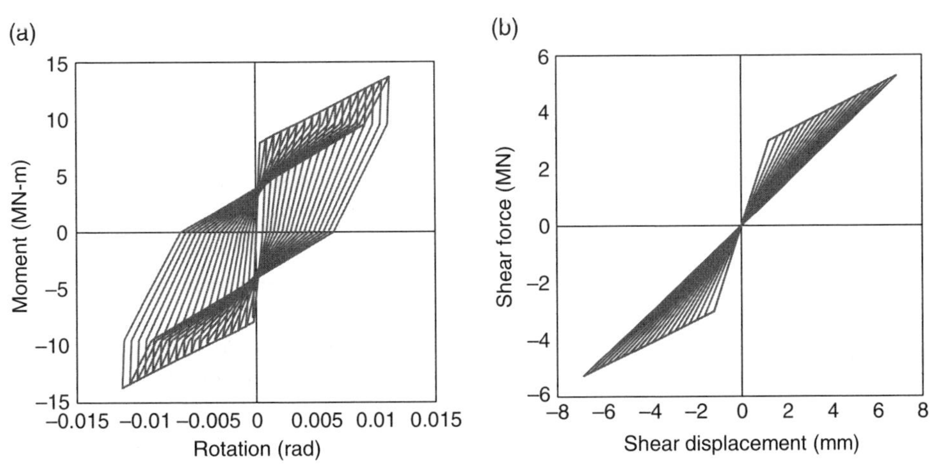

Figure 6.15 Hysteretic behaviors of the RC shear wall structure: (a) hysteretic behavior of moment versus rotation; (b) hysteretic behavior of shear force versus shear displacement.

Table 6.3 Numerical results of the shear wall subjected to cyclic loadings.

status	x (mm)	x'' (10^{-3} mm)	θ'' (10^{-6} rad)	Local plastic displacement (10^{-2} mm) δ_1''	δ_2''	δ_3''	Moment M (MN-m)	Shear force F (MN)
Loading	0	0	0	0	0	0	0	0
	0.1	7.62	2.93	1.76	0	0	0.369	0.142
	0.2	15.24	5.86	3.52	0	0	0.739	0.284
	⋮	⋮	⋮	⋮	⋮	⋮	⋮	⋮
	1.9	144.79	55.69	33.41	0	0	7.017	2.699
	2	152.41	58.62	35.17	0	0	7.386	2.841
Un-	1.9	148.96	57.29	34.38	0	0	7.000	2.692
loading	1.8	145.50	55.96	33.58	0	0	6.614	2.544
	⋮	⋮	⋮	⋮	⋮	⋮	⋮	⋮
	−1.9	−123.78	−47.61	5.25	33.81	0	−7.101	−2.731
	−2	−134.36	−51.68	4.51	35.52	0	−7.458	−2.869
Loading	−1.9	−127.67	−49.10	5.28	34.74	0	−7.086	−2.725
	−1.8	−120.97	−46.53	6.05	33.97	0	−6.712	−2.582
	⋮	⋮	⋮	⋮	⋮	⋮	⋮	⋮
	1.9	126.78	48.76	34.56	5.30	0	7.089	2.727
	2	134.27	51.64	35.52	4.53	0	7.459	2.867
Re-	2.1	147.85	55.58	37.16	3.81	0.33	7.804	3.002
loading	2.2	224.81	56.56	37.56	3.62	7.78	7.896	3.037
	⋮	⋮	⋮	⋮	⋮	⋮	⋮	⋮
	3.9	1851.91.	591.84	358.12	3.02	31.3	8.188	3.150
	4	1947.62	623.33	376.98	2.98	32.7	8.205	3.156
Un-	3.9	1928.78	621.06	376.29	3.66	31.40	7.880	3.031
loading	3.8	1909.94	618.78	375.60	4.33	30.11	7.556	2.906
	⋮	⋮	⋮	⋮	⋮	⋮	⋮	⋮
	−3.9	−1843.54	−578.21	48.03	394.96	−34.02	−8.221	−3.162
	−4	−1939.36	−609.88	47.40	413.33	−35.37	−8.238	−3.169
Loading	−3.9	−1903.64	−895.98	52.55	412.78	−34.26	−7.981	−3.070
	−3.8	−1867.92	−876.96	57.70	412.23	−33.16	−7.724	−2.971
	⋮	⋮	⋮	⋮	⋮	⋮	⋮	⋮
	3.9	1842.67	881.42	398.78	52.70	34.30	8.225	3.163
	4	1938.50	913.72	417.10	52.07	35.64	8.241	3.170
	⋮	⋮	⋮	⋮	⋮	⋮	⋮	⋮

where $\mathbf{f}_{k+1}$ is the discretized form of $\mathbf{f}(t)$. Each item on the right side of Eq. (6.105) is known and the objective is to calculate $\mathbf{f}_{k+1}$ and $d\mathbf{\Delta}''$. If the number of the VSHs and HSHs are $2n_v$ and n_h, respectively, there are totally $(2n_v + n_h)$ unknowns. However, Eq. (6.105) by itself only provides $(2n_v + n_h)$ equations, and $2n_v$ outer vertical spring forces versus axial plastic displacement and n_h shear spring force versus shear plastic displacement relationships for VSHs and HSHs should be incorporated for solving the $(2n_v + n_h)$ unknowns uniquely. Once $\mathbf{f}_{k+1}$ and $d\boldsymbol{\Delta}''$ are solved, the plastic deformation at time step $k+1$, $\boldsymbol{\Delta}''_{k+1}$, can be obtained using Eq. (6.104). Then, the plastic displacement $\mathbf{x}''_{k+1}$ can be calculated as:

$$\mathbf{x}''_{k+1} = \mathbf{K}^{-1}\mathbf{K}'\boldsymbol{\Delta}''_{k+1} \tag{6.106}$$

This completes the portion of using the FAM in computing the internal forces of the shear wall structures, and the inelastic displacement $\mathbf{x}''_{k+1}$ is passed back to the state space method to compute the structural response at the next step.

While for the RC frame-shear wall structure, the governing equation in the FAM can be expressed as:

$$\left[\begin{array}{c|cc} \mathbf{K} & \mathbf{K}'_{\mathbf{f}} & \mathbf{K}'_{\mathbf{w}} \\ \hline \mathbf{K}'^{\mathbf{T}}_{\mathbf{f}} & \mathbf{K}''_{\mathbf{f}} & \mathbf{0} \\ \mathbf{K}'^{\mathbf{T}}_{\mathbf{w}} & \mathbf{0} & \mathbf{K}''_{\mathbf{w}} \end{array}\right] \left\{\begin{array}{c} \mathbf{x} \\ \hline -\boldsymbol{\theta}'' \\ -\boldsymbol{\Delta}'' \end{array}\right\} = \left\{\begin{array}{c} \mathbf{F} \\ \hline \mathbf{m} \\ \mathbf{f} \end{array}\right\} \tag{6.107}$$

in which the second and third governing equations in Eq. (6.107) can be rewritten here in incremental form as:

$$\begin{cases} \mathbf{m}_{k+1} + \mathbf{K}''_{\mathbf{f}} d\boldsymbol{\theta}'' = \mathbf{K}'^{\mathbf{T}}_{\mathbf{f}} \mathbf{x}_{k+1} - \mathbf{K}''_{\mathbf{f}} \boldsymbol{\theta}''_k \\ \mathbf{f}_{k+1} + \mathbf{K}''_{\mathbf{w}} d\boldsymbol{\Delta}'' = \mathbf{K}'^{\mathbf{T}}_{\mathbf{w}} \mathbf{x}_{k+1} - \mathbf{K}''_{\mathbf{w}} \boldsymbol{\Delta}''_k \end{cases} \tag{6.108}$$

Then the plastic displacement $\mathbf{x}''_{k+1}$ of the structure can be calculated as

$$\mathbf{x}''_{k+1} = \mathbf{K}^{-1}\left(\mathbf{K}'_{\mathbf{f}}\boldsymbol{\theta}''_{k+1} + \mathbf{K}'_{\mathbf{w}}\boldsymbol{\Delta}''_{k+1}\right) \tag{6.109}$$

The plastic displacement $\mathbf{x}''_{k+1}$ is passed back to the state space method to compute the response at the next step.

For a given input displacement at time step $k+1$, $\mathbf{x}_{k+1}$, assuming that the structure is unloading, which means all the local plastic mechanisms are unloading. The unloading points of VSHs and HSHs are related with internal forces f and corresponding plastic deformations δ'' at time step k. Substituting these values back to the corresponding equations from Eq. (6.15) to Eq. (6.26), the unloading tangent stiffnesses of the local plastic mechanisms VSHs and HSHs, at time step $k+1$ can be obtained. As soon as the above tangent unloading stiffnesses are obtained, the additional equations for solving Eq. (6.105) are ready, and $\mathbf{f}_{k+1}$ and $d\boldsymbol{\Delta}''$ can be calculated. Then, comparing the internal force values between f_k and f_{k+1} for each local plastic mechanism, if $|f_{k+1}| < |f_k|$, the unloading condition of the VSHs or HSHs is satisfied, which means the assumption that these VSHs and HSHs are unloading at time step $k+1$ is correct; otherwise the VSHs and HSHs of the shear wall members are loading or reloading at time step $k+1$.

Example 6.5 Nonlinear dynamic analysis of the two-story RC frame-shear wall structure subjected to cyclic increasing excitation

Consider again the 2-story RC frame-wall structure expressed in Example 6.3. Now, the structure is subjected to the cyclic increasing excitation with peak acceleration 8 m/s^2 and time step 0.01 s as shown in Figure 6.16.

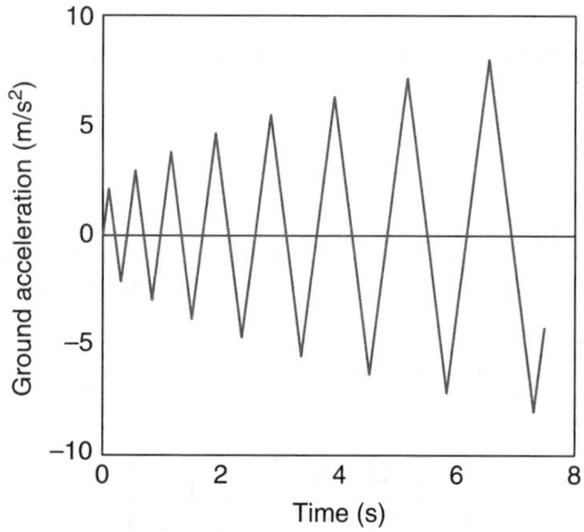

Figure 6.16 The cyclic increasing ground acceleration.

After static condensation, the global stiffness matrix can be written as:

$$\bar{\mathbf{K}} = \begin{bmatrix} 4230.2 & -1646.9 \\ -1646.9 & 1037.3 \end{bmatrix} \text{MN/m} \tag{6.110}$$

In addition, the $\bar{\text{K}}'$ and $\bar{\text{K}}''$ matrices can be similarly calculated according to the expression of the static condensation procedure in Chapter 2. The mass matrix $\mathbf{M}_{dd}$ can be written as:

$$\mathbf{M}_{dd} = \begin{bmatrix} 550 & 0 \\ 0 & 550 \end{bmatrix} \text{Mg} \tag{6.111}$$

With the damping ratio 2%, the damping matrix $\mathbf{C}_{dd}$ can be written as:

$$\mathbf{C}_{dd} = \begin{bmatrix} 1.849 & -0.5512 \\ -0.5512 & 0.7804 \end{bmatrix} \text{MN s/m} \tag{6.112}$$

Then the equations of motion for the structure can be written as:

$$\mathbf{M}_{dd}\ddot{\mathbf{X}}_d(t) + \mathbf{C}_{dd}\dot{\mathbf{X}}_d(t) + \bar{\mathbf{K}}\mathbf{x}'_d(t) = -\mathbf{M}_{dd}\ddot{\mathbf{g}}(t) \tag{6.113}$$

The transition matrices **A**, **H**, and **G** can be calculated using Eq. (3.139), and the discrete form, $\mathbf{F}_d$, $\mathbf{H}_d$, and $\mathbf{G}_d$, can be obtained using Eq. (3.138).

The displacement responses of the floors are shown in Figures 6.17. The hysteretic loops of moment versus rotation and shear force versus shear displacement of the first floor are plotted in Figure 6.18. Figure 6.19 shows the responses of selected hinges. Figure 6.20 shows the hysteretic loops of selected hinges.

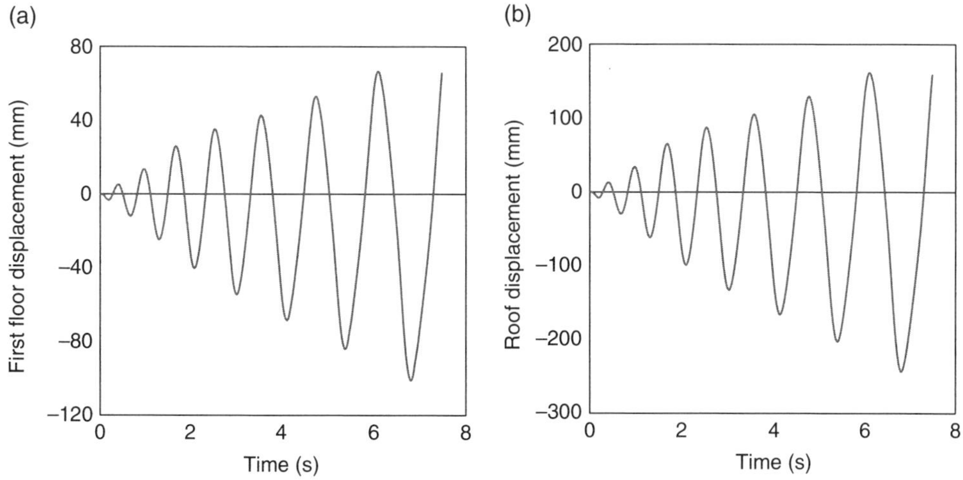

Figure 6.17 Global displacement responses of the two-story RC frame-shear wall structure: (a) response of the first floor; (b) response of the second floor.

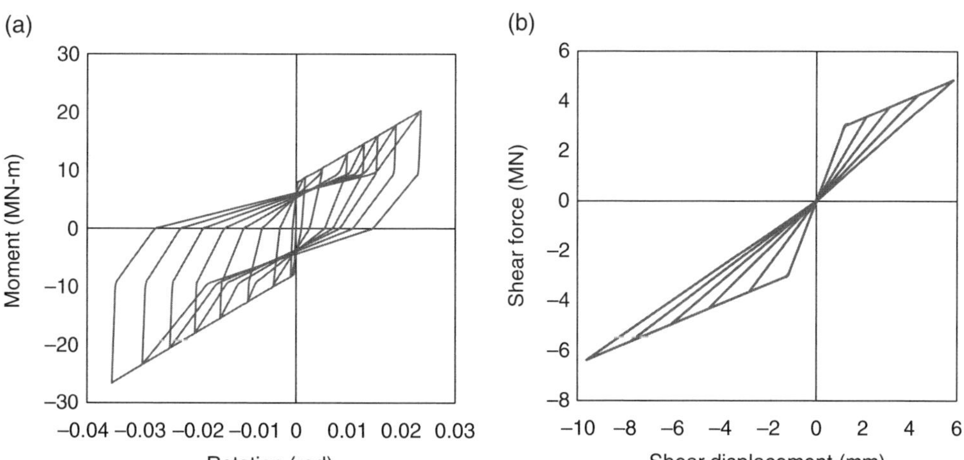

Figure 6.18 Hysteretic behaviors of the first floor: (a) hysteretic behavior of moment versus rotation; (b) hysteretic behavior of shear force versus shear displacement.

Example 6.6 Nonlinear dynamic analysis of a RC frame-shear wall structure subjected to earthquake excitation

Consider again the two-story RC frame-shear wall structure as expressed in Example 6.3. Now the structure is subjected to the El Centro earthquake ground motion with peak

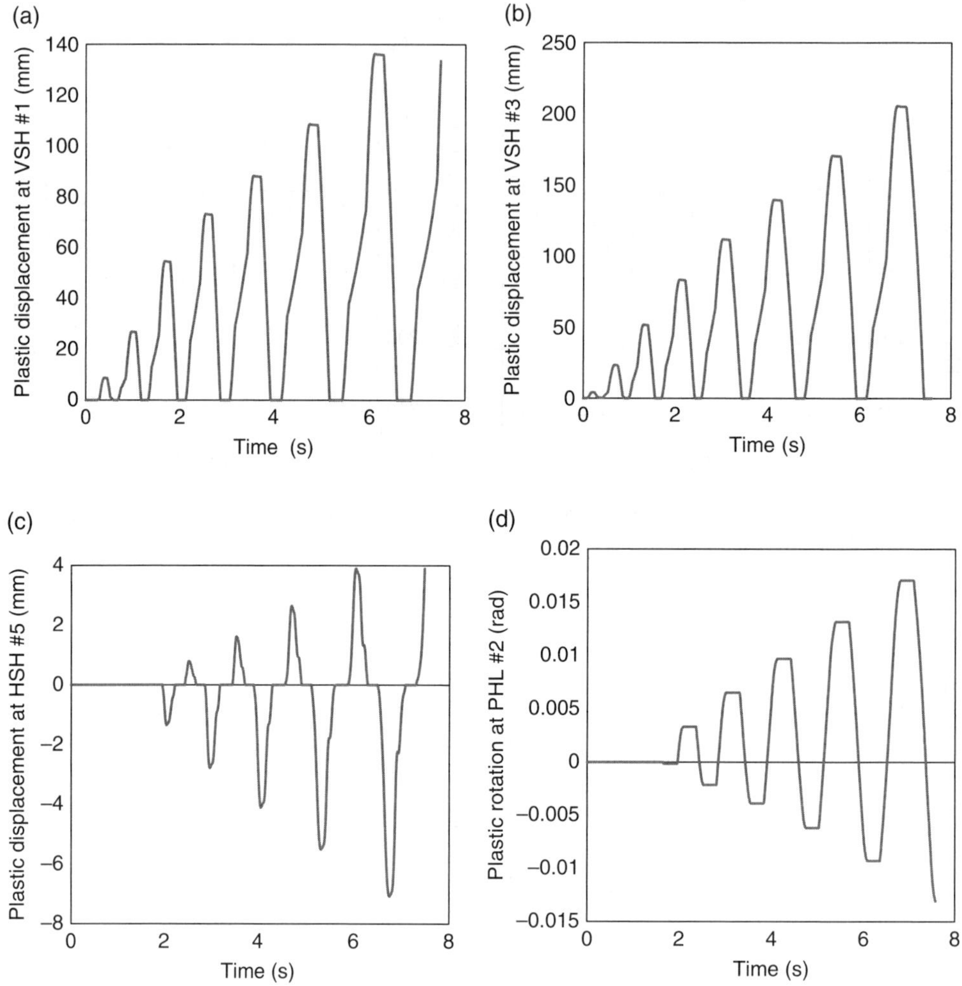

Figure 6.19 Plastic responses at selected local plastic mechanisms: (a) plastic response at VSH #1; (b) plastic response at VSH #3; (c) plastic response at HSH #5; (d) plastic response at PHL #2.

acceleration scaled to $9.5\,m/s^2$. The displacement and velocity responses of the floors are shown in Figures 6.21 and 6.22. The hysteretic loops of moment versus rotation and shear force versus shear displacement of the first floor are plotted in Figure 6.23. Figure 6.24 shows the responses of selected hinges. Figure 6.25 shows the hysteretic behaviors of selected hinges.

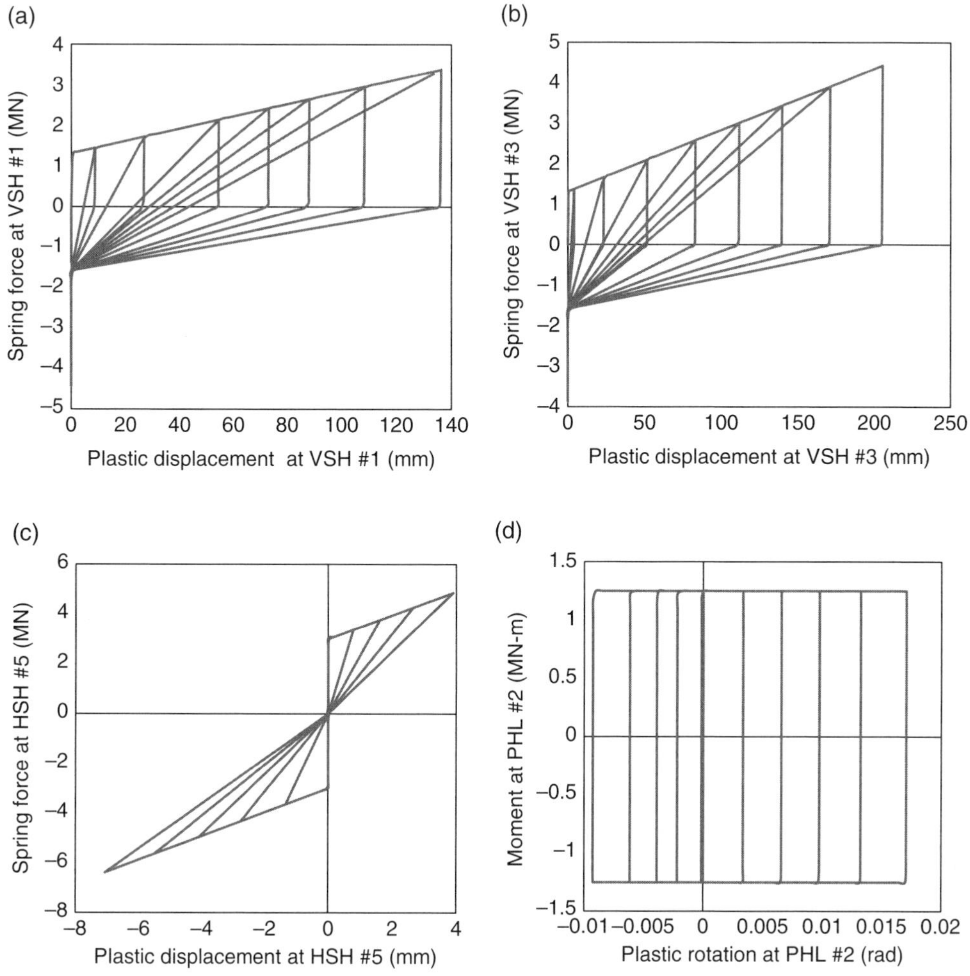

Figure 6.20 Hysteretic behaviors at selected local plastic mechanisms: (a) hysteretic behavior at VSH #1; (b) hysteretic behavior at VSH #3; (c) hysteretic behavior at HSH #5; (d) hysteretic behavior at PHL #2.

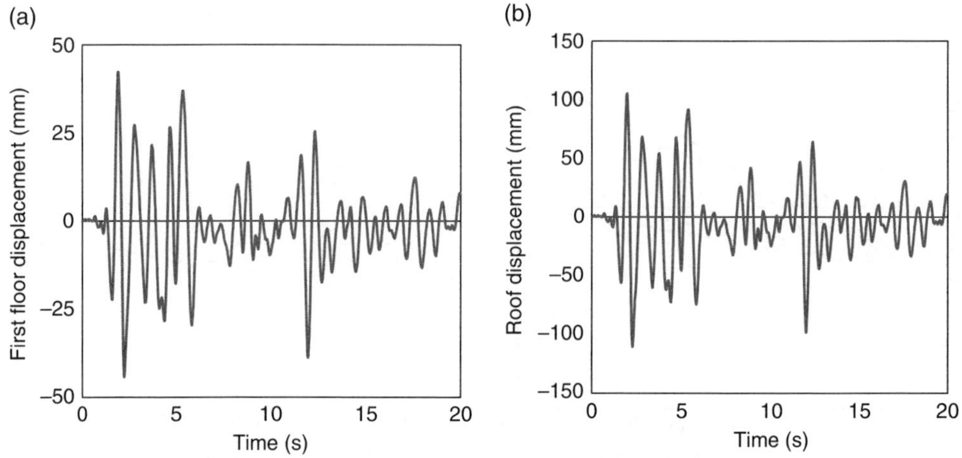

Figure 6.21 Global displacement responses: (a) response of the first floor; (b) response of the second floor.

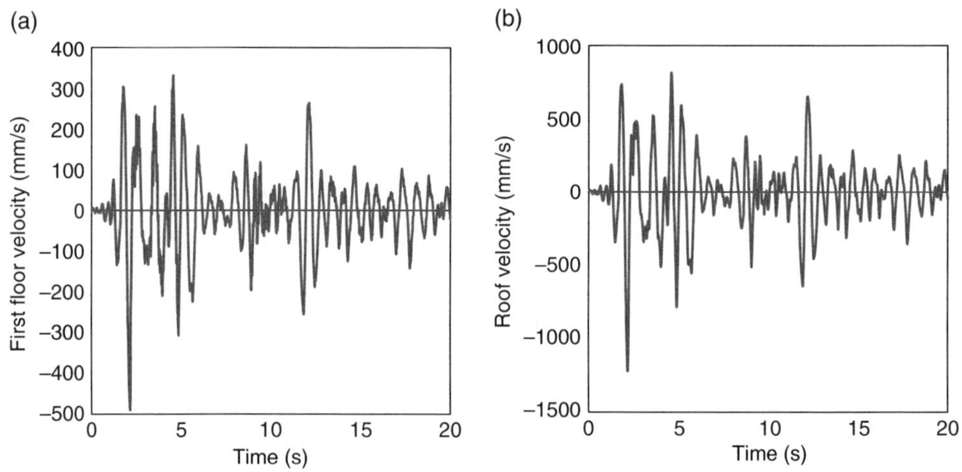

Figure 6.22 Global velocity responses: (a) response of the first floor; (b) response of the second floor.

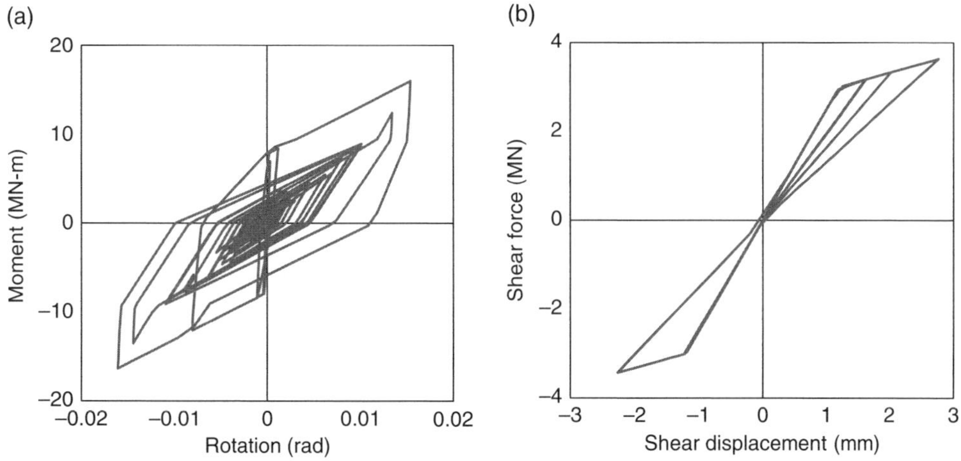

Figure 6.23 Hysteretic behaviors of the first floor: (a) hysteretic behavior of moment versus rotation; (b) hysteretic behavior of shear force versus shear displacement.

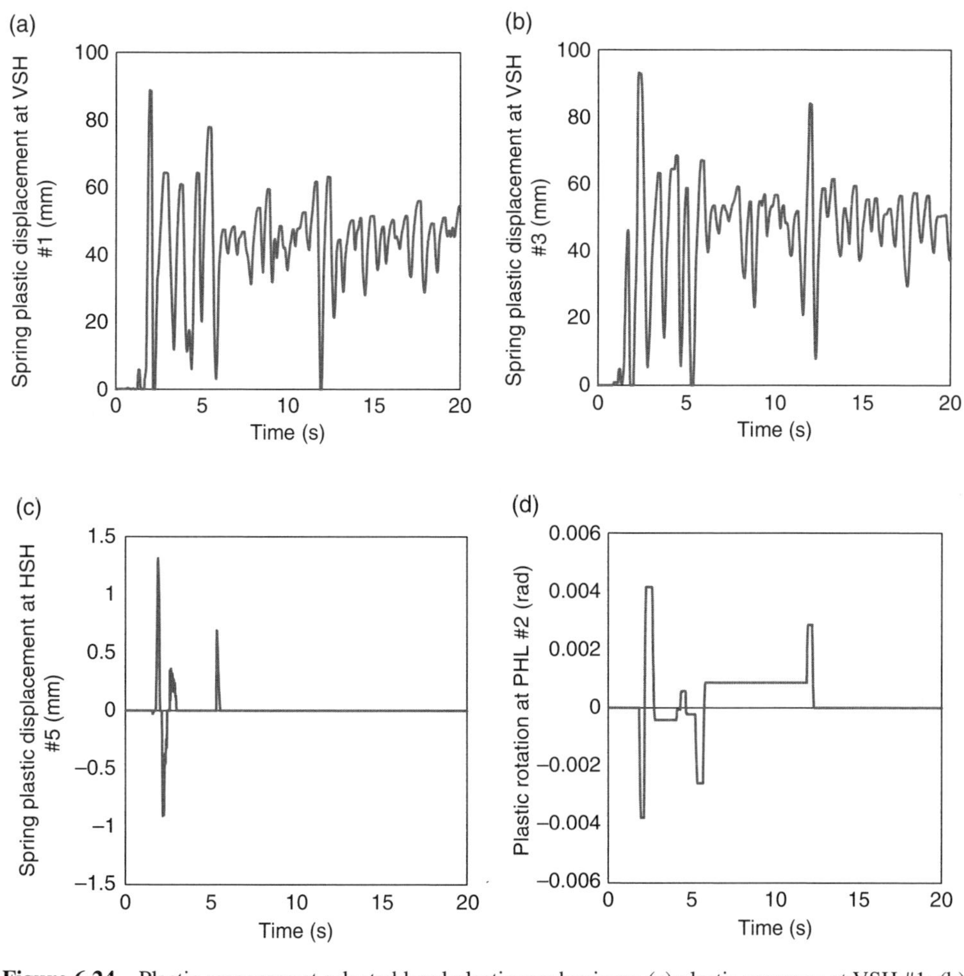

Figure 6.24 Plastic responses at selected local plastic mechanisms: (a) plastic response at VSH #1; (b) plastic response at VSH #3; (c) plastic response at HSH #5; (d) plastic response at PHL #2.

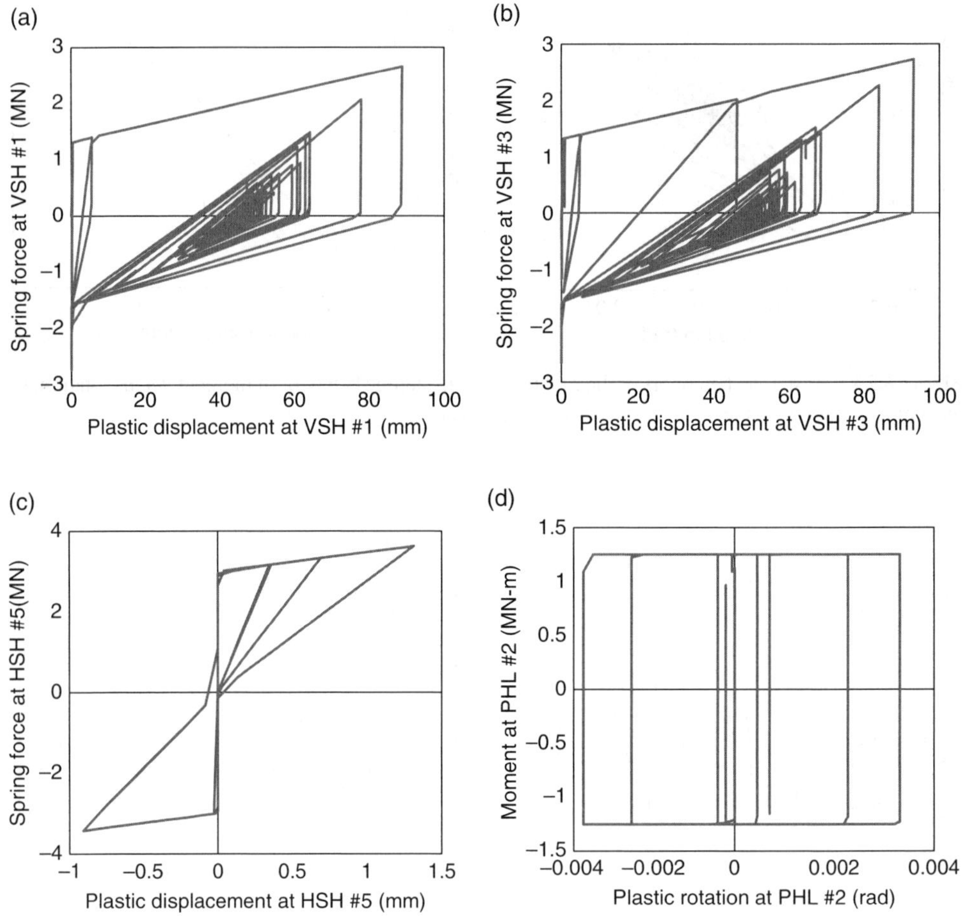

Figure 6.25 Hysteretic behaviors at selected local plastic mechanisms: (a) hysteretic behavior at VSH #1; (b) hysteretic behavior at VSH #3; (c) hysteretic behavior at HSH #5; (d) hysteretic behavior at PHL #2.

References

Linde P. and Bachmann H. (1994). "Dynamic modeling and design of earthquake-resistant walls". *Earthquake Engineering & Structural Dynamics*, 23(12):1331–50.

7

Geometric Nonlinearity

The theory of nonlinear static analysis of moment-resisting frames using the Force Analogy Method (FAM) was presented in Chapter 2, and the theory of nonlinear dynamic analysis of moment-resisting frames using the combination of the FAM and the state space method was presented in Chapter 3. In this chapter, the FAM combined with the state space method for solving nonlinear dynamic problems is extended to include geometric nonlinearity in the analysis. While the traditional method of nonlinear analysis with material nonlinearity uses a changing stiffness to quantify the stiffness force in the structure, it creates a conflict with geometric nonlinearity in defining the appropriate stiffness matrix that can represent both geometric nonlinearity and material nonlinearity. However, in the FAM, because it uses initial stiffness with changing displacements in the formulation to quantify the stiffness force in the structure, it works well with geometric nonlinearity that requires updating of stiffness matrix due to changes in axial compressive forces in the column members. The end result is that the stiffness force is computed by simply multiplying a nonlinear time-varying stiffness matrix with a nonlinear time-varying displacement vector.

There are several approaches that can be used to address geometric nonlinearity with changing stiffness. Since geometric nonlinearity must be updated at every time step of the analysis due to changes in axial compressive forces in the columns, no computational advantage is expected to be gained when any one of the geometric nonlinearity approaches is used with the state space method in the nonlinear dynamic analysis. But previous attempts have suggested that the geometric stiffness approach with its linear approximation is too sensitive in the FAM, which reduced its potential applications (Zhao and Wong 2006). Therefore, a more rigorous approach of using the stability functions is adopted in this analysis.

Theory of Nonlinear Structural Analysis: The Force Analogy Method for Earthquake Engineering,
First Edition. Gang Li and Kevin K.F. Wong.

In this chapter, different approaches of addressing geometric nonlinearity are first presented in Sections 7.1 and 7.2. This is followed by a detailed presentation on combining the FAM with the stability functions approach of geometric nonlinearity for solving fully nonlinear static problems in Section 7.3. The result is then further combined with the state space method of dynamic analysis in Section 7.4 for solving fully nonlinear dynamic problems. Static condensation is included in the derivation of the fully nonlinear dynamic analysis procedure for solving moment-resisting frame problems in Section 7.5. Finally, Section 7.6 presents additional examples of analyzing the nonlinear dynamics of real moment-resisting frames.

7.1 Classical Stiffness Matrices with Geometric Nonlinearity

Two commonly used methods of analyzing structures with geometric nonlinearity today are the P-Δ approach and the geometric stiffness approach. Both of these approaches use the lateral deflected shapes $v(x)$ in the form of cubic polynomials (sometimes called the Hermite polynomials) drawn according to their corresponding shapes shown in Figure 7.1. The equations for the four cases of deflected shapes are as follows:

- *Case 1*: $v(0)=1$, $v'(0)=0$, $v(L)=0$, and $v'(L)=0$

$$v(x)=1-3\left(\frac{x}{L}\right)^2+2\left(\frac{x}{L}\right)^3 \tag{7.1}$$

- *Case 2*: $v(0)=0$, $v'(0)=1$, $v(L)=0$, and $v'(L)=0$

$$v(x)=L\times\left[\left(\frac{x}{L}\right)-2\left(\frac{x}{L}\right)^2+\left(\frac{x}{L}\right)^3\right] \tag{7.2}$$

- *Case 3*: $v(0)=0$, $v'(0)=0$, $v(L)=1$, and $v'(L)=0$

$$v(x)=3\left(\frac{x}{L}\right)^2-2\left(\frac{x}{L}\right)^3 \tag{7.3}$$

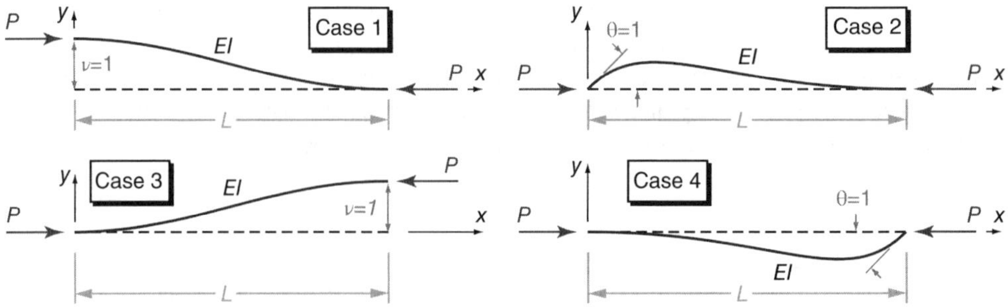

Figure 7.1 Four cases of deflected shapes.

- *Case 4*: $v(0) = 0$, $v'(0) = 0$, $v(L) = 0$, and $v'(L) = 1$

$$v(x) = L \times \left[-\left(\frac{x}{L}\right)^2 + \left(\frac{x}{L}\right)^3 \right] \tag{7.4}$$

where L is the length of the member. The member is subjected to an axial compressive load P in all four cases.

7.1.1 The P-Δ Approach

In this approach, the Bernoulli–Euler beam theory with "plane sections remain plane" is assumed. The strain at any point along the member (x-direction) and cross section (y-direction) can be expressed as:

$$\varepsilon(x, y) = \frac{du}{dx} - y\frac{d^2v}{dx^2} \tag{7.5}$$

where u is the displacement at any point of the member in the x-direction (or horizontal displacement), and v is the displacement at any point of the member in the y-direction (or vertical displacement). The displacement in the x-direction can be linearly interpolated and expressed as

$$u(x) = \left(1 - \frac{x}{L}\right)u(0) + \frac{x}{L}u(L) \tag{7.6}$$

where $u(0)$ is the horizontal displacement at the left-end of the member, and $u(L)$ is the horizontal displacement at the right-end of the member.

Equation (7.5) shows that the P-Δ approach begins with the same equation for kinematic relationships as the classical beam theory with curvature calculated as $\phi(x) = d^2v/dx^2$. Geometric nonlinearity is then included when equilibrium is considered, namely:

$$m(x) = EI\frac{d^2v}{dx^2}, \quad V(x) = EI\frac{d^3v}{dx^3} + P \times \Theta \tag{7.7}$$

where $m(x)$ and $V(x)$ the moment and shear along the member length. The term Θ is usually set to be $\Theta = dv/dx$ in the classical beam theory with both large P-Δ and small P-δ effects of geometric nonlinearity included, but it is here simplified to capture only the large P-Δ effect as:

$$\Theta = \frac{v(L) - v(0)}{L} \tag{7.8}$$

Substituting Eqs. (7.1) to (7.4) for the four cases into Eq. (7.7) and evaluate the shear and moment at the two ends of the member gives the stiffness matrix of the ith member as:

$$\mathbf{K}_i = \begin{bmatrix} \frac{12EI}{L^3} & \frac{6EI}{L^2} & -\frac{12EI}{L^3} & \frac{6EI}{L^2} \\ \frac{6EI}{L^2} & \frac{4EI}{L} & -\frac{6EI}{L^2} & \frac{2EI}{L} \\ -\frac{12EI}{L^3} & -\frac{6EI}{L^2} & \frac{12EI}{L^3} & -\frac{6EI}{L^2} \\ \frac{6EI}{L^2} & \frac{2EI}{L} & -\frac{6EI}{L^2} & \frac{4EI}{L} \end{bmatrix} + \begin{bmatrix} -\frac{P}{L} & 0 & \frac{P}{L} & 0 \\ 0 & 0 & 0 & 0 \\ \frac{P}{L} & 0 & -\frac{P}{L} & 0 \\ 0 & 0 & 0 & 0 \end{bmatrix} \tag{7.9}$$

Because the geometric nonlinear stiffness matrix in Eq. (7.9) is a simple linear function of the axial force per unit length with other terms being set equal to zero, it has been used extensively in many commercial software packages (e.g., Perform-3D (CSI 2011) and OpenSees (Regents of UC 2000)). However, keep in mind that Eq. (7.9) is the stiffness matrix that considers only the large P-Δ effect but ignores the small P-δ effect. This approach for handling geometric nonlinearity will be difficult to incorporate into the force analogy method, which requires a detailed expression to capture the moment magnification due to the small P-δ effect along any point on the member with material nonlinearity.

7.1.2 The Geometric Stiffness Approach

In this approach, both large P-Δ and small P-δ are considered by assuming that the strain at any point along the member and cross section is second-order with all higher-order terms truncated. This gives

$$\varepsilon(x,y) = \frac{du}{dx} - y\frac{d^2v}{dx^2} + \frac{1}{2}\left[\left(\frac{du}{dx}\right)^2 + \left(\frac{dv}{dx}\right)^2\right] \tag{7.10}$$

Because u and v are functions of x only, and the strain $\varepsilon(x,y)$ has a linear variation of y along the cross section as shown in the second term of Eq. (7.10), the assumption "plane sections remain plane" is again employed. The minimum energy method can be used to compute the stiffness matrix:

$$W = \int_V \sigma(x,y)\varepsilon(x,y)dV \tag{7.11}$$

where W is the work done due to the internal strain energy, $\sigma(x,y)$ is the stress at any point along the member and cross section, and the integration is done over the entire volume of the beam member. Substituting Eq. (7.10) into Eq. (7.11) and using the interpolation functions for the displacements in Eqs. (7.1) to (7.4) and Eq. (7.6) rewritten as:

$$u(x) = \left(1 - \frac{x}{L}\right)u(0) + \frac{x}{L}u(L) \tag{7.12a}$$

$$v(x)=\left[1-3\left(\frac{x}{L}\right)^2+2\left(\frac{x}{L}\right)^3\right]v(0)+\left[\left(\frac{x}{L}\right)-2\left(\frac{x}{L}\right)^2+\left(\frac{x}{L}\right)^3\right]Lv'(0)$$

$$+\left[3\left(\frac{x}{L}\right)^2-2\left(\frac{x}{L}\right)^3\right]v(L)+\left[-\left(\frac{x}{L}\right)^2+\left(\frac{x}{L}\right)^3\right]Lv'(L) \tag{7.12b}$$

The resulting integration and minimization of Eq. (7.11) gives the stiffness matrix of the *i*th member as:

$$\mathbf{K}_i=\begin{bmatrix} \frac{12EI}{L^3} & \frac{6EI}{L^2} & -\frac{12EI}{L^3} & \frac{6EI}{L^2} \\ \frac{6EI}{L^2} & \frac{4EI}{L} & -\frac{6EI}{L^2} & \frac{2EI}{L} \\ -\frac{12EI}{L^3} & -\frac{6EI}{L^2} & \frac{12EI}{L^3} & -\frac{6EI}{L^2} \\ \frac{6EI}{L^2} & \frac{2EI}{L} & -\frac{6EI}{L^2} & \frac{4EI}{L} \end{bmatrix}+\begin{bmatrix} -\frac{6P}{5L} & -\frac{P}{10} & \frac{6P}{5L} & -\frac{P}{10} \\ -\frac{P}{10} & -\frac{2PL}{15} & \frac{P}{10} & \frac{PL}{30} \\ \frac{6P}{5L} & \frac{P}{10} & -\frac{6P}{5L} & \frac{P}{10} \\ -\frac{P}{10} & \frac{PL}{30} & \frac{P}{10} & -\frac{2PL}{15} \end{bmatrix} \tag{7.13}$$

Because the geometric nonlinear stiffness matrix in Eq. (7.13) has a simple linear relationship with the axial force, it has been adopted in many commercial software packages (e.g., SAP2000 (CSI 2010)) and is widely accepted in the structural engineering textbooks as a common practice for including second-order effects (e.g., McGuire *et al.* (2000)). However, it needs to be emphasized that with the second-order approximation in which the deflected shapes are fitted with cubic polynomials, the stiffness matrix presented in Eq. (7.13) is not exact and therefore the static condensation only gives approximate results. This means, just like the P-Δ approach, the geometric stiffness approach also has difficulty in being incorporated into the force analogy method, but to a lesser extent.

7.2 Stability Functions

The stability theory of using stability functions was developed for elastic structures in the 1960s (Horne and Merchant 1965, Bazant and Cedolin 2003), but it has found limited applications because of its complexity in the closed-form solution. Even with the advance in computing technology, only one research publication was found in the recent literature on the analysis of framed structures using stability functions (Park and Kim 2008). This is an unfortunate trend, especially at a time when significant research emphasis is moving toward modeling and analyzing structures near collapse as a performance goal in the performance-based seismic engineering. For structures with significant lateral deflection, large geometric nonlinearity is expected, and linear or second-order approximation of the geometric stiffness may not be able to capture the near-collapse behavior accurately. Therefore, stability functions are here used to incorporate with the Force Analogy Method (FAM) to analyze structures with both geometric and material nonlinearities.

To incorporate geometric nonlinearity into the analysis with material nonlinearity based on the FAM, the stiffness matrices $\mathbf{K}_i$, $\mathbf{K}'_i$, and $\mathbf{K}''_i$ of the ith column member must be derived by taking into consideration the applied axial compressive force on the member. Though the theory of using stability functions in analyzing elastic frame buckling was fully developed half a century ago, the derivation of the elastic stiffness matrix $\mathbf{K}_i$ is repeated here mainly because the equations of the deflected shapes are important to the derivation of inelastic $\mathbf{K}'_i$ and $\mathbf{K}''_i$ matrices. In addition, the derivation of the deflected shapes also highlights the difference between the stability functions approach and the current approximate method of geometric nonlinear analysis using the geometric stiffness approach and the P-Δ approach.

By recognizing that only column members will be subjected to the axial compression while beam members will experience insignificant axial force due to the presence of the slab, distributed loading along the members due to gravity will therefore be ignored throughout the derivation of the stiffness matrices. In addition, the columns are rotated 90 degrees in the derivation using the classical beam theory with an applied axial compressive load, and therefore the word 'beam' will be used interchangeably with the word 'column' in this section.

7.2.1 Stiffness Matrix $[\boldsymbol{K}_i]$

Consider separately the four cases of a beam deflection subjected to various displacement patterns as shown in Figure 7.2 in deriving the stiffness matrices. Here, V_{1i}, m_{1i}, V_{2i}, and m_{2i} represent the fixed-end shears and moments of the beam, and $i = 1, \ldots, 4$ represents the four cases of unit displacement patterns of beam deflection.

Using the classical Bernoulli–Euler beam theory with "plane sections remain plane", where the moment is proportional to the curvature, the governing equilibrium equation describing the deflection of the beam member can be written as

$$(EIv'')'' + Pv'' = 0 \tag{7.14}$$

where E is the Young's modulus, I is the moment of inertia, v is the lateral deflection, and P is the axial compressive force of the member, and prime represents taking derivatives of the corresponding variable with respect to the x-direction of the member. By assuming EI is constant along the member, the solution to the fourth-order ordinary differential equation is:

$$v = A\sin kx + B\cos kx + Cx + D \tag{7.15}$$

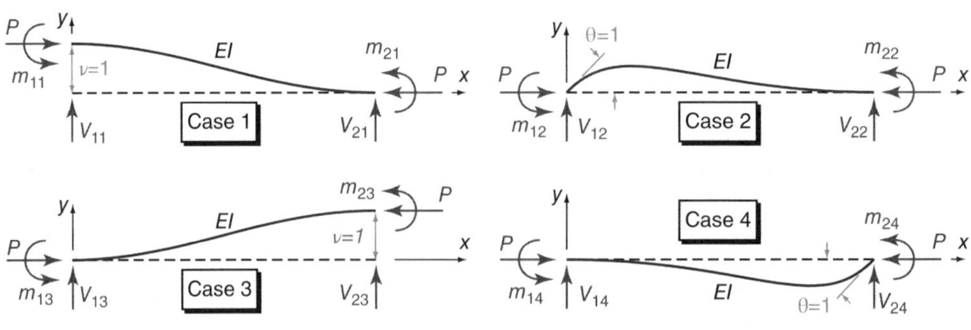

Figure 7.2 Displacement patterns and the corresponding fixed-end forces.

where $k=\sqrt{P/EI}$. Let L denote the length of the beam member, and let $\lambda = kL$ to simplify the subsequent derivations. The following four cases of boundary conditions (not in sequence) are considered.

Case 2: Imposing the boundary conditions $v(0)=0$, $v'(0)=1$, $v(L)=0$, and $v'(L)=0$ gives

$$v(0)=0: \quad B+D=0 \tag{7.16a}$$

$$v'(0)=1: \quad kA+C=1 \tag{7.16b}$$

$$v(L)=0: \quad A\sin\lambda+B\cos\lambda+CL+D=0 \tag{7.16c}$$

$$v'(L)=0: \quad kA\cos\lambda-kB\sin\lambda+C=0 \tag{7.16d}$$

Solving simultaneously for the constants in Eq. (7.16) gives

$$A=\frac{L(\lambda\sin\lambda+\cos\lambda-1)}{\lambda(\lambda\sin\lambda+2\cos\lambda-2)}, \quad B=\frac{L(\lambda\cos\lambda-\sin\lambda)}{\lambda(\lambda\sin\lambda+2\cos\lambda-2)}, \quad C=1-kA, \quad D=-B \tag{7.17}$$

Therefore, Eq. (7.15) along with the constants in Eq. (7.17) gives the deflected shape for Case 2.

The shears and moments at the two ends of the member (see Figure 7.2) are now evaluated using the classical beam theory formula:

$$m(x)=EI\frac{d^2v}{dx^2}, \quad V(x)=EI\frac{d^3v}{dx^3}+P\frac{dv}{dx} \tag{7.18}$$

Substituting Eq. (7.15) into Eq. (7.18) and using the constants calculated in Eq. (7.17), the fixed-end forces as labeled in Figure 7.2 (Case 2) are calculated as:

$$m_{12}=-EIv''(0)=EIk^2B=\left(\frac{EI}{L}\right)\frac{\lambda(\lambda\cos\lambda-\sin\lambda)}{\lambda\sin\lambda+2\cos\lambda-2}=\frac{sEI}{L} \tag{7.19a}$$

$$V_{12}=EIv'''(0)+Pv'(0)=-EIk^3A+P(1)=-\left(\frac{EI}{L^2}\right)\frac{\lambda^2(\lambda\sin\lambda+\cos\lambda-1)}{\lambda\sin\lambda+2\cos\lambda-2}+\frac{EI}{L^2}\lambda^2=\frac{\bar{s}EI}{L^2} \tag{7.19b}$$

$$m_{22}=EIv''(L)=-EIk^2(A\sin\lambda+B\cos\lambda)=\left(\frac{EI}{L}\right)\frac{\lambda(\sin\lambda-\lambda)}{\lambda\sin\lambda+2\cos\lambda-2}=\frac{scEI}{L} \tag{7.19c}$$

$$V_{22}=-EIv'''(L)-Pv'(L)=EIk^3(A\cos\lambda-B\sin\lambda)-P\times 0=-\frac{\bar{s}EI}{L^2} \tag{7.19d}$$

where

$$s=\frac{\lambda(\sin\lambda-\lambda\cos\lambda)}{2-2\cos\lambda-\lambda\sin\lambda}, \quad c=\frac{\lambda-\sin\lambda}{\sin\lambda-\lambda\cos\lambda}, \quad \bar{s}=s+sc=s(1+c)=\frac{\lambda^2(1-\cos\lambda)}{2-2\cos\lambda-\lambda\sin\lambda} \tag{7.20}$$

Note that the minus signs in front of the calculations for m_{12} and V_{22} are used because of the differences in sign convention between the classical beam theory and the theory for stiffness method of structural analysis.

Case 4: Imposing the boundary conditions $v(0)=0$, $v'(0)=0$, $v(L)=0$, and $v'(L)=1$ gives

$$v(0)=0: \quad B+D=0 \tag{7.21a}$$

$$v'(0)=0: \quad kA+C=0 \tag{7.21b}$$

$$v(L)=0: \quad A\sin\lambda+B\cos\lambda+CL+D=0 \tag{7.21c}$$

$$v'(L)=1: \quad kA\cos\lambda-kB\sin\lambda+C=1 \tag{7.21d}$$

Solving simultaneously for the constants in Eq. (7.21) gives

$$A=\frac{L(1-\cos\lambda)}{\lambda(\lambda\sin\lambda+2\cos\lambda-2)}, \quad B=\frac{L(\sin\lambda-\lambda)}{\lambda(\lambda\sin\lambda+2\cos\lambda-2)}, \quad C=-kA, \quad D=-B \tag{7.22}$$

These constants are used to give the deflected shape in Eq. (7.15) for Case 4.

Substituting Eq. (7.15) into Eq. (7.18) and using the constants calculated in Eq. (7.22), the fixed-end forces as labeled in Figure 7.2 (Case 4) are calculated as:

$$m_{14}=EIk^2B=\left(\frac{EI}{L}\right)\frac{\lambda(\sin\lambda-\lambda)}{\lambda\sin\lambda+2\cos\lambda-2}=\frac{scEI}{L} \tag{7.23a}$$

$$V_{14}=EIv'''(0)+Pv'(0)=-EIk^3A+P(0)=-\left(\frac{EI}{L^2}\right)\frac{\lambda^2(1-\cos\lambda)}{\lambda\sin\lambda+2\cos\lambda-2}=\frac{\bar{s}EI}{L^2} \tag{7.23b}$$

$$m_{24}=-EIk^2(A\sin\lambda+B\cos\lambda)=-\left(\frac{EI}{L}\right)\frac{\lambda[(1-\cos\lambda)\sin\lambda+(\sin\lambda-\lambda)\cos\lambda]}{\lambda\sin\lambda+2\cos\lambda-2}=\frac{sEI}{L} \tag{7.23c}$$

$$V_{24}=-EIv'''(L)-Pv'(L)=EIk^3(A\cos\lambda-B\sin\lambda)-P\times 1=\frac{\bar{s}EI}{L^2} \tag{7.23d}$$

where s, c, and $\bar{s}$ are given in Eq. (7.20).

Case 1: Imposing the boundary conditions $v(0)=1$, $v'(0)=0$, $v(L)=0$, and $v'(L)=0$ gives

$$v(0)=1: \quad B+D=1 \tag{7.24a}$$

$$v'(0)=0: \quad kA+C=0 \tag{7.24b}$$

$$v(L)=0: \quad A\sin\lambda+B\cos\lambda+CL+D=0 \tag{7.24c}$$

$$v'(L)=0: \quad kA\cos\lambda-kB\sin\lambda+C=0 \tag{7.24d}$$

Solving simultaneously for the constants in Eq. (7.24) gives

$$A=\frac{\sin\lambda}{\lambda\sin\lambda+2\cos\lambda-2}, \quad B=\frac{\cos\lambda-1}{\lambda\sin\lambda+2\cos\lambda-2}, \quad C=-kA, \quad D=1-B \tag{7.25}$$

These constants are used to give the deflected shape in Eq. (7.15) for Case 1.

Substituting Eq. (7.15) into Eq. (7.18) and using the constants calculated in Eq. (7.25), the fixed-end forces as labeled in Figure 7.2 (Case 1) are calculated as:

$$m_{11} = -EIv''(0) = EIk^2B = \left(\frac{EI}{L^2}\right)\frac{\lambda^2(\cos\lambda - 1)}{\lambda\sin\lambda + 2\cos\lambda - 2} = \frac{\bar{s}EI}{L^2} \tag{7.26a}$$

$$V_{11} = EIv'''(0) + Pv'(0) = -EIk^3A + P(0) = -\left(\frac{EI}{L^3}\right)\frac{\lambda^3\sin\lambda}{\lambda\sin\lambda + 2\cos\lambda - 2} = \frac{s'EI}{L^3} \tag{7.26b}$$

$$m_{21} = -EIk^2(A\sin\lambda + B\cos\lambda) = -\left(\frac{EI}{L^2}\right)\frac{\lambda^2[\sin^2\lambda + (\cos\lambda - 1)\cos\lambda]}{\lambda\sin\lambda + 2\cos\lambda - 2} = \frac{\bar{s}EI}{L^2} \tag{7.26c}$$

$$V_{21} = -EIv'''(L) - Pv'(L) = EIk^3(A\cos\lambda - B\sin\lambda) - P\times 0 = -\frac{s'EI}{L^3} \tag{7.26d}$$

where

$$s' = 2\bar{s} - \lambda^2 = \frac{\lambda^3\sin\lambda}{2 - 2\cos\lambda - \lambda\sin\lambda} \tag{7.27}$$

and $\bar{s}$ is given in Eq. (7.20).

Case 3: Finally, imposing the boundary conditions $v(0) = 0$, $v'(0) = 0$, $v(L) = 1$, and $v'(L) = 0$ gives

$$v(0) = 0: \quad B + D = 0 \tag{7.28a}$$

$$v'(0) = 0: \quad kA + C = 0 \tag{7.28b}$$

$$v(L) = 1: \quad A\sin\lambda + B\cos\lambda + CL + D = 1 \tag{7.28c}$$

$$v'(L) = 0: \quad kA\cos\lambda - kB\sin\lambda + C = 0 \tag{7.28d}$$

Solving simultaneously for the constants in Eq. (7.28) gives

$$A = -\frac{\sin\lambda}{\lambda\sin\lambda + 2\cos\lambda - 2}, \quad B = \frac{1 - \cos\lambda}{\lambda\sin\lambda + 2\cos\lambda - 2}, \quad C = -kA, \quad D = -B \tag{7.29}$$

These constants are used to give the deflected shape in Eq. (7.15) for Case 3.

Substituting Eq. (7.15) into Eq. (7.18) and using the constants calculated in Eq. (7.29), the fixed-end forces as labeled in Figure 7.2 (Case 3) are calculated as:

$$m_{13} = -EIv''(0) = EIk^2B = \left(\frac{EI}{L^2}\right)\frac{\lambda^2(1 - \cos\lambda)}{\lambda\sin\lambda + 2\cos\lambda - 2} = -\frac{\bar{s}EI}{L^2} \tag{7.30a}$$

$$V_{13} = EIv''(0) + Pv'(0) = -EIk^3A + P(0) = \left(\frac{EI}{L^3}\right)\frac{\lambda^3\sin\lambda}{\lambda\sin\lambda + 2\cos\lambda - 2} = -\frac{s'EI}{L^3} \tag{7.30b}$$

$$m_{23} = -EIk^2(A\sin\lambda + B\cos\lambda) = -\left(\frac{EI}{L^2}\right)\frac{\lambda^2[-\sin^2\lambda + (1-\cos\lambda)\cos\lambda]}{\lambda\sin\lambda + 2\cos\lambda - 2} = -\frac{\bar{s}EI}{L^2} \tag{7.30c}$$

$$V_{23} = -EIv'''(L) - Pv'(L) = \left(\frac{EI}{L^3}\right)\frac{\lambda^3(-\sin\lambda\cos\lambda - (1-\cos\lambda)\sin\lambda)}{\lambda\sin\lambda + 2\cos\lambda - 2} = \frac{s'EI}{L^3} \tag{7.30d}$$

where $\bar{s}$ is given in Eq. (7.20) and s' is given in Eq. (7.27).

In summary, based on the above four cases, the stiffness matrix of the ith member $\mathbf{K}_i$ after incorporating axial deformation becomes:

$$\mathbf{K}_i = \begin{bmatrix} s'EI/L^3 & \bar{s}EI/L^2 & -s'EI/L^3 & \bar{s}EI/L^2 \\ \bar{s}EI/L^2 & sEI/L & -\bar{s}EI/L^2 & scEI/L \\ -s'EI/L^3 & -\bar{s}EI/L^2 & s'EI/L^3 & -\bar{s}EI/L^2 \\ \bar{s}EI/L^2 & scEI/L & -\bar{s}EI/L^2 & sEI/L \end{bmatrix} \tag{7.31}$$

Note that if the stability coefficients in Eqs. (7.20) and (7.27) are expanded using Taylor series expansion, the results become

$$s = \frac{\lambda(\sin\lambda - \lambda\cos\lambda)}{2 - 2\cos\lambda - \lambda\sin\lambda} = 4 - \frac{2}{15}\lambda^2 + \ldots \tag{7.32a}$$

$$sc = \frac{\lambda(\lambda - \sin\lambda)}{2 - 2\cos\lambda - \lambda\sin\lambda} = 2 + \frac{1}{30}\lambda^2 + \ldots \tag{7.32b}$$

$$\bar{s} = \frac{\lambda^2(1-\cos\lambda)}{2 - 2\cos\lambda - \lambda\sin\lambda} = 6 - \frac{1}{10}\lambda^2 + \ldots \tag{7.32c}$$

$$s' = \frac{\lambda^3\sin\lambda}{2 - 2\cos\lambda - \lambda\sin\lambda} = 12 - \frac{6}{5}\lambda^2 + \ldots \tag{7.32d}$$

Recall that $\lambda = \sqrt{P/EI} \times L$. Substituting Eq. (7.32) into Eq. (7.31) and truncating the higher-order terms gives exactly the stiffness matrix in Eq. (7.13). Also, note that if the axial force is zero (i.e. $P = 0$), the resulting stiffness matrix becomes the classical member stiffness matrix of the form shown in Eq. (1.21):

$$\mathbf{K}_i = \begin{bmatrix} 12EI/L^3 & 6EI/L^2 & -12EI/L^3 & 6EI/L^2 \\ 6EI/L^2 & 4EI/L & -6EI/L^2 & 2EI/L \\ -12EI/L^3 & -6EI/L^2 & 12EI/L^3 & -6EI/L^2 \\ 6EI/L^2 & 2EI/L & -6EI/L^2 & 4EI/L \end{bmatrix} \tag{7.33}$$

7.2.2 Stiffness Matrix $[\mathbf{K}_i']$

The stiffness matrix $\mathbf{K}_i'$ relates the plastic rotations at the PHLs with the restoring forces applied at the degrees of freedom (DOFs). However, computing the restoring forces at the two ends of the member subjected to a unit plastic rotation at the PHL is a rather tedious process. To get

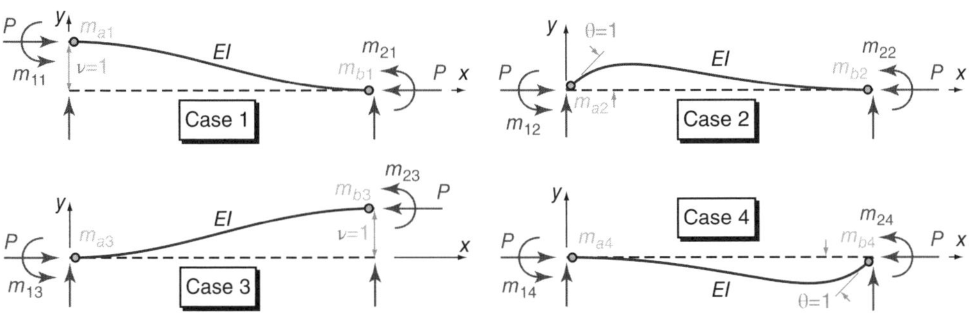

Figure 7.3 Displacement patterns for computation of moments at the plastic hinge locations.

around this problem, the $\mathbf{K}'^{T}_{i}$ matrix relating the displacements at the two ends of the beam member with the moment at the PHLs is therefore constructed.

Consider the same four cases of unit displacements of the beam member as shown in Figure 7.3, where the moment at the plastic hinges 'a' and 'b' (i.e. m_{ai} and m_{bi}, $i = 1, \ldots, 4$) are the desired quantities. Note that the fixed end moments (i.e. m_{1i} and m_{2i}, $i = 1, \ldots, 4$) have already been calculated and summarized in the second and fourth rows of the stiffness matrix given in Eq. (7.31). Based on Figure 7.3, the moments at the two plastic hinges for each of the four cases become:

Case 1: Imposing the boundary conditions $v(0) = 1$, $v'(0) = 0$, $v(L) = 0$, and $v'(L) = 0$ gives

$$m_{a1} = m_{11} = \frac{\bar{s}EI}{L^2}, \quad m_{b1} = m_{21} = \frac{\bar{s}EI}{L^2} \tag{7.34}$$

Case 2: Imposing the boundary conditions $v(0) = 0$, $v'(0) = 1$, $v(L) = 0$, and $v'(L) = 0$ gives

$$m_{a2} = m_{12} = \frac{sEI}{L}, \quad m_{b2} = m_{22} = \frac{scEI}{L} \tag{7.35}$$

Case 3: Imposing the boundary conditions $v(0) = 0$, $v'(0) = 0$, $v(L) = 1$, and $v'(L) = 0$ gives

$$m_{a3} = m_{13} = -\frac{\bar{s}EI}{L^2}, \quad m_{b3} = m_{23} = -\frac{\bar{s}EI}{L^2} \tag{7.36}$$

Case 4: Imposing the boundary conditions $v(0) = 0$, $v'(0) = 0$, $v(L) = 0$, and $v'(L) = 1$ gives

$$m_{a4} = m_{14} = \frac{scEI}{L}, \quad m_{b4} = m_{24} = \frac{sEI}{L} \tag{7.37}$$

From Eqs. (7.34) to (7.37), the transpose of stiffness matrix $\mathbf{K}'_i$ for the ith member becomes

$$\mathbf{K}_i'^T = \begin{bmatrix} \bar{s}EI/L^2 & sEI/L & -\bar{s}EI/L^2 & scEI/L \\ \bar{s}EI/L^2 & scEI/L & -\bar{s}EI/L^2 & sEI/L \end{bmatrix} \tag{7.38}$$

Once the $\mathbf{K}_i'^T$ matrix is derived, the $\mathbf{K}_i'$ matrix can be written as:

$$\mathbf{K}_i' = \begin{bmatrix} M_{a1} & M_{b1} \\ M_{a2} & M_{b2} \\ M_{a3} & M_{b3} \\ M_{a4} & M_{b4} \end{bmatrix} = \begin{bmatrix} \bar{s}EI/L^2 & \bar{s}EI/L^2 \\ sEI/L & scEI/L \\ -\bar{s}EI/L^2 & -\bar{s}EI/L^2 \\ scEI/L & sEI/L \end{bmatrix} = \begin{bmatrix} V_{1a} & V_{1b} \\ M_{1a} & M_{1b} \\ V_{2a} & V_{2b} \\ M_{2a} & M_{2b} \end{bmatrix} \tag{7.39}$$

and it represents the fixed-end forces due to unit plastic rotations at PHL 'a' and PHL 'b' as shown in Eq. (7.39) based on Maxwell's reciprocity theorem. Note that if the axial force is zero, the stability coefficients reduce to $s = 4$, $c = 0.5$, $\bar{s} = 6$, and $s' = 12$, and the stiffness matrix $\mathbf{K}_i'^T$ in Eq. (7.38) becomes

$$\mathbf{K}_i'^T = \begin{bmatrix} 6EI/L^2 & 4EI/L & -6EI/L^2 & 2EI/L \\ 6EI/L^2 & 2EI/L & -6EI/L^2 & 4EI/L \end{bmatrix} \tag{7.40}$$

7.2.3 *Stiffness Matrix* $[\mathbf{K}_i'']$

The stiffness matrix $\mathbf{K}_i''$ relates the moments at the PHLs with a corresponding unit plastic rotation at the PHL. To determine the $\mathbf{K}_i''$ matrix, the goal is to compute m_{aa}, m_{ab}, m_{ba}, and m_{bb} as shown in Figure 7.4.

The fixed-end forces derived in the $\mathbf{K}_i'$ matrix can be used to calculate the stiffness matrix $\mathbf{K}_i''$. For example, the first column of the 4×2 $\mathbf{K}_i'$ matrix in Eq. (7.39), $V_{1a} = \bar{s}EI/L^2$, $m_{1a} = sEI/L$, $V_{2a} = -\bar{s}EI/L^2$, and $m_{2a} = scEI/L$, represents the shear and moment at the ends of the beam member due to a unit plastic rotation at the left plastic hinge (i.e. PHL 'a'), as shown in Figure 7.4. Similarly, the second column of 4×2 $\mathbf{K}_i'$ matrix in Eq. (7.39), $V_{1b} = \bar{s}EI/L^2$, $m_{1b} = scEI/L$, $V_{2b} = -\bar{s}EI/L^2$, and $m_{2b} = sEI/L$, represents the shear and moment at the ends of the beam member due to a unit plastic rotation at the right plastic hinge (i.e. PHL 'b'), as shown in Figure 7.4. Then moments at the two plastic hinges for each of the two cases become:

Case '*a*': Imposing a unit plastic rotation $\theta_a'' = 1$ and $\theta_b'' = 0$ gives

$$m_{aa} = m_{1a} = \frac{sEI}{L}, \qquad m_{ba} = m_{2a} = \frac{scEI}{L} \tag{7.41}$$

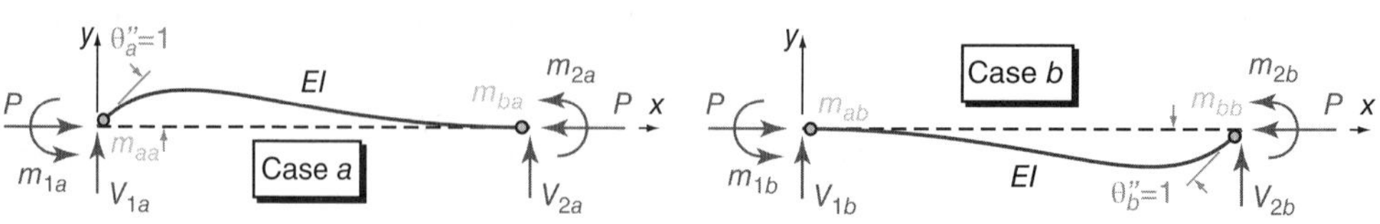

Figure 7.4 Displacement patterns for computation of moments due to unit plastic rotations.

Case '*b*': Imposing a unit plastic rotation $\theta_a''=0$ and $\theta_b''=1$ gives

$$m_{ab}=m_{1b}=\frac{scEI}{L},\qquad m_{bb}=m_{2b}=\frac{sEI}{L} \tag{7.42}$$

From Eqs. (7.41) and (7.42), the stiffness matrix $\mathbf{K}_i''$ for the ith member becomes

$$\mathbf{K}_i''=\begin{bmatrix} sEI/L & scEI/L \\ scEI/L & sEI/L \end{bmatrix} \tag{7.43}$$

and it represents the moments at the plastic hinges due to unit plastic rotations at PHL 'a' and PHL 'b'. Note that if the axial force is zero, the stability coefficients reduce to $s=4$, $c=0.5$, $\bar{s}=6$, and $s'=12$, and the stiffness matrix $\mathbf{K}_i''$ in Eq. (7.43) becomes

$$\mathbf{K}_i''=\begin{bmatrix} 4EI/L & 2EI/L \\ 2EI/L & 4EI/L \end{bmatrix} \tag{7.44}$$

Example 7.1 Static Condensation

Consider a beam member of length $3L$ with elastic modulus E and moment of inertia I subjected to an axial compressive force P as shown in Figure 7.5. The stiffness matrix using DOFs #1 to #4 with geometric nonlinearity considered can be written as

$$\mathbf{K}_{P\Delta}=\begin{bmatrix} \frac{12EI}{(3L)^3}-\frac{P}{3L} & \frac{6EI}{(3L)^2} & -\frac{12EI}{(3L)^3}+\frac{P}{3L} & \frac{6EI}{(3L)^2} \\ \frac{6EI}{(3L)^2} & \frac{4EI}{3L} & -\frac{6EI}{(3L)^2} & \frac{2EI}{3L} \\ -\frac{12EI}{(3L)^3}+\frac{P}{3L} & -\frac{6EI}{(3L)^2} & \frac{12EI}{(3L)^3}-\frac{P}{3L} & -\frac{6EI}{(3L)^2} \\ \frac{6EI}{(3L)^2} & \frac{2EI}{3L} & -\frac{6EI}{(3L)^2} & \frac{4EI}{3L} \end{bmatrix} \tag{7.45a}$$

Figure 7.5 Beam member subdivided into three elements.

$$\mathbf{K}_{GS}=\begin{bmatrix} \frac{12EI}{(3L)^3}-\frac{6P}{5(3L)} & \frac{6EI}{(3L)^2}-\frac{P}{10} & -\frac{12EI}{(3L)^3}+\frac{6P}{5(3L)} & \frac{6EI}{(3L)^2}-\frac{P}{10} \\ \frac{6EI}{(3L)^2}-\frac{P}{10} & \frac{4EI}{3L}-\frac{2P(3L)}{15} & -\frac{6EI}{(3L)^2}+\frac{P}{10} & \frac{2EI}{3L}+\frac{P(3L)}{30} \\ -\frac{12EI}{(3L)^3}+\frac{6P}{5(3L)} & -\frac{6EI}{(3L)^2}+\frac{P}{10} & \frac{12EI}{(3L)^3}-\frac{6P}{5(3L)} & -\frac{6EI}{(3L)^2}+\frac{P}{10} \\ \frac{6EI}{(3L)^2}-\frac{P}{10} & \frac{2EI}{3L}+\frac{P(3L)}{30} & -\frac{6EI}{(3L)^2}+\frac{P}{10} & \frac{4EI}{3L}-\frac{2P(3L)}{15} \end{bmatrix} \tag{7.45b}$$

$$\mathbf{K}_{SF}=\begin{bmatrix} \frac{s'EI}{(3L)^3} & \frac{\bar{s}EI}{(3L)^2} & -\frac{s'EI}{(3L)^3} & \frac{\bar{s}EI}{(3L)^2} \\ \frac{\bar{s}EI}{(3L)^2} & \frac{sEI}{3L} & -\frac{\bar{s}EI}{(3L)^2} & \frac{scEI}{3L} \\ -\frac{s'EI}{(3L)^3} & -\frac{\bar{s}EI}{(3L)^2} & \frac{s'EI}{(3L)^3} & -\frac{\bar{s}EI}{(3L)^2} \\ \frac{\bar{s}EI}{(3L)^2} & \frac{scEI}{3L} & -\frac{\bar{s}EI}{(3L)^2} & \frac{sEI}{3L} \end{bmatrix} \tag{7.45c}$$

where $\mathbf{K}_{P\Delta}$ is the stiffness matrix using the P-Δ approach (see Eq. (7.9)), $\mathbf{K}_{GS}$ is the stiffness matrix using the geometric stiffness approach (see Eq. (7.13)), and $\mathbf{K}_{SF}$ is the stiffness matrix using the stability functions approach (see Eq. (7.31)). The stability coefficients s, c, $\bar{s}$, and s' in Eq. (7.45c) are defined in Eqs. (7.20) and (7.27) with $\lambda=\sqrt{P/EI}\times 3L$. As a numerical illustration, let $E=200$ GPa, $I=20\times10^6$ mm^4, $L=4$ m, and $P=150$ kN. This gives

$$\mathbf{K}_{P\Delta}=\begin{bmatrix} 15.278 & 166.67 & -15.278 & 166.67 \\ 166.67 & 1333.3 & -166.67 & 666.67 \\ -15.278 & -166.67 & 15.278 & -166.67 \\ 166.67 & 666.67 & -166.67 & 1333.3 \end{bmatrix} \begin{matrix} \leftarrow \text{DOF\#1} \\ \leftarrow \text{DOF\#2} \\ \leftarrow \text{DOF\#3} \\ \leftarrow \text{DOF\#4} \end{matrix} \tag{7.46a}$$

$$\mathbf{K}_{GS}=\begin{bmatrix} 12.778 & 151.67 & -12.778 & 151.67 \\ 151.67 & 1093.3 & -151.67 & 726.67 \\ -12.778 & -151.67 & 12.778 & -151.67 \\ 151.67 & 726.67 & -151.67 & 1093.3 \end{bmatrix} \begin{matrix} \leftarrow \text{DOF\#1} \\ \leftarrow \text{DOF\#2} \\ \leftarrow \text{DOF\#3} \\ \leftarrow \text{DOF\#4} \end{matrix} \tag{7.46b}$$

$$\mathbf{K}_{SF}=\begin{bmatrix} 12.675 & 151.05 & -12.675 & 151.05 \\ 151.05 & 1074.1 & -151.05 & 738.48 \\ -12.675 & -151.05 & 12.675 & -151.05 \\ 151.05 & 738.48 & -151.05 & 1074.1 \end{bmatrix} \begin{matrix} \leftarrow \text{DOF\#1} \\ \leftarrow \text{DOF\#2} \\ \leftarrow \text{DOF\#3} \\ \leftarrow \text{DOF\#4} \end{matrix} \tag{7.46c}$$

Now assume that the same beam is subdivided into three elements of equal lengths. The stiffness matrix using DOFs #1 to #8 with geometric nonlinearity considered can be written as

$$\mathbf{K}_{P\Delta}=\left[\begin{array}{cccc:cccc} 712.5 & 1500 & 0 & 0 & -712.5 & 1500 & 0 & 0 \\ 1500 & 4000 & 0 & 0 & -1500 & 2000 & 0 & 0 \\ 0 & 0 & 712.5 & -1500 & 0 & 0 & -712.5 & -1500 \\ 0 & 0 & -1500 & 4000 & 0 & 0 & 1500 & 2000 \\ \hdashline -712.5 & -1500 & 0 & 0 & 1425 & 0 & -712.5 & 1500 \\ 1500 & 2000 & 0 & 0 & 0 & 8000 & -1500 & 2000 \\ 0 & 0 & -712.5 & 1500 & -712.5 & -1500 & 1425 & 0 \\ 0 & 0 & -1500 & 2000 & 1500 & 2000 & 0 & 8000 \end{array}\right] \tag{7.47a}$$

$$\mathbf{K}_{GS}=\left[\begin{array}{cccc:cccc} 705 & 1485 & 0 & 0 & -705 & 1485 & 0 & 0 \\ 1485 & 3920 & 0 & 0 & -1485 & 2020 & 0 & 0 \\ 0 & 0 & 705 & -1485 & 0 & 0 & -705 & -1485 \\ 0 & 0 & -1485 & 3920 & 0 & 0 & 1485 & 2020 \\ \hdashline -705 & -1485 & 0 & 0 & 1410 & 0 & -705 & 1485 \\ 1485 & 2020 & 0 & 0 & 0 & 7840 & -1485 & 2020 \\ 0 & 0 & -705 & 1485 & -705 & -1485 & 1410 & 0 \\ 0 & 0 & -1485 & 2020 & 1485 & 2020 & 0 & 7840 \end{array}\right] \tag{7.47b}$$

$$\mathbf{K}_{SF}=\left[\begin{array}{cccc:cccc} 704.97 & 1484.9 & 0 & 0 & -704.97 & 1484.9 & 0 & 0 \\ 1484.9 & 3919.4 & 0 & 0 & -1484.9 & 2020.4 & 0 & 0 \\ 0 & 0 & 704.97 & -1484.9 & 0 & 0 & -704.97 & -1484.9 \\ 0 & 0 & -1484.9 & 3919.4 & 0 & 0 & 1484.9 & 2020.4 \\ \hdashline -704.97 & -1484.9 & 0 & 0 & 1409.9 & 0 & -704.97 & 1484.9 \\ 1484.9 & 2020.4 & 0 & 0 & 0 & 7838.7 & -1484.9 & 2020.4 \\ 0 & 0 & -704.97 & 1484.9 & -704.97 & -1484.9 & 1409.9 & 0 \\ 0 & 0 & -1484.9 & 2020.4 & 1484.9 & 2020.4 & 0 & 7838.7 \end{array}\right] \tag{7.47c}$$

Comparing Eq. (7.47b) with Eq. (7.47c) shows that $\mathbf{K}_{GS} \approx \mathbf{K}_{SF}$ to within four significant digits when a length of L is used to construct the stiffness matrices, but comparing Eq. (7.46b) with Eq. (7.46c) shows that the two matrices differ by a larger magnitude when a length of $3L$ is used to construct the stiffness matrices.

Static condensation using Eq. (2.149) is then performed to eliminate DOFs #5 to #8 from the stiffness matrices, i.e.

$$\bar{\mathbf{K}} = \mathbf{K}_{dd} - \mathbf{K}_{dr}\mathbf{K}_{rr}^{-1}\mathbf{K}_{rd} \tag{7.48}$$

where $\mathbf{K}_{dd}$, $\mathbf{K}_{dr}$, $\mathbf{K}_{rd}$, and $\mathbf{K}_{rr}$ are partitioned sub-matrices according to those full stiffness matrices shown in Eq. (7.47). Now, substituting the sub-matrices in Eq. (7.47) into Eq. (7.48) and performing the matrix calculations gives

$$\bar{\mathbf{K}}_{P\Delta} = \begin{bmatrix} 13.971 & 158.82 & -13.971 & 158.82 \\ 158.82 & 1138.1 & -158.82 & 767.76 \\ -13.971 & -158.82 & 13.971 & -158.82 \\ 158.82 & 767.76 & -158.82 & 1138.1 \end{bmatrix} \begin{matrix} \leftarrow \text{DOF\#1} \\ \leftarrow \text{DOF\#2} \\ \leftarrow \text{DOF\#3} \\ \leftarrow \text{DOF\#4} \end{matrix} \tag{7.49a}$$

$$\bar{\mathbf{K}}_{GS} = \begin{bmatrix} 12.688 & 151.13 & -12.688 & 151.13 \\ 151.13 & 1074.8 & -151.13 & 738.77 \\ -12.688 & -151.13 & 12.688 & -151.13 \\ 151.13 & 738.77 & -151.13 & 1074.8 \end{bmatrix} \begin{matrix} \leftarrow \text{DOF\#1} \\ \leftarrow \text{DOF\#2} \\ \leftarrow \text{DOF\#3} \\ \leftarrow \text{DOF\#4} \end{matrix} \tag{7.49b}$$

$$\bar{\mathbf{K}}_{SF} = \begin{bmatrix} 12.675 & 151.05 & -12.675 & 151.05 \\ 151.05 & 1074.1 & -151.05 & 738.48 \\ -12.675 & -151.05 & 12.675 & -151.05 \\ 151.05 & 738.48 & -151.05 & 1074.1 \end{bmatrix} \begin{matrix} \leftarrow \text{DOF\#1} \\ \leftarrow \text{DOF\#2} \\ \leftarrow \text{DOF\#3} \\ \leftarrow \text{DOF\#4} \end{matrix} \tag{7.49c}$$

Comparing Eq. (7.46) with Eq. (7.49) shows that only the stability functions approach gives exactly the same stiffness matrix regardless of whether one long element is used or three subdivided elements are used. This indicates that while all three geometric nonlinearity approaches address large P-Δ appropriately, only stability functions approach captures small P-δ exactly. Note that the differences between the statically condensed matrices (i.e. $\bar{\mathbf{K}}_{P\Delta}$ and $\bar{\mathbf{K}}_{GS}$ relative to $\bar{\mathbf{K}}_{SF}$ in Eq. (7.49)) is smaller than that between the original matrices (i.e. $\mathbf{K}_{P\Delta}$ and $\mathbf{K}_{GS}$ relative to $\mathbf{K}_{SF}$ in Eq. (7.46)). This indicates that one way of capturing a small P-δ effect in the P-Δ approach or the geometric stiffness approach is to subdivide the member into several shorter elements.

7.3 Force Analogy Method with Stability Functions

Based on the stiffness matrices derived in Eqs. (7.31), (7.39), and (7.43) using the stability functions approach, the Force Analogy Method (FAM) can be used to statically analyze framed structures with both geometric and material nonlinearities. The governing equation of the FAM was presented in Eq. (2.63) for moment-resisting frame, which is

$$\begin{bmatrix} \mathbf{K} & \mathbf{K}' \\ \mathbf{K}'^T & \mathbf{K}'' \end{bmatrix} \begin{Bmatrix} \mathbf{x} \\ -\mathbf{\Theta}'' \end{Bmatrix} = \begin{Bmatrix} \mathbf{F}_a \\ \mathbf{m} \end{Bmatrix} \tag{7.50}$$

where $\mathbf{x}$ is the total displacement vector at the degrees of freedom (DOFs), $\mathbf{F}_a$ is the statically applied force vector, $\mathbf{m}$ is the total moment vector at the plastic hinge locations (PHLs), and $\mathbf{\Theta}''$ is the plastic rotation vector at the PHLs. The global stiffness matrices in Eq. (7.50) are constructed by assembling individual stiffness matrices of each member, and these individual stiffness matrices are presented in Eq. (7.31) for $\mathbf{K}_i$, Eq. (7.39) for $\mathbf{K}'_i$, and Eq. (7.43) for $\mathbf{K}''_i$, where i represents the ith member of the moment-resisting frame.

Example 7.2 Pushover of One-Story Frame with Updating Geometric Nonlinearity
Consider a one-story one-bay moment-resisting frame as shown in Figure 7.6. Assume the three members as labeled are axially rigid, this results in a system with three DOFs ($n = 3$) and six

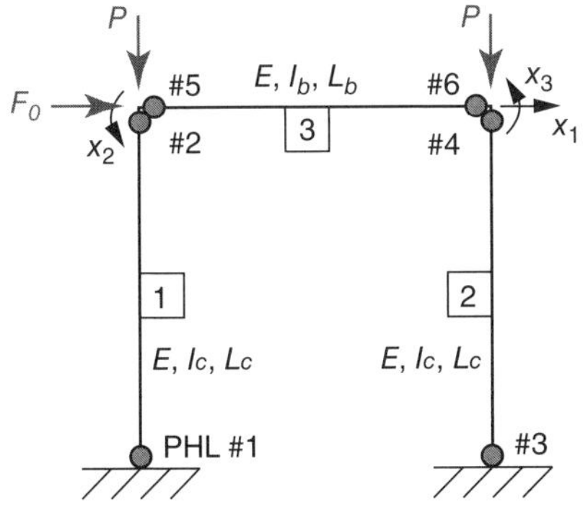

Figure 7.6 One-story one-bay moment-resisting frame.

PHLs ($q = 6$). For simplicity, let $I_c = I_b = I$ and $L_c = L_b = L$. Assume that only a lateral force of F_o is applied at the horizontal degree of freedom x_1, this gives $F_1 = F_o$ and $F_2 = F_3 = 0$.

Since the axial compressive force in Member 1 (denote as P_1) can be different from that of Member 2 (denote as P_2) due to overturning caused the lateral applied load F_o, the resulting stability coefficients can be different as well. The axial compressive force in Member 3 is assumed to be negligible (i.e. $P_3 \approx 0$) due to the presence of a slab, even though it is not shown in the figure. Therefore, let

$$\lambda_1 = \sqrt{\frac{P_1}{EI}} \times L, \qquad \lambda_2 = \sqrt{\frac{P_2}{EI}} \times L \tag{7.51}$$

It then follows that the stiffness matrices of the FAM for this one-story frame become similar to those in Eq. (2.54), i.e.

$$\mathbf{K} = \begin{bmatrix} s_1' EI/L^3 + s_2' EI/L^3 & \bar{s}_1 EI/L^2 & \bar{s}_2 EI/L^2 \\ \bar{s}_1 EI/L^2 & s_1 EI/L + 4EI/L & 2EI/L \\ \bar{s}_2 EI/L^2 & 2EI/L & s_2 EI/L + 4EI/L \end{bmatrix} \begin{matrix} \leftarrow x_1 \\ \leftarrow x_2 \\ \leftarrow x_3 \end{matrix} \tag{7.52a}$$

$$\mathbf{K}' = \left[\begin{array}{cc:cc:cc} \bar{s}_1 EI/L^2 & \bar{s}_1 EI/L^2 & \bar{s}_2 EI/L^2 & \bar{s}_2 EI/L^2 & 0 & 0 \\ s_1 c_1 EI/L & s_1 EI/L & 0 & 0 & 4EI/L & 2EI/L \\ 0 & 0 & s_2 c_2 EI/L & s_2 EI/L & 2EI/L & 4EI/L \end{array}\right] \begin{matrix} \leftarrow x_1 \\ \leftarrow x_2 \\ \leftarrow x_3 \end{matrix} \tag{7.52b}$$

$$\mathbf{K}'' = \left[\begin{array}{cc:cc:cc} s_1 EI/L & s_1 c_1 EI/L & 0 & 0 & 0 & 0 \\ s_1 c_1 EI/L & s_1 EI/L & 0 & 0 & 0 & 0 \\ \hdashline 0 & 0 & s_2 EI/L & s_2 c_2 EI/L & 0 & 0 \\ 0 & 0 & s_2 c_2 EI/L & s_2 EI/L & 0 & 0 \\ \hdashline 0 & 0 & 0 & 0 & 4EI/L & 2EI/L \\ 0 & 0 & 0 & 0 & 2EI/L & 4EI/L \end{array}\right] \begin{matrix} \leftarrow \theta_1 \\ \leftarrow \theta_2 \\ \leftarrow \theta_3 \\ \leftarrow \theta_4 \\ \leftarrow \theta_5 \\ \leftarrow \theta_6 \end{matrix} \tag{7.52c}$$

where s_1, c_1, $\bar{s}_1$, and s_1' are the stability coefficients of Member 1 as functions of λ_1 and P_1, and s_2, c_2, $\bar{s}_2$, and s_2' are the stability coefficients of Member 2 as functions of λ_2 and P_2. The governing equation then becomes

$$\left[\begin{array}{c:c} \mathbf{K} & \mathbf{K}' \\ \hdashline \mathbf{K}'^T & \mathbf{K}'' \end{array}\right] \left\{\begin{array}{c} x_1 \\ x_2 \\ x_3 \\ \hdashline -\theta_1'' \\ -\theta_2'' \\ -\theta_3'' \\ -\theta_4'' \\ -\theta_5'' \\ -\theta_6'' \end{array}\right\} = \left\{\begin{array}{c} F_o \\ 0 \\ 0 \\ \hdashline m_1 \\ m_2 \\ m_3 \\ m_4 \\ m_5 \\ m_6 \end{array}\right\} \tag{7.53}$$

As a numerical illustration, let $P = 200$ kN, $E = 200$ GPa, $I_b = I_c = I = 20 \times 10^6$ mm^4, and $L_b = L_c = L = 4$ m for computing the pushover curve. Assume that the interaction between axial force and moment is included in determining the yield surface with the relationships:

$$\text{Columns:} \quad \left(\frac{P}{P_c}\right)^2 + \left(\frac{m}{m_c}\right)^2 \leq 1 \tag{7.54a}$$

$$\text{Beam:} \quad \left|\frac{m}{m_b}\right| \leq 1 \tag{7.54b}$$

where $P_c = 600$ kN, $m_c = 100$ kN m, and $m_b = 90$ kN m. Let the post-yield hardening stiffness of the columns be $K_{t,c} = 800$ kN/m and that of the beam be $K_{t,b} = 600$ kN/m. This gives

$$\textit{if} \begin{cases} m_i \leq m_{yc,i} \\ m_i > m_{yc,i} \end{cases}, \quad \textit{then} \quad \begin{cases} \theta_i'' = 0 \\ m_i = m_{yc,i} + 800 \times \theta_i'' \end{cases} \quad i = 1,2,3,4 \tag{7.55a}$$

$$\textit{if} \begin{cases} m_i \leq 90 \\ m_i > 90 \end{cases}, \quad \textit{then} \quad \begin{cases} \theta_i'' = 0 \\ m_i = 90 + 600 \times \theta_i'' \end{cases} \quad i = 5,6 \tag{7.55b}$$

where $m_{yc,i}$ is the yield moment of the ith column hinge computed using the yield surface equation given in Eq. (7.54a) after the axial compressive force in the column is specified, i.e.

$$m_{yc,i} = m_c \sqrt{1 - (P/P_c)^2} \tag{7.56}$$

A pushover curve is now constructed for the one-story one-bay frame. The frame initially responds in the linearly elastic range. At an applied lateral force of $F_o = 73.28$ kN and $P_1 = P_2 = P = 200$ kN, extracting the first three equations of Eq. (7.53) gives

$$\begin{bmatrix} 1379.9 & 1479.9 & 1479.9 \\ 1479.9 & 7892.2 & 2000 \\ 1479.9 & 2000 & 7892.2 \end{bmatrix} \begin{Bmatrix} x_1 \\ x_2 \\ x_3 \end{Bmatrix} = \begin{Bmatrix} 73.28 \\ 0 \\ 0 \end{Bmatrix} \tag{7.57}$$

Solving for the displacements at the DOFs gives

$$\begin{Bmatrix} x_1 \\ x_2 \\ x_3 \end{Bmatrix} = \begin{Bmatrix} 0.0782 \\ -0.0117 \\ -0.0117 \end{Bmatrix} \tag{7.58}$$

Then, substituting Eq. (7.58) back into the last six equations of Eq. (7.53) gives the moments

$$\begin{Bmatrix} m_1 \\ m_2 \\ m_3 \\ m_4 \\ m_5 \\ m_6 \end{Bmatrix} = \begin{bmatrix} 1479.9 & 2027.3 & 0 \\ 1479.9 & 3892.2 & 0 \\ 1479.9 & 0 & 2027.3 \\ 1479.9 & 0 & 3892.2 \\ 0 & 4000 & 2000 \\ 0 & 2000 & 4000 \end{bmatrix} \begin{Bmatrix} 0.0782 \\ -0.0117 \\ -0.0117 \end{Bmatrix} = \begin{Bmatrix} 92.00 \\ 70.19 \\ 92.00 \\ 70.19 \\ -70.19 \\ -70.19 \end{Bmatrix} \tag{7.59}$$

Based on the setup of the frame as shown in Figure 7.6, the column axial force can be updated by computing the shear force at the two ends of the beam using equilibrium of the beam, i.e.

$$P_1 = P + \frac{m_5 + m_6}{L_b} = 200 + \frac{-70.19 - 70.19}{4} = 164.9\,\text{kN} \tag{7.60a}$$

$$P_2 = P - \frac{m_5 + m_6}{L_b} = 200 - \frac{-70.19 - 70.19}{4} = 235.1\,\text{kN} \tag{7.60b}$$

Now, to check whether any plastic hinge has yielded, it appears that PHL #3 has the largest combination of axial force and moment. Using Eq. (7.54a) for PHL #3 gives

$$\text{PHL\#3:} \quad \left(\frac{235.1}{600}\right)^2 + \left(\frac{92.00}{100}\right)^2 = 1.00 \tag{7.61}$$

which indicates that PHL #3 has reached yielding at a lateral force of $F_o = 73.28$ kN that causes a lateral displacement of $x_1 = 7.82$ cm.

The analysis continues with PHL #3 yielded. The lateral force is now applied up to $F_o = 75.42$ kN. Since updated information on the column axial forces has been obtained in Eq. (7.60), the stiffness matrices are updated accordingly and Rows 1, 2, 3, and 6 are now extracted from Eq. (7.53) while making use of Eq. (7.55a) with an addition of stiffness $K_{t,c} = 800$ kN/m to the appropriate diagonal term:

$$\begin{bmatrix} 1379.9 & 1483.4 & 1476.3 & 1476.3 \\ 1483.4 & 7911.3 & 2000 & 0 \\ 1476.3 & 2000 & 7873.0 & 2032.3 \\ 1476.3 & 0 & 2032.3 & 4673.0 \end{bmatrix} \begin{Bmatrix} x_1 \\ x_2 \\ x_3 \\ -\theta_3'' \end{Bmatrix} = \begin{Bmatrix} 75.42 \\ 0 \\ 0 \\ 92.00 \end{Bmatrix} \tag{7.62}$$

Note that in Eq. (7.62), $m_{yc,3}$ is computed by using Eq. (7.56) as

$$m_{yc,3} = 100 \times \sqrt{1 - (235.1/600)^2} = 92.00\ \text{kN m} \tag{7.63}$$

Solving for the displacements at the DOFs and plastic rotation in Eq. (7.62) gives

$$\begin{Bmatrix} x_1 \\ x_2 \\ x_3 \\ -\theta_3'' \end{Bmatrix} = \begin{Bmatrix} 0.0816 \\ -0.0123 \\ -0.0120 \\ -0.0009 \end{Bmatrix} \tag{7.64}$$

Then substituting Eq. (7.64) back into the last six equations of Eq. (7.53) gives the moments

$$\begin{Bmatrix} m_1 \\ m_2 \\ m_3 \\ m_4 \\ m_5 \\ m_6 \end{Bmatrix} = \begin{bmatrix} 1483.4 & 2022.4 & 0 & 0 \\ 1483.4 & 3911.3 & 0 & 0 \\ 1476.3 & 0 & 2032.3 & 3873.0 \\ 1476.3 & 0 & 3873.0 & 2032.3 \\ 0 & 4000 & 2000 & 0 \\ 0 & 2000 & 4000 & 0 \end{bmatrix} \begin{Bmatrix} 0.0816 \\ -0.0123 \\ -0.0120 \\ -0.0009 \end{Bmatrix} = \begin{Bmatrix} 96.21 \\ 73.02 \\ 92.72 \\ 72.36 \\ -73.02 \\ -72.36 \end{Bmatrix} \tag{7.65}$$

and the column axial forces are updated as

$$P_1 = P + \frac{m_5 + m_6}{L_b} = 200 + \frac{-73.02 - 72.36}{4} = 163.7 \text{ kN} \tag{7.66a}$$

$$P_2 = P - \frac{m_5 + m_6}{L_b} = 200 - \frac{-73.02 - 72.36}{4} = 236.3 \text{ kN} \tag{7.66b}$$

Equation (7.54a) is now used to ensure PHL #3 continues to yield and PHL #1 has reached yielding:

$$\text{PHL\#1:} \left(\frac{163.7}{600}\right)^2 + \left(\frac{96.21}{100}\right)^2 = 1.00 \tag{7.67a}$$

$$\text{PHL\#3:} \left(\frac{236.3}{600}\right)^2 + \left(\frac{92.72}{100}\right)^2 = 1.01 \tag{7.67b}$$

which indicates that both PHLs #1 and #3 have reached or exceeded yielding at a lateral force of $F_o = 75.42$ kN that causes a lateral displacement of $x_1 = 8.16$ cm.

The analysis continues with both PHLs #1 and #3 yielded. The lateral force is now applied up to $F_o = 84.17$ kN. Since updated information on the column axial forces was obtained in Eq. (7.66), the stiffness matrices are updated accordingly and Rows 1, 2, 3, 4, and 6 are now extracted from Eq. (7.53) while making use of Eq. (7.55a) by adding stiffness to the appropriate diagonal terms:

$$\begin{bmatrix} 1379.9 & 1483.6 & 1476.2 & 1483.6 & 1476.2 \\ 1483.6 & 7912.0 & 2000 & 2022.3 & 0 \\ 1476.2 & 2000 & 7872.4 & 0 & 2032.5 \\ 1483.6 & 2022.3 & 0 & 4712.0 & 0 \\ 1476.2 & 0 & 2032.5 & 0 & 4672.4 \end{bmatrix} \begin{Bmatrix} x_1 \\ x_2 \\ x_3 \\ -\theta_1'' \\ -\theta_3'' \end{Bmatrix} = \begin{Bmatrix} 84.17 \\ 0 \\ 0 \\ 96.21 \\ 91.92 \end{Bmatrix} \tag{7.68}$$

where

$$m_{yc,1} = 100 \times \sqrt{1-(163.7/600)^2} = 96.21 \text{ kN m} \tag{7.69a}$$

$$m_{yc,3} = 100 \times \sqrt{1-(236.3/600)^2} = 91.92 \text{ kN m} \tag{7.69b}$$

Solving for the displacements at the DOFs and plastic rotations in Eq. (7.68) gives

$$\begin{Bmatrix} x_1 \\ x_2 \\ x_3 \\ -\theta_1'' \\ -\theta_3'' \end{Bmatrix} = \begin{Bmatrix} 0.1119 \\ -0.0151 \\ -0.0148 \\ -0.0083 \\ -0.0093 \end{Bmatrix} \tag{7.70}$$

Then substituting Eq. (7.70) back into the last six equations of Eq. (7.53) gives the moments

$$\begin{Bmatrix} m_1 \\ m_2 \\ m_3 \\ m_4 \\ m_5 \\ m_6 \end{Bmatrix} = \begin{bmatrix} 1483.6 & 2022.3 & 0 & 3912.0 & 0 \\ 1483.6 & 3912.0 & 0 & 2022.3 & 0 \\ 1476.2 & 0 & 2032.5 & 0 & 3872.4 \\ 1476.2 & 0 & 3872.4 & 0 & 2032.5 \\ 0 & 4000 & 2000 & 0 & 0 \\ 0 & 2000 & 4000 & 0 & 0 \end{bmatrix} \begin{Bmatrix} 0.1119 \\ -0.0151 \\ -0.0148 \\ -0.0083 \\ -0.0093 \end{Bmatrix} = \begin{Bmatrix} 102.86 \\ 90.00 \\ 99.33 \\ 89.24 \\ -90.00 \\ -89.24 \end{Bmatrix} \tag{7.71}$$

and the column axial force are updated as

$$P_1 = P + \frac{m_5 + m_6}{L_b} = 200 + \frac{-90.00 - 89.24}{4} = 155.2 \text{ kN} \tag{7.72a}$$

$$P_2 = P - \frac{m_5 + m_6}{L_b} = 200 - \frac{-90.00 - 89.24}{4} = 244.8 \text{ kN} \tag{7.72b}$$

Equation (7.54) is now used to ensure PHLs #1 and #3 continues to yield and PHL #5 has reached yielding:

$$\text{PHL\#1: } \left(\frac{155.2}{600}\right)^2 + \left(\frac{102.86}{100}\right)^2 = 1.12 \tag{7.73a}$$

$$\text{PHL\#3: } \left(\frac{244.8}{600}\right)^2 + \left(\frac{99.33}{100}\right)^2 = 1.15 \tag{7.73b}$$

$$\text{PHL\#5: } \left|\frac{90}{90}\right| = 1.00 \tag{7.73c}$$

which indicates that PHLs #1, #3, and #5 have reached or exceeded yielding at a lateral force of $F_o = 84.17$ kN that causes a lateral displacement of $x_1 = 11.19$ cm.

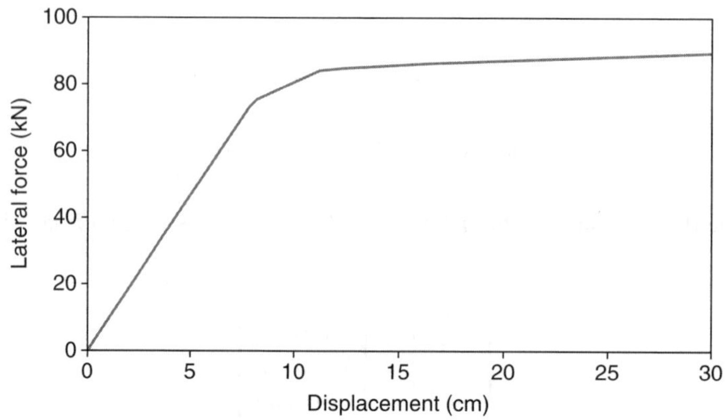

Figure 7.7 Pushover curve of the one-story frame with update on geometric nonlinearity.

At this point,

$$m_{yc,1} = 100 \times \sqrt{1-(155.2/600)^2} = 96.60\ \text{kN m} \tag{7.74a}$$

$$m_{yc,3} = 100 \times \sqrt{1-(244.8/600)^2} = 91.30\ \text{kN m} \tag{7.74b}$$

The calculation process continues, and the final pushover curve is shown in Figure 7.7.

Example 7.3 Pushover of One-Story Frame with No Update on Geometric Nonlinearity
Consider again the one-story one-bay moment-resisting frame as shown in Figure 7.6 with three DOFs ($n = 3$) and six PHLs ($q = 6$). Let $I_c = I_b = I$ and $L_c = L_b = L$. Assume again that only a lateral force of F_o is applied at the horizontal degree of freedom x_1, this gives $F_1 = F_o$ and $F_2 = F_3 = 0$. It then follows that the global stiffness matrices of the FAM for this one-story frame are given in Eq. (7.52), and the governing equation is given in Eq. (7.53).

For the numerical illustration, again let $E = 200$ GPa, $I_b = I_c = I = 20 \times 10^6$ mm^4, $L_b = L_c = L = 4$ m, and $P = 200$ kN. Now assume that geometric nonlinearity is not updated as the new axial force is calculated (i.e. the global stiffness matrices remain constant due to the initial load of $P = 200$ kN), then Eq. (7.53) becomes

$$\left[\begin{array}{ccc|cccccc} 1379.9 & 1479.9 & 1479.9 & 1479.9 & 1479.9 & 1479.9 & 1479.9 & 0 & 0 \\ 1479.9 & 7892.2 & 2000 & 2027.3 & 3892.2 & 0 & 0 & 4000 & 2000 \\ 1479.9 & 2000 & 7892.2 & 0 & 0 & 2027.3 & 3892.2 & 2000 & 4000 \\ \hline 1479.9 & 2027.3 & 0 & 3892.2 & 2027.3 & 0 & 0 & 0 & 0 \\ 1479.9 & 3892.2 & 0 & 2027.3 & 3892.2 & 0 & 0 & 0 & 0 \\ 1479.9 & 0 & 2027.3 & 0 & 0 & 3892.2 & 2027.3 & 0 & 0 \\ 1479.9 & 0 & 3892.2 & 0 & 0 & 2027.3 & 3892.2 & 0 & 0 \\ 0 & 4000 & 2000 & 0 & 0 & 0 & 0 & 4000 & 2000 \\ 0 & 2000 & 4000 & 0 & 0 & 0 & 0 & 2000 & 4000 \end{array}\right] \left\{\begin{array}{c} x_1 \\ x_2 \\ x_3 \\ \hline -\theta_1'' \\ -\theta_2'' \\ -\theta_3'' \\ -\theta_4'' \\ -\theta_5'' \\ -\theta_6'' \end{array}\right\} = \left\{\begin{array}{c} F_o \\ 0 \\ 0 \\ \hline m_1 \\ m_2 \\ m_3 \\ m_4 \\ m_5 \\ m_6 \end{array}\right\} \tag{7.75}$$

Assume that the interaction between the axial force and moment is included in determining the yield surface with the relationships as given in Eq. (7.54) and moment versus plastic rotation relationships given in Eq. (7.55).

A pushover curve is now constructed for the one-story one-bay frame. The frame initially responds in the linearly elastic range. At an applied lateral force of $F_o = 73.28$ kN and $P_1 = P_2 = P = 200$ kN, extracting the first three equations of Eq. (7.75) gives

$$\begin{bmatrix} 1379.9 & 1479.9 & 1479.9 \\ 1479.9 & 7892.2 & 2000 \\ 1479.9 & 2000 & 7892.2 \end{bmatrix} \begin{Bmatrix} x_1 \\ x_2 \\ x_3 \end{Bmatrix} = \begin{Bmatrix} 73.28 \\ 0 \\ 0 \end{Bmatrix} \tag{7.76}$$

Solving for the displacements at the DOFs gives

$$\begin{Bmatrix} x_1 \\ x_2 \\ x_3 \end{Bmatrix} = \begin{Bmatrix} 0.0782 \\ -0.0117 \\ -0.0117 \end{Bmatrix} \tag{7.77}$$

Then substituting Eq. (7.77) back into the last six equations of Eq. (7.75) gives the moments

$$\begin{Bmatrix} m_1 \\ m_2 \\ m_3 \\ m_4 \\ m_5 \\ m_6 \end{Bmatrix} = \begin{bmatrix} 1479.9 & 2027.3 & 0 \\ 1479.9 & 3892.2 & 0 \\ 1479.9 & 0 & 2027.3 \\ 1479.9 & 0 & 3892.2 \\ 0 & 4000 & 2000 \\ 0 & 2000 & 4000 \end{bmatrix} \begin{Bmatrix} 0.0782 \\ -0.0117 \\ -0.0117 \end{Bmatrix} = \begin{Bmatrix} 92.00 \\ 70.19 \\ 92.00 \\ 70.19 \\ -70.19 \\ -70.19 \end{Bmatrix} \tag{7.78}$$

which are exactly the same as Eqs. (7.58) and (7.59), since both examples begin with a linear analysis.

The column axial force are updated by computing the shear force at the two ends of the beam using equilibrium of the beam, i.e.

$$P_1 = P + \frac{m_5 + m_6}{L_b} = 200 + \frac{-70.19 - 70.19}{4} = 164.9 \text{ kN} \tag{7.79a}$$

$$P_2 = P - \frac{m_5 + m_6}{L_b} = 200 - \frac{-70.19 - 70.19}{4} = 235.1 \text{ kN} \tag{7.79b}$$

Now, to check whether any plastic hinge has yielded, it appears that PHL #3 has the largest combination of axial force and moment. Using Eq. (7.54a) for PHL #3 gives

$$\text{PHL\#3:} \left(\frac{235.1}{600}\right)^2 + \left(\frac{92.00}{100}\right)^2 = 1.00 \tag{7.80}$$

which indicates that PHL #3 has reached yielding at a lateral force of $F_o = 73.28$ kN that causes a lateral displacement of $x_1 = 7.82$ cm.

The analysis continues with PHL #3 yielded. The lateral force is now applied up to $F_o = 75.51$ kN. Using the same stiffness matrix in Eq. (7.75) by assuming no update to the geometric nonlinearity even when the axial compressive force in the column changes, extracting Rows 1, 2, 3, and 6, while making use of Eq. (7.55a) with an addition of stiffness $K_{t,c} = 800$ kN/m to the appropriate diagonal term, gives:

$$\begin{bmatrix} 1379.9 & 1479.9 & 1479.9 & 1479.9 \\ 1479.9 & 7892.2 & 2000 & 0 \\ 1479.9 & 2000 & 7892.2 & 2027.3 \\ 1479.9 & 0 & 2027.3 & 4692.2 \end{bmatrix} \begin{Bmatrix} x_1 \\ x_2 \\ x_3 \\ -\theta_3'' \end{Bmatrix} = \begin{Bmatrix} 75.51 \\ 0 \\ 0 \\ 92.00 \end{Bmatrix} \tag{7.81}$$

where $m_{yc,3}$ in Eq. (7.81) is computed using Eq. (7.56) as

$$m_{yc,3} = 100 \times \sqrt{1 - (235.1/600)^2} = 92.00 \text{ kN m} \tag{7.82}$$

Solving for the displacements at the DOFs and plastic rotation in Eq. (7.81) gives

$$\begin{Bmatrix} x_1 \\ x_2 \\ x_3 \\ -\theta_3'' \end{Bmatrix} = \begin{Bmatrix} 0.0819 \\ -0.0123 \\ -0.0120 \\ -0.0011 \end{Bmatrix} \tag{7.83}$$

Then, substituting Eq. (7.83) back into the last six equations of Eq. (7.75) gives the moments

$$\begin{Bmatrix} m_1 \\ m_2 \\ m_3 \\ m_4 \\ m_5 \\ m_6 \end{Bmatrix} = \begin{bmatrix} 1479.9 & 2027.3 & 0 & 0 \\ 1479.9 & 3892.2 & 0 & 0 \\ 1479.9 & 0 & 2027.3 & 3892.2 \\ 1479.9 & 0 & 3892.2 & 2027.3 \\ 0 & 4000 & 2000 & 0 \\ 0 & 2000 & 4000 & 0 \end{bmatrix} \begin{Bmatrix} 0.0819 \\ -0.0123 \\ -0.0120 \\ -0.0011 \end{Bmatrix} = \begin{Bmatrix} 96.21 \\ 73.23 \\ 92.85 \\ 72.50 \\ -73.23 \\ -72.50 \end{Bmatrix} \tag{7.84}$$

and the column axial forces are updated as

$$P_1 = P + \frac{m_5 + m_6}{L_b} = 200 + \frac{-73.23 - 72.50}{4} = 163.6 \text{ kN} \tag{7.85a}$$

$$P_2 = P - \frac{m_5 + m_6}{L_b} = 200 - \frac{-73.23 - 72.50}{4} = 236.4 \text{ kN} \tag{7.85b}$$

Equation (7.54a) is now used to ensure PHL #3 continues to yield and PHL #1 has reached yielding:

$$\text{PHL\#1:} \left(\frac{163.6}{600}\right)^2 + \left(\frac{96.21}{100}\right)^2 = 1.00 \tag{7.86a}$$

$$\text{PHL\#3:}\ \left(\frac{236.4}{600}\right)^2 + \left(\frac{92.85}{100}\right)^2 = 1.02 \qquad (7.86b)$$

which indicates that both PHLs #1 and #3 have reached or exceeded yielding at a lateral force of $F_o = 75.51$ kN that causes a lateral displacement of $x_1 = 8.19$ cm.

The analysis continues with both PHLs #1 and #3 yielded. The lateral force is now applied up to $F_o = 84.17$ kN. Using the same stiffness matrix in Eq. (7.75) by assuming no update to the geometric nonlinearity even when the axial compressive force in the column changes, extracting Rows 1, 2, 3, 4, and 6, while making use of Eq. (7.55a), gives:

$$\begin{bmatrix} 1379.9 & 1479.9 & 1479.9 & 1479.9 & 1479.9 \\ 1479.9 & 7892.2 & 2000 & 2027.3 & 0 \\ 1479.9 & 2000 & 7892.2 & 0 & 2027.3 \\ 1479.9 & 2027.3 & 0 & 4692.2 & 0 \\ 1479.9 & 0 & 2027.3 & 0 & 4692.2 \end{bmatrix} \begin{Bmatrix} x_1 \\ x_2 \\ x_3 \\ -\theta_1'' \\ -\theta_3'' \end{Bmatrix} = \begin{Bmatrix} 84.17 \\ 0 \\ 0 \\ 96.21 \\ 91.91 \end{Bmatrix} \qquad (7.87)$$

where

$$m_{yc,1} = 100 \times \sqrt{1-(163.6/600)^2} = 96.21\ \text{kN m} \qquad (7.88a)$$

$$m_{yc,3} = 100 \times \sqrt{1-(236.4/600)^2} = 91.91\ \text{kN m} \qquad (7.88b)$$

Solving for the displacements at the DOFs and plastic rotations in Eq. (7.87) gives

$$\begin{Bmatrix} x_1 \\ x_2 \\ x_3 \\ -\theta_1'' \\ -\theta_3'' \end{Bmatrix} = \begin{Bmatrix} 0.1119 \\ -0.0151 \\ -0.0148 \\ -0.0083 \\ -0.0093 \end{Bmatrix} \qquad (7.89)$$

Then, substituting Eq. (7.89) back into the last six equations of Eq. (7.75) gives the moments

$$\begin{Bmatrix} m_1 \\ m_2 \\ m_3 \\ m_4 \\ m_5 \\ m_6 \end{Bmatrix} = \begin{bmatrix} 1479.9 & 2027.3 & 0 & 3892.2 & 0 \\ 1479.9 & 3892.2 & 0 & 2027.3 & 0 \\ 1479.9 & 0 & 2027.3 & 0 & 3892.2 \\ 1479.9 & 0 & 3892.2 & 0 & 2027.3 \\ 0 & 4000 & 2000 & 0 & 0 \\ 0 & 2000 & 4000 & 0 & 0 \end{bmatrix} \begin{Bmatrix} 0.1119 \\ -0.0151 \\ -0.0148 \\ -0.0083 \\ -0.0093 \end{Bmatrix} = \begin{Bmatrix} 102.81 \\ 90.00 \\ 99.37 \\ 89.26 \\ -90.00 \\ -89.26 \end{Bmatrix} \qquad (7.90)$$

and the column axial force are updated as

$$P_1 = P + \frac{m_5 + m_6}{L_b} = 200 + \frac{-90.00 - 89.26}{4} = 155.2\ \text{kN} \qquad (7.91a)$$

$$P_2 = P - \frac{m_5 + m_6}{L_b} = 200 - \frac{-90.00 - 89.26}{4} = 244.8 \text{ kN} \tag{7.91b}$$

Equation (7.54) is now used to ensure PHLs #1 and #3 continue to yield and PHL #5 has reached yielding:

$$\text{PHL\#1:} \left(\frac{155.2}{600}\right)^2 + \left(\frac{102.86}{100}\right)^2 = 1.12 \tag{7.92a}$$

$$\text{PHL\#3:} \left(\frac{244.8}{600}\right)^2 + \left(\frac{99.33}{100}\right)^2 = 1.15 \tag{7.92b}$$

$$\text{PHL\#5:} \left|\frac{90}{90}\right| = 1.00 \tag{7.92c}$$

which indicates that PHLs #1, #3, and #5 have reached or exceeded yielding at a lateral force of $F_o = 84.17$ kN that causes a lateral displacement of $x_1 = 11.19$ cm.

At this point,

$$m_{yc,1} = 100 \times \sqrt{1 - (155.2/600)^2} = 96.60 \text{ kN m} \tag{7.93a}$$

$$m_{yc,3} = 100 \times \sqrt{1 - (244.8/600)^2} = 91.30 \text{ kN m} \tag{7.93b}$$

The calculation process continues, and the final pushover curve is shown in Figure 7.8. The pushover curve that was obtained in Example 7.2 with updates to geometric nonlinearity (see Figure 7.7) is also shown in Figure 7.8 as a comparison.

Figure 7.8 shows that the difference between whether or not to update the geometric nonlinearity at every step due to changes in column axial force on the pushover curve is negligible. One assumption often made among various software packages is that the effect of changing axial force on the geometric stiffness matrices is ignored, and therefore the geometric

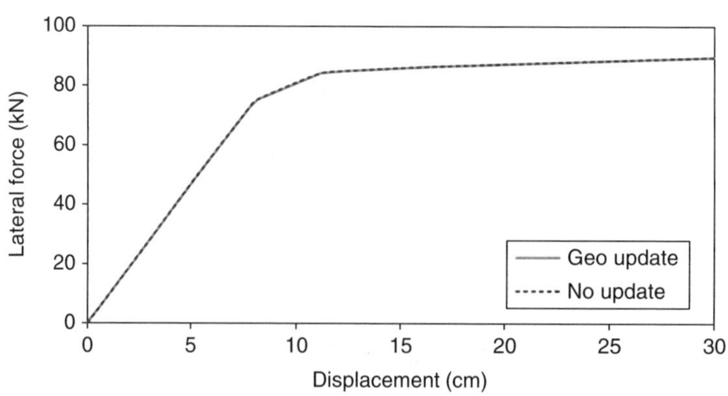

Figure 7.8 Comparison of pushover curves on geometric nonlinearity with and without updates.

nonlinearity need not be updated. The idea behind this assumption is that when lateral displacement imposes an overturning moment on the entire structure, global equilibrium requires that there will be an increase in column compression on one side of the frame and an equal amount of reduction in column compression on the opposite side of the frame. While an increase in column compression on one side of the frame reduces the lateral stiffness of these columns, a reduction in column compression on the opposite side of the frame increases the lateral stiffness of those columns by a similar amount. The result is that the net change in total lateral stiffness of the entire frame becomes negligible, and this can be seen when the stiffness matrix in Eq. (7.62) is compared with that in Eq. (7.81). Similar observations can be made when the stiffness matrix in Eq. (7.68) is compared with that in Eq. (7.87) – there is practically no difference in the stiffness matrices in the comparisons. Therefore, an assumption to keep the geometrically nonlinear stiffness matrices unchanged as loading increases is justifiable.

7.4 Nonlinear Dynamic Analysis Using Stability Functions

7.4.1 Force Analogy Method

In a structural dynamic analysis, the axial forces in columns are often varying with time (i.e. $P(t)$), and therefore the stiffness matrices due to geometric nonlinearity will also be functions of time. The stiffness matrices of moment-resisting frames incorporating material nonlinearity based on the FAM can then be written as $\mathbf{K}(t)$, $\mathbf{K}'(t)$, and $\mathbf{K}''(t)$. The concept of the FAM for dynamic analysis was presented in detail in Chapter 3 without using any geometric nonlinearity. The details of the method are briefly summarized here with the incorporation of geometric nonlinearity in the equations.

For a moment-resisting frame with n degrees of freedom (DOFs) and q plastic hinge locations (PHLs), the displacement relationship can be expressed as

$$\mathbf{x}(t) = \mathbf{x}'(t) + \mathbf{x}''(t) \tag{7.94}$$

where $\mathbf{x}(t)$ is the total displacement vector, $\mathbf{x}'(t)$ is the elastic displacement vector, and $\mathbf{x}''(t)$ is the inelastic displacement vector. Similar to the total displacements in Eq. (7.94), the total moment $\mathbf{m}(t)$ at the PHLs of a typical moment-resisting frame can be separated into elastic moment and inelastic moment, i.e.

$$\mathbf{m}(t) = \mathbf{m}'(t) + \mathbf{m}''(t) \tag{7.95}$$

where $\mathbf{m}'(t)$ is the elastic moment due to the elastic displacement $\mathbf{x}'(t)$, and $\mathbf{m}''(t)$ is the inelastic moment due to the inelastic displacement $\mathbf{x}''(t)$. The displacements in Eq. (7.94) and moments in Eq. (7.95) are related by the following equations:

$$\mathbf{m}'(t) = \mathbf{K}'(t)^T \mathbf{x}'(t) \tag{7.96}$$

$$\mathbf{m}''(t) = -\left[\mathbf{K}''(t) - \mathbf{K}'(t)^T \mathbf{K}(t)^{-1} \mathbf{K}'(t)\right] \mathbf{\Theta}''(t) \tag{7.97}$$

where $\mathbf{\Theta}''(t)$ is the plastic rotation at the PHLs.

Substituting Eqs. (7.96) and (7.97) into Eq. (7.95) and rearranging the terms, the first governing equation of the FAM for dynamic analysis is obtained as follows:

$$\mathbf{m}(t)+\mathbf{K}''(t)\mathbf{\Theta}''(t)=\mathbf{K}'(t)^T\mathbf{x}(t) \tag{7.98}$$

The second governing equation of the FAM relates the inelastic displacement $\mathbf{x}''(t)$ at the DOFs and the plastic rotation $\mathbf{\Theta}''(t)$ in the PHLs. This equation can be written as

$$\mathbf{x}''(t)=\mathbf{K}(t)^{-1}\mathbf{K}'(t)\mathbf{\Theta}''(t) \tag{7.99}$$

7.4.2 Nonlinear Dynamic Analysis with the Force Analogy Method

When the FAM is used, the stiffness force in the equation of motion is calculated by multiplying the stiffness matrix $\mathbf{K}(t)$ with the elastic displacement $\mathbf{x}'(t)$. For an n-DOF system subjected to earthquake ground motions, this equation can be written as

$$\mathbf{M}\ddot{\mathbf{x}}(t)+\mathbf{C}\dot{\mathbf{x}}(t)+\mathbf{K}(t)\mathbf{x}'(t)=-\mathbf{M}\ddot{\mathbf{g}}(t)-\mathbf{F}_f(t) \tag{7.100}$$

where $\mathbf{M}$ is the $n\times n$ invertible mass matrix, $\mathbf{C}$ is the $n\times n$ damping matrix, $\dot{\mathbf{x}}(t)$ is the $n\times 1$ velocity vector, $\ddot{\mathbf{x}}(t)$ is the $n\times 1$ acceleration vector, $\ddot{\mathbf{g}}(t)$ is the $n\times 1$ earthquake ground acceleration vector corresponding to the effect of ground motion at each DOF, and $\mathbf{F}_f(t)$ is the $n\times 1$ vector of additional forces imposed on the frame due to geometric nonlinearity (mainly the P-Δ effect) of the gravity columns. This effect can often be modeled using a leaning column (or a P-Δ column) in a two-dimensional analysis but requires a detailed modeling of the gravity columns in a three-dimensional analysis. The relationship between this lateral force $\mathbf{F}_f(t)$ and the lateral displacement can be written in the form:

$$\mathbf{F}_f(t)=\mathbf{K}_f\mathbf{x}(t) \tag{7.101}$$

where $\mathbf{K}_f$ is an $n\times n$ stiffness matrix that is a function of the gravity loads on the leaning column and the corresponding story height, but it is not a function of time. In a two-dimensional analysis, this $\mathbf{K}_f$ matrix usually takes the form similar to assembling the second term of Eq. (7.9) for each element stiffness matrix:

$$\mathbf{K}_f=\begin{bmatrix} -Q_1/h_1-Q_2/h_2 & Q_2/h_2 & 0 & \cdots & 0 \\ Q_2/h_2 & -Q_2/h_2-Q_3/h_3 & \ddots & \ddots & \vdots \\ 0 & \ddots & \ddots & Q_{n-1}/h_{n-1} & 0 \\ \vdots & \ddots & Q_{n-1}/h_{n-1} & -Q_{n-1}/h_{n-1}-Q_n/h_n & Q_n/h_n \\ 0 & \cdots & 0 & Q_n/h_n & -Q_n/h_n \end{bmatrix} \tag{7.102}$$

where Q_i is the axial force due to gravity on the leaning column of the ith floor, and h_i is the story height of the ith floor.

While the $\mathbf{K}_f$ matrix takes care of the geometric nonlinear effects from all the gravity columns, the stiffness matrix $\mathbf{K}(t)$ in Eq. (7.100) must consider both large P-Δ and small P-δ effects for the moment-resisting frame itself. Let this time-dependent global stiffness matrix $\mathbf{K}(t)$ be represented in the form:

$$\mathbf{K}(t) = \mathbf{K}_o + \mathbf{K}_g(t) \tag{7.103}$$

where $\mathbf{K}_o$ denotes the initial stiffness of the frame due to the gravity loads only, and $\mathbf{K}_g(t)$ denotes the change in stiffness due to the change in axial load on members during the dynamic loading. Note that the $\mathbf{K}_o$ matrix is a constant stiffness matrix computed by using only the gravity loads on the columns (which means $\mathbf{K}_o = \mathbf{K}(t_0) = \mathbf{K}(0)$, the stiffness matrix computed at time step 0) and is not a function of time.

First, solving for the elastic displacement $\mathbf{x}'(t)$ in Eq. (7.94) and substituting the result into Eq. (7.100) gives

$$\mathbf{M}\ddot{\mathbf{x}}(t) + \mathbf{C}\dot{\mathbf{x}}(t) + \mathbf{K}(t)\mathbf{x}(t) = -\mathbf{M}\ddot{\mathbf{g}}(t) - \mathbf{F}_f(t) + \mathbf{K}(t)\mathbf{x}''(t) \tag{7.104}$$

Then substituting Eqs. (7.101) and (7.103) into Eq. (7.104), the equation of motion after considering both P-δ and P-Δ effects of geometric nonlinearity of the entire structure becomes

$$\mathbf{M}\ddot{\mathbf{x}}(t) + \mathbf{C}\dot{\mathbf{x}}(t) + \mathbf{K}_o\mathbf{x}(t) = -\mathbf{M}\ddot{\mathbf{g}}(t) - \mathbf{K}_f\mathbf{x}(t) - \mathbf{K}_g(t)\mathbf{x}(t) + \mathbf{K}(t)\mathbf{x}''(t) \tag{7.105}$$

Define

$$\mathbf{K}_e = \mathbf{K}_o + \mathbf{K}_f \tag{7.106}$$

where $\mathbf{K}_e$ represents the elastic stiffness of the structure. With a negative definite $\mathbf{K}_f$ matrix as shown in Eq. (7.102), it can be seen from Eq. (7.106) that $\det(\mathbf{K}_e) < \det(\mathbf{K}_o)$. Therefore, the natural periods of vibration of the structure based on $\mathbf{K}_e$ (i.e. reduced stiffness) are always longer than those calculated based on $\mathbf{K}_o$ (i.e. initial stiffness). Finally, on substituting Eq. (7.106) into Eq. (7.105), it follows that

$$\mathbf{M}\ddot{\mathbf{x}}(t) + \mathbf{C}\dot{\mathbf{x}}(t) + \mathbf{K}_e\mathbf{x}(t) = -\mathbf{M}\ddot{\mathbf{g}}(t) - \mathbf{K}_g(t)\mathbf{x}(t) + \mathbf{K}(t)\mathbf{x}''(t) \tag{7.107}$$

An alternate equation of Eq. (7.107) can be written by first pre-multiplying Eq. (7.99) by the stiffness matrix $\mathbf{K}(t)$ and then substituting the result into the last term of Eq. (7.107). This gives

$$\mathbf{M}\ddot{\mathbf{x}}(t) + \mathbf{C}\dot{\mathbf{x}}(t) + \mathbf{K}_e\mathbf{x}(t) = -\mathbf{M}\ddot{\mathbf{g}}(t) - \mathbf{K}_g(t)\mathbf{x}(t) + \mathbf{K}'(t)\mathbf{\Theta}''(t) \tag{7.108}$$

7.4.3 State Space Analysis with Geometric and Material Nonlinearities

Let the material nonlinearity term (i.e. $\mathbf{K}(t)\mathbf{x}''(t)$) and geometric nonlinearity term (i.e. $-\mathbf{K}_g(t)\mathbf{x}(t)$) shown on the right-hand side of Eq. (7.107) be treated as the equivalent forces applied to the structure, the equation can now be solved using the state space method. To represent Eq. (7.107) in state space form, let the state vector $\mathbf{z}(t)$ be defined as

$$\mathbf{z}(t) = \begin{Bmatrix} \mathbf{x}(t) \\ \dot{\mathbf{x}}(t) \end{Bmatrix} \tag{7.109}$$

which is a $2n \times 1$ vector with a collection of states of the responses. It follows from Eq. (7.107) that

$$\dot{\mathbf{z}}(t) = \begin{Bmatrix} \dot{\mathbf{x}}(t) \\ \ddot{\mathbf{x}}(t) \end{Bmatrix} = \begin{bmatrix} \mathbf{0} & \mathbf{I} \\ -\mathbf{M}^{-1}\mathbf{K}_e & -\mathbf{M}^{-1}\mathbf{C} \end{bmatrix} \begin{Bmatrix} \mathbf{x}(t) \\ \dot{\mathbf{x}}(t) \end{Bmatrix} + \begin{bmatrix} \mathbf{0} \\ -\mathbf{h} \end{bmatrix} \mathbf{a}(t) + \begin{bmatrix} \mathbf{0} \\ \mathbf{M}^{-1} \end{bmatrix} \left(-\mathbf{K}_g(t)\mathbf{x}(t) + \mathbf{K}(t)\mathbf{x}''(t)\right) \tag{7.110}$$

where $\mathbf{h}$ is an $n \times 3$ matrix that relates the directions of each DOF with the global x-, y-, and z-directions (i.e. a collection of 0 s and 1 s in all entries), and $\mathbf{a}(t)$ is the 3×1 ground acceleration vector in the three global directions of $\ddot{g}_x(t)$, $\ddot{g}_y(t)$, and $\ddot{g}_z(t)$. The relationship between the ground acceleration vector $\ddot{\mathbf{g}}(t)$ for each DOF in Eq. (7.108) and the three-component ground acceleration vector $\mathbf{a}(t)$ can be expressed as

$$\ddot{\mathbf{g}}(t) = \mathbf{h}\mathbf{a}(t) = \mathbf{h} \begin{Bmatrix} \ddot{g}_x(t) \\ \ddot{g}_y(t) \\ \ddot{g}_z(t) \end{Bmatrix} \tag{7.111}$$

To simplify Eq. (7.110), let

$$\mathbf{A} = \begin{bmatrix} \mathbf{0} & \mathbf{I} \\ -\mathbf{M}^{-1}\mathbf{K}_e & -\mathbf{M}^{-1}\mathbf{C} \end{bmatrix}, \quad \mathbf{H} = \begin{bmatrix} \mathbf{0} \\ -\mathbf{h} \end{bmatrix}, \quad \mathbf{B} = \begin{bmatrix} \mathbf{0} \\ -\mathbf{M}^{-1} \end{bmatrix} \tag{7.112a}$$

$$\mathbf{f}_g(t) = -\mathbf{K}_g(t)\mathbf{x}(t), \qquad \mathbf{f}_m(t) = \mathbf{K}(t)\mathbf{x}''(t) \tag{7.112b}$$

where $\mathbf{A}$ is the $2n \times 2n$ state transition matrix in the continuous form, $\mathbf{H}$ is the $2n \times 3$ ground motion transition matrix in the continuous form, $\mathbf{B}$ is the $2n \times n$ geometric nonlinearity transition matrix in the continuous form, $\mathbf{f}_g(t)$ is the $n \times 1$ equivalent force vector due to geometric nonlinearity, and $\mathbf{f}_m(t)$ is the $n \times 1$ equivalent force vector due to material nonlinearity. Then, Eq. (7.110) becomes

$$\dot{\mathbf{z}}(t) = \mathbf{A}\mathbf{z}(t) + \mathbf{H}\mathbf{a}(t) + \mathbf{B}\mathbf{f}_g(t) + \mathbf{B}\mathbf{f}_m(t) \tag{7.113}$$

Solving for the first-order linear differential equation in Eq. (7.113) gives

$$\mathbf{z}(t) = \mathbf{e}^{\mathbf{A}(t-t_o)}\mathbf{z}(t_o) + \mathbf{e}^{\mathbf{A}t} \int_{t_o}^{t} \mathbf{e}^{-\mathbf{A}s}\left[\mathbf{H}\mathbf{a}(s) + \mathbf{B}\mathbf{f}_g(s) + \mathbf{B}\mathbf{f}_m(s)\right] ds \tag{7.114}$$

where t_o is the time of reference when the integration begins, which is typically the time when the states $\mathbf{z}(t_o)$ are known, such as the initial conditions.

To integrate Eq. (7.114) numerically, let $t_{k+1} = t$, $t_k = t_o$, and $\Delta t = t_{k+1} - t_k$, and the subscript k denotes the kth time step, then it follows from Eq. (7.114) that

$$\mathbf{z}_{k+1} = \mathbf{e}^{\mathbf{A}\Delta t}\mathbf{z}_k + \mathbf{e}^{\mathbf{A}t_{k+1}} \int_{t_k}^{t_{k+1}} \mathbf{e}^{-\mathbf{A}s}\left[\mathbf{Ha}(s) + \mathbf{Bf}_g(s) + \mathbf{Bf}_m(s)\right] ds \tag{7.115}$$

Using the delta function approximation for the variables in the integral, where the ground acceleration vector $\mathbf{a}(s)$, equivalent geometric nonlinear force vector $\mathbf{f}_g(s)$, and equivalent material nonlinear force vector $\mathbf{f}_m(s)$ take the form:

$$\mathbf{a}(s) = \mathbf{a}_k\delta(s - t_k)\Delta t, \quad t_k \le s < t_{k+1} \tag{7.116a}$$

$$\mathbf{f}_g(s) = \mathbf{f}_{g,k}\delta(s - t_k)\Delta t, \quad t_k \le s < t_{k+1} \tag{7.116b}$$

$$\mathbf{f}_m(s) = \mathbf{f}_{m,k}\delta(s - t_k)\Delta t, \quad t_k \le s < t_{k+1} \tag{7.116c}$$

Substituting Eq. (7.116) into Eq. (7.115) and performing the integration gives

$$\mathbf{z}_{k+1} = \mathbf{e}^{\mathbf{A}\Delta t}\mathbf{z}_k + \Delta t\mathbf{e}^{\mathbf{A}\Delta t}\mathbf{Ha}_k + \Delta t\mathbf{e}^{\mathbf{A}\Delta t}\mathbf{Bf}_{g,k} + \Delta t\mathbf{e}^{\mathbf{A}\Delta t}\mathbf{Bf}_{m,k} \tag{7.117}$$

where $\mathbf{z}_k$, $\mathbf{a}_k$, $\mathbf{f}_{g,k}$, and $\mathbf{f}_{m,k}$ are the discretized forms of $\mathbf{z}(t)$, $\mathbf{a}(t)$, $\mathbf{f}_g(t)$, and $\mathbf{f}_m(t)$, respectively. Let

$$\mathbf{F}_d = \mathbf{e}^{\mathbf{A}\Delta t}, \quad \mathbf{H}_d = \mathbf{e}^{\mathbf{A}\Delta t}\mathbf{H}\Delta t, \quad \mathbf{B}_d = \mathbf{e}^{\mathbf{A}\Delta t}\mathbf{B}\Delta t \tag{7.118}$$

Then, Eq. (7.117) becomes

$$\mathbf{z}_{k+1} = \mathbf{F}_d\mathbf{z}_k + \mathbf{H}_d\mathbf{a}_k + \mathbf{B}_d\mathbf{f}_{g,k} + \mathbf{B}_d\mathbf{f}_{m,k} \tag{7.119}$$

Equation (7.119) represents the recursive equation for calculating the dynamic response of moment-resisting framed structures while considering updates on geometric nonlinearity as the axial compressive force in columns changes with time.

In Examples 7.2 and 7.3, it has been demonstrated that ignoring updates on geometric nonlinearity results in minor differences in the responses. Therefore, for the case when updates on geometric nonlinearity are ignored in the nonlinear dynamic analysis, the $\mathbf{K}_g(t)$ matrix as given in Eq. (7.103) becomes $\mathbf{K}_g(t) = \mathbf{0}$. Therefore, from the same equation, $\mathbf{K}(t) = \mathbf{K}_o$. Then, it follows from Eq. (7.112b) that

$$\mathbf{f}_g(t) = [\mathbf{0}]\cdot\mathbf{x}(t) = \mathbf{0} \tag{7.120a}$$

$$\mathbf{f}_m(t) = \mathbf{K}_o\mathbf{x}''(t) \tag{7.120b}$$

and Eq. (7.119) becomes

$$\mathbf{z}_{k+1} = \mathbf{F}_d\mathbf{z}_k + \mathbf{H}_d\mathbf{a}_k + \mathbf{G}_d\mathbf{x}_k'' \tag{7.121}$$

where, similar to Eq. (3.56),

$$\mathbf{G}_d = \mathbf{B}_d\mathbf{K}_o = \mathbf{e}^{\mathbf{A}\Delta t}\begin{bmatrix}\mathbf{0}\\ \mathbf{M}^{-1}\end{bmatrix}\mathbf{K}_o\Delta t = \mathbf{e}^{\mathbf{A}\Delta t}\begin{bmatrix}\mathbf{0}\\ \mathbf{M}^{-1}\mathbf{K}_o\end{bmatrix}\Delta t = \mathbf{e}^{\mathbf{A}\Delta t}\mathbf{G}\Delta t \tag{7.122}$$

with the $\mathbf{G}$ matrix similarly defined in Eq. (3.50) and $\mathbf{x}_k''$ is the discretized forms of $\mathbf{x}''(t)$. Equation (7.121) is similar to Eq. (3.57), and it represents the recursive equation for calculating the nonlinear dynamic response of moment-resisting framed structures without updates on geometric nonlinearity.

To perform nonlinear dynamic analysis, either Eq. (7.119) or Eq. (7.121) is used in conjunction with the governing equations of the FAM (see Eqs. (7.98) and (7.99)), which are written here in discretized forms as:

$$\mathbf{m}_{k+1} + \mathbf{K}_{k+1}''\Delta\mathbf{\Theta}'' = \mathbf{K}'^{T}_{k+1}\mathbf{x}_{k+1} - \mathbf{K}_{k+1}''\mathbf{\Theta}_k'' \tag{7.123}$$

$$\mathbf{x}_{k+1}'' = \mathbf{K}_{k+1}^{-1}\mathbf{K}_{k+1}'\mathbf{\Theta}_{k+1}'' \tag{7.124}$$

where $\mathbf{x}_k$, $\mathbf{\Theta}_k''$, and $\mathbf{m}_k$ are the discretized forms of $\mathbf{x}(t)$, $\mathbf{\Theta}''(t)$, and $\mathbf{m}(t)$, respectively, $\Delta\mathbf{\Theta}'' = \mathbf{\Theta}_{k+1}'' - \mathbf{\Theta}_k''$, and $\mathbf{K}_k$, $\mathbf{K}_k'$, and $\mathbf{K}_k''$ are the stiffness matrices at time step k computed using the axial forces in columns at the same time step. Note that the stiffness matrices in Eqs. (7.123) and (7.124) are written in terms of time step $k+1$. However, the axial forces in columns at time step $k+1$ are unknown prior to the calculation of moments and change in plastic rotations, which means Eq. (7.123) requires an iterative procedure and may be difficult to execute. Therefore, the stiffness matrices in these two equations may often be replaced by those at time step k without loss of accuracy, i.e.

$$\mathbf{m}_{k+1} + \mathbf{K}_k''\Delta\mathbf{\Theta}'' = \mathbf{K}'^{T}_{k}\mathbf{x}_{k+1} - \mathbf{K}_k''\mathbf{\Theta}_k'' \tag{7.125}$$

$$\mathbf{x}_{k+1}'' = \mathbf{K}_k^{-1}\mathbf{K}_k'\mathbf{\Theta}_{k+1}'' \tag{7.126}$$

If updates to geometric nonlinearity are ignored, these stiffness matrices become:

$$\mathbf{K}_k = \mathbf{K}(0) = \mathbf{K}_o, \quad \mathbf{K}_k' = \mathbf{K}'(0) = \mathbf{K}_o', \quad \mathbf{K}_k'' = \mathbf{K}''(0) = \mathbf{K}_o'' \tag{7.127}$$

Then, Eqs. (7.123) and (7.124) become

$$\mathbf{m}_{k+1} + \mathbf{K}_o''\Delta\mathbf{\Theta}'' = \mathbf{K}'^{T}_{o}\mathbf{x}_{k+1} - \mathbf{K}_o''\mathbf{\Theta}_k'' \tag{7.128}$$

$$\mathbf{x}_{k+1}'' = \mathbf{K}_o^{-1}\mathbf{K}_o'\mathbf{\Theta}_{k+1}'' \tag{7.129}$$

Finally, the absolute acceleration vector (i.e. $\ddot{\mathbf{y}}(t) = \ddot{\mathbf{x}}(t) + \ddot{\mathbf{g}}(t)$) can be computed by rewriting Eq. (7.100) in discretized form as

$$\ddot{\mathbf{y}}_k = -\mathbf{M}^{-1}\mathbf{C}\dot{\mathbf{x}}_k - \mathbf{M}^{-1}\mathbf{K}_k\left(\mathbf{x}_k - \mathbf{x}_k''\right) - \mathbf{M}^{-1}\mathbf{K}_f\mathbf{x}_k \tag{7.130}$$

where $\mathbf{x}_k$, $\dot{\mathbf{x}}_k$, and $\ddot{\mathbf{y}}_k$ are the discretized forms of $\mathbf{x}(t)$, $\dot{\mathbf{x}}(t)$, and $\ddot{\mathbf{y}}(t)$, respectively.

Example 7.4 Single Degree of Freedom Column

Consider the single degree of freedom (SDOF) column as shown in Figure 7.9 subjected to a constant axial compressive force P. Let the mass be M, damping be C, elastic modulus be E, moment of inertia be I, and length be L. The degree of freedom is set up so that the mass is restrained from rotation, but it is free to translate in the horizontal direction.

If geometric nonlinearity is totally ignored in this problem, the lateral stiffness related to the horizontal translation degree of freedom at the column top is simply:

$$K_{OR} = \frac{12EI}{L^3} \tag{7.131}$$

where K_{OR} represents the original stiffness of the SDOF system without considering any geometric nonlinearity. If geometric nonlinearity is considered through the P-Δ approach, the lateral stiffness of the SDOF column becomes:

$$K_{P\Delta} = \frac{12EI}{L^3} - \frac{P}{L} \tag{7.132}$$

where $K_{P\Delta}$ denotes the geometrically nonlinear stiffness of the column using the P-Δ approach. If geometric nonlinearity is considered through the geometric stiffness approach, the lateral stiffness of the SDOF column becomes:

$$K_{GS} = \frac{12EI}{L^3} - \frac{6P}{5L} \tag{7.133}$$

where K_{GS} denotes the geometrically nonlinear stiffness of the column using the geometric stiffness approach. By comparing K_{GS} in Eq. (7.133) with $K_{P\Delta}$ in Eq. (7.132), it can be seen that the stiffness is increased by $P/5L$ when the small P-δ effect is ignored. Finally, if geometric nonlinearity is considered through the use of stability functions, the stiffness matrix becomes:

$$K_{SF} = \frac{s'EI}{L^3} \tag{7.134}$$

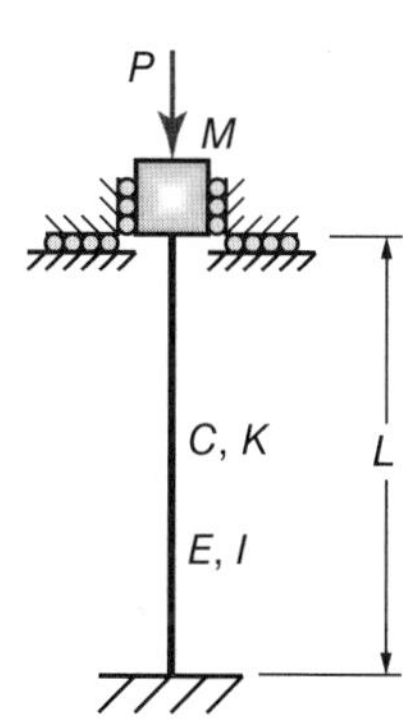

Figure 7.9 A single-degree-of-freedom column with geometric nonlinearity.

Table 7.1 Comparison of geometric nonlinear stiffnesses, periods, and maximum responses.

	Stiffness (kN)	Period (s)	Max Displacement (cm)	Max Velocity (cm/s)	Max Acceleration (g)
K_{OR}	375.0	1.000	42.26	261.7	1.704
$K_{P\Delta}$	250.0	1.225	40.74	220.8	1.095
K_{GS}	225.0	1.291	50.35	260.6	1.218
K_{SF}	224.3	1.293	50.56	261.0	1.218

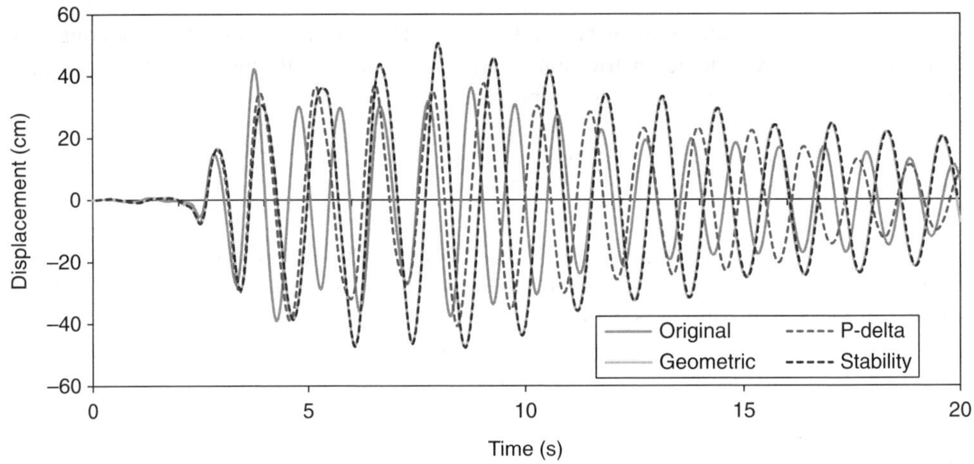

Figure 7.10 Displacement responses of the SDOF column using different stiffnesses.

where K_{SF} denotes the geometrically nonlinear stiffness computed using the stability functions approach, and s' is one of the stability coefficients that is given in Eq. (7.27).

In this study, let $E = 100$ GPa, $I = 20 \times 10^6$ mm^4, $L = 4$ m, and $P = 500$ kN in compression. Using a mass of $M = 9.5$ Mg, the calculated stiffnesses and the corresponding periods of vibration are summarized in Table 7.1. Note that the critical buckling load for the SDOF column is $P_{cr} = \pi^2 EI/L^2 = 1,234$ kN, and therefore the applied load is at 40.5% of the critical buckling load, i.e. $P/P_{cr} = 0.405$. At this axial compressive force level, it is observed that $K_{GS} \approx K_{SF}$, but there is an 11.5% difference in stiffness between $K_{P\Delta}$ and K_{SF}. This means ignoring the small P-δ effect results in an increase in stiffness of 11.5%. Table 7.1 also shows that the difference between using $K_{P\Delta}$ and using K_{SF} results in a 5.3% difference in the calculated period of vibration of the SDOF column.

Assume the SDOF column has a 2% damping. By subjecting the SDOF column to the 1995 Kobe earthquake presented in Figure 3.2, the displacement responses of using the original stiffness, P-Δ stiffness, geometric stiffness, and stability function stiffness are presented in Figure 7.10. Similarly, the velocity responses using the four different stiffnesses are presented in Figure 7.11, and the absolute acceleration responses are presented in Figure 7.12. Finally, Table 7.1 also summarizes the maximum displacement, velocity, and absolute acceleration responses.

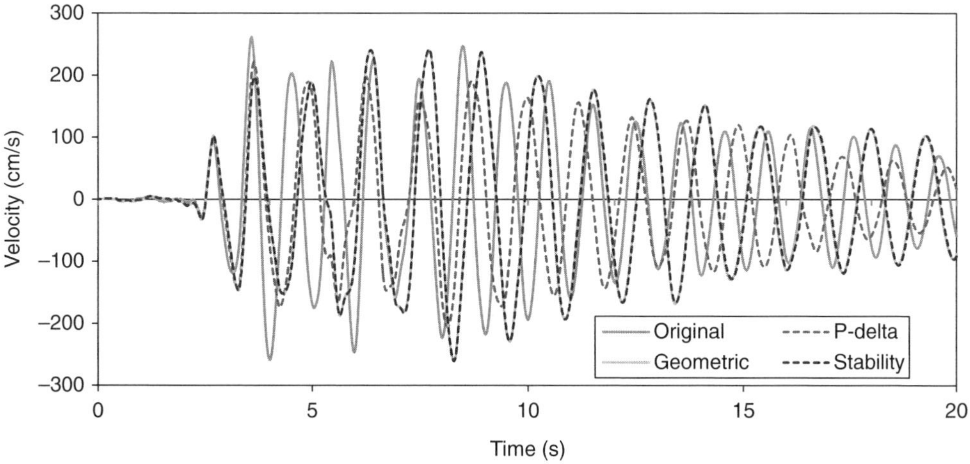

Figure 7.11 Velocity responses of the SDOF column using different stiffnesses.

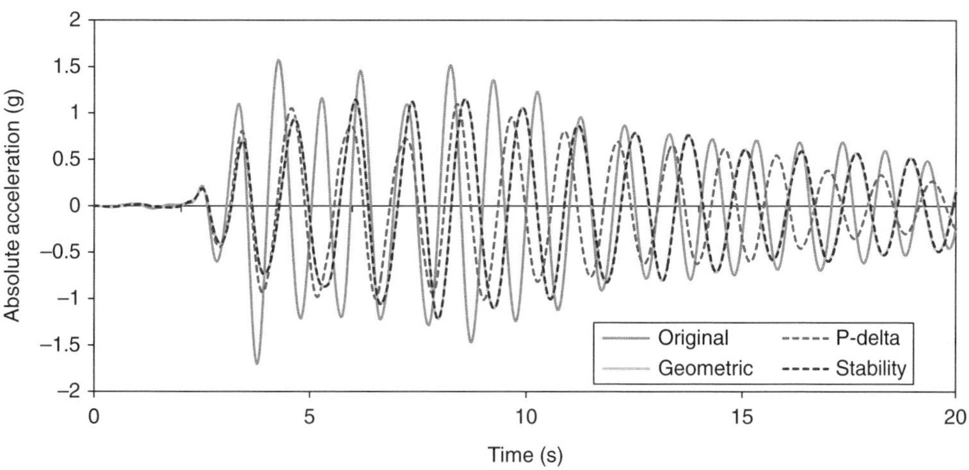

Figure 7.12 Acceleration responses of the SDOF column using different stiffnesses.

It can be seen that the responses using K_{GS} match those using K_{SF} very well, and Table 7.1 shows that the difference in maximum displacement response is only 0.4%. However, there are more noticeable differences between the responses using $K_{P\Delta}$ and those using K_{SF}, where the difference in maximum displacement response goes up to 19.4%.

Example 7.5 Single-Degree-of-Freedom Column with Plastic Hinges

Consider a similar SDOF column used in Figure 7.9 of Example 7.4 but with two plastic hinges (i.e. $q = 2$) as shown in Figure 7.13. Let $E = 100$ GPa, $I = 20 \times 10^6$ mm^4, $L = 4$ m, and $P =$ 500 kN in compression, $M = 9.5$ Mg, and a damping ratio of 2%. The equation used to calculate

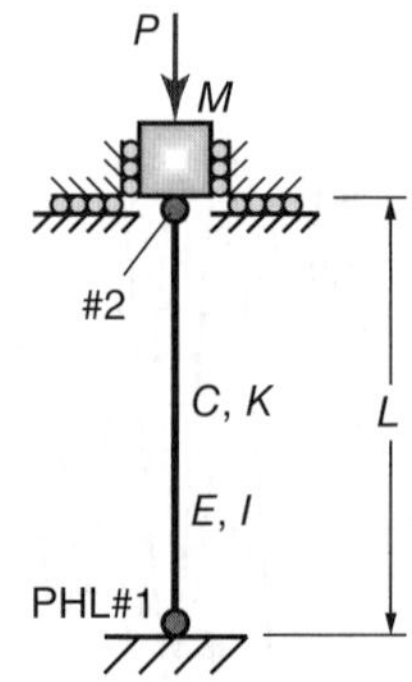

Figure 7.13 A single degree of freedom column with two plastic hinges.

the stiffness K_{SF} using the stability functions approach was presented in Eq. (7.134) and it is $K_{SF} = 224.3$ kN/m as shown in Table 7.1.

Since the axial compressive force of $P = 500$ kN remains constant throughout the duration of the earthquake ground motion, no update of the geometric stiffness is needed. This means Eq. (7.121) can be used in the nonlinear dynamic analysis. The structural properties are evaluated as follows:

$$\omega = \sqrt{K_{SF}/M} = \sqrt{224.3/9.5} = 4.859 \text{ rad/s} \tag{7.135a}$$

$$T = 2\pi\sqrt{M/K} = 2\pi\sqrt{9.5/224.3} = 1.293 \text{ s} \tag{7.135b}$$

$$C = 2\zeta\omega M = 2\times 0.02\times 4.859\times 9.5 = 1.846 \text{ kN s/m} \tag{7.135c}$$

$$\omega_d = \omega\sqrt{1-\zeta^2} = 4.859\times\sqrt{1-(0.02)^2} = 4.858 \text{ rad/s} \tag{7.135d}$$

Using a time step size of $\Delta t = 0.01$ s, the transition matrices become

$$\mathbf{A} = \begin{bmatrix} 0 & 1 \\ -23.606 & -0.1944 \end{bmatrix}, \quad \mathbf{H} = \begin{bmatrix} 0 \\ -1 \end{bmatrix}, \quad \mathbf{G} = \begin{bmatrix} 0 \\ 23.606 \end{bmatrix} \tag{7.136a}$$

$$\mathbf{F}_d = \begin{bmatrix} 0.998821 & 0.009986 \\ -0.23573 & 0.996880 \end{bmatrix}, \quad \mathbf{H}_d = \begin{bmatrix} 0.00010 \\ -0.00997 \end{bmatrix}, \quad \mathbf{G}_d = \begin{bmatrix} 0.00236 \\ 0.23532 \end{bmatrix} \tag{7.136b}$$

and therefore Eq. (7.121) becomes

$$\begin{Bmatrix} x_{k+1} \\ \dot{x}_{k+1} \end{Bmatrix} = \begin{bmatrix} 0.998821 & 0.009986 \\ -0.23573 & 0.996880 \end{bmatrix} \begin{Bmatrix} x_k \\ \dot{x}_k \end{Bmatrix} + \begin{bmatrix} 0.00010 \\ -0.00997 \end{bmatrix} a_k + \begin{bmatrix} 0.00236 \\ 0.23532 \end{bmatrix} x_k'' \tag{7.137}$$

The two remaining stiffness matrices related to the FAM can be formulated as:

$$\mathbf{K}_o' = \left[\bar{s}EI/L^2 \quad \bar{s}EI/L^2\right] = [698.5 \quad 698.5] \tag{7.138a}$$

$$\mathbf{K}_o'' = \begin{bmatrix} sEI/L & scEI/L \\ scEI/L & sEI/L \end{bmatrix} = \begin{bmatrix} 1718.1 & 1076.0 \\ 1076.0 & 1718.1 \end{bmatrix} \tag{7.138b}$$

Then, the governing equations of the FAM in Eqs. (7.128) and (7.129) become

$$\begin{Bmatrix} m_1 \\ m_2 \end{Bmatrix}_{k+1} + \begin{bmatrix} 1718.1 & 1076.0 \\ 1076.0 & 1718.1 \end{bmatrix} \begin{Bmatrix} \Delta\theta_1'' \\ \Delta\theta_2'' \end{Bmatrix} = \begin{bmatrix} 698.5 \\ 698.5 \end{bmatrix} x_{k+1} - \begin{bmatrix} 1718.1 & 1076.0 \\ 1076.0 & 1718.1 \end{bmatrix} \begin{Bmatrix} \theta_1'' \\ \theta_2'' \end{Bmatrix}_k \tag{7.139}$$

$$x_{k+1}'' = \frac{1}{224.3} [698.5 \quad 698.5] \begin{Bmatrix} \theta_1'' \\ \theta_2'' \end{Bmatrix}_{k+1} \tag{7.140}$$

At the axial compressive force of $P = 500$ kN, assume that the yield moments are $m_{y1} = 200$ kN m and $m_{y2} = 220$ kN m. Let both plastic hinges exhibit bilinear kinematic hardening behavior with $K_{t1} = 300$ kN m/rad and $K_{t2} = 200$ kN m/rad, then the corresponding moment versus plastic rotation relationships can be written as:

$$if \begin{cases} m_1 \leq 200 \\ m_1 > 200 \end{cases}, \quad then \quad \begin{cases} \theta_1'' = 0 \\ m_1 = 200 + 300 \times \theta_1'' \end{cases} \tag{7.141a}$$

$$if \begin{cases} m_2 \leq 220 \\ m_2 > 220 \end{cases}, \quad then \quad \begin{cases} \theta_2'' = 0 \\ m_2 = 220 + 200 \times \theta_2'' \end{cases} \tag{7.141b}$$

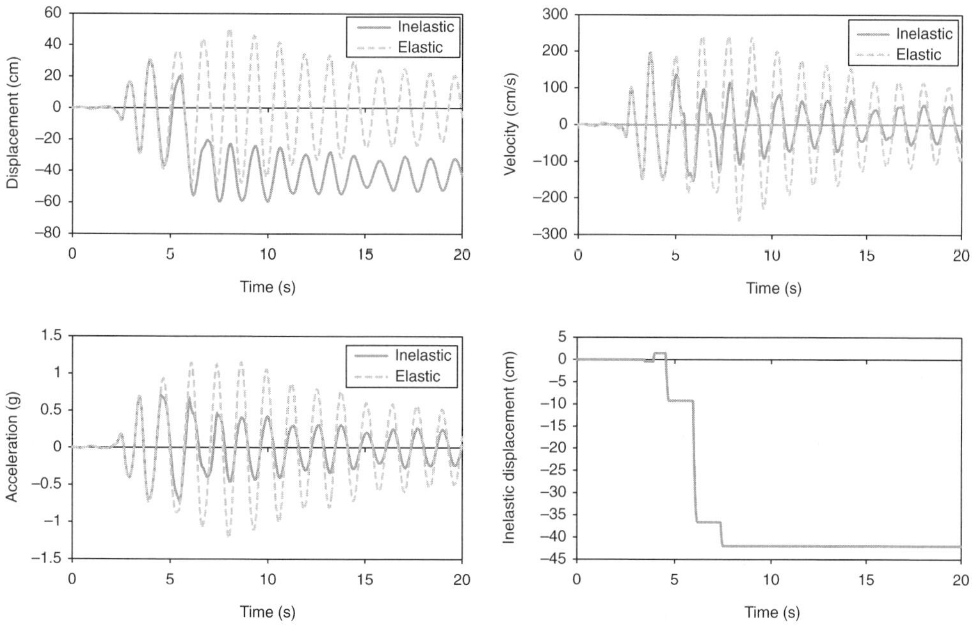

Figure 7.14 Global responses of the SDOF column with geometric nonlinearity.

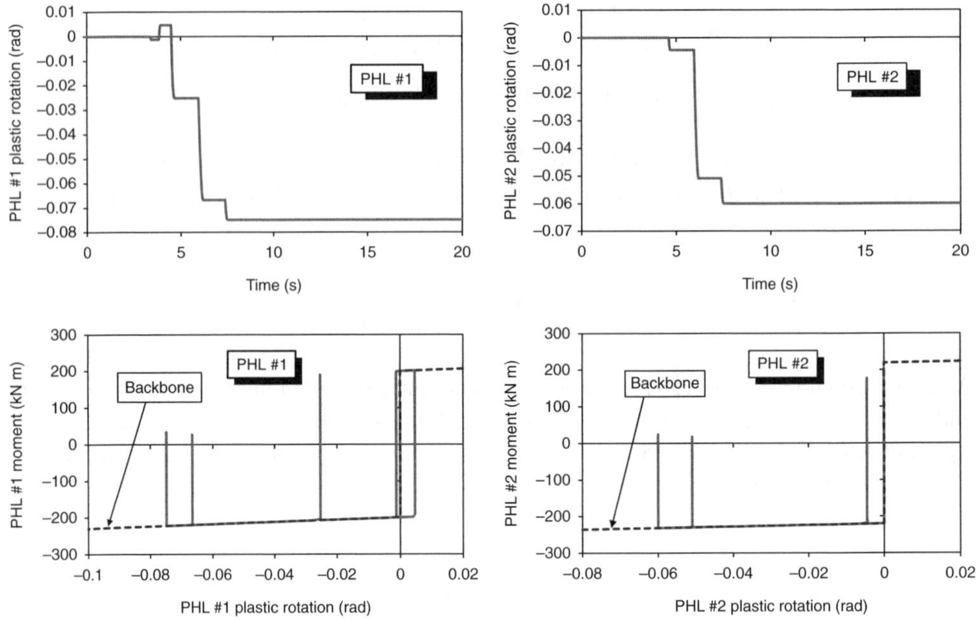

Figure 7.15 Local plastic hinge responses of the SDOF column with geometric nonlinearity.

By subjecting the SDOF column to the 1995 Kobe earthquake as shown in Figure 3.2, the displacement, velocity, absolute acceleration, and inelastic displacement responses are shown in Figure 7.14. The geometric nonlinear elastic responses obtained in Example 7.4 are also plotted in the figure for comparison. The plastic rotation responses at the two PHLs are shown in Figure 7.15. In addition, the moment versus plastic rotation responses of the two PHLs along with the corresponding backbone curves are also shown in Figure 7.15.

7.5 Nonlinear Dynamic Analysis with Static Condensation Using Stability Functions

In Section 3.3, the nonlinear dynamic analysis procedure was derived with static condensation on framed structures. In this section, the knowledge gained from the previous sections on the dynamic analysis of structures with static condensation is combined with geometric nonlinearity. Similar to Section 3.3, the goal here is to condense the stiffness matrix when the mass moment of inertia is ignored at certain DOFs. In this case, static condensation is needed in order to eliminate those DOFs without mass moment of inertia, resulting in a mass matrix that is invertible.

Consider a moment-resisting frame with n DOFs and q PHLs, the equation of motion can be written in the matrix form as

$$\begin{bmatrix} \mathbf{M}_{dd} & \mathbf{0} \\ \mathbf{0} & \mathbf{0} \end{bmatrix} \begin{Bmatrix} \ddot{\mathbf{x}}_d(t) \\ \ddot{\mathbf{x}}_r(t) \end{Bmatrix} + \begin{bmatrix} \mathbf{C}_{dd} & \mathbf{0} \\ \mathbf{0} & \mathbf{0} \end{bmatrix} \begin{Bmatrix} \dot{\mathbf{x}}_d(t) \\ \dot{\mathbf{x}}_r(t) \end{Bmatrix} + \begin{bmatrix} \mathbf{K}_{dd}(t) & \mathbf{K}_{dr}(t) \\ \mathbf{K}_{rd}(t) & \mathbf{K}_{rr}(t) \end{bmatrix} \begin{Bmatrix} \mathbf{x}'_d(t) \\ \mathbf{x}'_r(t) \end{Bmatrix} = -\begin{bmatrix} \mathbf{M}_{dd} & \mathbf{0} \\ \mathbf{0} & \mathbf{0} \end{bmatrix} \begin{Bmatrix} \ddot{\mathbf{g}}(t) \\ \mathbf{0} \end{Bmatrix} - \begin{Bmatrix} \mathbf{F}_f(t) \\ \mathbf{0} \end{Bmatrix} \tag{7.142}$$

where $\mathbf{M}_{dd}$ is the mass matrix associated with translational DOFs only, $\mathbf{C}_{dd}$ is the damping matrix associated with translational DOFs only, and $\mathbf{K}_{dd}(t)$, $\mathbf{K}_{dr}(t)$, $\mathbf{K}_{rd}(t)$, and $\mathbf{K}_{rr}(t)$ are the stiffness sub-matrices partitioned according to the DOFs with mass and those without mass. Note that the stiffness sub-matrices are functions of time, since these matrices depend on the axial compressive force in the columns that are also functions of time. The vector $\mathbf{x}'(t)$ is the elastic displacement response, $\dot{\mathbf{x}}(t)$ is the velocity response, $\ddot{\mathbf{x}}(t)$ is the acceleration response, $\mathbf{F}_f(t)$ is the $d \times 1$ vector of additional forces imposed on the translational DOFs due to geometric nonlinearity of gravity columns, and the earthquake ground acceleration vector $\ddot{\mathbf{g}}(t)$ corresponds to the effect of ground motion at each DOF associated with nonzero mass. The subscript d denotes the number of degrees of freedom that have nonzero mass, and subscript r denotes the number of degrees of freedom that have zero mass moment of inertia. This gives $n = d + r$ in an n-DOF system.

The governing equations of the FAM presented in Eqs. (7.98) and (7.99) can similarly be partitioned and written as:

$$\mathbf{m}(t) + \mathbf{K}''(t)\boldsymbol{\Theta}''(t) = \left[\mathbf{K}'_d(t)^T \;\vdots\; \mathbf{K}'_r(t)^T\right]\begin{Bmatrix}\mathbf{x}_d(t)\\ \hdashline \mathbf{x}_r(t)\end{Bmatrix} \tag{7.143}$$

$$\begin{Bmatrix}\mathbf{x}''_d(t)\\ \hdashline \mathbf{x}''_r(t)\end{Bmatrix} = \left[\begin{array}{c:c}\mathbf{K}_{dd}(t) & \mathbf{K}_{dr}(t)\\ \hdashline \mathbf{K}_{rd}(t) & \mathbf{K}_{rr}(t)\end{array}\right]^{-1}\begin{bmatrix}\mathbf{K}'_d(t)\\ \hdashline \mathbf{K}'_r(t)\end{bmatrix}\boldsymbol{\Theta}''(t) \tag{7.144}$$

where $\mathbf{m}(t)$ is the $q \times 1$ total moment vector of the PHLs, $\boldsymbol{\Theta}''(t)$ is the $q \times 1$ plastic rotation vector of the PHLs, and

$$\mathbf{K}'(t) = \begin{bmatrix}\mathbf{K}'_d(t)\\ \hdashline \mathbf{K}'_r(t)\end{bmatrix}, \quad \mathbf{x}''(t) = \begin{Bmatrix}\mathbf{x}''_d(t)\\ \hdashline \mathbf{x}''_r(t)\end{Bmatrix}, \quad \begin{Bmatrix}\mathbf{x}_d(t)\\ \hdashline \mathbf{x}_r(t)\end{Bmatrix} = \begin{Bmatrix}\mathbf{x}'_d(t)\\ \hdashline \mathbf{x}'_r(t)\end{Bmatrix} + \begin{Bmatrix}\mathbf{x}''_d(t)\\ \hdashline \mathbf{x}''_r(t)\end{Bmatrix} \tag{7.145}$$

Static condensation is now performed on Eqs. (7.142) to (7.144) by following the same procedure that was presented in Section 3.3. The resulting equations in the condensed form are:

$$\mathbf{M}_{dd}\ddot{\mathbf{x}}_d(t) + \mathbf{C}_{dd}\dot{\mathbf{x}}_d(t) + \bar{\mathbf{K}}(t)\mathbf{x}'_d(t) = -\mathbf{M}_{dd}\ddot{\mathbf{g}}(t) - \mathbf{F}_f(t) \tag{7.146a}$$

$$\mathbf{m}(t) + \bar{\mathbf{K}}''(t)\boldsymbol{\Theta}''(t) = \bar{\mathbf{K}}'(t)^T\mathbf{x}_d(t) \tag{7.146b}$$

$$\mathbf{x}''_d(t) = \bar{\mathbf{K}}(t)^{-1}\bar{\mathbf{K}}'(t)\boldsymbol{\Theta}''(t) \tag{7.146c}$$

where

$$\bar{\mathbf{K}}(t) = \mathbf{K}_{dd}(t) - \mathbf{K}_{dr}(t)\mathbf{K}_{rr}(t)^{-1}\mathbf{K}_{rd}(t) \tag{7.147a}$$

$$\bar{\mathbf{K}}'(t) = \mathbf{K}'_d(t) - \mathbf{K}_{dr}(t)\mathbf{K}_{rr}(t)^{-1}\mathbf{K}'_r(t) \tag{7.147b}$$

$$\bar{\mathbf{K}}''(t) = \mathbf{K}''(t) - \mathbf{K}'_r(t)^T\mathbf{K}_{rr}(t)^{-1}\mathbf{K}'_r(t) \tag{7.147c}$$

The state space method of nonlinear dynamic analysis is then performed on Eq. (7.146a) by following the same procedure that was presented in Section 7.4. Define

$$\mathbf{F}_f(t)=\mathbf{K}_f\mathbf{x}_d(t), \quad \bar{\mathbf{K}}(t)=\bar{\mathbf{K}}_o+\bar{\mathbf{K}}_g(t), \quad \bar{\mathbf{K}}_e=\bar{\mathbf{K}}_o+\mathbf{K}_f \tag{7.148}$$

where $\mathbf{K}_f$ is the $d \times d$ stiffness matrix that is a function of the gravity loads on the leaning column and the corresponding story height, $\bar{\mathbf{K}}_o$ is the $d \times d$ initial stiffness matrix of the frame computed by using the gravity loads on the frame only, $\bar{\mathbf{K}}_g(t)$ is the change in stiffness matrix due to the change in axial load on members during the dynamic loading, and $\bar{\mathbf{K}}_e$ is the $d \times d$ elastic stiffness matrix of the structure. Let

$$\mathbf{z}(t)=\begin{Bmatrix}\mathbf{x}_d(t)\\ \dot{\mathbf{x}}_d(t)\end{Bmatrix}, \quad \mathbf{A}=\begin{bmatrix}\mathbf{0} & \mathbf{I}\\ -\mathbf{M}_{dd}^{-1}\bar{\mathbf{K}}_e & -\mathbf{M}_{dd}^{-1}\mathbf{C}_{dd}\end{bmatrix}, \quad \ddot{\mathbf{g}}(t)=\mathbf{h}\mathbf{a}(t)=\mathbf{h}\begin{Bmatrix}\ddot{g}_x(t)\\ \ddot{g}_y(t)\\ \ddot{g}_z(t)\end{Bmatrix} \tag{7.149a}$$

$$\mathbf{H}=\begin{bmatrix}\mathbf{0}\\ -\mathbf{h}\end{bmatrix}, \quad \mathbf{B}=\begin{bmatrix}\mathbf{0}\\ \mathbf{M}_{dd}^{-1}\end{bmatrix}, \quad \mathbf{f}_g(t)=\bar{\mathbf{K}}_g(t)\mathbf{x}_d(t), \quad \mathbf{f}_m(t)=\bar{\mathbf{K}}(t)\mathbf{x}''_d(t) \tag{7.149b}$$

$$\mathbf{F}_d=\mathbf{e}^{\mathbf{A}\Delta t}, \quad \mathbf{H}_d=\mathbf{e}^{\mathbf{A}\Delta t}\mathbf{H}\Delta t, \quad \mathbf{B}_d=\mathbf{e}^{\mathbf{A}\Delta t}\mathbf{B}\Delta t \tag{7.149c}$$

where Δt is the time step size. Then, it follows from Eq. (7.119) that

$$\mathbf{z}_{k+1}=\mathbf{F}_d\mathbf{z}_k+\mathbf{H}_d\mathbf{a}_k+\mathbf{B}_d\mathbf{f}_{g,k}+\mathbf{B}_d\mathbf{f}_{m,k} \tag{7.150}$$

where $\mathbf{z}_k$, $\mathbf{a}_k$, $\mathbf{f}_{g,k}$, and $\mathbf{f}_{m,k}$ are the discretized forms of $\mathbf{z}(t)$, $\mathbf{a}(t)$, $\mathbf{f}_g(t)$, and $\mathbf{f}_m(t)$, respectively. The discretized forms of the governing equations of the FAM follow directly from Eqs. (7.125) and (7.126), which are

$$\mathbf{m}_{k+1}+\bar{\mathbf{K}}''_k\Delta\boldsymbol{\Theta}''=\bar{\mathbf{K}}'^{T}_k\mathbf{x}_{d,k+1}-\bar{\mathbf{K}}''_k\boldsymbol{\Theta}''_k \tag{7.151}$$

$$\mathbf{x}''_{d,k+1}=\bar{\mathbf{K}}_k^{-1}\bar{\mathbf{K}}'_k\boldsymbol{\Theta}''_{k+1} \tag{7.152}$$

where $\mathbf{m}_k$, $\boldsymbol{\Theta}''_k$, $\mathbf{x}_{d,k}$, and $\mathbf{x}''_{d,k}$ are the discretized forms of $\mathbf{m}(t)$, $\boldsymbol{\Theta}''(t)$, $\mathbf{x}_d(t)$, and $\mathbf{x}''_d(t)$, respectively, $\Delta\boldsymbol{\Theta}''=\boldsymbol{\Theta}''_{k+1}-\boldsymbol{\Theta}''_k$, and $\bar{\mathbf{K}}_k$, $\bar{\mathbf{K}}'_k$, and $\bar{\mathbf{K}}''_k$ are the condensed geometric nonlinear stiffness matrices at time step k. The displacement vector associated with the DOFs with zero mass moment of inertia (i.e. $\mathbf{x}_r(t)$) can be written using Eq. (3.130) as:

$$\mathbf{x}_{r,k}=-\mathbf{K}_{rr}^{-1}\mathbf{K}_{rd}\mathbf{x}_{d,k}+\mathbf{K}_{rr}^{-1}\mathbf{K}'_r\boldsymbol{\Theta}''_k \tag{7.153}$$

where $\mathbf{x}_{r,k}$ is the discretized form of $\mathbf{x}_r(t)$.

If updates to geometric nonlinearity are ignored, then the condensed geometric nonlinear stiffness matrices become:

$$\bar{\mathbf{K}}_k=\bar{\mathbf{K}}(t_k)=\bar{\mathbf{K}}(0)=\bar{\mathbf{K}}_o, \quad \bar{\mathbf{K}}'_k=\bar{\mathbf{K}}'(t_k)=\bar{\mathbf{K}}'(0)=\bar{\mathbf{K}}'_o \tag{7.154a}$$

$$\bar{\mathbf{K}}''_k=\bar{\mathbf{K}}''(t_k)=\bar{\mathbf{K}}''(0)=\bar{\mathbf{K}}''_o, \quad \bar{\mathbf{K}}_g(t)=\mathbf{0} \tag{7.154b}$$

According to Eq. (7.149b), this gives

$$\mathbf{f}_g, k=[\mathbf{0}]\cdot\mathbf{x}_{d,k}=\mathbf{0}, \quad \mathbf{f}_{m,k}=\bar{\mathbf{K}}_o\mathbf{x}''_k \tag{7.155}$$

Then, Eqs. (7.150) to (7.152) become

$$\mathbf{z}_{k+1} = \mathbf{F}_d\mathbf{z}_k + \mathbf{H}_d\mathbf{a}_k + \mathbf{G}_d\mathbf{x}_k'' \tag{7.156}$$

$$\mathbf{m}_{k+1} + \bar{\mathbf{K}}_o''\Delta\boldsymbol{\Theta}'' = \bar{\mathbf{K}}_o'^T\mathbf{x}_{d,k+1} - \bar{\mathbf{K}}_o''\boldsymbol{\Theta}_k'' \tag{7.157}$$

$$\mathbf{x}_{d,k+1}'' = \bar{\mathbf{K}}_o^{-1}\bar{\mathbf{K}}_o'\boldsymbol{\Theta}_{k+1}'' \tag{7.158}$$

where

$$\mathbf{G}_d = \mathbf{B}_d\bar{\mathbf{K}}_o = \mathbf{e}^{\mathbf{A}\Delta t}\begin{bmatrix}\mathbf{0}\\ \mathbf{M}_{dd}^{-1}\end{bmatrix}\bar{\mathbf{K}}_o\Delta t = \mathbf{e}^{\mathbf{A}\Delta t}\begin{bmatrix}\mathbf{0}\\ \mathbf{M}_{dd}^{-1}\bar{\mathbf{K}}_o\end{bmatrix}\Delta t = \mathbf{e}^{\mathbf{A}\Delta t}\mathbf{G}\Delta t \tag{7.159}$$

Finally, the absolute acceleration vector (i.e. $\ddot{\mathbf{y}}_d(t) = \ddot{\mathbf{x}}_d(t) + \ddot{\mathbf{g}}(t)$) can be computed in a way similar to Eq. (7.130), i.e.

$$\ddot{\mathbf{y}}_{d,k} = -\mathbf{M}_{dd}^{-1}\mathbf{C}_{dd}\dot{\mathbf{x}}_{d,k} - \mathbf{M}_{dd}^{-1}\bar{\mathbf{K}}_k\left(\mathbf{x}_{d,k} - \mathbf{x}_{d,k}''\right) - \mathbf{M}_{dd}^{-1}\mathbf{K}_f\mathbf{x}_{d,k} \tag{7.160}$$

where $\dot{\mathbf{x}}_{d,k}$ and $\ddot{\mathbf{y}}_{d,k}$ are the discretized forms of $\dot{\mathbf{x}}_d(t)$ and $\ddot{\mathbf{y}}_d(t)$, respectively.

Example 7.6 One-story Moment-Resisting Frame

Consider a one-story one-bay moment-resisting frame as shown in Figure 7.16 with members assumed to be axially rigid. This gives a total of 3 DOFs (i.e. $n = 3$) and 6 PHLs (i.e. $q = 6$) for the frame as labeled in the figure. Let $I_b = I_c = I$ and $L_b = L_c = L$. Assume no update of geometric nonlinearity due to changes in axial forces in columns is performed, then the global stiffness matrices $\mathbf{K}$, $\mathbf{K}'$, and $\mathbf{K}''$ become

$$\mathbf{K} = \begin{bmatrix} 2s'EI/L^3 & \bar{s}EI/L^2 & \bar{s}EI/L^2 \\ \bar{s}EI/L^2 & (s+4)EI/L & 2EI/L \\ \bar{s}EI/L^2 & 2EI/L & (s+4)EI/L \end{bmatrix} \begin{matrix} \leftarrow x_1 \\ \leftarrow x_2 \\ \leftarrow x_3 \end{matrix} \tag{7.161a}$$

$$\mathbf{K}' = \left[\begin{array}{cc:cc:cc} \bar{s}EI/L^2 & \bar{s}EI/L^2 & \bar{s}EI/L^2 & \bar{s}EI/L^2 & 0 & 0 \\ scEI/L & sEI/L & 0 & 0 & 4EI/L & 2EI/L \\ 0 & 0 & scEI/L & sEI/L & 2EI/L & 4EI/L \end{array}\right] \begin{matrix} \leftarrow x_1 \\ \leftarrow x_2 \\ \leftarrow x_3 \end{matrix} \tag{7.161b}$$

$$\mathbf{K}'' = \left[\begin{array}{cc:cc:cc} sEI/L & scEI/L & 0 & 0 & 0 & 0 \\ scEI/L & sEI/L & 0 & 0 & 0 & 0 \\ \hdashline 0 & 0 & sEI/L & scEI/L & 0 & 0 \\ 0 & 0 & scEI/L & sEI/L & 0 & 0 \\ \hdashline 0 & 0 & 0 & 0 & 4EI_b/L_b & 2EI_b/L_b \\ 0 & 0 & 0 & 0 & 2EI_b/L_b & 4EI_b/L_b \end{array}\right] \begin{matrix} \leftarrow \theta_1'' \\ \leftarrow \theta_2'' \\ \leftarrow \theta_3'' \\ \leftarrow \theta_4'' \\ \leftarrow \theta_5'' \\ \leftarrow \theta_6'' \end{matrix} \tag{7.161c}$$

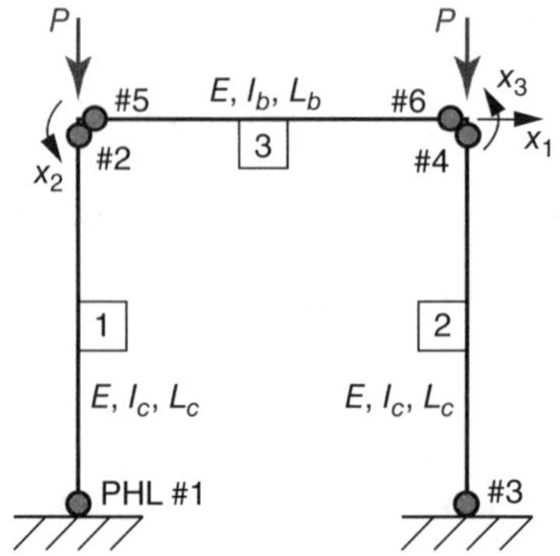

Figure 7.16 One-story one-bay moment-resisting frame with gravity loads.

where s, c, $\bar{s}$, and s' are stability coefficients computed using Eqs. (7.20) and (7.27) based on $\lambda = \sqrt{P/EI} \times L$, where P is the axial compressive force on the columns due to gravity loads. Note that the axial compressive force P due to gravity load only is the same for both columns.

Assume that the frame has a mass of $M_{dd} = 5.0$ Mg and a damping $C_{dd} = 2.0$ Mg s/m at DOF #1, while the mass moment of inertia at DOFs #2 and #3 are ignored. Therefore, static condensation can be applied to eliminate the DOFs for x_2 and x_3. Let $E = 200.0$ GPa, $I = 6.0 \times 10^6$ mm^4, $L = 4$ m, and $P = 50$ kN. Using Eq. (7.147), the condensed stiffness matrices $\bar{\mathbf{K}}_o$, $\bar{\mathbf{K}}_o'$, and $\bar{\mathbf{K}}_o''$ become:

$$\bar{K}_o = \bar{K}_e = 286.8 \text{ kN/m} \tag{7.162a}$$

$$\bar{\mathbf{K}}_o' = [354.2 \quad 269.4 \quad 354.2 \quad 269.4 \quad -269.4 \quad -269.4] \text{ kN} \tag{7.162b}$$

$$\bar{\mathbf{K}}_o'' = \begin{bmatrix} 1007.3 & 286.4 & 41.9 & 81.0 & -286.4 & -81.0 \\ 286.4 & 553.6 & 81.0 & 156.6 & -553.6 & -156.6 \\ 41.9 & 81.0 & 1007.3 & 286.4 & -81.0 & -286.4 \\ 81.0 & 156.6 & 286.4 & 553.6 & -156.6 & -553.6 \\ -286.4 & -553.6 & -81.0 & -156.6 & 553.6 & 156.6 \\ -81.0 & -156.6 & -286.4 & -553.6 & 156.6 & 553.6 \end{bmatrix} \text{kN m/rad} \tag{7.162c}$$

The period of vibration T and the damping ratio ζ can also be calculated:

$$T = 2\pi\sqrt{M_{dd}/\bar{K}_o} = 2\pi\sqrt{5.0/286.8} = 0.830 \text{ s} \tag{7.163a}$$

$$\zeta = \frac{C}{2\sqrt{\bar{K}_o M_{dd}}} = \frac{2.0}{2\sqrt{286.8 \times 5.0}} = 0.0264 = 2.64\% \tag{7.163b}$$

With a time step size of $\Delta t = 0.01$ s, the transition matrices are calculated as:

$$\mathbf{A} = \begin{bmatrix} 0 & 1 \\ -57.36 & -0.4 \end{bmatrix}, \quad \mathbf{H} = \begin{bmatrix} 0 \\ -1 \end{bmatrix}, \quad \mathbf{G} = \begin{bmatrix} 0 \\ 57.36 \end{bmatrix} \tag{7.164a}$$

$$\mathbf{F}_d = \mathbf{e}^{\mathbf{A}\Delta t} = \begin{bmatrix} 0.997137 & 0.009970 \\ -0.57187 & 0.993150 \end{bmatrix}, \quad \mathbf{H}_d = \begin{bmatrix} -0.00010 \\ -0.00993 \end{bmatrix}, \quad \mathbf{G}_d = \begin{bmatrix} 0.00572 \\ 0.56967 \end{bmatrix} \tag{7.164b}$$

and the equation for performing the state space analysis becomes

$$\begin{Bmatrix} x_1 \\ \dot{x}_1 \end{Bmatrix}_{k+1} = \begin{bmatrix} 0.997137 & 0.009970 \\ -0.57187 & 0.993150 \end{bmatrix} \begin{Bmatrix} x_1 \\ \dot{x}_1 \end{Bmatrix}_k + \begin{bmatrix} -0.00010 \\ -0.00993 \end{bmatrix} a_k + \begin{bmatrix} 0.00572 \\ 0.56967 \end{bmatrix} x''_{d,k} \tag{7.165}$$

Assume that the plastic hinges exhibit elastic–plastic behavior with moment capacities of $m_b = 15.0$ kN m for the beam and $m_c = 20.0$ kN m for the two columns. Assume also that the interaction between axial force and moment on the yield surface of the columns is ignored. This gives

$$\text{if} \begin{cases} m_i \le 20.0 \\ m_i > 20.0 \end{cases}, \quad \text{then} \quad \begin{cases} \Delta\theta''_i = 0 \\ m_i = 20.0 \end{cases} \quad i = 1,2,3,4 \tag{7.166a}$$

$$\text{if} \begin{cases} m_i \le 15.0 \\ m_i > 15.0 \end{cases}, \quad \text{then} \quad \begin{cases} \Delta\theta''_i = 0 \\ m_i = 15.0 \end{cases} \quad i = 5,6 \tag{7.166b}$$

Consider now the frame is subjected to the 1994 Northridge earthquake ground motion as shown in Figure 3.3. The global displacement, velocity, acceleration, and inelastic displacement responses at DOF #1 are plotted in Figure 7.17. In addition, the same responses but for the case without considering any geometric nonlinearity (see Example 3.7) are presented in the figure for comparisons.

The local responses at PHLs #1 and #5 are plotted in Figure 7.18. These include the moment response and the plastic rotation response. Again, the same responses but for the case without considering any geometric nonlinearity (see Example 3.7) are also presented in the figure for comparisons. Finally, the moment versus plastic rotation responses for PHLs #1 and #5 for the case with geometric nonlinearity is also shown in Figure 7.18. As shown in the figure, it can be seen that when geometric nonlinearity is considered, the moment responses are skewed to one direction. This suggests that the tendency after yielding in one direction is to have further yielding in the same direction when geometric nonlinearity is considered. On the other hand, the moment responses continue to oscillate about the centerline when geometric nonlinearity is ignored, suggesting that yielding in one direction does not favor or prevent subsequent yielding in either the same or opposite direction.

Example 7.7 Two-story Moment-Resisting Frame

Consider a two-story one-bay moment-resisting frame as shown in Figure 7.19 with members assumed to be axially rigid. This frame has a total of 6 DOFs (i.e. $n = 6$) and 12 PHLs (i.e. $q = 12$) as labeled in the figure. Assume that the two-story frame has member properties of $E = 200.0$ GPa, $I = 6.0 \times 10^6$ mm^4, $L = 4.0$ m, and $P = 50$ kN on each of the column members. Also assume that updating of geometric nonlinearity due to changes in axial forces in columns is ignored, then the $\mathbf{K}$ (6×6), $\mathbf{K}'$ (6×12), and $\mathbf{K}''$ (12×12) global stiffness matrices can be constructed. Note that while Columns #3 and #4 are subjected to an axial force of 50 kN, Columns #1 and #2 are subjected to an axial force of 100 kN.

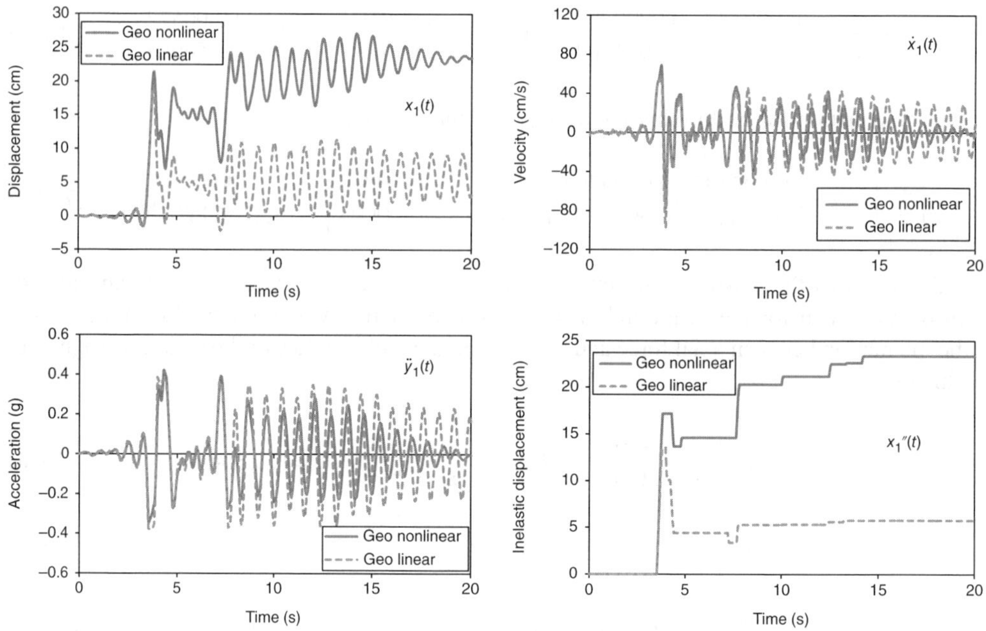

Figure 7.17 Global response comparisons of the one-story frame with geometric nonlinearity.

Assume that the mass moment of inertia are ignored for the rotational joints, the frame can then be statically condensed to a 2-DOF system by eliminating DOFs #3 to #6 while retaining only DOFs #1 and #2. By following Eq. (7.147a), the condensed global stiffness matrix becomes

$$\bar{\mathbf{K}}_o = \begin{bmatrix} 673.415 & -288.698 \\ -288.698 & 210.869 \end{bmatrix} \text{kN/m} \tag{7.167}$$

and the $2 \times 12\bar{\mathbf{K}}_o'$ and $12 \times 12\bar{\mathbf{K}}_o''$ matrices can similarly be calculated using Eqs. (7.147b) and (7.147c).

Let the mass be 2.5 Mg on each floor and the damping be 2% for the two modes of vibration. This gives

$$\mathbf{M}_{dd} = \begin{bmatrix} 2.5 & 0 \\ 0 & 2.5 \end{bmatrix} \text{Mg}, \quad \mathbf{C}_{dd} = \begin{bmatrix} 2.3479 & -0.7403 \\ -0.7403 & 1.1618 \end{bmatrix} \text{kN s/m} \tag{7.168}$$

and by using $\mathbf{M}_{dd}$ and $\bar{\mathbf{K}}_o$, the two periods of vibration are computed as 1.169 s and 0.349 s. Then using Eqs. (7.149a), (7.149b), and (7.159), the calculations of the transition matrices in the continuous form give

$$\mathbf{A} = \begin{bmatrix} 0.0 & 0.0 & 1.0 & 0.0 \\ 0.0 & 0.0 & 0.0 & 1.0 \\ -269.4 & 115.5 & -0.939 & 0.296 \\ 115.5 & -84.3 & 0.296 & -0.465 \end{bmatrix}, \quad \mathbf{H} = \begin{bmatrix} 0 \\ 0 \\ -1 \\ -1 \end{bmatrix}, \quad \mathbf{G} = \begin{bmatrix} 0.0 & 0.0 \\ 0.0 & 0.0 \\ 269.4 & -115.5 \\ -115.5 & 84.3 \end{bmatrix} \tag{7.169}$$

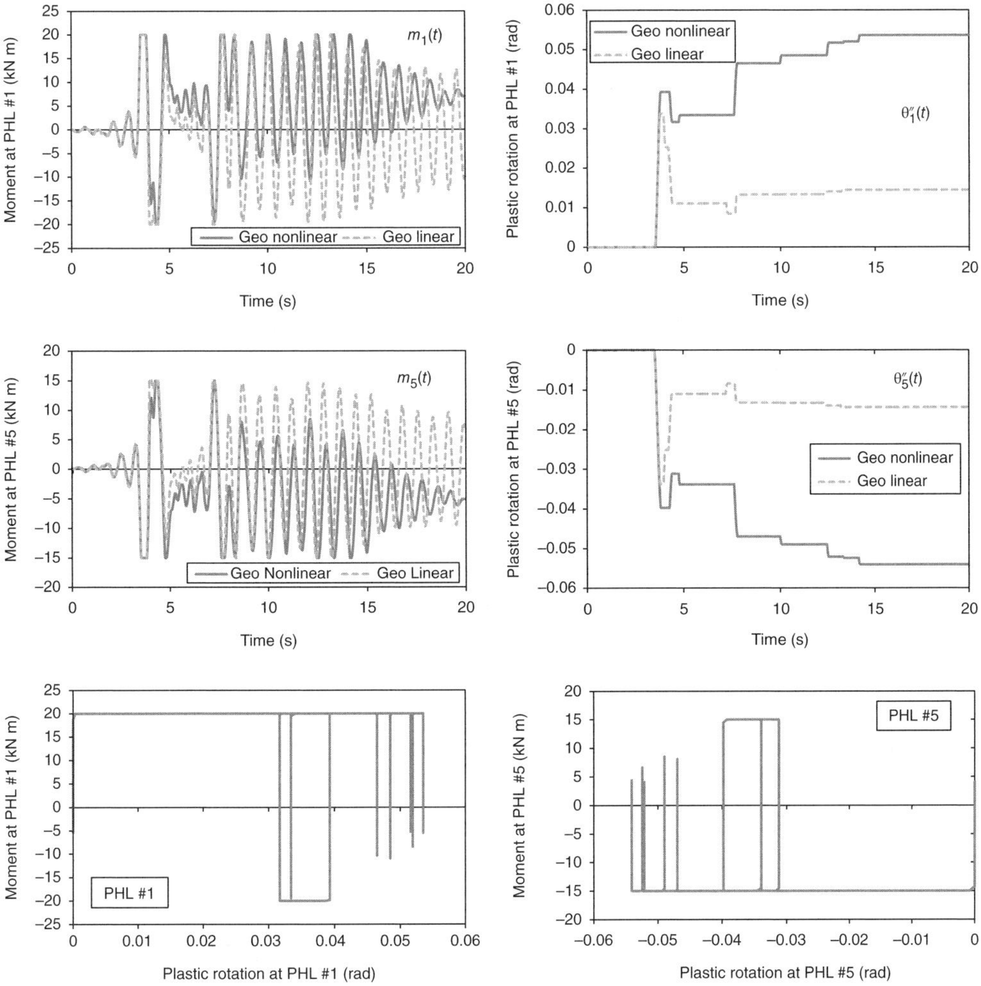

Figure 7.18 Local response comparisons of the one-story frame with geometric nonlinearity.

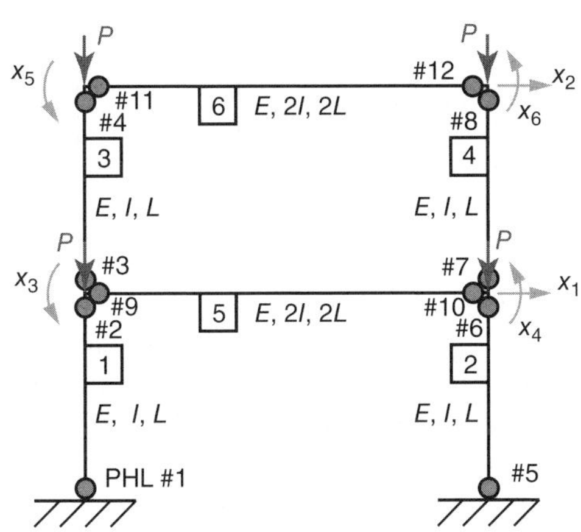

Figure 7.19 Two-story moment-resisting frame with gravity loads.

From Eq. (7.149c) with a time step size of $\Delta t = 0.01$ s, the transition matrices in the discrete form are calculated as

$$\mathbf{F}_d = \begin{bmatrix} 0.986615 & 0.005735 & 0.009909 & 0.000034 \\ 0.005735 & 0.995803 & 0.000034 & 0.009963 \\ -2.66514 & 1.14138 & 0.977319 & 0.008653 \\ 1.14138 & -0.836433 & 0.008653 & 0.991183 \end{bmatrix} \quad (7.170a)$$

$$\mathbf{H}_d = \begin{bmatrix} -0.000099 \\ -0.000100 \\ -0.009860 \\ -0.009998 \end{bmatrix}, \quad \mathbf{G}_d = \begin{bmatrix} 0.026651 & -0.011414 \\ -0.011414 & 0.008364 \\ 2.62257 & -1.12130 \\ -1.12130 & 0.826047 \end{bmatrix} \quad (7.170b)$$

Therefore, by using the matrices presented in Eq. (7.170), Eq. (7.156) becomes

$$\begin{Bmatrix} x_1 \\ x_2 \\ \dot{x}_1 \\ \dot{x}_2 \end{Bmatrix}_{k+1} = \mathbf{F}_d \begin{Bmatrix} x_1 \\ x_2 \\ \dot{x}_1 \\ \dot{x}_2 \end{Bmatrix}_k + \mathbf{H}_d a_k + \mathbf{G}_d \begin{Bmatrix} x_1'' \\ x_2'' \end{Bmatrix}_k \quad (7.171)$$

Assume that the plastic hinges exhibit elastic–plastic behavior with moment capacities of $m_b = 15.0$ kN m for the two beams and $m_c = 20.0$ kN m for the four columns. This gives

$$if \begin{cases} m_i \le 20.0 \\ m_i > 20.0, \end{cases} \quad then \quad \begin{cases} \Delta\theta_i'' = 0 \\ m_i = 20.0 \end{cases} \quad i = 1, \ldots, 8 \quad (7.172a)$$

$$if \begin{cases} m_i \le 15.0 \\ m_i > 15.0, \end{cases} \quad then \quad \begin{cases} \Delta\theta_i'' = 0 \\ m_i = 15.0 \end{cases} \quad i = 9, \ldots, 12 \quad (7.172b)$$

and Eqs. (7.157) and (7.158) become

$$\begin{Bmatrix} m_1 \\ m_2 \\ \vdots \\ m_{12} \end{Bmatrix}_{k+1} + \bar{\mathbf{K}}_o'' \begin{Bmatrix} \Delta\theta_1'' \\ \Delta\theta_2'' \\ \vdots \\ \Delta\theta_{12}'' \end{Bmatrix} = \bar{\mathbf{K}}_o'^T \begin{Bmatrix} x_1 \\ x_2 \end{Bmatrix}_k - \bar{\mathbf{K}}_o'' \begin{Bmatrix} \theta_1'' \\ \theta_2'' \\ \vdots \\ \theta_{12}'' \end{Bmatrix}_k, \quad \begin{Bmatrix} x_1'' \\ x_2'' \end{Bmatrix}_k = \bar{\mathbf{K}}_o^{-1} \bar{\mathbf{K}}_o' \begin{Bmatrix} \theta_1'' \\ \theta_2'' \\ \vdots \\ \theta_{12}'' \end{Bmatrix}_k \quad (7.173)$$

Consider now the two-story moment-resisting frame is subjected to the 1994 Northridge earthquake ground motion as shown in Figure 3.3. The nonlinear dynamic responses are calculated using Eqs. (7.171) and (7.173), and the displacement, velocity, acceleration, and inelastic displacement responses are plotted in Figures 7.20 to 7.23, respectively. In addition, the same responses but for the case without considering any geometric nonlinearity (see Example 3.8) are presented in the figures for comparisons.

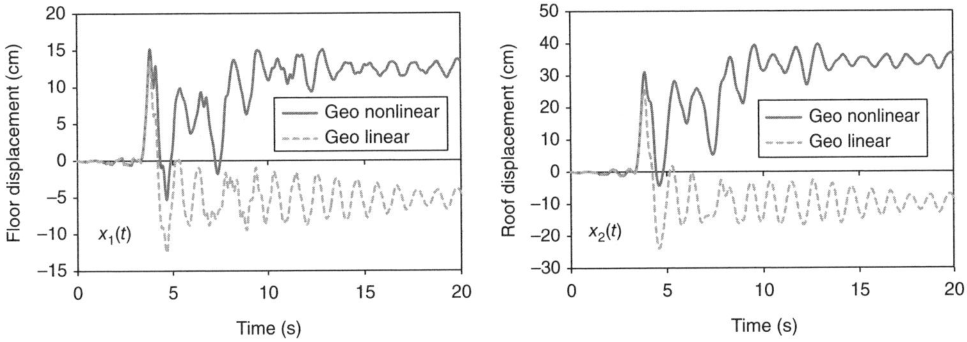

Figure 7.20 Global displacement responses of the two-story frame with geometric nonlinearity.

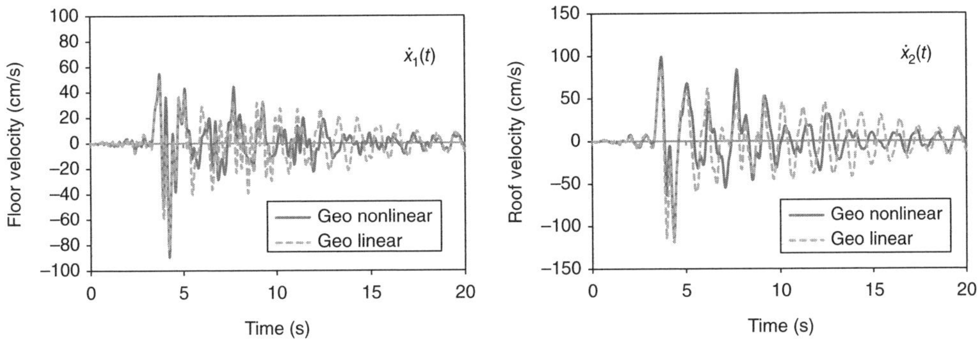

Figure 7.21 Global velocity responses of the two-story frame with geometric nonlinearity.

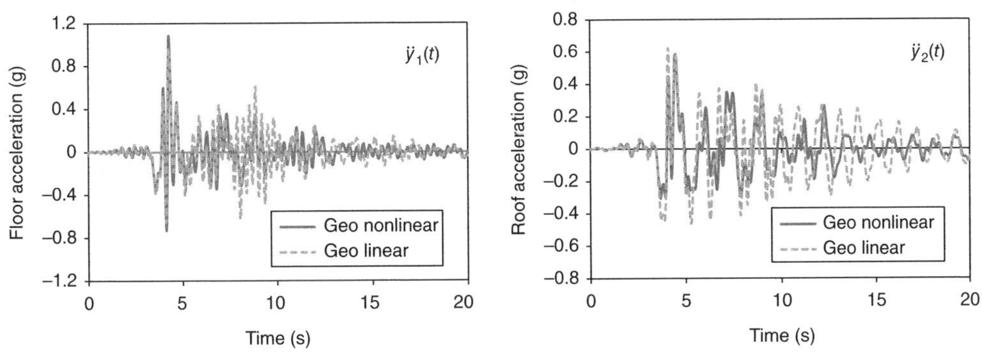

Figure 7.22 Global acceleration responses of the two-story frame with geometric nonlinearity.

The local moment and plastic rotation responses at PHLs #1 and #9 are plotted in Figures 7.24 and 7.25, respectively. In addition, the same responses but for the case without considering any geometric nonlinearity (see Example 3.8) are presented in the figures for comparisons. Finally, the corresponding hysteresis loops are plotted in Figure 7.26.

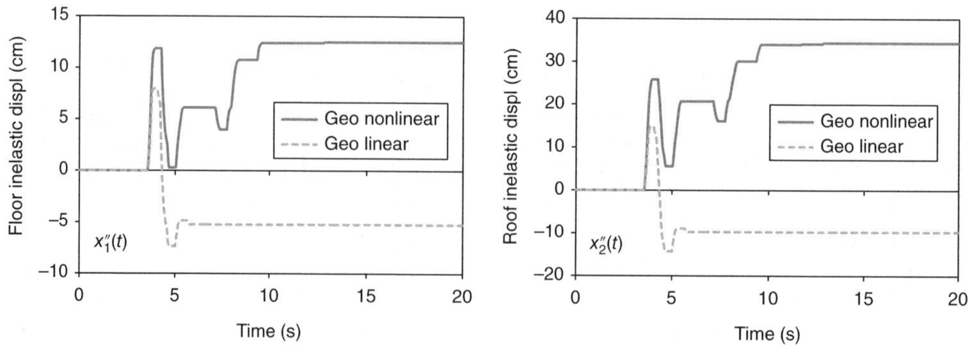

Figure 7.23 Global inelastic displacement responses of the frame with geometric nonlinearity.

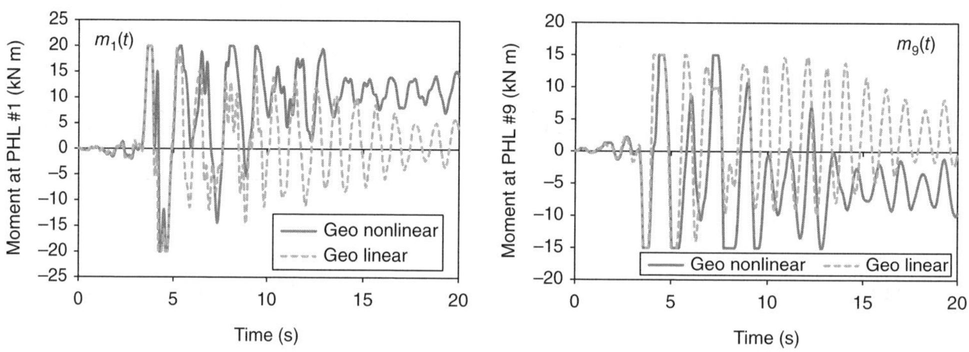

Figure 7.24 Moments at selected PHLs of the two-story frame with geometric nonlinearity.

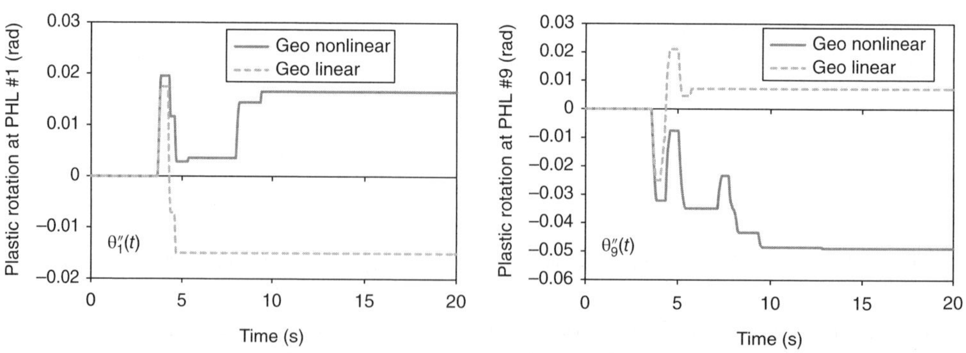

Figure 7.25 Plastic rotations at selected PHLs of the frame with geometric nonlinearity.

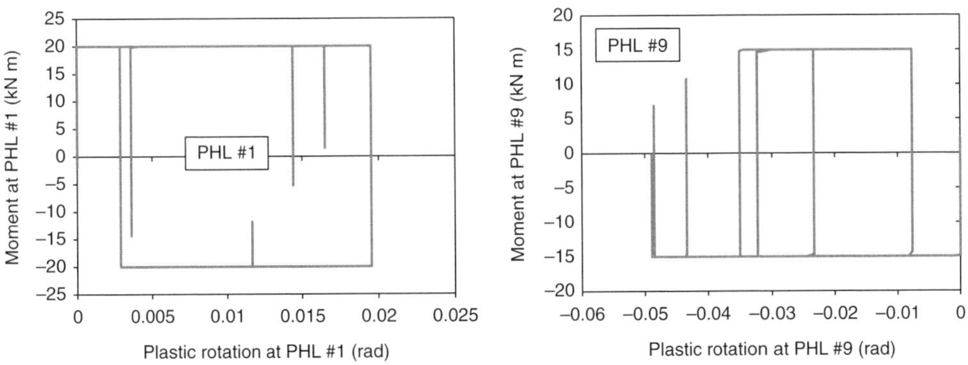

Figure 7.26 Hysteresis loops at selected PHLs of the frame with geometric nonlinearity.

7.6 Nonlinear Dynamic Examples

Additional examples are presented here to demonstrate the use of the FAM with static condensation, geometric nonlinearity, and the state space method for solving nonlinear dynamic problems of two-dimensional moment-resisting frames.

Example 7.8 Six-story Moment-Resisting Steel Frame

Consider the 6-story moment-resisting steel frame as shown in Figure 7.27. Assuming the members are axially rigid, this gives a total of 30 DOFs (i.e. $n = 30$) and 84 PHLs (i.e. $q = 84$). Assume that updating of geometric nonlinearity due to changes in axial forces in columns is ignored. A uniform gravity load of 21.89 kN/m is applied on all the beams, and this gives the axial compressive forces in the columns that are used to calculate the ith member stiffness matrices $\mathbf{K}_i$, $\mathbf{K}_i'$, and $\mathbf{K}_i''$ along with the standard properties of each member. Finally, the 30×30 $\mathbf{K}$ matrix, 30×84 $\mathbf{K}'$ matrix, and 84×84 $\mathbf{K}''$ matrix are constructed by assembling the individual stiffness matrices of each member.

No mass moment of inertia is assumed at the rotational joints, and therefore static condensation is used to eliminate the 24 rotational DOFs (i.e. $r = 24$), resulting in only 6 translational DOFs for the frame (i.e. $d = 6$). By applying the static condensation using Eq. (7.147), the $6 \times 6\,\bar{\mathbf{K}}_o$ matrix, $6 \times 84\,\bar{\mathbf{K}}_o'$ matrix, and $84 \times 84\,\bar{\mathbf{K}}_o''$ matrix are obtained, where these condensed stiffness matrices are constant because updates of geometric nonlinearity due to changes in axial force are ignored. As a numerical illustration, the resulting $\bar{\mathbf{K}}_o$ matrix is of the form

$$\bar{\mathbf{K}}_o = \begin{bmatrix} 462{,}400 & -251{,}400 & 39{,}160 & -3{,}716 & 346.0 & -29.4 \\ -251{,}400 & 444{,}100 & -265{,}800 & 38{,}140 & -3{,}551 & 305.4 \\ 39{,}160 & -265{,}800 & 449{,}700 & -252{,}200 & 36{,}170 & -3{,}106 \\ -3{,}716 & 38{,}140 & -252{,}200 & 383{,}600 & -190{,}600 & 24{,}430 \\ 346.0 & -3{,}551 & 36{,}170 & -190{,}600 & 282{,}400 & -124{,}800 \\ -29.4 & 305.4 & -3{,}106 & 24{,}430 & -124{,}800 & 103{,}200 \end{bmatrix} \text{kN/m} \tag{7.174}$$

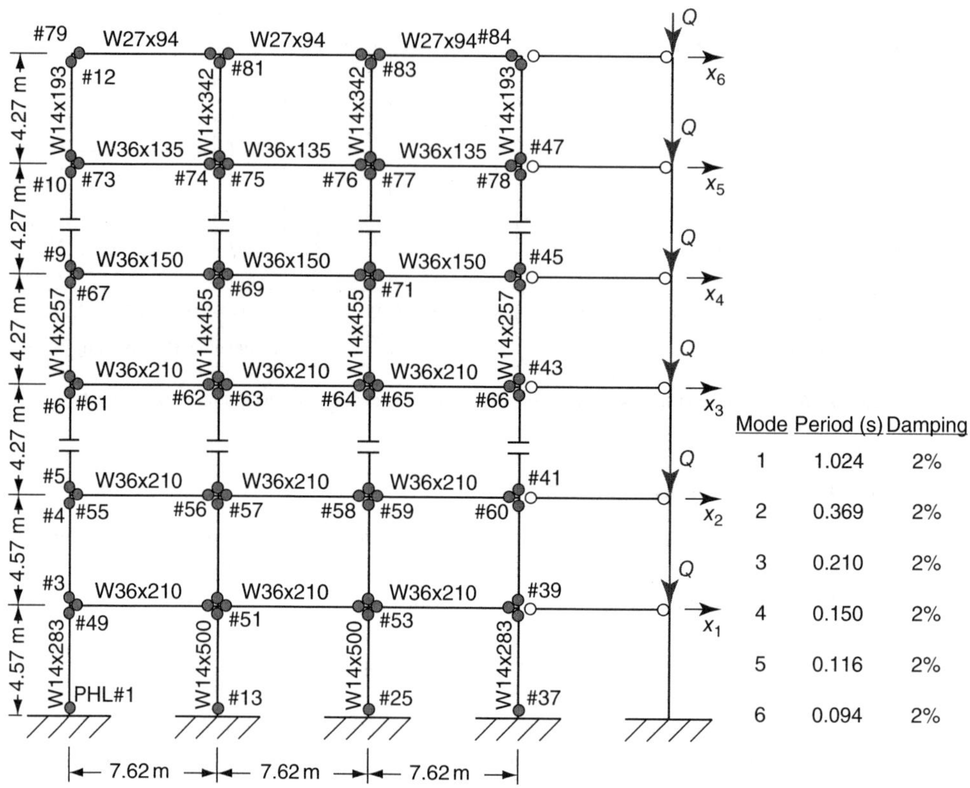

Figure 7.27 Six-story moment-resisting steel frame with a leaning column.

Assume that a gravity load of $Q = 5{,}978$ kN is applied on each floor of the leaning column as shown in Figure 7.27. Then, the $\mathbf{K}_f$ matrix becomes

$$\mathbf{K}_f = \begin{bmatrix} -14{,}390 & 6{,}540 & 0 & 0 & 0 & 0 \\ 6{,}540 & -12{,}140 & 5{,}600 & 0 & 0 & 0 \\ 0 & 5{,}600 & -9{,}800 & 4{,}200 & 0 & 0 \\ 0 & 0 & 4{,}200 & -7{,}000 & 2{,}800 & 0 \\ 0 & 0 & 0 & 2{,}800 & -4{,}200 & 1{,}400 \\ 0 & 0 & 0 & 0 & 1{,}400 & -1{,}400 \end{bmatrix} \text{kN/m} \tag{7.175}$$

Now, from the third equation of Eq. (7.148) where $\bar{\mathbf{K}}_e = \bar{\mathbf{K}}_o + \mathbf{K}_f$, the condensed elastic stiffness matrix $\bar{\mathbf{K}}_e$ becomes

$$\bar{\mathbf{K}}_e = \begin{bmatrix} 448{,}000 & -244{,}900 & 39{,}160 & -3{,}716 & 346.0 & -29.4 \\ -244{,}900 & 432{,}000 & -260{,}200 & 38{,}140 & -3{,}551 & 305.4 \\ 39{,}160 & -260{,}200 & 439{,}900 & -248{,}000 & 36{,}170 & -3{,}106 \\ -3{,}716 & 38{,}140 & -248{,}000 & 376{,}600 & -187{,}800 & 24{,}430 \\ 346.0 & -3{,}551 & 36{,}170 & -187{,}800 & 278{,}200 & -123{,}400 \\ -29.4 & 305.4 & -3{,}106 & 24{,}430 & -123{,}400 & 101{,}800 \end{bmatrix} \text{kN/m} \tag{7.176}$$

Let the mass be 200.0 Mg on each floor, giving a total mass of 1,200 Mg for the entire frame. The diagonal nonzero mass matrix $\mathbf{M}_{dd}$ can be written as

$$\mathbf{M}_{dd} = 200.0 \times \mathbf{I} = \begin{bmatrix} 200 & 0 & 0 & 0 & 0 & 0 \\ 0 & 200 & 0 & 0 & 0 & 0 \\ 0 & 0 & 200 & 0 & 0 & 0 \\ 0 & 0 & 0 & 200 & 0 & 0 \\ 0 & 0 & 0 & 0 & 200 & 0 \\ 0 & 0 & 0 & 0 & 0 & 200 \end{bmatrix} \text{Mg} \tag{7.177}$$

Using the condensed elastic stiffness matrix $\bar{\mathbf{K}}_e$ in Eq. (7.176) and the diagonal $\mathbf{M}_{dd}$ mass matrix in Eq. (7.177), the 6 periods of vibration are calculated and summarized in Figure 7.27. The damping is assumed to be 2% in all six modes of vibration. This gives a damping matrix of

$$\mathbf{C}_{dd} = \begin{bmatrix} 361 & -114 & -4.2 & -4.6 & -2.0 & -1.3 \\ -114 & 329 & -130 & -9.2 & -6.2 & -3.4 \\ -4.2 & -130 & 327 & -129 & -7.2 & -7.0 \\ -4.6 & -9.2 & -129 & 302 & -110 & -9.0 \\ -2.0 & -6.2 & -7.2 & -110 & 259 & -99.2 \\ -1.3 & -3.4 & -7.0 & -9.0 & -99.2 & 150 \end{bmatrix} \text{kN s/m} \tag{7.178}$$

Assume that all 84 plastic hinges exhibit elastic–plastic behavior with moment capacity of the ith plastic hinge, $m_{c,i}$, calculated as

$$m_{c,i} = f_y \times Z_i \quad i = 1, \ldots, 84 \tag{7.179}$$

where f_y is the yield stress of steel and Z_i is the plastic section modulus of the ith plastic hinge. This gives

$$\textit{if} \begin{cases} m_i(t) \le m_{c,i} \\ m_i(t) > m_{c,i} \end{cases}, \quad \textit{then} \quad \begin{cases} \Delta\theta_i''(t) = 0 \\ m_i(t) = m_{c,i} \end{cases} \quad i = 1, \ldots, 84 \tag{7.180}$$

By subjecting the 6-story frame to the 1995 Kobe earthquake ground motion as shown in Figure 3.2, Figures 7.28 to 7.30 show the displacement, velocity, and absolute acceleration responses, respectively, of selected floors. In addition, the same responses but for the case without considering any geometric nonlinearity (see Example 3.9) are presented in the figures for comparisons.

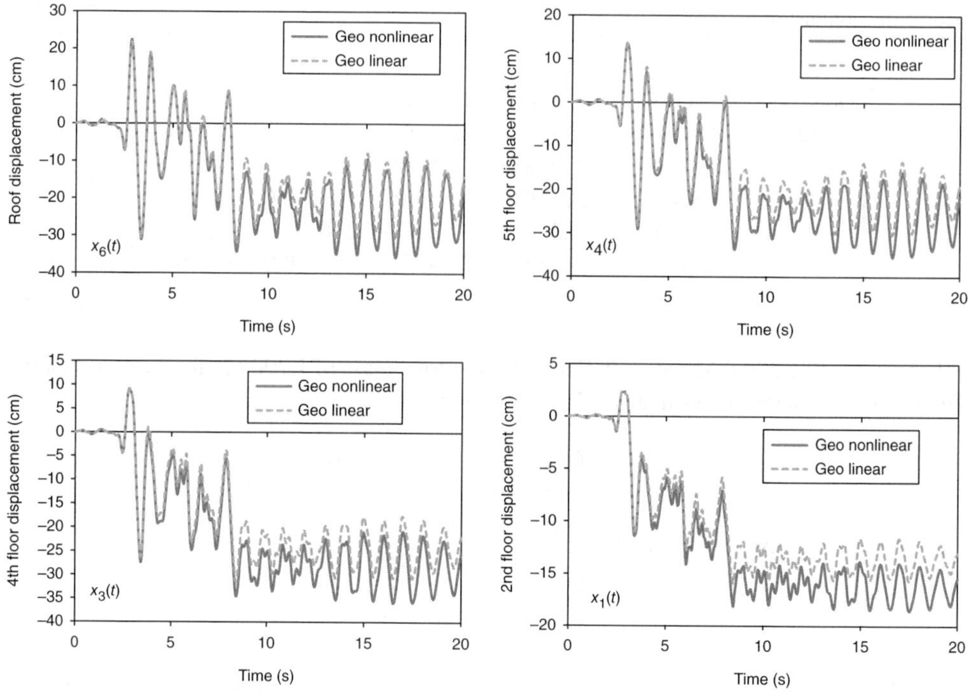

Figure 7.28 Global displacement responses of the 6-story frame with geometric nonlinearity.

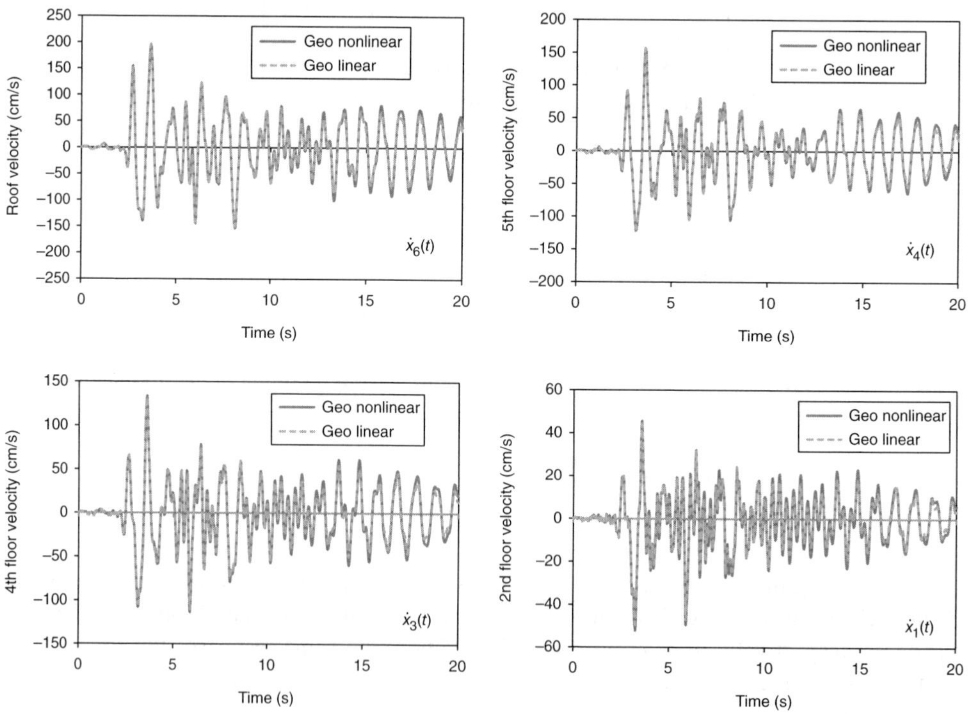

Figure 7.29 Global velocity responses of the 6-story frame with geometric nonlinearity.

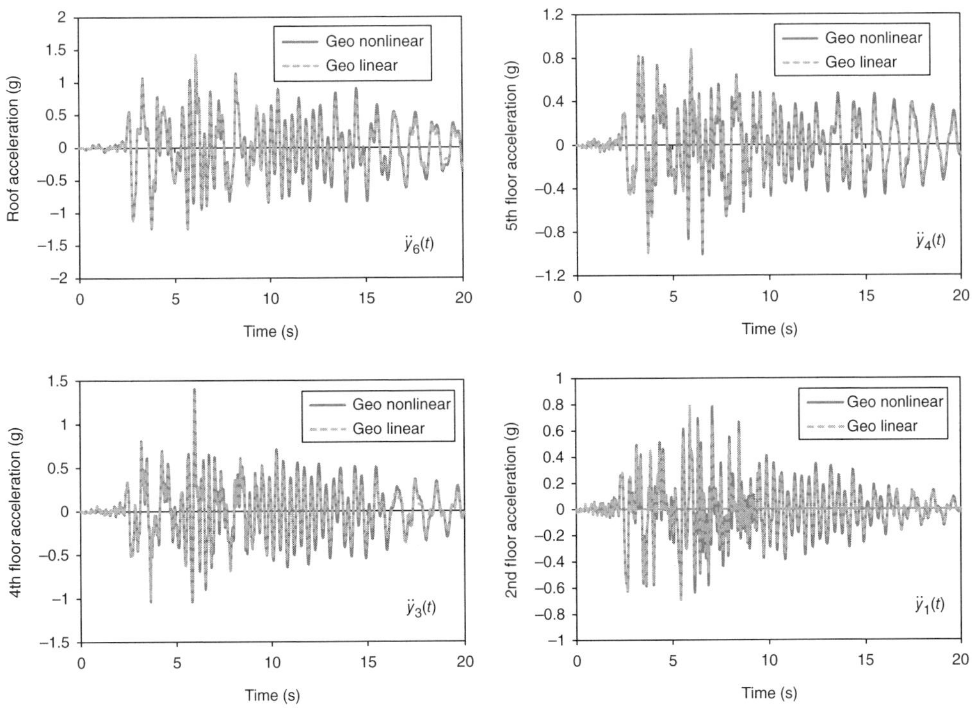

Figure 7.30 Global acceleration responses of the 6-story frame with geometric nonlinearity.

Figures 7.31 and 7.32 show the moment and plastic rotation responses, respectively, of selected plastic hinges. These selected plastic hinges include the base of the exterior column (PHL #1), base of the interior column (PHL #13), exterior hinge of the beam on the second floor (PHL #49), and interior hinge of the beam on the third floor (PHL #56). In addition, the same responses but for the case without considering any geometric nonlinearity (see Example 3.9) are presented in the figures for comparisons. Finally, the corresponding hysteresis loops are plotted in Figure 7.33.

Example 7.9 Ten-story Moment-Resisting Steel Frame

Consider the 10-story moment-resisting steel frame as shown in Figure 7.34. Assuming the members are axially rigid, this gives a total of 50 DOFs (i.e. $n = 50$) and 140 PHLs (i.e. $q = 140$). Also assume that updating of geometric nonlinearity due to changes in axial forces in columns is ignored. A uniform gravity load of 21.89 kN/m is applied on all the beams, and this gives the axial compressive forces in the columns that are used to calculate the ith member stiffness matrices $\mathbf{K}_i$, $\mathbf{K}'_i$, and $\mathbf{K}''_i$ along with the standard properties of each member. Finally, the 50×50 $\mathbf{K}$ matrix, 50×140 $\mathbf{K}'$ matrix, and 140×140 $\mathbf{K}''$ matrix are constructed by assembling the individual stiffness matrices of each member.

No mass moment of inertia is assumed at the rotational joints, and therefore static condensation is used to eliminate the 40 rotational DOFs (i.e. $r = 40$), resulting in only

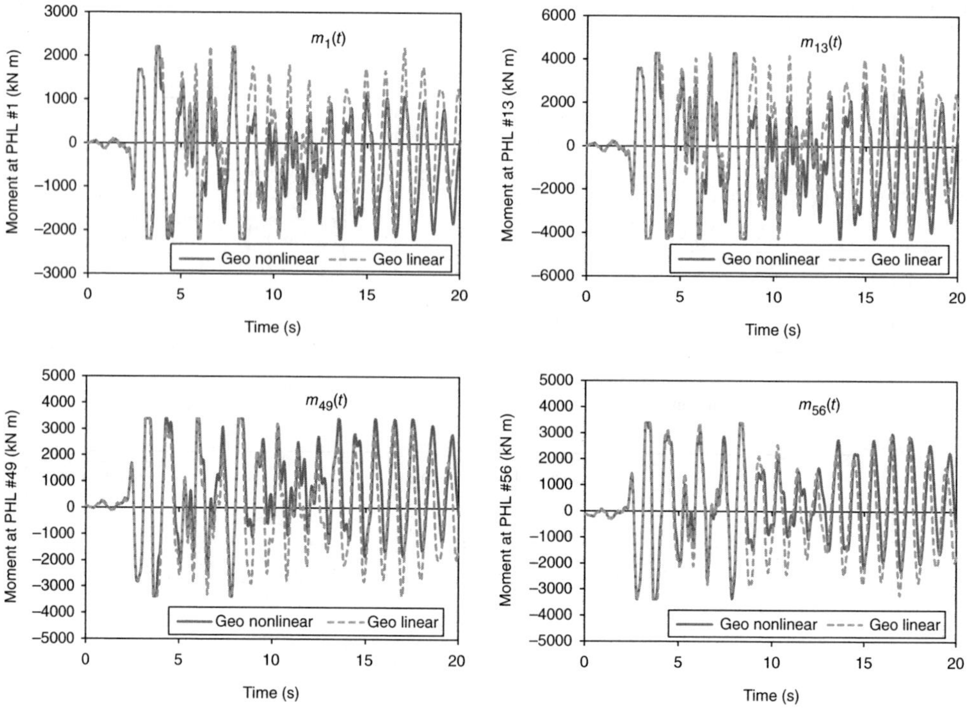

Figure 7.31 Selected PHL moment responses of the frame with geometric nonlinearity.

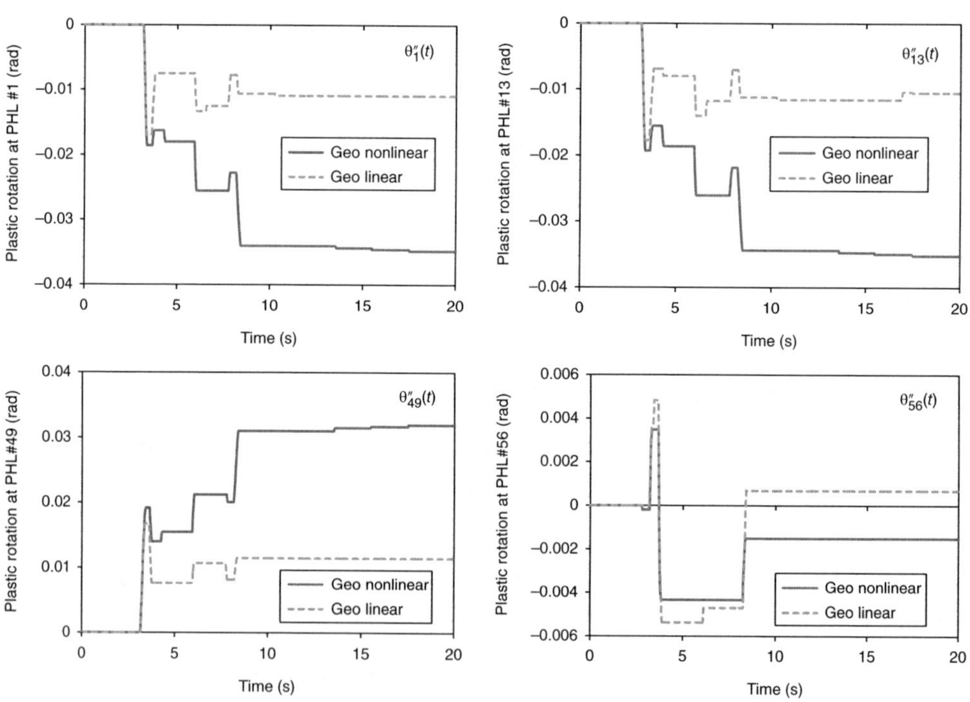

Figure 7.32 Selected PHL plastic rotation responses of the frame with geometric nonlinearity.

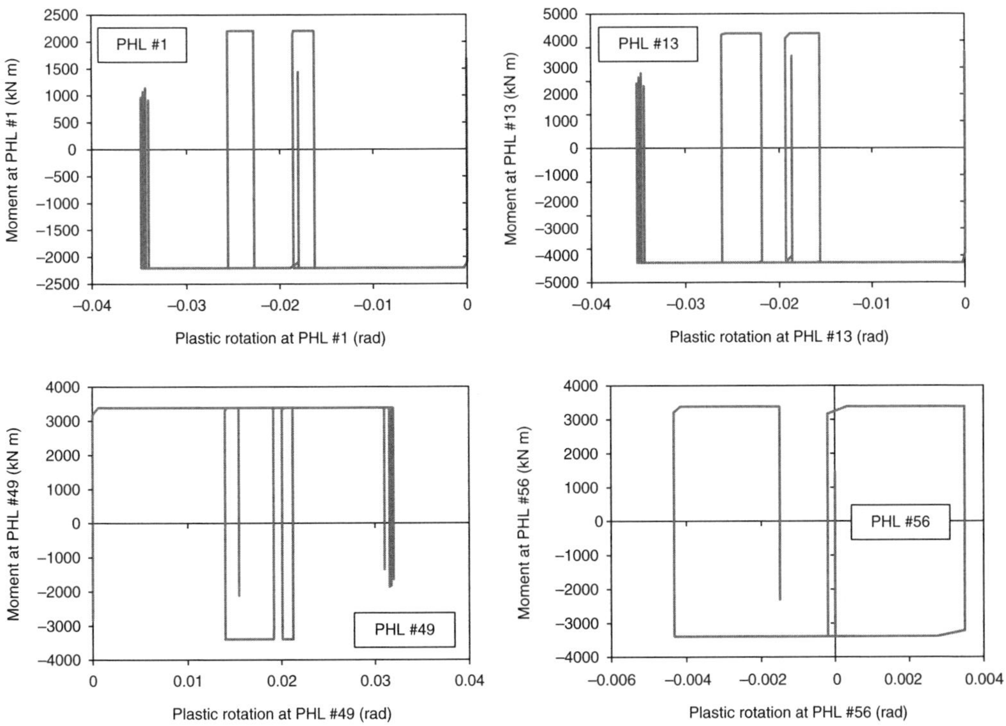

Figure 7.33 Hysteresis loops of selected PHLs of the frame with geometric nonlinearity.

10 translational DOFs for the frame (i.e. $d = 10$). By applying the static condensation using Eq. (7.147), the 10×10 $\bar{\mathbf{K}}_o$ matrix, 10×140 $\bar{\mathbf{K}}_o'$ matrix, and 140×140 $\bar{\mathbf{K}}_o''$ matrix are obtained, where these condensed stiffness matrices are constant because updates of geometric nonlinearity due to changes in axial force are ignored. Assume that a gravity load of $Q = 5{,}978$ kN is applied on each floor of the leaning column, as shown in Figure 7.34. Then, the $\mathbf{K}_f$ matrix becomes

$$\mathbf{K}_f = \begin{bmatrix} -26600 & 12600 & 0 & 0 & 0 & 0 & 0 & 0 & 0 & 0 \\ 12600 & -13800 & 11200 & 0 & 0 & 0 & 0 & 0 & 0 & 0 \\ 0 & 11200 & -21000 & 9800 & 0 & 0 & 0 & 0 & 0 & 0 \\ 0 & 0 & 9800 & -18200 & 8400 & 0 & 0 & 0 & 0 & 0 \\ 0 & 0 & 0 & 8400 & -15400 & 7000 & 0 & 0 & 0 & 0 \\ 0 & 0 & 0 & 0 & 7000 & -12600 & 5600 & 0 & 0 & 0 \\ 0 & 0 & 0 & 0 & 0 & 5600 & -9800 & 4200 & 0 & 0 \\ 0 & 0 & 0 & 0 & 0 & 0 & 4200 & -7000 & 2800 & 0 \\ 0 & 0 & 0 & 0 & 0 & 0 & 0 & 2800 & -4200 & 1400 \\ 0 & 0 & 0 & 0 & 0 & 0 & 0 & 0 & 1400 & -1400 \end{bmatrix} \text{kN/m} \tag{7.181}$$

and the condensed elastic stiffness matrix $\bar{\mathbf{K}}_e$ can be computed accordingly using the third equation of Eq. (7.148).

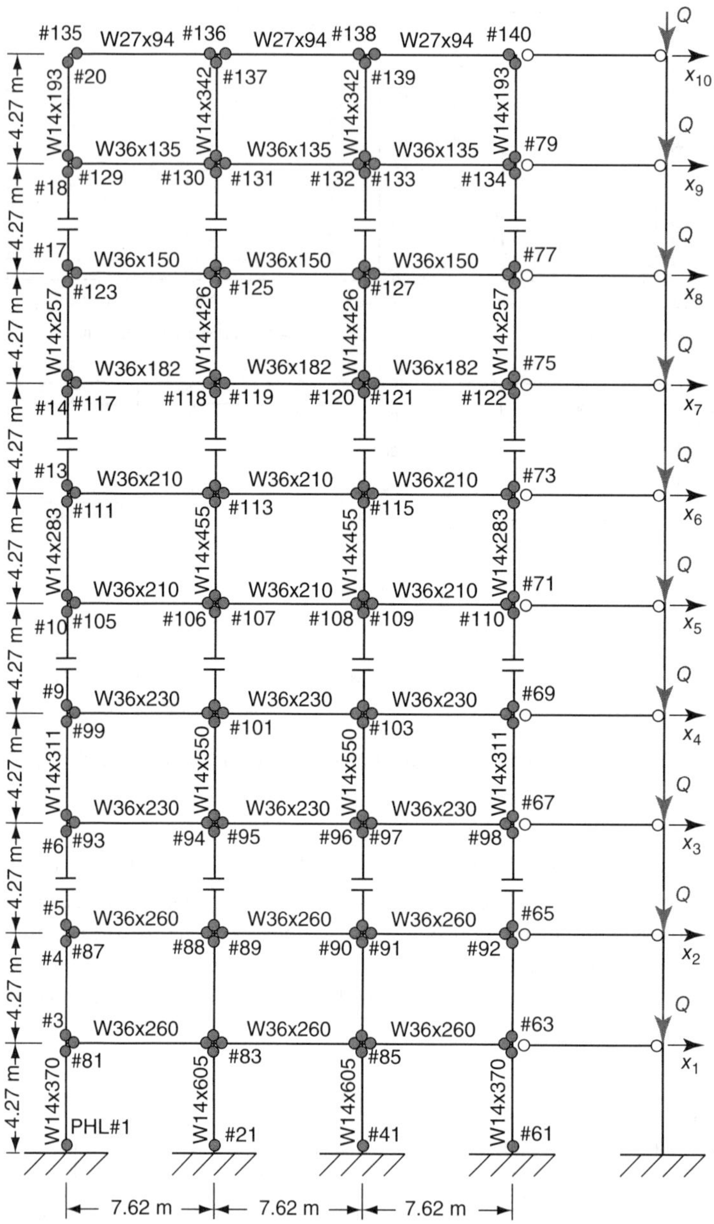

Mode	Period (s)	Damping
1	1.546	3%
2	0.559	3%
3	0.339	3%
4	0.237	3%
5	0.181	3%
6	0.148	3%
7	0.123	3%
8	0.105	3%
9	0.094	3%
10	0.082	3%

Figure 7.34 Ten-story moment-resisting steel frame with a leaning column.

Let the mass be 218.9 Mg on each floor, giving a total mass of 2,189 Mg for the entire frame. This gives

$$\mathbf{M}_{dd}=218.9\times\mathbf{I}=\begin{bmatrix}218.9 & 0 & \cdots & 0\\ 0 & 218.9 & \ddots & \vdots\\ \vdots & \ddots & \ddots & 0\\ 0 & \cdots & 0 & 218.9\end{bmatrix}\text{Mg} \quad (7.182)$$

Using the resulting condensed $\bar{\mathbf{K}}_e$ stiffness matrix and the diagonal $\mathbf{M}_{dd}$ mass matrix, the 10 periods of vibration are calculated and summarized in Figure 7.34. The damping is assumed to be 3% in all ten modes.

Assume that all 140 plastic hinges exhibit elastic–plastic behavior with moment capacity of the ith plastic hinge, $m_{c,i}$, calculated as

$$m_{c,i}=f_y\times Z_i \quad i=1,\ldots,140 \quad (7.183)$$

where f_y is the yield stress of steel and Z_i is the plastic section modulus of the ith plastic hinge. This gives

$$\textit{if}\ \begin{cases} m_i(t)\le m_{c,i} \\ m_i(t)>m_{c,i}\end{cases}, \quad \textit{then}\ \begin{cases}\Delta\theta_i''(t)=0\\ m_i(t)=m_{c,i}\end{cases} \quad i=1,\ldots,140 \quad (7.184)$$

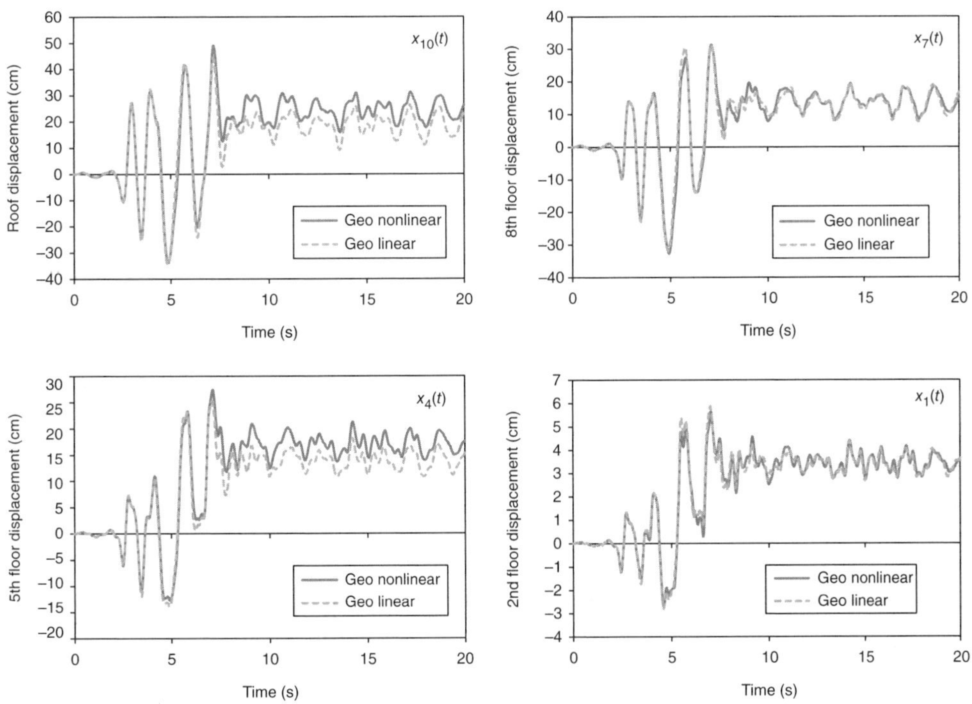

Figure 7.35 Global displacement responses of the 10-story frame with geometric nonlinearity.

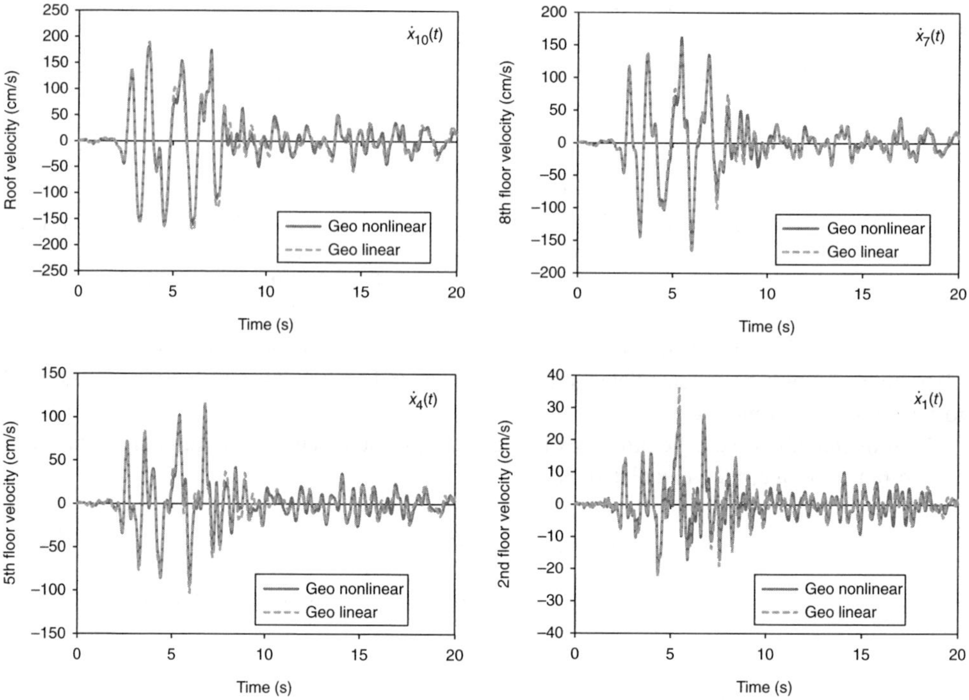

Figure 7.36 Global velocity responses of the 10-story frame with geometric nonlinearity.

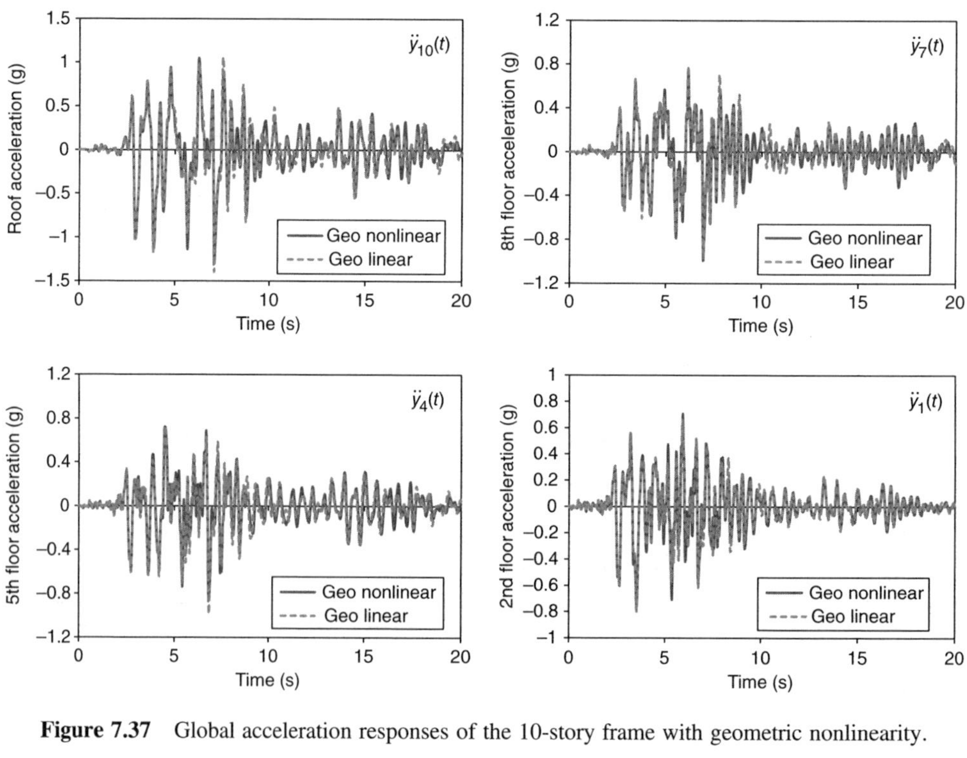

Figure 7.37 Global acceleration responses of the 10-story frame with geometric nonlinearity.

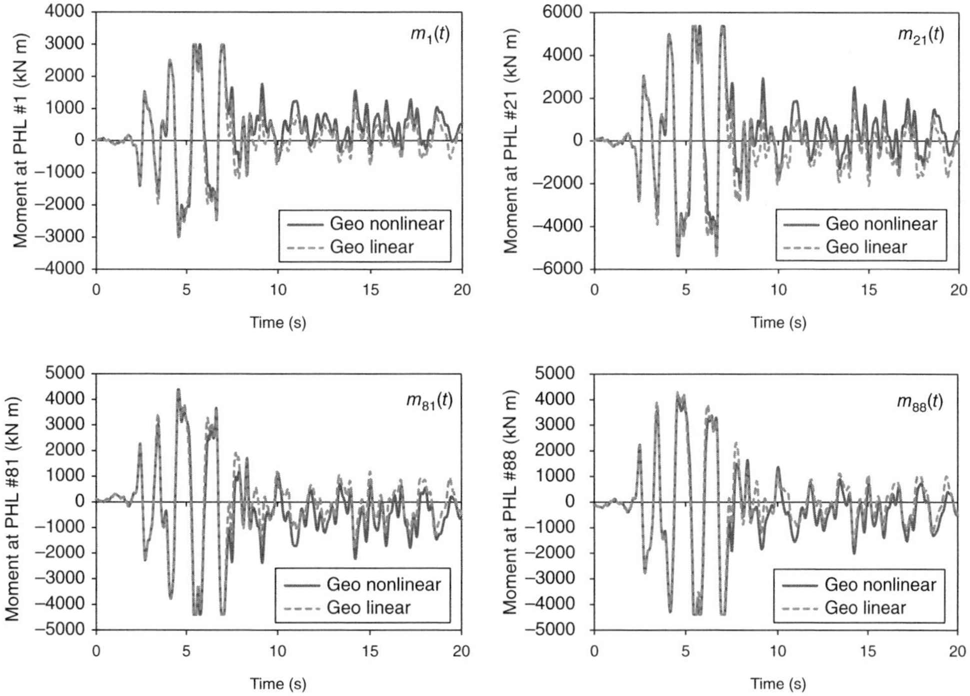

Figure 7.38 Selected PHL moment responses of the 10-story frame with geometric nonlinearity.

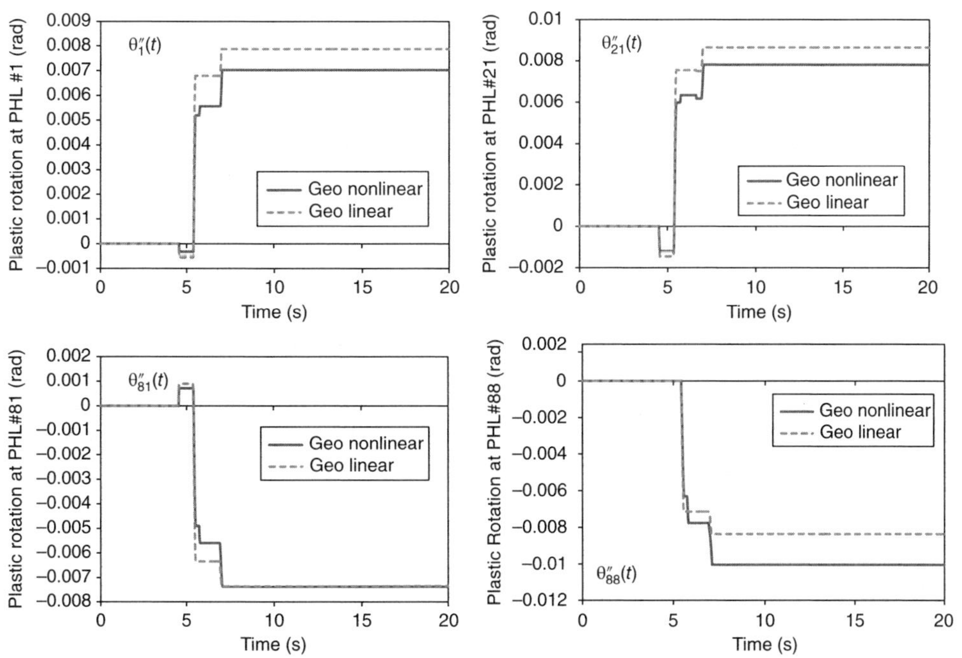

Figure 7.39 Selected PHL plastic rotation responses of the frame with geometric nonlinearity.

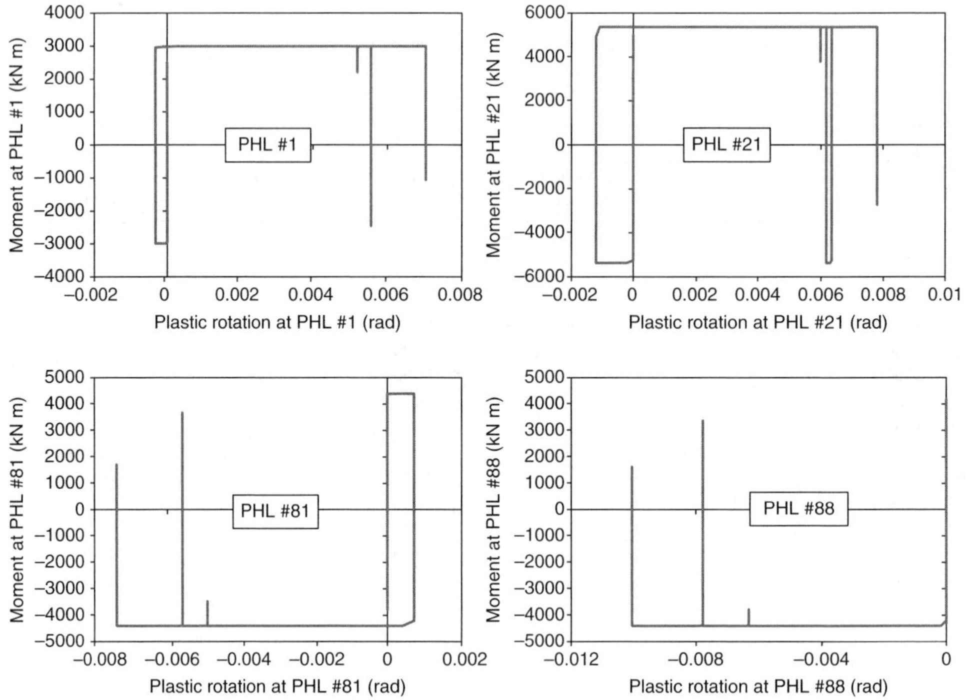

Figure 7.40 Hysteresis loops of selected PHLs of the frame with geometric nonlinearity.

By subjecting the 10-story frame to the 1995 Kobe earthquake ground motion as shown in Figure 3.2, Figures 7.35 to 7.37 show the displacement, velocity, and absolute acceleration responses, respectively, of selected floors. In addition, the same responses but for the case without considering any geometric nonlinearity (see Example 3.10) are presented in the figures for comparisons.

Figures 7.38 and 7.39 show the moment and plastic rotation responses, respectively, of selected plastic hinges. These selected plastic hinges include the base of the exterior column (PHL #1), base of the interior column (PHL #21), exterior hinge of the beam on the second floor (PHL #81), and interior hinge of the beam on the third floor (PHL #88). In addition, the same responses but for the case without considering any geometric nonlinearity (see Example 3.10) are presented in the figures for comparisons. Finally, the corresponding hysteresis loops are plotted in Figure 7.40.

References

Bazant Z.P. and Cedolin L. (2003). *Stability of Structures*, Dover Publication, New York.

Computers and Structures, Inc. (2010). SAP2000 Version 15.1.0, *Integrated Software for Structural Analysis and Design*, Berkeley, CA.

Computers and Structures, Inc. (2011). Perform-3D Version 5.0.0, *Nonlinear Analysis and Performance Assessment of 3D Structures*, Berkeley, CA.

Horne M.Z. and Merchant W. (1965). *The Stability of Frames*, Pergamon Press, New York.

McGuire W., Gallagher R.H. and Ziemian R.D. (2000). *Matrix Structural Analysis*, John Wiley and Sons, New York.

Park J.W. and Kim S.E. (2008). "Nonlinear Inelastic Analysis of Steel-Concrete Composite Beam-Columns Using the Stability Functions," *Structural Engineering and Mechanics*, 30(6): 763–785.

Regents of the University of California (2000). OpenSees Version 2.3.2, *Open System for Earthquake Engineering Simulation*, Berkeley, CA.

Zhao D.F. and Wong K.K.F. (2006). "New Approach for Seismic Nonlinear Analysis of Inelastic Framed Structures," *Journal of Engineering Mechanics ASCE*, 132(9): 959–966.

8

Application of the Force Analogy Method in Modal Superposition

The conventional modal superposition method (MSM) has been validated as a fast and simple way to calculate responses of multi-degree-of-freedom (MDOF) systems under environmental loads. Significant effort has been spent towards extending the usage of MSM in linear dynamic analysis to nonlinear dynamic analysis. In this chapter, the MSM for decoupling governing equation of motions in the force analogy method (FAM) is implemented. The main feasible reason is that the nonlinear deformation of the MDOF system can be separated to the elastic and plastic displacement so that the left-hand side of the equation of motion of nonlinear systems retains the linear properties of corresponding elastic system in the FAM. The nonlinear static pushover analysis (NSPA) procedure in the FAM is established and some examples are presented to demonstrate it. In addition, it is validated that the equation of motion for the MDOF systems in the FAM can be decoupled and two others for solving the force and moment of structural members cannot be decomposed directly. However, the NSPA procedure in the FAM provides a way to determine each equivalent single-degree-of-freedom (SDOF) system corresponding to various models such that the decomposition of these two equations can be achieved.

This chapter consists of three parts: Nonlinear static pushover analysis, modal decomposition method and response summation. They are presented in Section 8.1, 8.2 and 8.3, respectively.

Theory of Nonlinear Structural Analysis: The Force Analogy Method for Earthquake Engineering, First Edition. Gang Li and Kevin K.F. Wong.

8.1 Nonlinear Static Pushover Analysis in the FAM

The nonlinear static pushover analysis (NSPA) is an approximate, quick and convenient method to evaluate the ultimate load and deflection capability of structures under gradually increasing lateral loads. The equivalent static lateral loads approximately represent earthquake-induced forces. Local nonlinear structure effects, such as flexural hinges at member joints, are modeled and the structure is deformed or pushed until enough hinges form to develop a collapse mechanism or until the plastic deformation limit is reached at the hinge. The plot for depicting the relation of base shear versus top displacement of the structure is obtained by the method.

Consider the SDOF system used in Section 2.2 that is selected to address the NSPA procedure in the FAM. According to Eq. (2.22), there are

$$K = 3EI/L^3; \quad K' = 3EI/L^2; \quad K'' = 3EI/L \tag{8.1}$$

Assume that the column is subjected to a set of horizontal displacements at the roof and the maximum displacement is $x^{(l)}$. The maximum displacement is divided into l steps, the discrete displacement set can be written as:

$$\mathbf{x} = \left[x^{(1)}, \cdots, \quad x^{(k)}, \cdots x^{(l)}\right] \tag{8.2}$$

Now, we show how to perform the NSPA procedure for the SDOF in the FAM. First, consider that the responses of the column are known at the step k. Then, the lateral force and moment at the base of the column at step $k+1$is expressed as an incremental form as:

$$\begin{Bmatrix} F^{(k+1)} \\ m^{(k+1)} \end{Bmatrix} = \begin{bmatrix} 3EI/L^3 & 3EI/L^2 \\ 3EI/L^2 & 3EI/L \end{bmatrix} \begin{Bmatrix} x^{(k)} \\ m^{(k)} \end{Bmatrix} + \begin{bmatrix} 3EI/L^3 & 3EI/L^2 \\ 3EI/L^2 & 3EI/L \end{bmatrix} \begin{Bmatrix} \Delta x^{(k)} \\ -\Delta\theta''^{(k)} \end{Bmatrix} \tag{8.3}$$

There are totally three unknown quantities, $F^{(k+1)}$, $m^{(k+1)}$ and $\Delta\theta''^{(k)}$ at step $k+1$ in Eq. (8.3). First, assume that the incremental plastic rotation at the PHL at step k equals zero as:

$$\Delta\theta''^{(k)} = 0 \tag{8.4}$$

Thus, a trial lateral force and moment at step $k+1$ can be written as

$$\begin{Bmatrix} \tilde{F}^{(k+1)} \\ \tilde{m}^{(k+1)} \end{Bmatrix} = \begin{bmatrix} 3EI/L^3 & 3EI/L^2 \\ 3EI/L^2 & 3EI/L \end{bmatrix} \begin{Bmatrix} x^{(k)} \\ \theta''^{(k)} \end{Bmatrix} + \begin{bmatrix} 3EI/L^3 & 3EI/L^2 \\ 3EI/L^2 & 3EI/L \end{bmatrix} \begin{Bmatrix} \Delta x^{(k)} \\ 0 \end{Bmatrix} \tag{8.5}$$

If the trial moment $\tilde{m}^{(k+1)}$ at step $k+1$ exceeds the yield moment m_y, the final moment $m^{(k+1)}$ equals the yield moment m_y. Through using Eq. (8.3), the incremental plastic rotation $\Delta\theta''^{(k)}$ can be solved by

$$\Delta\theta''^{(k)} = \frac{L}{3EI}\left(-m_y + \frac{3EI}{L^2}x^{(k)}\right) + \frac{\Delta x^{(k)}}{L} \tag{8.6}$$

By substituting Eq. (8.6) into Eq. (8.3), the final lateral force and moment at step $k+1$ can be solved.

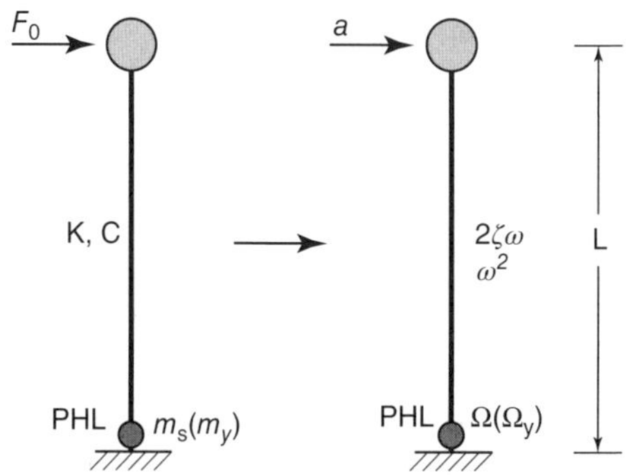

Figure 8.1 A SDOF system and corresponding normalized-mass system model.

8.1.1 NSPA for Mass-Normalized SDOF Systems

In the NSPA procedure, lateral static continuous forces are applied on structures in proportion. The distribution pattern of lateral loads broadly falls into three patterns: (1) vertical uniform distribution, (2) inverted triangle (first mode) load pattern and (3) vertical distribution proportional to the story shear distribution calculated by combining modal responses. Transfer a general SDOF system with a plastic hinge (PH) at the base of the column to a normalized SDOF system shown in Figure 8.1 and select the latter as an example to investigate how to implement the NSPA in the FAM.

The governing equations of motion for a general SDOF system in the FAM can be expressed as:

$$M\ddot{x}(t) + C\dot{x}(t) + Kx(t) = -M\ddot{x}_g(t) + Kx''(t) \tag{8.7}$$

$$F_s(t) = Kx(t) - Kx''(t) = Kx(t) - K'\theta''(t) \tag{8.8}$$

$$m(t) = K'x(t) - K''\theta''(t) \tag{8.9}$$

In addition, an additional relationship between the moment and plastic rotation can be written as

$$if \begin{cases} m(t) \le m_y \\ m(t) > m_y \end{cases}, \quad then \begin{cases} \theta''(\mathrm{t}) = 0 \\ m(t) = f(\theta''(\mathrm{t})) \end{cases} \tag{8.10}$$

Equations (8.7), (8.8), (8.9) and (8.10) are divided by m and define

$$\omega^2 = K/M; \quad C = 2\zeta\omega; \quad a = F_s/M; \quad \omega'^2 = K'/M \tag{8.11}$$
$$\omega''^2 = K''/M; \quad \Omega = m/M; \quad x'' = K'\theta''/K$$

Then, Eqs. (8.7) to (8.10) are changed as:

$$\ddot{x}(t) + 2\varsigma\omega\dot{x}(t) + \omega^2 x(t) = -\ddot{x}_g(t) + \omega^2 x''(t) \tag{8.12}$$

$$a(t) = \omega^2 x(t) - \omega'^2 \theta''(t) \tag{8.13}$$

$$\Omega(t) = \omega'^2 x(t) - \omega''^2 \theta''(t) \tag{8.14}$$

$$\text{if } \begin{cases} \Omega(t) \le \Omega_y \\ \Omega(t) > \Omega_y \end{cases}, \quad \text{then } \begin{cases} \theta''(t) = 0 \\ \Omega(t) = \dfrac{1}{m} f(\theta''(\mathrm{t})) \end{cases} \tag{8.15}$$

Equations (8.12) to (8.15) are the governing equations of the mass-normalized SDOF system in the FAM. Assume that the mass-normalized SDOF system is subjected to a continuous increasing horizontal displacement at the roof and there are also totally l steps, as shown in Eq. (8.2). Structural information at step $k+1$ can be predicted by that at step k, thus there is

$$\begin{Bmatrix} a^{(k+1)} \\ \Omega^{(k+1)} \end{Bmatrix} = \begin{bmatrix} \omega^2 & \omega'^2 \\ \omega'^2 & \omega''^2 \end{bmatrix} \begin{Bmatrix} x^{(k)} \\ \theta''^{(k)} \end{Bmatrix} + \begin{bmatrix} \omega^2 & \omega'^2 \\ \omega'^2 & \omega''^2 \end{bmatrix} \begin{Bmatrix} \Delta x^{(k)} \\ -\Delta\theta''^{(k)} \end{Bmatrix} \tag{8.16}$$

in which, $a^{(k+1)}$ and $\Omega^{(k+1)}$ are the normalized-mass force and moment at step $k+1$. Assume that the incremental plastic rotation at the plastic hinge location (PHL) equals zero as:

$$\Delta\theta^{(k)''} = 0 \tag{8.17}$$

Then, the trial normalized-mass force and moment at step $k+1$ is expressed by

$$\begin{Bmatrix} \tilde{a}^{(k+1)} \\ \tilde{\Omega}^{(k+1)} \end{Bmatrix} = \begin{bmatrix} \omega^2 & \omega'^2 \\ \omega'^2 & \omega''^2 \end{bmatrix} \begin{Bmatrix} x^{(k)} \\ \Omega^{(k)} \end{Bmatrix} + \begin{bmatrix} \omega^2 & \omega'^2 \\ \omega'^2 & \omega''^2 \end{bmatrix} \begin{Bmatrix} \Delta x^{(k)} \\ 0 \end{Bmatrix} \tag{8.18}$$

If the moment at step $k+1$exceeds the mass-normalized yield moment Ω_y, the final mass-normalized moment equals the yield moment Ω_y and the incremental plastic rotation $\Delta\theta''^{(k)}$ at step $k+1$ can be solved by

$$\Delta\theta''^{(k)} = \frac{1}{\omega''^2}\left(-\Omega_y + \omega^2 x^{(k)}\right) + \frac{\Delta x^{(k)}}{\omega'^2} \tag{8.19}$$

Substituting Eq. (8.19) back into Eq. (8.16), the final mass-normalized lateral force and moment at step $k+1$ can be expressed by

$$\begin{Bmatrix} a^{(k+1)} \\ \Omega^{(k+1)} \end{Bmatrix} = \begin{bmatrix} \omega^2 & \omega'^2 \\ \omega'^2 & \omega''^2 \end{bmatrix} \begin{Bmatrix} x^{(k)} \\ \theta''^{(k)} \end{Bmatrix} + \begin{bmatrix} \omega^2 & \omega'^2 \\ \omega'^2 & \omega''^2 \end{bmatrix} \begin{Bmatrix} \Delta x^{(k)} \\ -\Delta\theta''^{(k)} \end{Bmatrix} \tag{8.20}$$

Example 8.1 NSPA for a normalized SDOF system

Consider a mass-normalized SDOF system conversed from Example 3.6. By using Eq. (8.11), there are

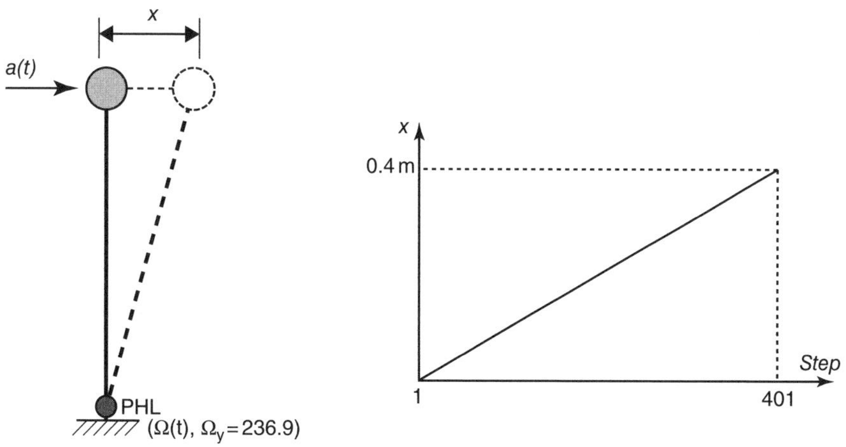

Figure 8.2 A normalized SDOF system and loads history.

$$\omega = 12.56; \quad \omega' = 39.74; \quad \omega'' = 125.67; \quad \Omega_y = 236.9 \tag{8.21}$$

Assume that the system shown in Figure 8.2 is subjected to a set of horizontal displacements, in which the maximum displacement is 0.4 m and the incremental displacement is 0.001 m so that the whole displacement set is divided into 400 displacement steps. In order to be consistent with the mass-normalized load $\ddot{x}_g$ in the right item of Eq. (8.12), the latter resistant load at step k is written as the form of $a^{(k)}$, in which $k = 1$ to 401.

Solution

***Step* 1**: $(k = 1, x = 0)$

Since the initial state of the system is static, there are

$$a^{(1)} = 0; \quad \Omega^{(1)} = 0 \tag{8.22}$$

***Step* 2**: $(k = 2, x = 0.001 \text{ m})$

Assuming the mass-normalized SDOF system at step $k = 2$ is linear yields

$$\Delta\theta''^{(1)} = 0 \tag{8.23}$$

$$\Delta\tilde{a}^{(1)} = \omega^2 \Delta x^1 = 157.92 \times 0.001 = 0.158 \text{ N/kg} \tag{8.24}$$

$$\Delta\tilde{\Omega}^{(1)} = \omega'^2 \Delta x^{(1)} = 1579.2 \times 0.001 = 1.579 \text{ N-m/kg} \tag{8.25}$$

The trial responses at step $k = 2$ is expressed by

$$\tilde{a}^{(2)} = a^{(2)} + \Delta\tilde{a}^{(2)} = 0 + 0.158 = 0.158 \text{ N/kg} \tag{8.26}$$

$$\tilde{\Omega}^{(2)} = \Omega^{(1)} + \Delta\tilde{\Omega}^{(1)} = 0 + 1.579 = 1.579 \text{ N-m/kg} \tag{8.27}$$

Since the trial mass-normalized moment, $\widetilde{\Omega}^{(2)} < \Omega_y = 236.9$, assuming $\Delta\theta''^{(k)} = 0$ at the beginning of this step is correct. Thus, the final mass-normalized force and moment at step $k = 2$ equal to the corresponding trial ones as

$$a^{(2)} = \widetilde{a}^{(2)} = 0.158 \text{ N/kg} \tag{8.28}$$

$$\Omega^{(2)} = \widetilde{\Omega}^{(2)} = 1.579 \text{ N-m/kg} \tag{8.29}$$

Steps 3 to 151: ($k = 3$ to 151, $x = 0.002$ to 0.150 m)
Similarly, the procedures for solving responses of the system at the steps from 3 to 151 are the same as the first two steps because of the system being linear in this range. The solution process is omitted.

Step 152: ($k = 152$, $x = 0.151$ m)
The hypothesis of the incremental plastic rotation being zero is still considered, there are

$$\Delta\theta''^{(151)} = 0 \tag{8.30}$$

$$\Delta\widetilde{a}^{(151)} = \omega^2 \Delta x^{(151)} = 157.92 \times 0.001 = 0.158 \text{ N/kg} \tag{8.31}$$

$$\Delta\widetilde{\Omega}^{(151)} = \omega'^2 \Delta x^{(151)} = 1579.2 \times 0.001 = 1.579 \text{ N-m/kg} \tag{8.32}$$

The trial normalized-mass moment at step $k = 151$ is expressed by

$$\widetilde{\Omega}^{(152)} = \Omega^{(151)} + \Delta\widetilde{\Omega}^{(151)} = 236.9 + 1.58 = 238.5 \text{ N-m/kg} \tag{8.33}$$

Since the trial mass-normalized moment, $\widetilde{\Omega}^{(152)} > \Omega_y = 236.9$, the corrected incremental plastic rotation at step $k = 152$ can be solved by

$$\Delta\theta''^{(151)} = \frac{\omega'^2}{\omega''^2} \Delta x^{(151)} = \frac{1579.2}{15792} \times 0.001 = 0.0001 \text{ rad} \tag{8.34}$$

Then, the incremental normalized-mass force and moment can be written as

$$\Delta a^{(151)} = \omega^2 \Delta x^{(151)} - \omega'^2 \Delta\theta''^{(151)} = 157.92 \times 0.001 - 1579.2 \times 0.0001 = 0 \tag{8.35}$$

$$\Delta\Omega^{(151)} = \omega'^2 \Delta x^{(151)} - \omega''^2 \Delta\theta''^{(151)} = 157.92 \times 0.001 - 1579.2 \times 0.0001 = 0 \tag{8.36}$$

The final mass-normalized force and moment at the step $k = 152$ are

$$a_{152} = a_{151} + \Delta a_{151} = 23.66 \tag{8.37}$$

$$\Omega_{152} = \Omega_{151} + \Delta\Omega_{151} = 236.9 \tag{8.38}$$

***Steps* 153 to 401**: ($x = 0.152$ m to 0.400 m)
The procedures for solving the variables at the steps from 153 to 401 are the same as the step of 152, because the normalized-mass SDOF is always in plasticity. The solution process is omitted.

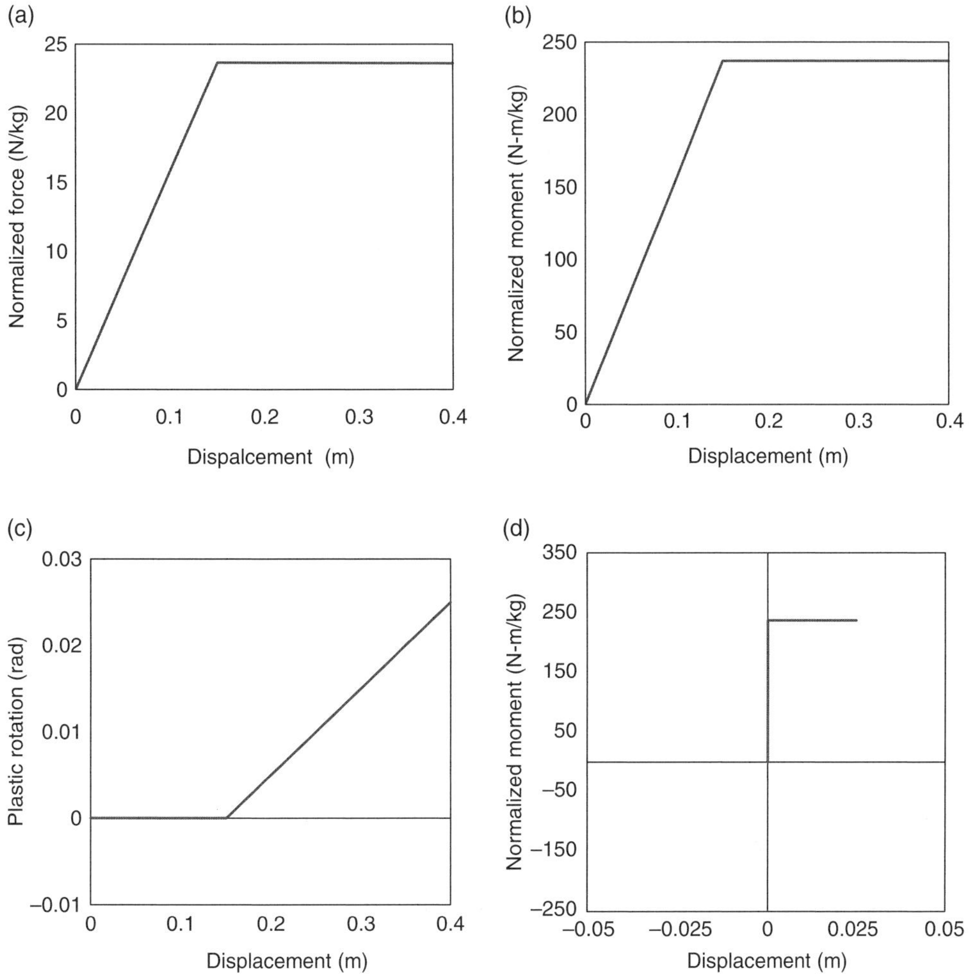

Figure 8.3 Mass-normalized force, moment and plastic rotation responses: (a) Normalized-mass force; (b) normalized-mass moment; (c) Plastic rotation; (d) normalized-mass moment and plastic rotation relation.

Finally, the relations of normalized force, moment and plastic rotation at the PHL versus displacement are shown in Figure 8.3(a) to 8.3(c), and the relation of moment versus plastic rotation of the plastic hinge is shown in Figure 8.3(d).

8.1.2 NSPA for Multi-Degree-of-Freedom Systems

Consider an n-degree-of-freedom (DOFs) system shown in Figure 8.4 as an example. Assume that the system is subjected to a set of horizontal continuous loads instead of previous displacements, and the whole load process is divided into l steps. The discrete loads set can be written as

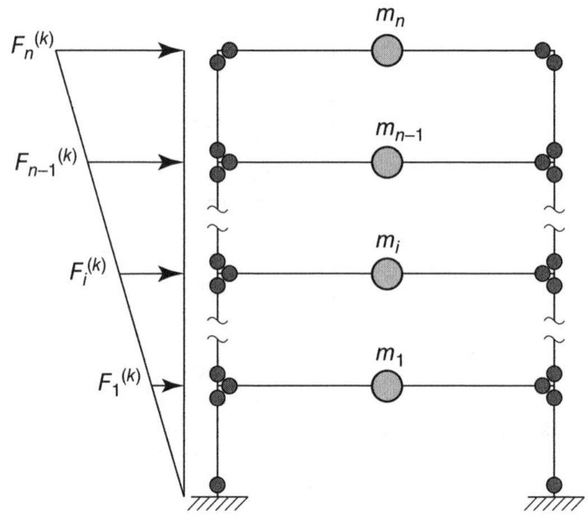

Figure 8.4 *n*-DOFs system model.

$$\mathbf{F} = \left[\mathbf{F}^{(1)} \ \cdots \ \mathbf{F}^{(k)} \ \cdots \ \mathbf{F}^{(l)}\right] \tag{8.39}$$

Then, the incremental lateral load is expressed by

$$\Delta\mathbf{F}^{(k)} = \mathbf{F}^{(k+1)} - \mathbf{F}^{(k)} \tag{8.40}$$

in which, $\mathbf{F}^{(k)} = [F_1^{(k)}, \cdots, F_i^{(k)}, \cdots, F_n^{(k)}]$ and $F_i^{(k)}$ represents the lateral load applied on the *i*th floor at step *k*. Let's consider how to carry out the NSPA procedure for the *n*-DOFs system in the FAM.

The similar discrete increments method is used here again. The roof displacement response and moments at the ends of each framed members of the *n*-DOFs system under the lateral loads at step of $k+1$ are

$$\mathbf{x}^{(k+1)} = \bar{\mathbf{K}}^{-1}\left(\mathbf{F}^{(k)} + \Delta\mathbf{F} - \bar{\mathbf{K}}'\boldsymbol{\theta}''^{(k)} - \bar{\mathbf{K}}'\Delta\boldsymbol{\theta}''^{(k)}\right) \tag{8.41}$$

$$\mathbf{m}^{(k+1)} = \bar{\mathbf{K}}'^{\mathrm{T}}\bar{\mathbf{K}}^{-1}\left(\mathbf{F}^{(k)} + \Delta\mathbf{F} - \bar{\mathbf{K}}'\boldsymbol{\theta}''^{(k)} - \bar{\mathbf{K}}'\Delta\boldsymbol{\theta}''^{(k)}\right) - \mathbf{K}''\boldsymbol{\theta}''^{(k)} - \mathbf{K}''\boldsymbol{\theta}''^{(k)} \tag{8.42}$$

Assuming that incremental pastic rotations of all members of the *n*-DOFs system at step $k+1$ equals zero gives

$$\Delta\boldsymbol{\theta}''^{(k)} = \mathbf{0} \tag{8.43}$$

By substituting Eq. (8.43) into Eqs. (8.41) and (8.42), the trial displacement and moment at step $k+1$ can be written as

$$\tilde{\mathbf{x}}^{(k+1)} = \bar{\mathbf{K}}^{-1}\left(\mathbf{F}^{(k)} + \Delta\mathbf{F} - \bar{\mathbf{K}}'\boldsymbol{\theta}''^{(k)}\right) \tag{8.44}$$

$$\widetilde{\mathbf{m}}^{(k+1)} = \bar{\mathbf{K}}'^{\mathrm{T}}\bar{\mathbf{K}}^{-1}\left(\mathbf{F}^{(k)} + \Delta\mathbf{F}\right) - \left(\bar{\mathbf{K}}'^{T}\bar{\mathbf{K}}^{-1}\bar{\mathbf{K}}' + \mathbf{K}''\right)\boldsymbol{\theta}''^{(k)} \tag{8.45}$$

If any trial moment at step $k+1$ exceeds the yield moment m_y, the final mass-normalized moment equals the yield moment. Then, the incremental displacement matrix $\mathbf{\Delta x}^{(k)}$ and moments $\mathbf{\Delta\theta}''^{(k)}$ of the n-DOFs system can be solved through

$$\begin{Bmatrix} \mathbf{F}^{(k+1)} \\ \mathbf{m}^{(k+1)} \end{Bmatrix} = \begin{bmatrix} \bar{\mathbf{K}} & \bar{\mathbf{K}}' \\ \bar{\mathbf{K}}'^{T} & \mathbf{K}'' \end{bmatrix} \begin{Bmatrix} \mathbf{x}^{(k)} \\ \boldsymbol{\theta}''^{(k)} \end{Bmatrix} + \begin{bmatrix} \bar{\mathbf{K}} & \bar{\mathbf{K}}' \\ \bar{\mathbf{K}}'^{T} & \mathbf{K}'' \end{bmatrix} \begin{Bmatrix} \Delta\mathbf{x}^{(k)} \\ -\Delta\boldsymbol{\theta}''^{(k)} \end{Bmatrix} \tag{8.46}$$

The final displacements $\mathbf{x}^{(k+1)}$ and moments $\mathbf{m}^{(k+1)}$ of the n-DOFs system at step $k+1$ can be solved. Finally, the relationship of base shear and roof displacement of the n-DOFs system applied the given lateral load patterns is obtained.

Example 8.2 NSPA for a two-story moment-resisting frame

Consider a two-story one-bay moment-resisting frame with members assumed to be axially rigid. The same member properties of $E = 200.0$ GPa, $I = 6.0 \times 10^6$ mm^4, and $L = 4$ m presented in Example 3.8 are selected. Assume that the plastic hinges have elastic–plastic behavior with moment capacities of $m_{b,y} = 15.0$ kN m for the two beams and $m_{c,y} = 20.0$ kN m for the four columns. This frame has a total of 6 DOFs (i.e. $n = 6$) and 12 PHLs (i.e. $q = 12$). The mass moment of inertial are ignored for the rotational joints, the frame can then be statically condensed to a 2-DOF system. The condensed global stiffness matrix gives

$$\bar{\mathbf{K}} = \begin{bmatrix} 761.03 & -330.88 \\ -330.88 & 251.47 \end{bmatrix} \text{kN/m} \tag{8.47}$$

Similarly, $\bar{\mathbf{K}}'$ and $\mathbf{K}''$ matrices can also be obtained.

The mass of 2.5 Mg on each floor gives

$$\mathbf{M}_{dd} = \begin{bmatrix} 2.5 & 0 \\ 0 & 2.5 \end{bmatrix} \text{Mg} \tag{8.48}$$

Two periods and corresponding model shapes are expressed as:

$$T_1 = 1.06s; \quad \mathbf{s}_1 = \begin{Bmatrix} 0.49 \\ 1.00 \end{Bmatrix}; \quad T_2 = 0.33s; \quad \mathbf{s}_2 = \begin{Bmatrix} -2.03 \\ 1.00 \end{Bmatrix} \tag{8.49}$$

The NSPA procedure for the 2-DOFs frame under loads with different patterns is discussed as below:

(1) Under 1st model shape-dependent lateral loads

Assume the lateral load distribution patterns are associated with the first model shape, so the lateral applied force vector is constructed as (Chopra, 2009)

$$\mathbf{F}_1 = \left[\mathbf{F}_1{}^{(1)} \;\cdots\; \mathbf{F}_1{}^{(k)} \;\cdots\; \mathbf{F}_1{}^{(l)}\right] = \left[\lambda_{1,1}\mathbf{M_{dd}}\,\mathbf{s}_1 \;\cdots\; \lambda_{1,k}\mathbf{M_{dd}}\mathbf{s}_1 \;\cdots\; \lambda_{1,l}\mathbf{M_{dd}}\mathbf{s}_1\right] \tag{8.50}$$

in which $\boldsymbol{\lambda}_1 = [0\ \ 0.05\ \ 0.10\ \ \cdots\ \ 3.95]_{80\times1}$ is a scaled factor vector for keeping the lateral loads gradually increasing. The applied lateral force vector with the same increment gives

$$\Delta\mathbf{F}_1 = \Delta\lambda\mathbf{M}_{dd}\mathbf{s}_1 = 0.05\times\begin{bmatrix}2500 & 0\\ 0 & 2500\end{bmatrix}\begin{Bmatrix}0.49\\ 1.00\end{Bmatrix} = \begin{Bmatrix}31.25\\ 62.50\end{Bmatrix}\mathrm{N} \tag{8.51}$$

Perform a discrete numerical solution for the 2-DOFs frame.

***Step* 1**: ($k=1$, $\mathbf{F}_1{}^{(1)} = [0, 0]^{\mathrm{T}}$)
Since the initial state is static, there are

$$\mathbf{m} = [\mathbf{0}]_{12\times1};\quad \mathbf{x} = \begin{Bmatrix}0\\ 0\end{Bmatrix} \tag{8.52}$$

***Step* 2**: ($k=2$, $\mathbf{F}_1{}^{(2)} = [31.25, 62.5]^{\mathrm{T}}$)
Assuming the 2-DOFs system at step $k=2$ is linear yields

$$\tilde{\mathbf{x}}^{(2)} = \bar{\mathbf{K}}^{-1}\left(\mathbf{F}_1{}^{(2)} + \Delta\mathbf{F}_1, -\bar{\mathbf{K}}'\boldsymbol{\theta}''^{(k)}\right) = \begin{Bmatrix}0.0007\\ 0.0014\end{Bmatrix}\mathrm{m} \tag{8.53}$$

$$\tilde{\mathbf{m}}^{(2)} = \bar{\mathbf{K}}'^{T}\bar{\mathbf{K}}^{-1}\left(\mathbf{F}_1{}^{(1)} + \Delta\mathbf{F}_1\right) = \left[\mathbf{m}_c^{(2)}\ \ \mathbf{m}_c^{(2)}\ \ \mathbf{m}_b^{(2)}\right]^T \tag{8.54}$$

in which

$$\mathbf{m}_c^{(2)} = [-228.4\ \ -144.6\ \ -106.9\ \ -143.1]\,\mathrm{N\text{-}m}$$

$$\mathbf{m}_b^{(2)} = [225.5\ \ 225.5\ \ 143.1\ \ 143.1]\,\mathrm{N\text{-}m}$$

$$\Delta\boldsymbol{\theta}_1'' = [\mathbf{0}]_{12\times1} \tag{8.55}$$

***Steps* 3–60**: ($k=3$ to 60)
Similarly, procedures for solving the variables at the steps from 3 to 60 are the same as the second step is always in elastic. Thus, the solution process is omitted.

***Step* 61**: ($k=61$, $\mathbf{F}_1{}^{(1)} = [3629, 7375]^{\mathrm{T}}$)
The hypothesis of incremental plastic rotations of all members being zeros is considered first, the trial displacement and moment at this step can be expressed by

$$\tilde{\mathbf{x}}^{(61)} = \bar{\mathbf{K}}^{-1}\left(\mathbf{F}^{(61)} + \Delta\mathbf{F} - \bar{\mathbf{K}}'\boldsymbol{\theta}''^{(60)}\right) = \begin{Bmatrix}0.042\\ 0.085\end{Bmatrix}\mathrm{m} \tag{8.56}$$

$$\tilde{\mathbf{m}}^{(61)} = \bar{\mathbf{K}}'^{T}\bar{\mathbf{K}}^{-1}\left(\mathbf{F}^{(60)} + \Delta\mathbf{F}\right) = \left[\tilde{\mathbf{m}}_c^{(61)}\ \ \tilde{\mathbf{m}}_c^{(61)}\ \ \tilde{\mathbf{m}}_b^{(61)}\right]^T \tag{8.57}$$

in which

$$\tilde{\mathbf{m}}_c^{(61)} = [-13706\ \ -8676\ \ -6416\ \ -8584]\,\mathrm{N\text{-}m}$$

$$\tilde{\mathbf{m}}_b^{(61)} = [15091\ \ 15091\ \ 85844\ \ 85844]\,\mathrm{N\text{-}m}$$

Since the moments at PHLs #9 and #10, $m_9 = m_{10} = 15091$ N-m, exceeds the yield moment $m_{b,y} = 15000$ N m, the corrected incremental plastic rotation and the final incremental moment at step $k = 61$ is

$$\mathbf{m}^{(61)} = \left[\mathbf{m}_c^{(61)} \quad \mathbf{m}_c^{(61)} \quad \mathbf{m}_b^{(61)}\right]^T \tag{8.58}$$

in which

$$\mathbf{m}_c^{(61)} = [-13706 \quad -8676 \quad -6416 \quad -8584]\,\text{N-m}$$

$$\mathbf{m}_b^{(61)} = [15000 \quad 15000 \quad 85844 \quad 85844]\,\text{N-m}$$

By using Eq. (8.46), the incremental plastic rotation and displacement at this step are calculated as

$$\Delta\mathbf{x}^{(60)} = \begin{Bmatrix} 0.0008 \\ 0.0017 \end{Bmatrix} \text{m} \tag{8.59}$$

and

$$\Delta\boldsymbol{\theta}''_{60} = [\mathbf{0}_{4\times 1} \quad \mathbf{0}_{4\times 1} \quad 0.0003 \quad 0.0003 \quad 0 \quad 0]^{\mathrm{T}} \tag{8.60}$$

Substituting Eqs. (8.59) and (8.60) into Eq. (8.46) yields

$$\mathbf{x}^{(61)} = \begin{Bmatrix} 0.042 \\ 0.085 \end{Bmatrix} \text{m} \tag{8.61}$$

$$\boldsymbol{\theta}''^{(61)} = \boldsymbol{\theta}''^{(60)} + \Delta\boldsymbol{\theta}''^{(60)} = [\mathbf{0}_{4\times 1} \quad \mathbf{0}_{4\times 1} \quad 0.0003 \quad 0.0003 \quad 0 \quad 0]^{\mathrm{T}}\text{rad} \tag{8.62}$$

***Steps* 62–106**: ($k = 62$ to 106)
The procedures for solving responses at the steps from 62 to 106 are the same as that at step 61. The solution process is omitted. Finally, plastic hinge distribution at step 106 is shown in Figure 8.5. The base shear force and moments at the PHLs versus roof displacement and hysteretic loop of selected plastic hinges are shown in Figure 8.6.

(2) Under 2nd model shape-dependent lateral force
Assume that the lateral load distribution patterns are associated with the second model shape, so the lateral applied force vector is constructed as

$$\mathbf{F}_2 = \left[\mathbf{F}_2{}^{(1)} \ \cdots \ \mathbf{F}_2{}^{(k)} \ \cdots \ \mathbf{F}_2{}^{(l)}\right] = \left[\lambda_2^{(1)}\mathbf{M_{dd}}\,\mathbf{s}_2^{(1)} \ \cdots \ \lambda_2^{(k)}\mathbf{M_{dd}}\mathbf{s}_2^{(k)} \ \cdots \ \lambda_2^{(l)}\mathbf{M_{dd}}\mathbf{s}_2^{(l)}\right] \tag{8.63}$$

in which $\boldsymbol{\lambda}_2 = [0 \quad 0.05 \quad 0.10 \quad \cdots \quad 7.95]_{160\times 1}$ is a scaled factor vector for keeping the lateral loads gradually increasing. The applied lateral force vector with the same incremental force gives

$$\Delta\mathbf{F}_2 = \Delta\lambda\mathbf{M}_{dd}\mathbf{s}_1 = 0.05 \times \begin{bmatrix} 2500 & 0 \\ 0 & 2500 \end{bmatrix} \begin{Bmatrix} -2.03 \\ 1.00 \end{Bmatrix} = \begin{Bmatrix} -254 \\ 125 \end{Bmatrix} \text{N} \tag{8.64}$$

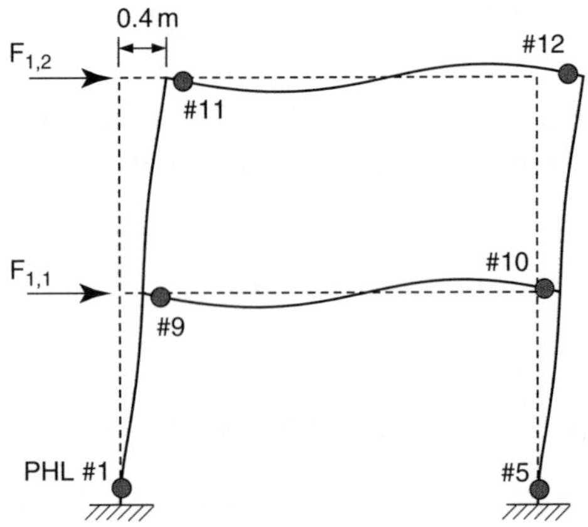

Figure 8.5 Plastic hinges distribution at last displacement step.

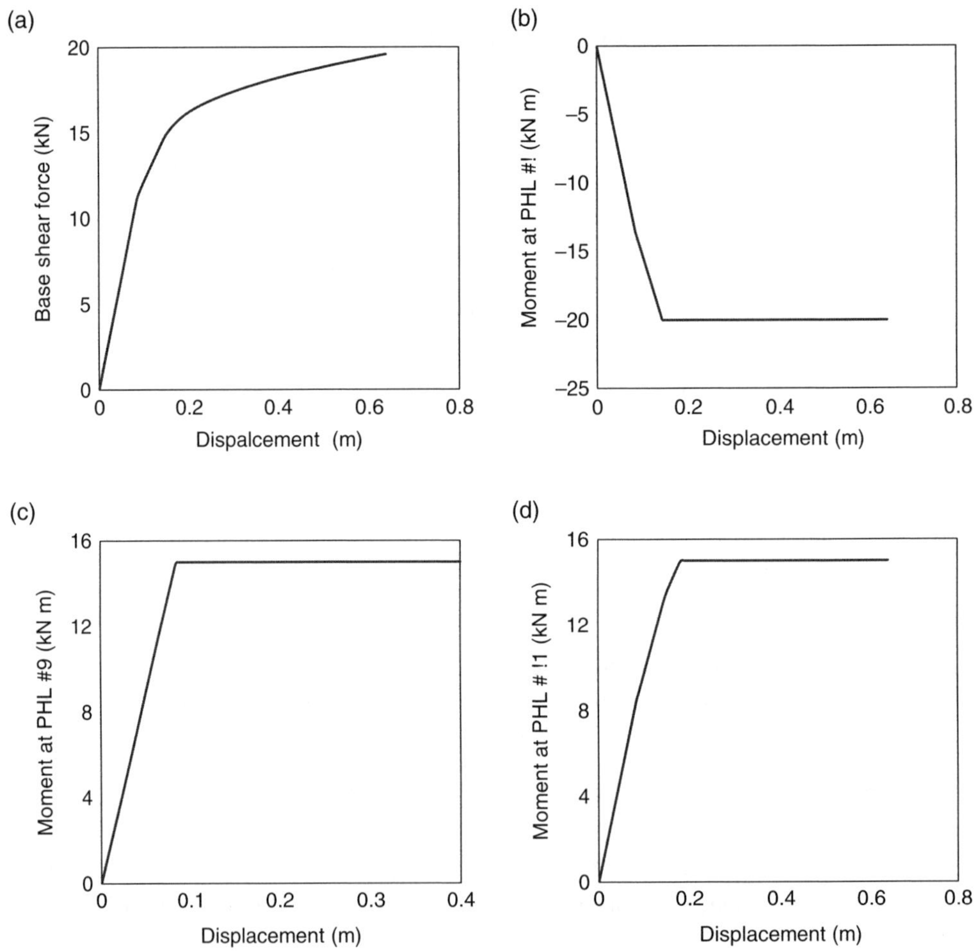

Figure 8.6 Responses of structure under 1st model shape-dependent lateral loads: (a) Base shear; (b) moment at PHL #1; (c) moment at PHL #9; (d) moment at PHL #11;

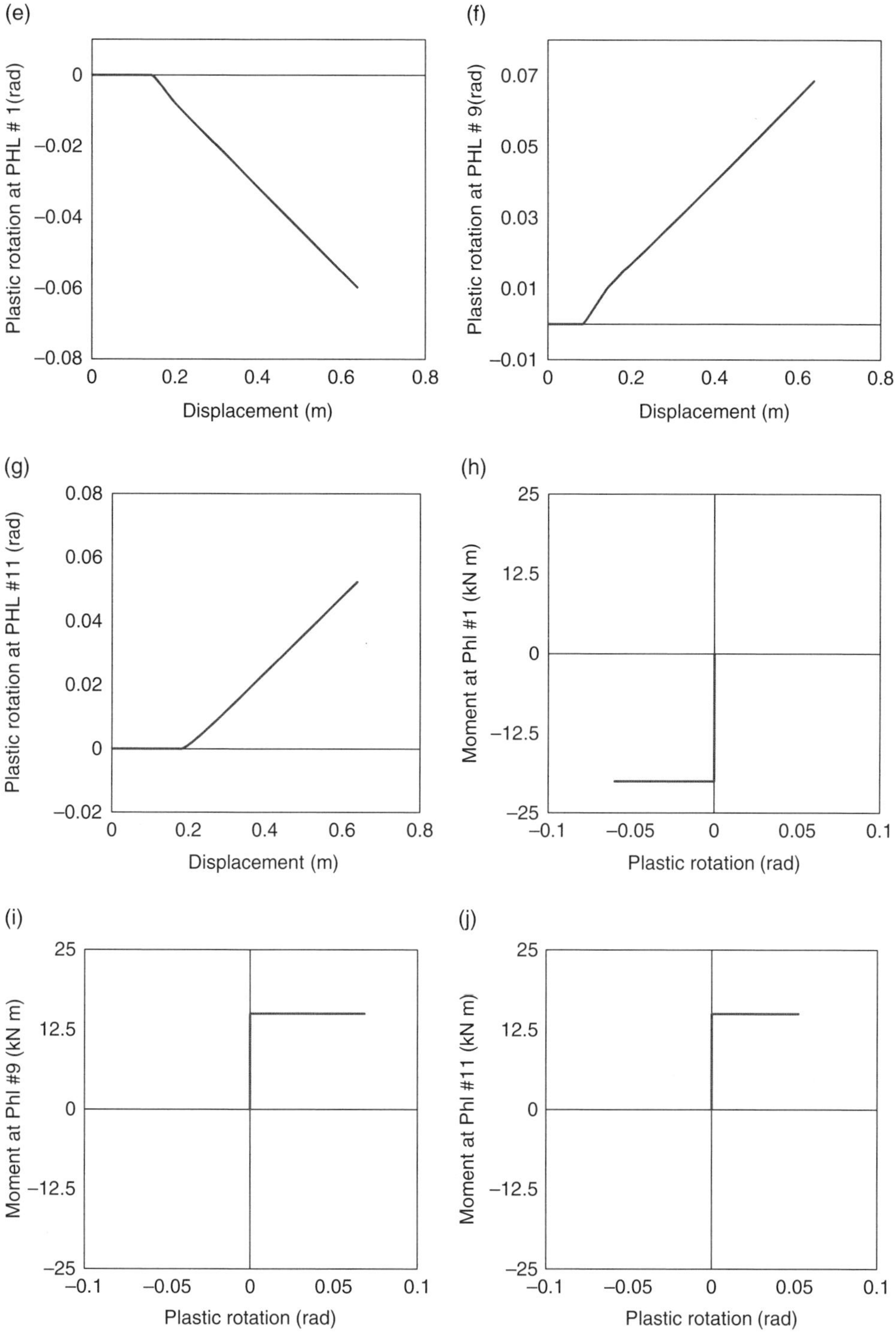

Figure 8.6 (*Continued*) (e) plastic rotation at PHL #1; (f) plastic rotation at PHL #9; (g) plastic rotation at PHL #11; (h) moment and plastic rotation relation of PH #1; (i) moment and plastic rotation relation of PH #9; (j) moment and plastic rotation relation of PH #11.

Perform a discrete numerical solution for the 2-DOFs frame.

***Step* 1**: ($k = 1$, $\mathbf{F}_2^{(1)} = [0\ 0]^{\mathrm{T}}$)
Since the initial state is still static, there are

$$\mathbf{m} = [\mathbf{0}]_{12\times 1}; \quad \mathbf{x} = \begin{Bmatrix} 0 \\ 0 \end{Bmatrix} \tag{8.65}$$

***Step* 2**: ($k = 2$, $\mathbf{F}_2^{(2)} = [-254,\ 125]^{\mathrm{T}}$)
Assuming the SDOF at step $k = 2$ is linear yields

$$\tilde{\mathbf{x}}^{(2)} = \bar{\mathbf{K}}^{-1}\left(\mathbf{F}_2^{(2)} + \Delta\mathbf{F}_2, - \bar{\mathbf{K}}'\boldsymbol{\theta}''^{(1)} = \begin{Bmatrix} -0.0003 \\ 0.0001 \end{Bmatrix}\text{m} \tag{8.66}$$

$$\tilde{\mathbf{m}}^{(2)} = \bar{\mathbf{K}}'^{T}\bar{\mathbf{K}}^{-1}\left(\mathbf{F}_2^{(1)} + \Delta\mathbf{F}_2\right) = \left[\tilde{\mathbf{m}}_c^{(2)}\ \ \tilde{\mathbf{m}}_c^{(2)}\ \ \tilde{\mathbf{m}}_b^{(2)}\right]^{\mathrm{T}} \tag{8.67}$$

in which

$$\tilde{\mathbf{m}}_c^{(2)} = [127.3\ \ 130.8\ \ -141.3\ \ -108.7]\text{N-m}$$

$$\tilde{\mathbf{m}}_b^{(2)} = [10.6\ \ 10.6\ \ 108.7\ \ 108.7]\text{N-m}$$

$$\Delta\boldsymbol{\theta}_1'' = [\mathbf{0}]_{12\times 1} \tag{8.68}$$

Since the response at the step is elastic, the final displacement and moment are satisfied by

$$\mathbf{x}^{(2)} = \tilde{\mathbf{x}}^{(2)} = \begin{Bmatrix} -0.0003 \\ 0.0001 \end{Bmatrix}\text{m} \tag{8.69}$$

$$\mathbf{m}^{(2)} = \tilde{\mathbf{m}}^{(2)} \tag{8.70}$$

***Steps* 3–139**: ($k = 3$ to 139)
Similarly, the procedures for solving the variables at the steps from 3 to 139 are the same as the first two steps as the system is always elastic. The solution process is omitted.

***Step* 140**: ($k = 140$, $\mathbf{F}_2^{(140)} = [-35308,\ 17375]^{\mathrm{T}}$)
The hypothesis of the 2-DOFs frame in elastic still is considered, there are

$$\tilde{\mathbf{x}}^{(140)} = \bar{\mathbf{K}}^{-1}\left(\mathbf{F}_2^{(140)} + \Delta\mathbf{F}_2, - \bar{\mathbf{K}}'\boldsymbol{\theta}''^{(139)} = \begin{Bmatrix} -0.038 \\ 0.019 \end{Bmatrix}\text{m} \tag{8.71}$$

$$\tilde{\mathbf{m}}^{(140)} = \bar{\mathbf{K}}'^{\mathrm{T}}\bar{\mathbf{K}}^{-1}\left(\mathbf{F}_2^{(139)} + \Delta\mathbf{F}_2\right) = \left[\tilde{\mathbf{m}}_c^{(140)}\ \ \tilde{\mathbf{m}}_c^{(140)}\ \ \tilde{\mathbf{m}}_b^{(140)}\right]^{\mathrm{T}} \tag{8.72}$$

in which

$$\tilde{\mathbf{m}}_c^{(140)} = [17688\ \ 18178\ \ -19647\ \ -15103]\text{N-m}$$

$$\tilde{\mathbf{m}}_b^{(140)} = [14695\ \ 14695\ \ 15103\ \ 15103]\text{N-m}$$

Since the moment at PHLs #11and #12, $m_{11} = m_{12} = 15103$ N-m, exceeds the yield moment $m_{b,y} = 15000$ N-m, the corrected incremental plastic rotation and the final incremental moment at step $k = 140$ is

$$\mathbf{m}^{(140)} = \left[\mathbf{m}_c^{(140)} \ \ \mathbf{m}_c^{(140)} \ \ \mathbf{m}_b^{(140)}\right]^{\mathrm{T}} \tag{8.73}$$

In which

$$\mathbf{m}_c^{(140)} = [17688 \ \ 18178 \ \ -19647 \ \ -15103]\text{N-m}$$

$$\mathbf{m}_c^{(140)} = [14695 \ \ 14695 \ \ 15000 \ \ 15000]\text{N-m}$$

Then, the incremental plastic rotation and displacement at this step are calculated by Eq. (8.46) as

$$\Delta\mathbf{x}^{(139)} = \begin{Bmatrix} -0.0003 \\ 0.0001 \end{Bmatrix} \text{m} \tag{8.74}$$

$$\Delta\boldsymbol{\theta}''^{(139)} = [\mathbf{0}_{4\times1} \ \ \mathbf{0}_{4\times1} \ \ 0.00015 \ \ 0.00015 \ \ 0 \ \ 0]^{\mathrm{T}} \tag{8.75}$$

Substituting Eqs. (8.59) and (8.60) into Eqs. (8.46) yields

$$\mathbf{x}^{(140)} = \begin{Bmatrix} -0.038 \\ 0.019 \end{Bmatrix} \text{m} \tag{8.76}$$

$$\boldsymbol{\theta}''^{(140)} = \boldsymbol{\theta}''^{(139)} + \Delta\boldsymbol{\theta}''^{(139)} = [\mathbf{0}_{4\times1} \ \ \mathbf{0}_{4\times1} \ \ 0.00015 \ \ 0.00015 \ \ 0 \ \ 0]^{\mathrm{T}}\text{rad} \tag{8.77}$$

Steps 141–160: ($k = 141$ to 160)

The procedures for solving the variables at the steps from 141 to 160 are the same as step 141 for the 2-DOFs system is always in plastic. The solution process is omitted. Thus, the solution process is omitted. Finally, the plastic hinge distribution is shown in Figure 8.7. The

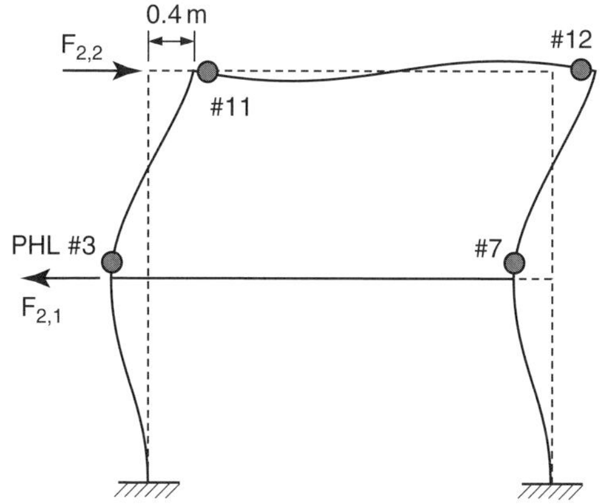

Figure 8.7 Plastic hinges distribution.

base shear force and moment and plastic rotation at selected PHLs versus roof displacement, and the relationship between moment and plastic rotation of selected plastic hinges are shown in Figure 8.8.

8.2 Modal Decomposition in the FAM

The governing equation of motion for MDOF systems with static condensation is written again as:

$$\mathbf{M}\ddot{\mathbf{X}}(\mathbf{t}) + \mathbf{C}\dot{\mathbf{X}}(\mathbf{t}) + \bar{\mathbf{K}}\mathbf{X}(\mathbf{t}) = -\mathbf{M}\iota\ddot{x}_g(t) + \bar{\mathbf{K}}\mathbf{X}''(\mathbf{t}) \tag{8.78}$$

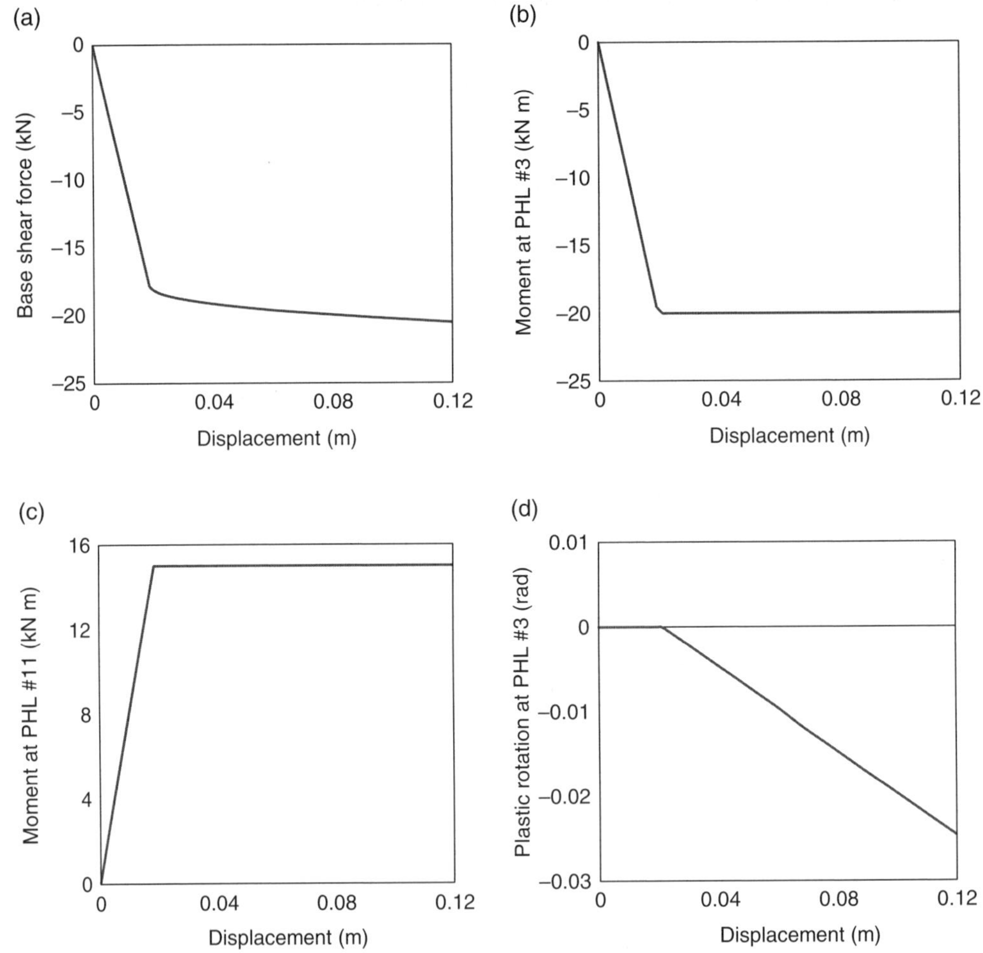

Figure 8.8 Responses of structure under 1st model shape-dependent lateral loads: (a) Base shear; (b) moment at PHL #3; (c) moment at PHL #11; (d) plastic rotation at PHL #3;

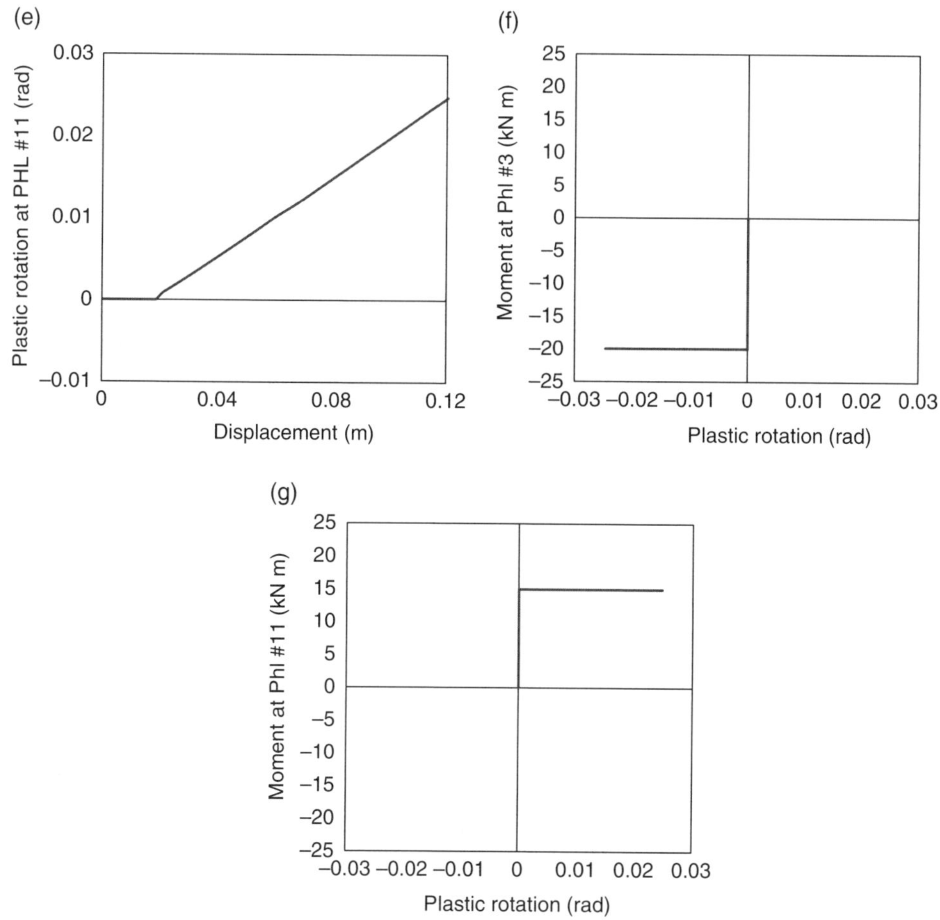

Figure 8.8 (*Continued*) (e) plastic rotation at PHL #11; (f) moment and plastic rotation relation of PH #3; (g) moment and plastic rotation relation of PH #111.

where **M** denotes the mass matrix, **C** is the structural damping matrix constructed based on the Rayleigh damping; $\mathbf{X}$, $\dot{\mathbf{X}}$ and $\ddot{\mathbf{X}}$ are the displacement, velocity and acceleration vectors, respectively; $\boldsymbol{\iota}$ is the unit vector.

Additional governing equations of the MDOF system based on the FAM can be expressed by:

$$\mathbf{F}(t) = \bar{\mathbf{K}}\mathbf{X}(t) - \bar{\mathbf{K}}'\boldsymbol{\theta}''(t) \tag{8.79}$$

$$\mathbf{m}_s(t) = \bar{\mathbf{K}}'^{\mathrm{T}}\mathbf{X}(t) - \bar{\mathbf{K}}''\boldsymbol{\theta}''(t) \tag{8.80}$$

$$\text{if} \quad \begin{cases} m_{s,i}(t) \le m_y \\ m_{s,i}(t) > m_y \end{cases}, \quad \text{then} \quad \begin{cases} \theta_i''(t) = 0 \\ m_{s,i}(t) = f\left(\theta_i''(t)\right) \end{cases} \tag{8.81}$$

where $\bar{\mathbf{K}}$ is the initial stiffness matrix of the MDOF system, and $\bar{\mathbf{K}}'$ means the factor matrix relating rotations with horizontal forces; $\bar{\mathbf{K}}'^{\mathrm{T}}$ means the transposed matrix of $\bar{\mathbf{K}}'$; $\bar{\mathbf{K}}''$ denotes the coefficient matrix that reflects the relationship between moments and rotations; $m_{s,i}$ represents plastic rotation at the ith PHL.

Define

$$\mathbf{X}(\mathbf{t}) = \mathbf{\Phi}\mathbf{Y}(\mathbf{t}) \tag{8.82}$$

$$\mathbf{X}''(\mathbf{t}) = \mathbf{\Phi}\mathbf{Y}''(\mathbf{t}) \tag{8.83}$$

where $\mathbf{\Phi}$ means the modal matrix of the MDOF system and $\mathbf{Y(t)}$ denotes a vector of normal coordinate. Substitute Eqs. (8.82) and (8.83) into Eq. (8.78), and multiply $\mathbf{\Phi}^{\mathrm{T}}$ on the left hand of each item of Eq. (8.78), then Eq. (8.78) can be decoupled to n different separated equations. The jth modal equation of motion can be written as

$$\ddot{Y}_j(t) + 2\varsigma_j\omega_j\dot{Y}_j(t) + \omega_j^2 Y_j(t) = -\gamma_j x_g(t) + \omega_j^2 Y_j''(t) \tag{8.84}$$

in which

$$M_j^* = \mathbf{\Phi}_j^T\mathbf{M}\mathbf{\Phi}_j; \quad K_j^* = \mathbf{\Phi}_j^T\mathbf{M}\mathbf{\Phi}_j; \quad C_j^* = \mathbf{\Phi}_j^T\mathbf{C}\mathbf{\Phi}_j$$

$$\omega_j^2 = \frac{K_j^*}{M_j^*}; \quad \varsigma_j = \frac{C_j^*}{2\omega_j M_j^*}; \quad \gamma_j = \frac{\mathbf{MI}}{M_j^*}$$

Although Eq. (8.78) can be decoupled to n different equations, Eqs. (8.79) and Eq. (8.80) still cannot be decoupled through the simple modal decomposition due to the presence of plastic rotation. A suitable method for establishing an equivalent SDOF system that is corresponding to the jth modal SDOF system can be established through the NSPA procedure presented in section 8.1. Detailed process are addressed as follows:

Step 1: Define that the lateral load pattern is associated with the jth modal shape, as shown in Figure 8.9(a) and written as

$$\mathbf{F}_j = \left[\lambda^{(1)}\mathbf{M_{dd}}\,\mathbf{s}_j \quad \cdots \quad \lambda^{(k)}\mathbf{M_{dd}}\mathbf{s}_j \quad \cdots \quad \lambda^{(l)}\mathbf{M_{dd}}\mathbf{s}_j\right] \tag{8.85}$$

Step 2: Perform the NSPA procedure through Eqs. (8.41) to (8.46) and obtain a curve reflecting the relationship of the base shear and roof displacement, as shown in Figure 8.9(b).

Step 3: Assume the jth modal SDOF with unite height and transfer the relation of base shear and roof displacement to the moment and displacement of the jth SDOF system through

$$\Omega_j = V_j H = V_j \tag{8.86}$$

$$Y_j = \frac{\mathbf{X}_{j,n}^{(l+1)}}{\gamma_j s_{n,j}} \tag{8.87}$$

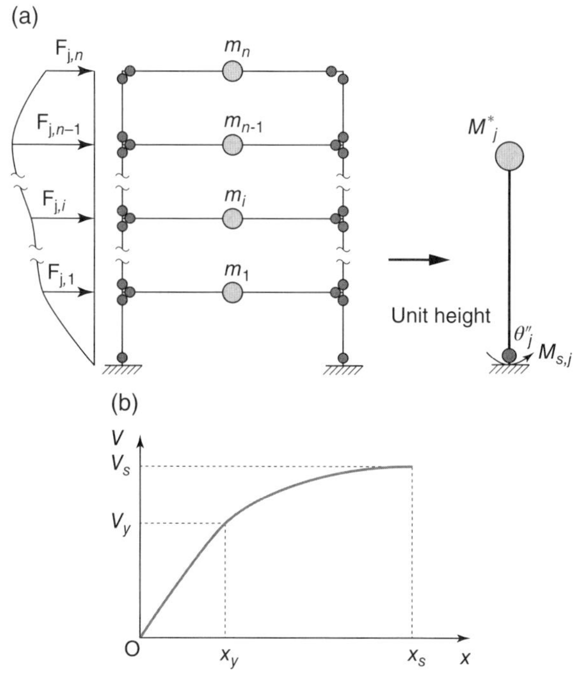

Figure 8.9 System model and response: (a) MDOF system model and corresponding jth modal SDOF system; (b) relation of base shear and roof displacement.

Step 4: Plot the moment versus displacement of the *j*th modal SDOF curve and determine the yielding moment $\Omega_{y,j}$.

Step 5: As for the *j*th modal SDOF system, there are

$$a_{s,j} = \omega_j^2 Y_j - \omega_j''^2 \theta_j'' \tag{8.88}$$

$$\Omega_j = \omega_j'^2 Y_j - \omega_j''^2 \theta_j'' \tag{8.89}$$

$$\theta_j'' = \begin{cases} \dfrac{1}{\omega_j'^2}\left(\omega_j'^2 Y_j - \Omega_{y,j}\right) & \Omega_j \geq \Omega_{j,y} \\ 0 & \Omega_j < \Omega_{j,y} \end{cases} \tag{8.90}$$

Equations (8.88) to (8.90) can be interpreted as two governing equations of the *j*th modal SDOF system after performing the modal decomposition for the MDOF system.

Example 8.3 Modal decomposition of the two-story moment-resisting frame
Consider that the one-bay two-story framed structure used in Example 8.2 is now subjected to two different lateral load patterns based on two modal shapes.

(1) 1st modal system
According to Eq. (8.49), there are

$$M_1^* = \mathbf{\Phi}_1^T \mathbf{M} \mathbf{\Phi}_1 = \{0.49 \quad 1.00\} \begin{bmatrix} 25000 & 0 \\ 0 & 25000 \end{bmatrix} \begin{Bmatrix} 0.49 \\ 1.00 \end{Bmatrix} = 3105.4\,\text{kg}$$

$$K_1^* = \mathbf{\Phi}_1^T \mathbf{K} \mathbf{\Phi}_1 = \{0.49 \quad 1.00\} \begin{bmatrix} 761030 & -330880 \\ -330880 & 251470 \end{bmatrix} \begin{Bmatrix} 0.49 \\ 1.00 \end{Bmatrix} = 110110\,\text{N/m}$$

$$C_1^* = \mathbf{\Phi}_1^T \mathbf{C} \mathbf{\Phi}_1 = \{0.49 \quad 1.00\} \begin{bmatrix} 1163.7 & -525.7 \\ -525.7 & 854.1 \end{bmatrix} \begin{Bmatrix} 0.49 \\ 1.00 \end{Bmatrix} = 739.7\,\text{N-s/m}$$

$$\omega_1^2 = \frac{K_1^*}{M_1^*} = \frac{110110}{3105.4} = 35.46$$

$$\varsigma_1 = \frac{C_1^*}{2\omega_1 M_1^*} = \frac{739.7}{2 \times 5.95 \times 3105.4} = 0.02$$

$$\gamma_1 = \frac{\mathbf{MI}}{M_1^*} = 1.2$$

The NPAS were performed and through using the graphic method and Eqs. (8.86) and (8.87), the yield moment is satisfied by

$$\Omega_{1,y} = \frac{\Gamma_1 V_{1,y}}{M_1^*} = \frac{1.2 \times 16000}{3105.4} = 6.18\,\text{N-m} \tag{8.91}$$

Thus, governing equations of the 1st modal system in the FAM is written as

$$\ddot{Y}_1(t) + 0.24\dot{Y}_1(t) + 35.46Y_1(t) = -1.2\ddot{x}_g(t) + 35.46Y_1''(t) \tag{8.92}$$

$$a_{s,1}(t) = 35.46Y_1(t) - 35.46\,\theta_1''(t) \tag{8.93}$$

$$\Omega_1(t) = 35.46Y_1(t) - 35.46\,\theta_1''(t) \tag{8.94}$$

$$\theta_1''(t) = \begin{cases} \dfrac{1}{35.46}\left(35.46Y_1(t) - \Omega_{y,1}\right) & \Omega_1(t) \geq 6.18 \\ 0 & \Omega_1(t) < 6.18 \end{cases} \tag{8.95}$$

Perform the state space formulation presented in Chapter 2.2 on the 1st modal system. Displacement and acceleration time history of each floor, and plastic rotation in the 1st modal coordinate time history is shown in Figure 8.10.

(2) 2nd modal system

According to Eq. (8.49), there are

$$M_2^* = \mathbf{\Phi}_2^T \mathbf{M} \mathbf{\Phi}_2 = \{-2.03 \quad 1.00\} \begin{bmatrix} 25000 & 0 \\ 0 & 25000 \end{bmatrix} \begin{Bmatrix} -2.03 \\ 1.00 \end{Bmatrix} = 12824\,\text{kg}$$

$$K_2^* = \mathbf{\Phi}_2^T \mathbf{K} \mathbf{\Phi}_2 = \{-2.03 \quad 1.00\} \begin{bmatrix} 761030 & -330880 \\ -330880 & 251470 \end{bmatrix} \begin{Bmatrix} 2.03 \\ 1.00 \end{Bmatrix} = 4738900\,\text{N/m}$$

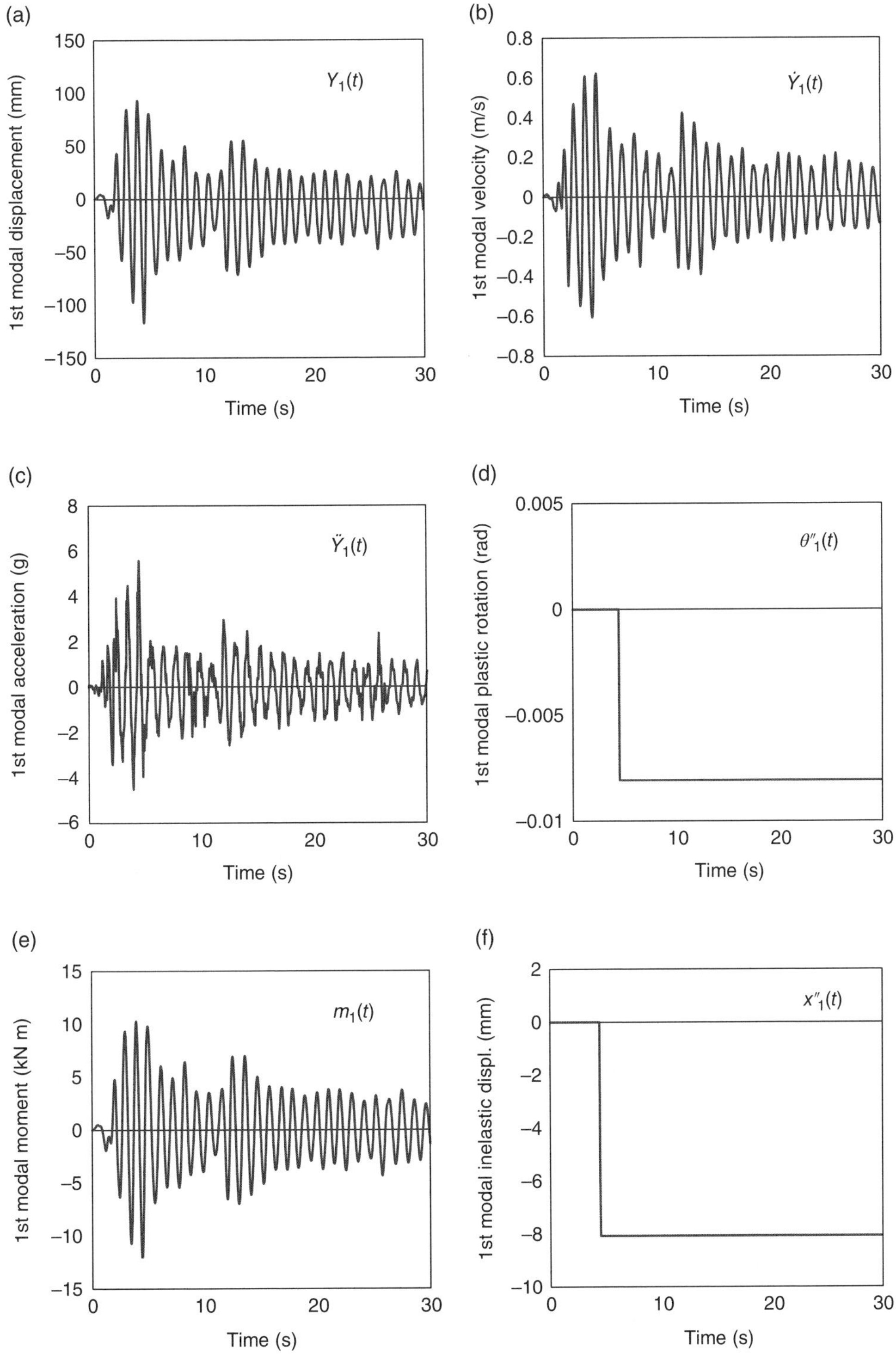

Figure 8.10 Responses of 1st modal SDOF system: (a) 1st modal displacement; (b) 1st modal velocity; (c) 1st modal acceleration; (d) 1st modal plastic rotation; (e) 1st modal moment; (f) 1st modal inelastic displacement.

$$C_2^* = \mathbf{\Phi}_2^T \mathbf{C} \mathbf{\Phi}_2 = \{-2.03 \;\; 1.00\} \begin{bmatrix} 1163.7 & -525.7 \\ -525.7 & 854.1 \end{bmatrix} \begin{Bmatrix} -2.03 \\ 1.00 \end{Bmatrix} = 9860.6 \text{ Ns/m}$$

$$\omega_2^2 = \frac{K_2^*}{M_2^*} = \frac{4738900}{12824} = 369.5$$

$$\varsigma_2 = \frac{C_2^*}{2\omega_2 M_2^*} = \frac{9860.6}{2 \times 19.22 \times 12824} = 0.02$$

$$\gamma_2 = \frac{\mathbf{MI}}{M_2^*} = -0.2$$

The yield moment can be determined as

$$\Omega_{2,y} = \frac{\Gamma_2 V_{2,y}}{M_2^*} = \frac{-0.2 \times (-18000)}{12824} = 0.28 \text{ N/m} \tag{8.96}$$

Thus, governing equations of the 2nd modal SDOF system in the FAM is written by

$$\ddot{Y}_2(t) + 0.77\dot{Y}_2(t) + 369.5Y_2(t) = 0.2x_g(t) + 369.5Y_2''(t) \tag{8.97}$$

$$a_{s,2}(t) = 369.5Y_2(t) - 369.5\theta_2''(t) \tag{8.98}$$

$$\Omega_2(t) = 369.5Y_2(t) - 369.5\theta_2''(t) \tag{8.99}$$

$$\theta_2''(t) = \begin{cases} \dfrac{1}{369.5}\left(369.5Y_2(t) - \Omega_{y,2}\right) & \Omega_2(t) \geq 0.28 \\ 0 & \Omega_2(t) < 0.28 \end{cases} \tag{8.100}$$

Then, the displacement response and plastic rotation of the 2nd modal system in modal coordinate are shown in Figure 8.11.

8.3 Modal Response Summation

Responses of the MDOF system using the modal superposition method can be obtained by summing all the responses of the SDOF systems in modal coordinates. Thus, Eqs. (8.82) and (8.83) can also be written in other forms as

$$\mathbf{X}(\mathbf{t}) = \sum_{j=1}^{n} \gamma_j \mathbf{\Phi}_j Y_j(t) \tag{8.101}$$

$$\dot{\mathbf{X}}(\mathbf{t}) = \sum_{j=1}^{n} \gamma_j \mathbf{\Phi}_j \dot{Y}_j(t) \tag{8.102}$$

$$\mathbf{X}''(\mathbf{t}) = \sum_{j=1}^{n} \gamma_j \mathbf{\Phi}_j Y_j''(t) \tag{8.103}$$

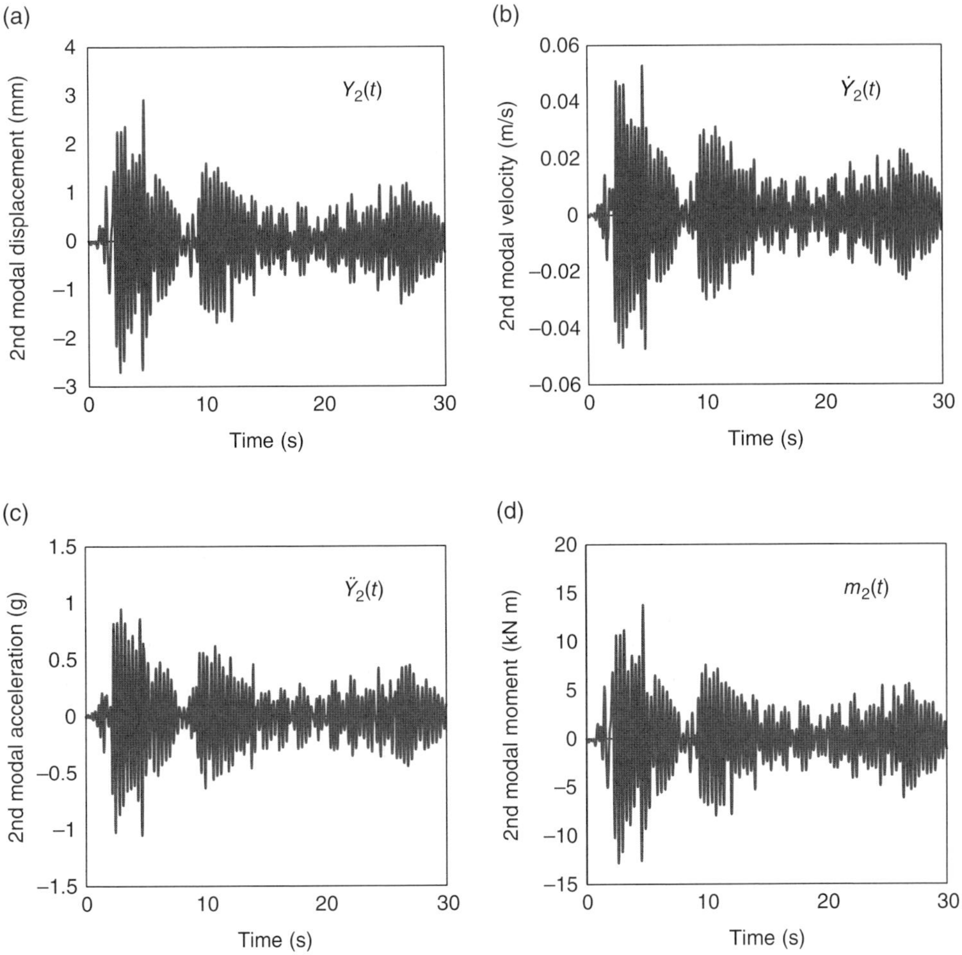

Figure 8.11 Responses of 2nd modal SDOF system: (a) 2nd modal displacement; (b) 2nd modal velocity; (c) 2nd modal acceleration; (e) 2nd modal moment.

Example 8.4 Responses summation of the two-story frame

Consider the one-bay two-story framed structure used in Example 8.2 again, the base and roof displacement responses using the modal superposition method (MSM) and general solution procedure of state space method (SSM) for the FAM, respectively, are plotted together, as shown in Figure 8.12. It can be seen from these figures that the displacement, velocity and acceleration response using MSM agree well with that using the SSM.

8.4 Nonlinear Modal Superposition Method Example

An additional example is presented to verify the use of the MSM in the FAM for solving nonlinear dynamics problems of a moment-resistant frame.

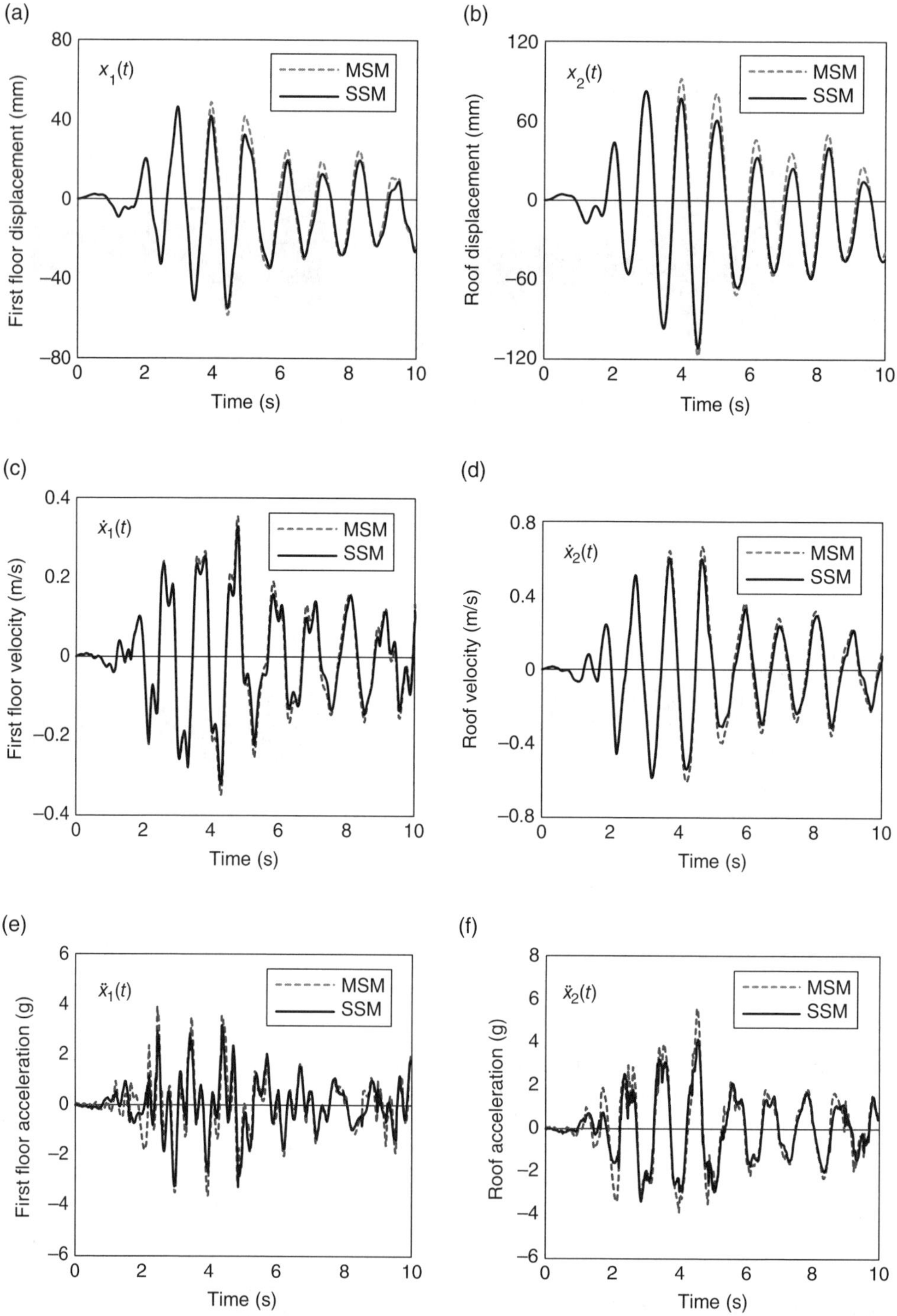

Figure 8.12 Response comparison of the MDOF system using the MSM and SSM: (a) First, floor displacement; (b) Roof displacement; (c) first floor velocity; (d) roof velocity; (e) first floor acceleration; (f) roof acceleration.

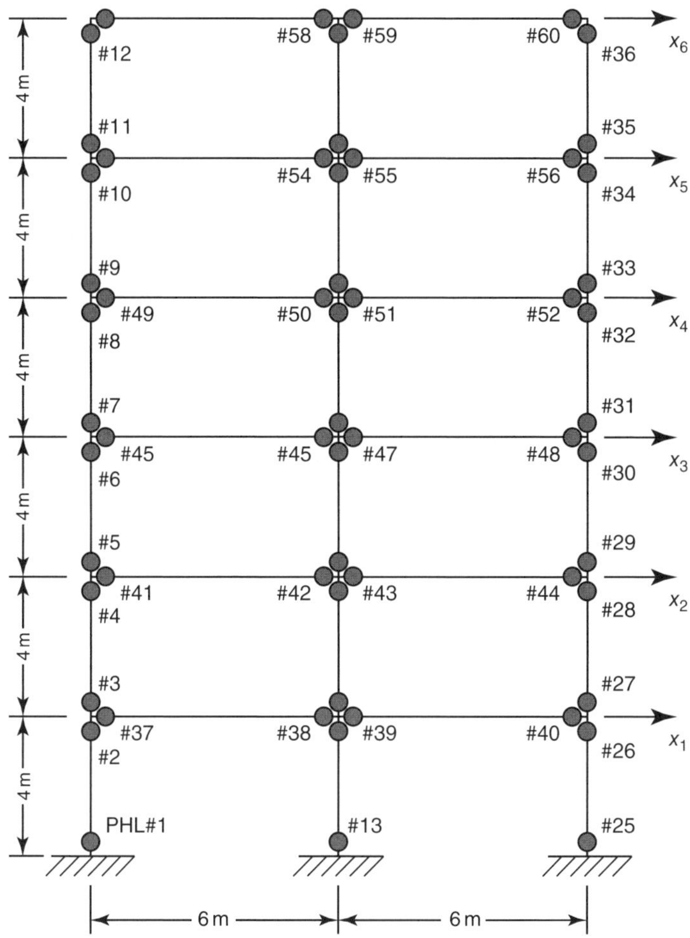

Figure 8.13 A 6-story 2-bay steel moment-resisting frame.

Examples 8.5 A Six-story Moment-Resistant Steel Framed Structure

Consider the 6-story moment-resistant steel framed structure shown in Figure 8.13. Assuming the members are axially rigid, this gives a total of 24 DOFs and 60 PHLs. No mass moment of inertia is assumed at the rotational joints, and therefore static condensation is used to eliminate the 18 rotational DOFs, resulting in only 6 translational DOFs for the frame.

Static condensation using Eq. (3.135) is used to obtain the $\bar{\mathbf{K}}$ matrix as

$$\bar{\mathbf{K}} = \begin{bmatrix} 457100 & -269340 & 79556 & -15828 & 3109 & -468 \\ -269340 & 383670 & -254730 & 76534 & -14838 & 2217 \\ 79556 & -254730 & 380650 & -253740 & 74555 & -11066 \\ -15828 & 76534 & -253740 & 378670 & -224730 & 56032 \\ 3109 & -14838 & 74555 & -22473 & 334450 & -151930 \\ -468 & 2217 & -11066 & -151930 & -151930 & 105130 \end{bmatrix} \text{kN/m} \qquad (8.104)$$

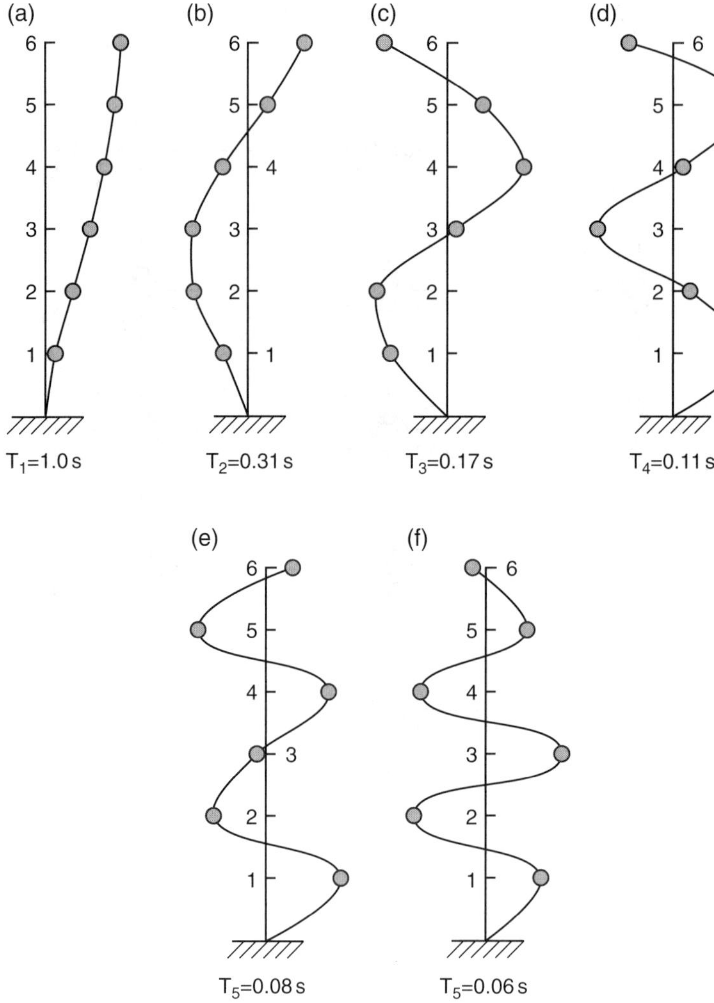

Figure 8.14 First, six model shapes: (a) 1st model; (b) 2nd model; (c) 3rd model; (d) 4th model; (e) 5th model; (f) 6th model.

The mass matrix $\mathbf{M}_{\mathrm{dd}}$ can be written as

$$\mathbf{M}_{dd} = \begin{bmatrix} 90 & 0 & 0 & 0 & 0 & 0 \\ 0 & 90 & 0 & 0 & 0 & 0 \\ 0 & 0 & 90 & 0 & 0 & 0 \\ 0 & 0 & 0 & 90 & 0 & 0 \\ 0 & 0 & 0 & 0 & 90 & 0 \\ 0 & 0 & 0 & 0 & 0 & 90 \end{bmatrix} \text{Mg} \tag{8.105}$$

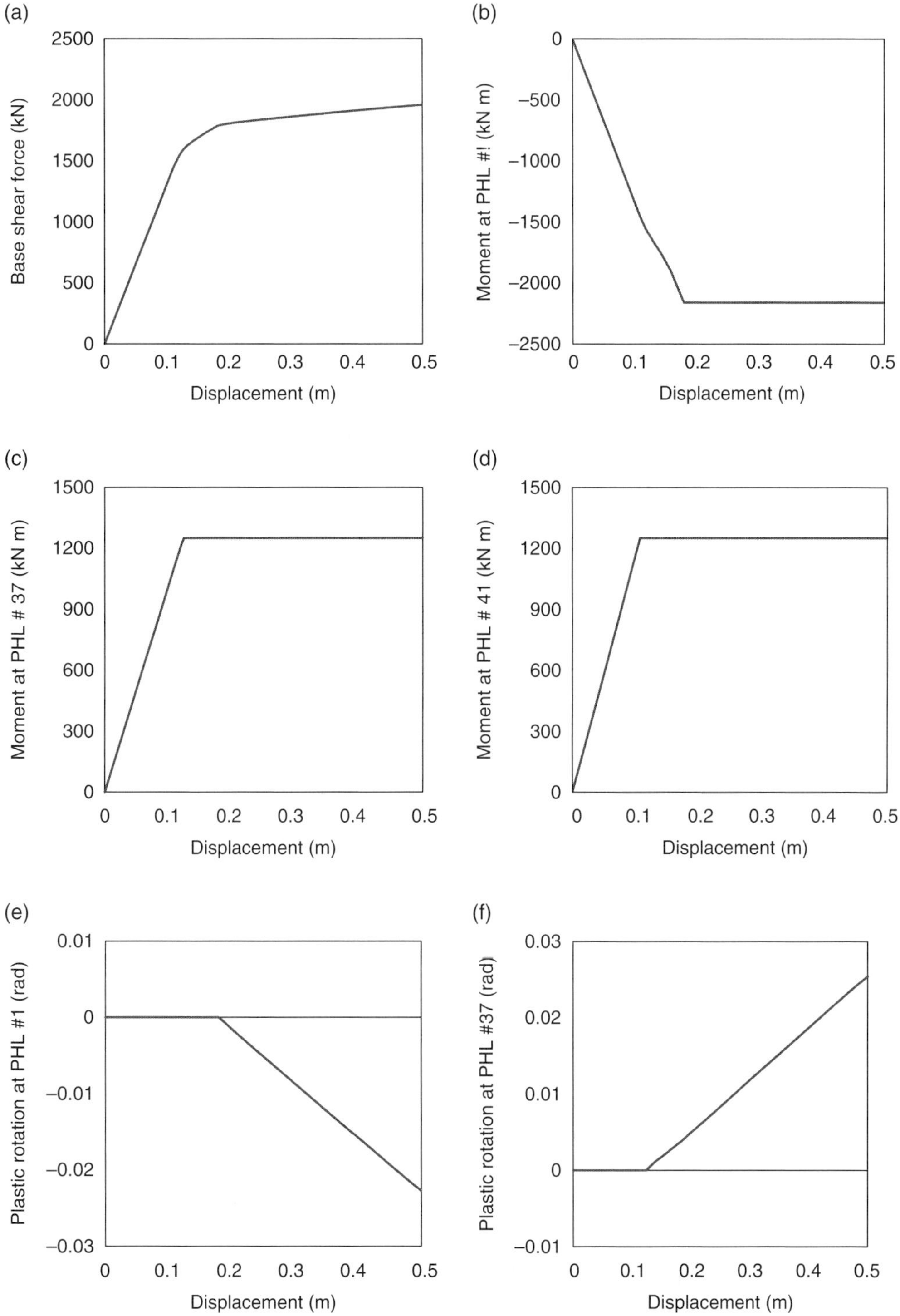

Figure 8.15 Responses of 1st modal SDOF system: (a) Base shear; (b) moment at PHL #1; (c) moment at PHL #37; (d) moment at PHL #41; (e) plastic rotation at PHL #1; (f) plastic rotation at PHL #37;

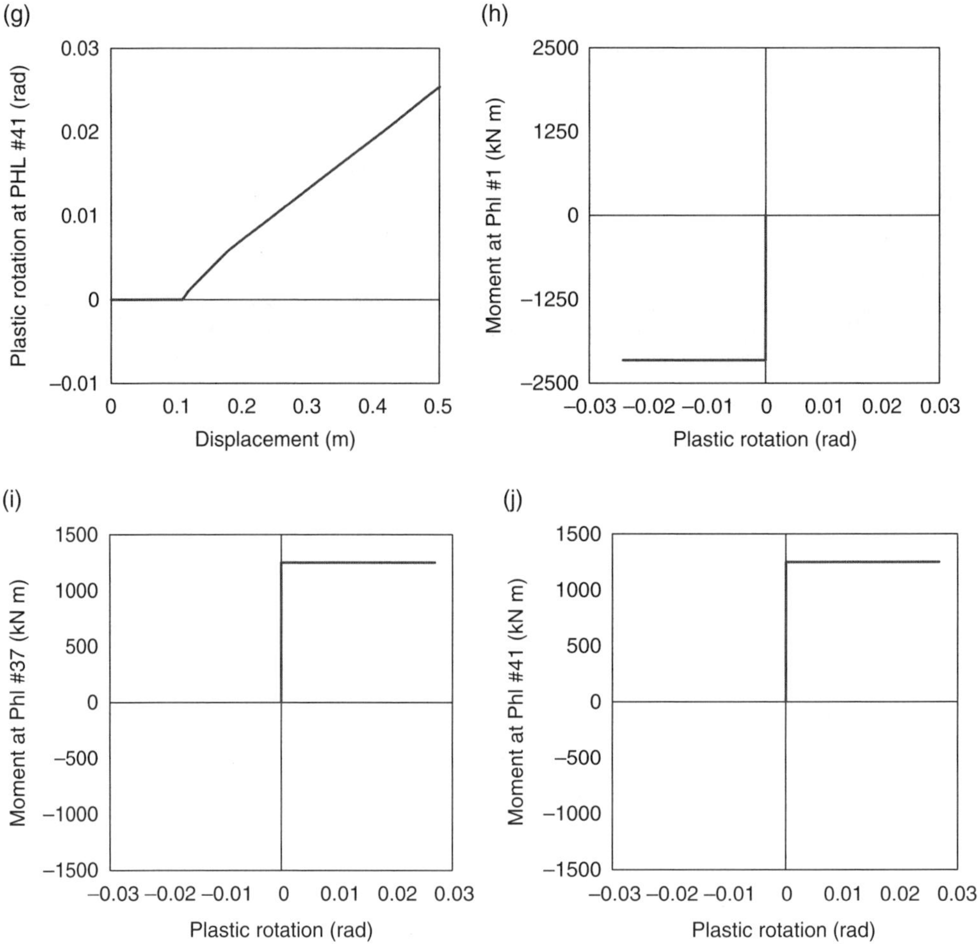

Figure 8.15 (*Continued*) (g) plastic rotation at PHL #41; (h) moment and plastic rotation relation of PH #1; (i) moment and plastic rotation relation of PH #37; (j) moment and plastic rotation relation of PH #41.

Assume that all plastic hinges at beams and columns exhibit elastic–plastic behavior with moment capacity of 1250kN m and 2160kN m, and elastic modules (*EI*) of beam and column sections are 240000kN m^2 and 480000kN m^2. The six periods and corresponding modal shape of vibration are calculated and summarized in Figure 8.14 through the condensed $\bar{\mathbf{K}}$ stiffness matrix in Eq. (8.104) and the mass matrix $\mathbf{M}_{dd}$ in Eq. (8.105).

(1) Nonlinear static pushover analysis

We now perform the NSPA procedure on the 6-story structure and the first three models are considered here. Figure 8.15 shows the relation of the base shear force, moment and plastic rotation of selected two plastic hinges (PHL #1 and PHL #9) versus roof displacement, and

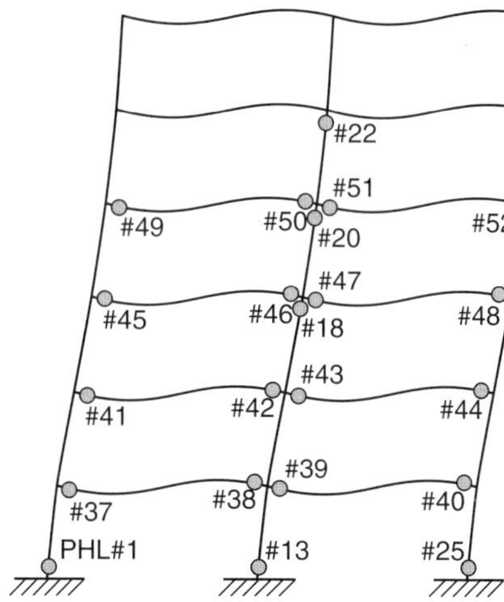

Figure 8.16 PHs distributions of the 1st modal SDOF system in the NSPA.

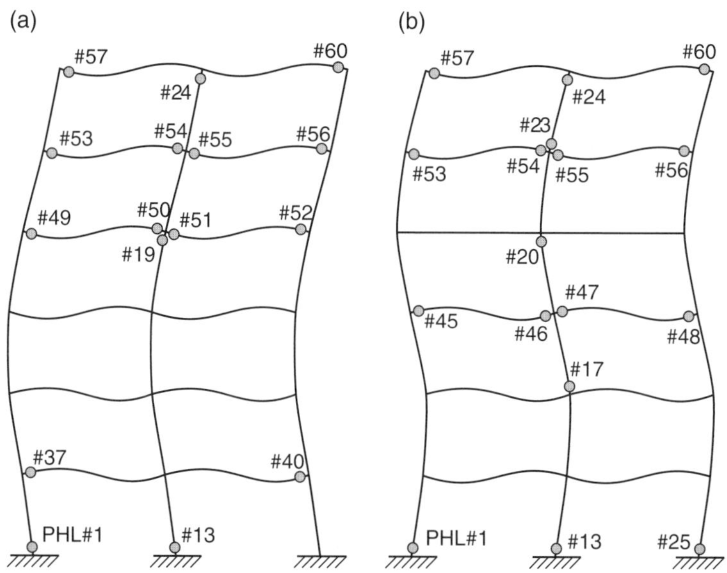

Figure 8.17 PH distributions of modal SDOF systems: (a) 2nd model; (b) 3rd model.

the relationship between the moment and plastic rotation of PHL #1. The PHLs distribution is shown in Figure 8.16 at the last step.

The PHLs distribution corresponding to other two models is shown in Figure 8.17 at the last displacement step. Figure 8.18 shows the relations of the base shear force versus roof displacement under 2nd and 3rd modal systems.

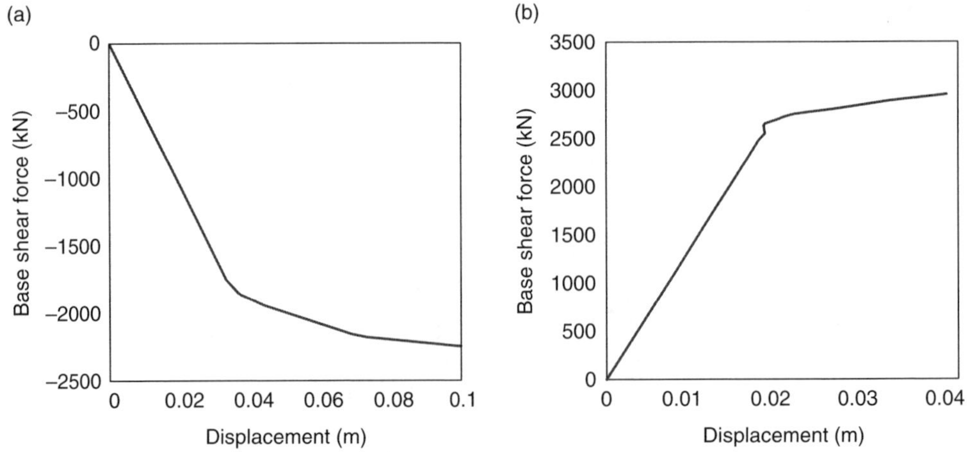

Figure 8.18 Base shear of modal SDOF systems: (a) 2nd model; (b) 3rd model.

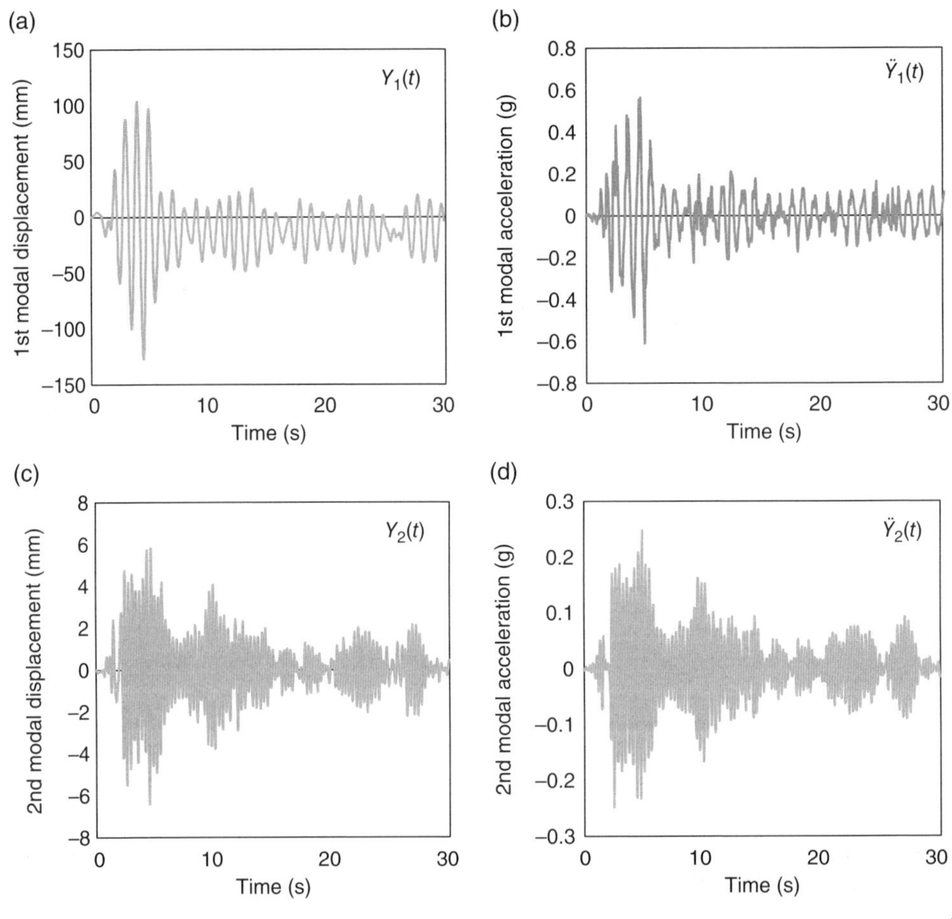

Figure 8.19 Displacement and acceleration responses of the first three modal SDOF systems: (a) 1st modal displacement; (b) 1st modal acceleration; (c) 2nd modal displacement; (d) 2nd modal acceleration;

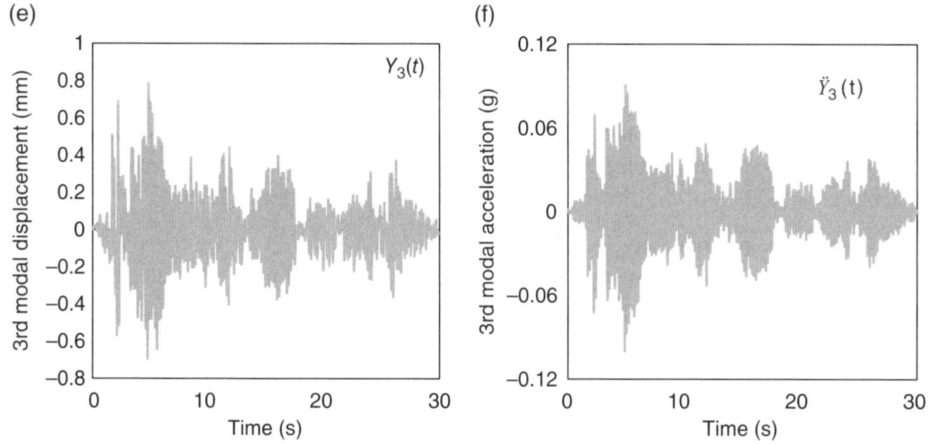

Figure 8.19 (*Continued*) (e) 3rd modal displacement; (f) 3rd modal acceleration.

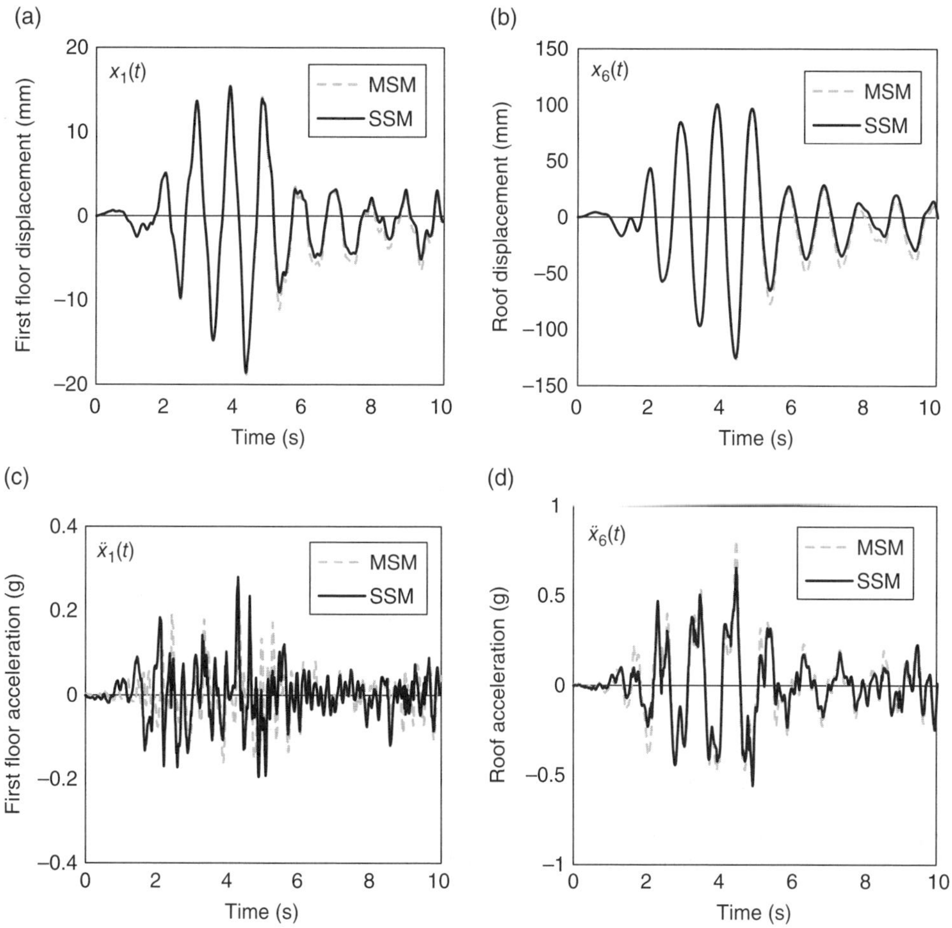

Figure 8.20 Response comparison of the structure using MSM and SSM: (a) first floor displacement; (b) roof displacement; (c) first floor acceleration; (d) roof acceleration.

(2) Modal decomposition

Now consider the structure is excited by the El Centro north–south ground acceleration record from the 1940 Imperial Valley Earthquake, the peak ground acceleration of which equals 0.2 g. By using the modal decomposition method in Section 8.2, the displacement, velocity and acceleration responses of the first three models are shown in Figures 8.19.

(3) Responses summation

Base and top floor displacement and acceleration of the structure are computed by the modal superposition method and the state space method, and are plotted together in Figure 8.20. First,

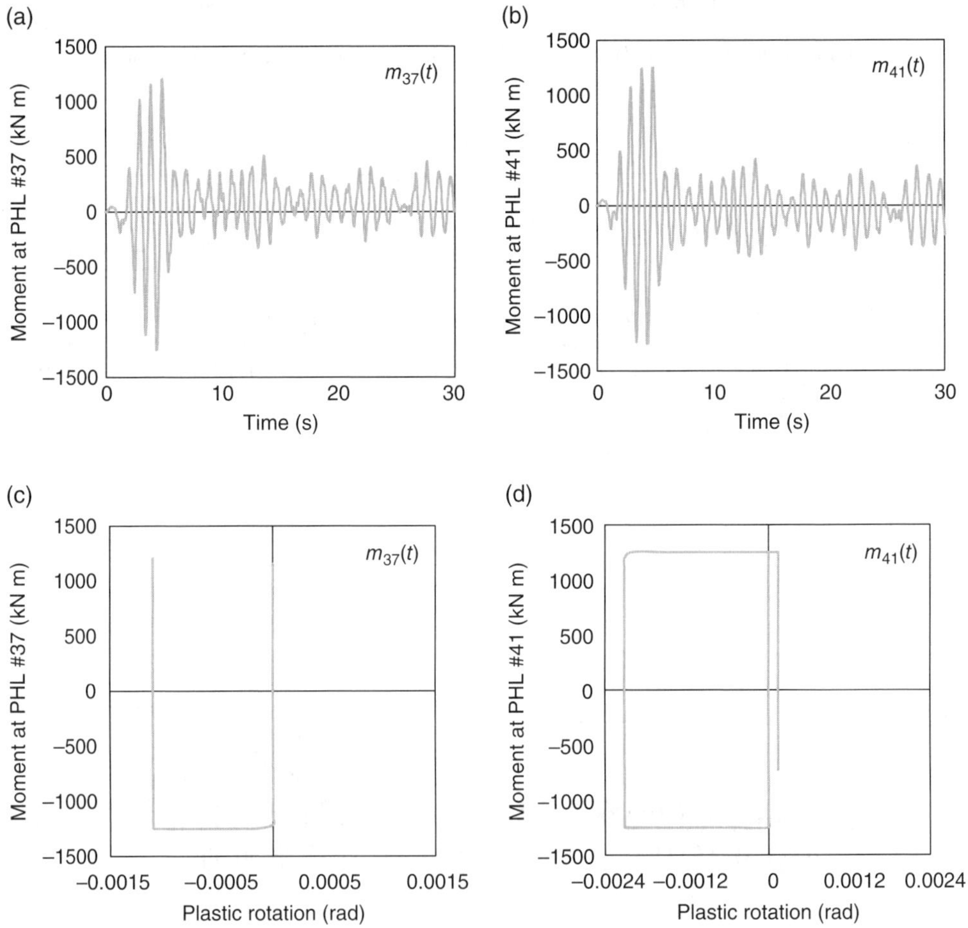

Figure 8.21 Global responses of the frame: (a) moment response at PHL #37; (b) moment response at PHL #41; (c) moment and plastic rotation relation of PH #37; (d) moment and plastic rotation relation of PH #41.

ten-second responses are shown in these figures for clear comparison. Figure 8.21 shows the moment history and relation of the moments and plastic rotation of the selected plastic hinge #37 and #41.

References

Chopra A.K. (2001). *Dynamics of Structures: Theory and Applications to Earthquake Engineering*. Prentice-Hall, New Jersey.

9

Application: Structural Vibration Control

The structural vibration control began in mechanical engineering in the early 20th century, and subsequently developed in civil engineering. At present, the structural vibration control technique is verified as an effective way to reduce responses of structures under natural hazards such as earthquakes and strong winds. Over the past few years, a number of structural control techniques have been developed and applied in practice, specifically in seismic regions. Structural control systems, in general, fall into four categories: passive, active, semi-active and hybrid control based on whether the power source is necessary. For the active and semi-active control technique, the control force is often determined by control strategies, but the passive control technique is through energy-dissipation devices to dissipate the input energy. In this chapter, the FAM is used to simulate the seismic dynamical analysis of structures with passive and active controlling devices.

9.1 Passive Control Technique

The passive control technique is through leading the input energy from environmental loads to special elements or devices to reduce energy-dissipating demand on primary structural members and minimize possible structural damage. These special elements or devices are so-called passive energy dissipation devices (PEDD) due to their energy-dissipation capacity.

9.1.1 Model of Passive Energy-Dissipation Devices

The PEDDs are mainly categorized into velocity-dependent dampers and displacement-dependent dampers.

Theory of Nonlinear Structural Analysis: The Force Analogy Method for Earthquake Engineering,
First Edition. Gang Li and Kevin K.F. Wong.

(1) Velocity-dependent PEDDs

Viscous and viscoelastic dampers are two typical kinds of velocity-dependent PEDDs. The restoring force of the dampers is associated with the inter-story velocity. Thus, the mathematical model of a viscous damper can be written as

$$F_d(t) = C_d \dot{x}(t) \tag{9.1}$$

where C_d represents the damping coefficient of linear viscous damper and $\dot{x}(t)$ denotes the velocity.

For the viscoelastic damper under the sinusoidal load with frequency ω_L, the force–displacement relationship of the viscoelastic damper is satisfied by

$$F_d(t) = k_d x(t) + \frac{\eta k_d}{\omega_L} \dot{x}(t) \tag{9.2}$$

where k_d is defined as the shear storage modulus of the viscoelastic material, η denotes the loss factor and $x(t)$ denotes the relative displacement of the viscoelastic damper. Figure 9.1 show the force–displacement relations of the viscous damper and viscoelastic damper, respectively.

(2) Displacement-dependent PEDDs

Friction dampers and metallic yielding dampers belong to the range of displacement-dependent PEDDs.

As for the friction damper, the mathematical model can be expressed by

$$F_d(t) = F_y \operatorname{sgn}(x(t)) \tag{9.3}$$

where F_y represents the slip load; sgn(.) is the sign function.

For the metallic yielding damper, the bilinear model is often used and written by

$$F(t) = \begin{cases} k_d x(t) & x(t) \le x_y \\ \alpha k_d x(t) & x(t) > x_y \end{cases} \tag{9.4}$$

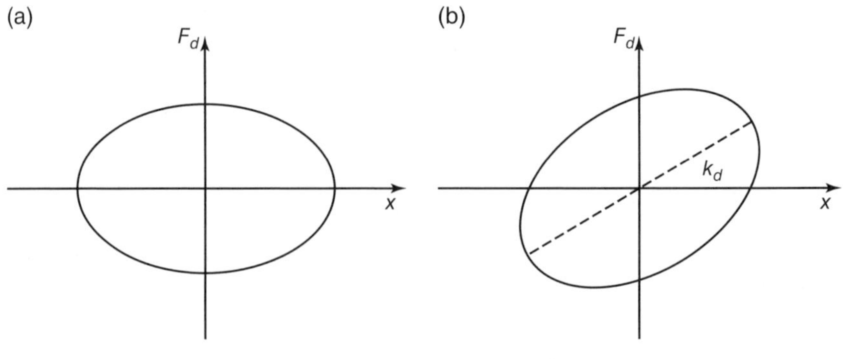

Figure 9.1 Force–displacement relations of: (a) viscous damper; (b) viscoelastic damper. Reproduced with permission from John Wiley and Sons, (License No. 3132491366031).

where k_d denotes the initial stiffness of metallic damper, α denotes the second stiffness coefficient and x_y is the yielding displacement. Figure 9.2 show force–displacement relations of these two dampers.

9.1.2 Model of Framed Structures with PEDDs

To dissipate a great deal of input earthquake energy, the PEDD is usually connected with braces or diagonal bars in series. In the FAM, assume that the force offered by the PEDD is distributed on the joints of the frame and ignore the axial deformation of the bar, as shown in Figure 9.3 and Figure 9.4 (Li and Li, 2011a), respectively.

When the PEDD is mounted eccentrically on the joint of chevron brace, the horizontal force generated by the PEDD can be transmitted at the support joints as the horizontal restoring forces, as illustrated in Figure 9.5 and is satisfied by

$$F_{d,A}=F_d\frac{b}{L};\quad F_{d,B}=F_d\frac{a}{L} \tag{9.5}$$

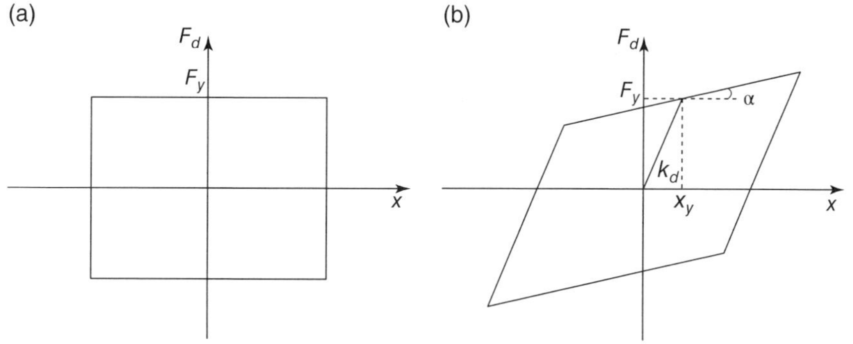

Figure 9.2 Force–displacement relations of: (a) friction damper; (b) metallic yielding damper. Reproduced with permission from John Wiley and Sons, (License No. 3132491366031).

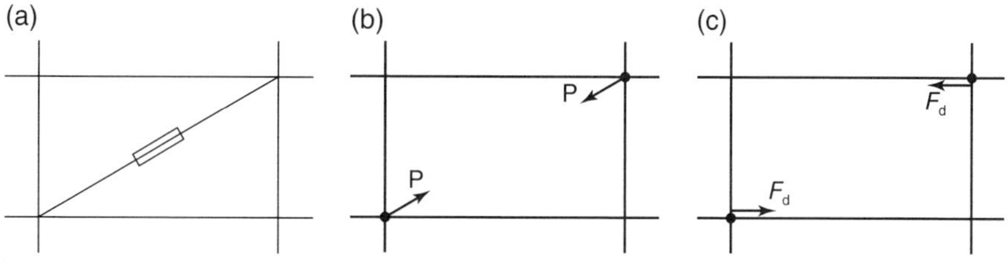

Figure 9.3 Model of a diagonal brace equipped with a PEDD (F_d is the horizontal force generated by PEDD; P is the axial force of diagonal brace): (a) diagonal brace model; (b) simplified axial force; (c) simplified horizontal force. Reproduced with permission from John Wiley and Sons, (License No. 3132491366031).

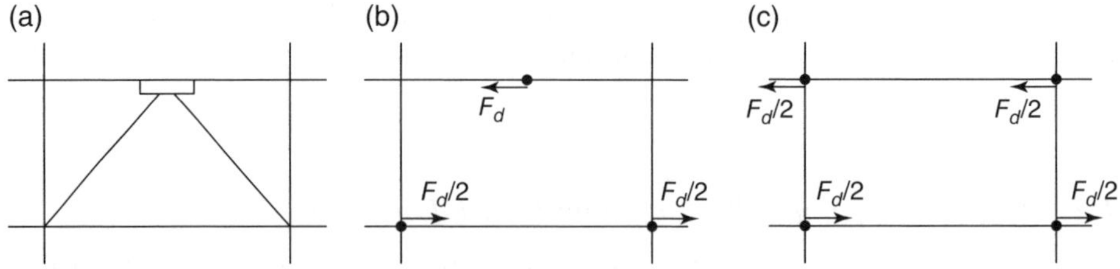

Figure 9.4 Model of a chevron brace equipped with a PEDD: (a) chevron brace model; (b) simplified braced joint forces; (c) simplified framed joint forces. Reproduced with permission from John Wiley and Sons, (License No. 3132491366031).

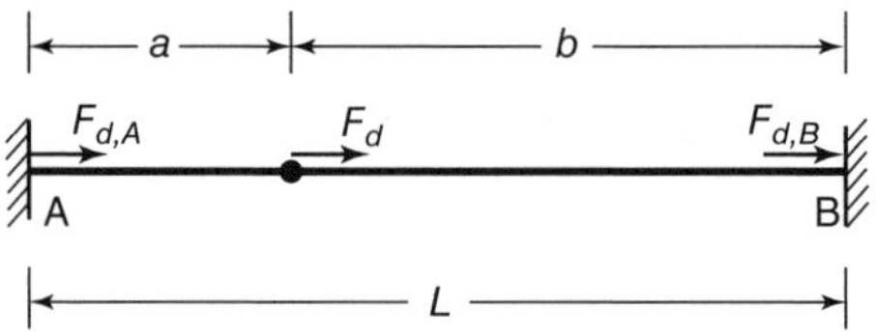

Figure 9.5 Damping force distributed on framed joints.

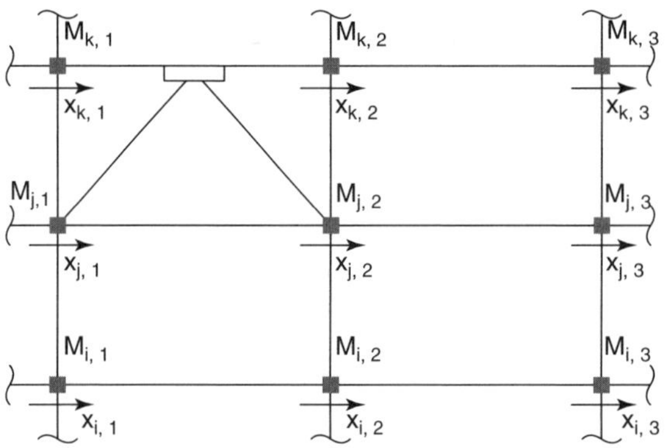

Figure 9.6 A MDOF structure with PEDDs model.

9.1.3 Force Analogy Method for Dynamical Analysis of Multi-Degree-Freedom Systems

Consider a multi-degree-of-freedom framed structure with PEDDs, which are connected to the frame through a chevron brace as shown in Figure 9.6.

The equation of motion of the MDOF system with the PEDD can be written as

$$\mathbf{M\ddot{X}(t)+C\dot{X}(t)+KX'(t)}=-\mathbf{M}\iota x_g(t)-\mathbf{F_d(t)} \tag{9.6}$$

Where **M** and **K** denote the mass, and stiffness matrices; $\mathbf{X(t)}$, $\dot{\mathbf{X}}\mathbf{(t)}$ and $\ddot{\mathbf{X}}\mathbf{(t)}$ are the displacement, velocity and acceleration vectors, respectively; ι is the unit vector.

The restoring force vector of the dissipation devices is satisfied by

$$\mathbf{F_d}(t) = \sum_{i=1}^{s} F_{d,i}(t)\mathbf{L_i} \tag{9.7}$$

in which $F_{d,i}$(t) is the joint force offered by the PEDD, $\mathbf{L}_\mathrm{i}$ implies the joint locations vector, for example, if there are five freedoms in the structure, the joint locations vector of the second joint is $\mathbf{L}_2 = [0\ 1\ 0\ 0\ 0]^\mathrm{T}$, s is the amount of the joints that are connected with the PEDDs.

According to Eq. (2.4), Eq. (9.6) is changed to (Li and Li, 2011a)

$$\mathbf{M}\ddot{\mathbf{X}}(\mathbf{t}) + \mathbf{C}\dot{\mathbf{X}}(\mathbf{t}) + \mathbf{K}\mathbf{X}(\mathbf{t}) = -\mathbf{M}\iota\ddot{x}_g(t) - \mathbf{F_d} + \mathbf{K}\mathbf{X}''(\mathbf{t}) \tag{9.8}$$

Equation (9.8) can be solved by state space method presented in Chapter 3.

Example 9.1 Eight-story Braced Frame with PEDDs

An eight-story steel framed structure is selected to illustrate the application of FAM in passive structural control technique. Its mass from first to eighth floor is 9.0×10^4 kg and other parameters of structural members are listed in Table 9.1. The location of the PH is shown in Figure 9.7 depicts joint numbers of structural members. The structure is excited by the El Centro north–south ground acceleration record from the 1940 Imperial Valley Earthquake, scaled by 1.74 such that the peak ground acceleration is equal to 0.6 g. Eight PEDDs, with stiffness and yielding displacement of 500 kN/mm and 5 mm, were installed on each floor.

The nonlinear dynamic responses are calculated using Eqs. (9.7) and (9.8), and the displacement, velocity and acceleration responses are shown in Figure 9.8.

The local plastic hinge moment and plastic rotation responses at selected PHLs are illustrated in Figure 9.9. Additionally, the relationship between moment and plastic rotation of selected five PHs are shown in Figure 9.10.

Figure 9.11 show that the hysteretic loops of seven dampers, which are located from the first floor to the seventh floors. In addition, the damper in the eighth floor keeps elasticity during the whole period.

Table 9.1 Parameters of the structure.

Structural member	Elastic Module EI (N-m^2)	Yielding moment M (N-m)
Columns	4.8×10^8	2.00×10^6
Left and right beams	2.4×10^8	1.20×10^6
Middle beams	2.4×10^8	1.20×10^6

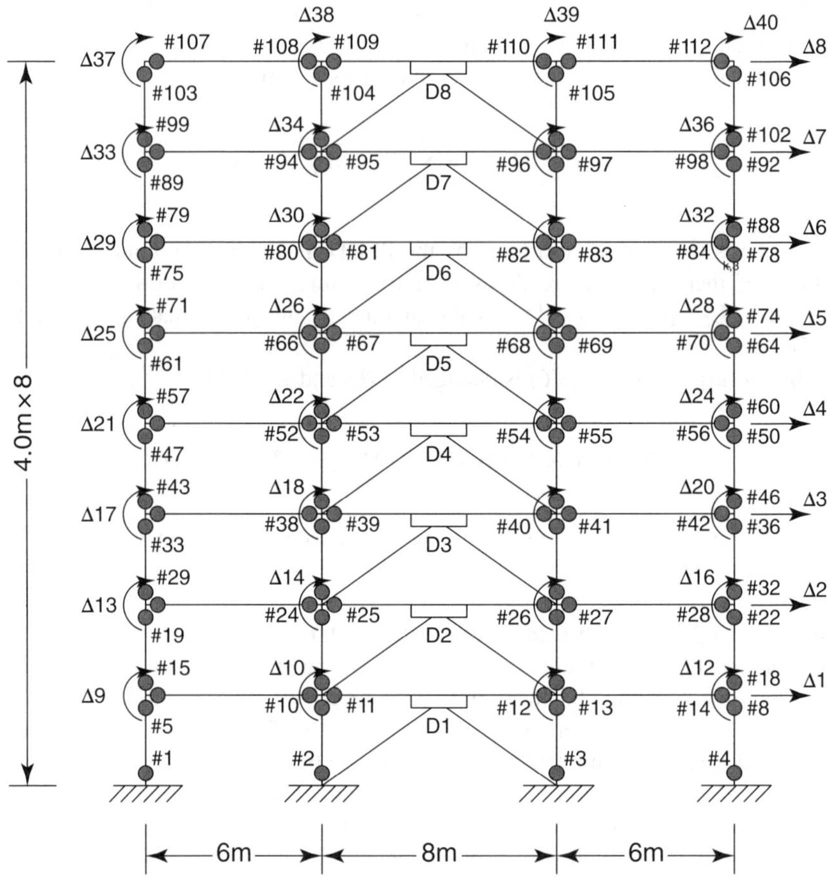

Figure 9.7 Model of the structure.

9.2 Application of the FAM in Active or Semi-Active Structural Control

For active and semi-active control techniques, the control force is often determined by the control laws. Some typical control laws, i.e. linear quadratic regulator, modal control, smart control, H_2 and H∞, etc., have become hot research topics in recent years. As one of control laws, the market-based control (MBC) is a multi-objective distributed control approach that provides effective control for a structure by means of simulating the activities of free market. Here, we will exhibit how to develop the FAM with the MBC theory for considering the inelastic behavior of a structure in which the MBC is used to reduce the structural vibration response and the FAM is proposed to perform the inelastic analysis (Li *et al*, 2011b).

9.2.1 Background of MBC

The MBC was proposed firstly by Clearwater in 1996. Lynch and Law (2002) first applied the MBC strategy to the structural vibration control in civil engineering and proposed

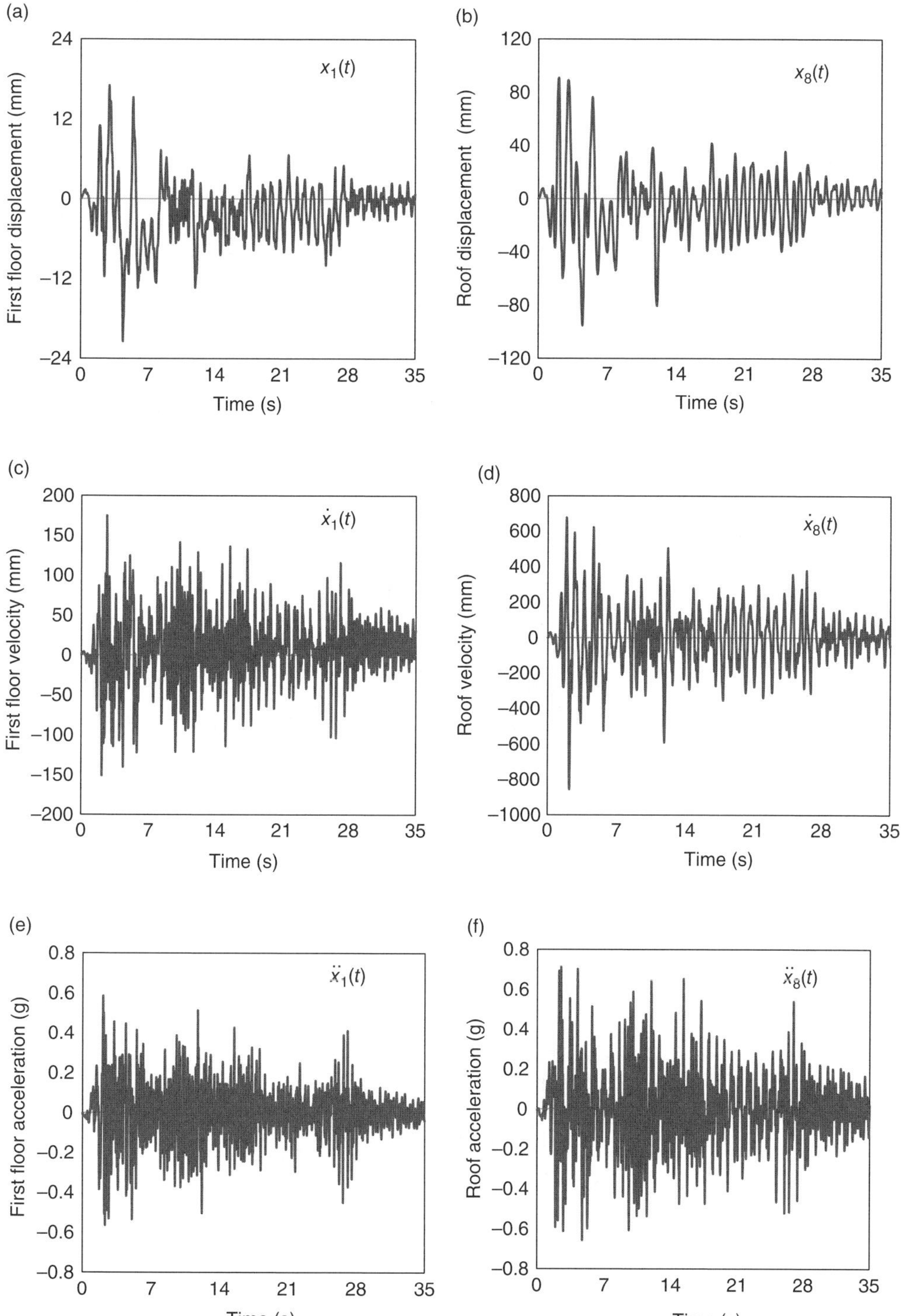

Figure 9.8 Displacement, velocity and acceleration responses of the structures: (a) first floor displacement; (b) roof displacement; (c) first floor velocity; (d) roof velocity; (e) first floor acceleration; (f) roof acceleration.

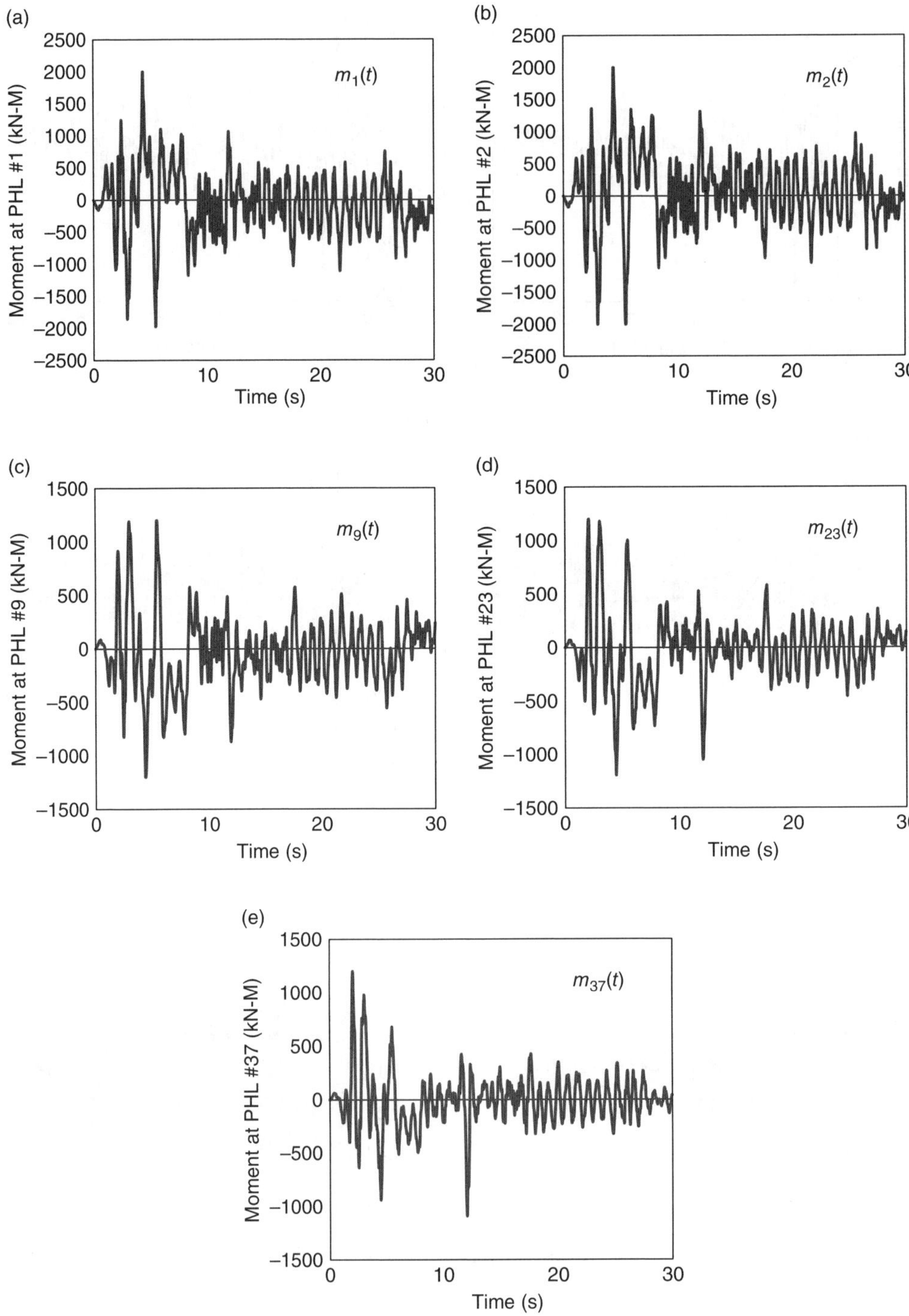

Figure 9.9 Moments at selected PHLs responses: (a) PHL #1; (b) PHL #2; (c) PHL #9; (d) PHL #23; (e) PHL #37.

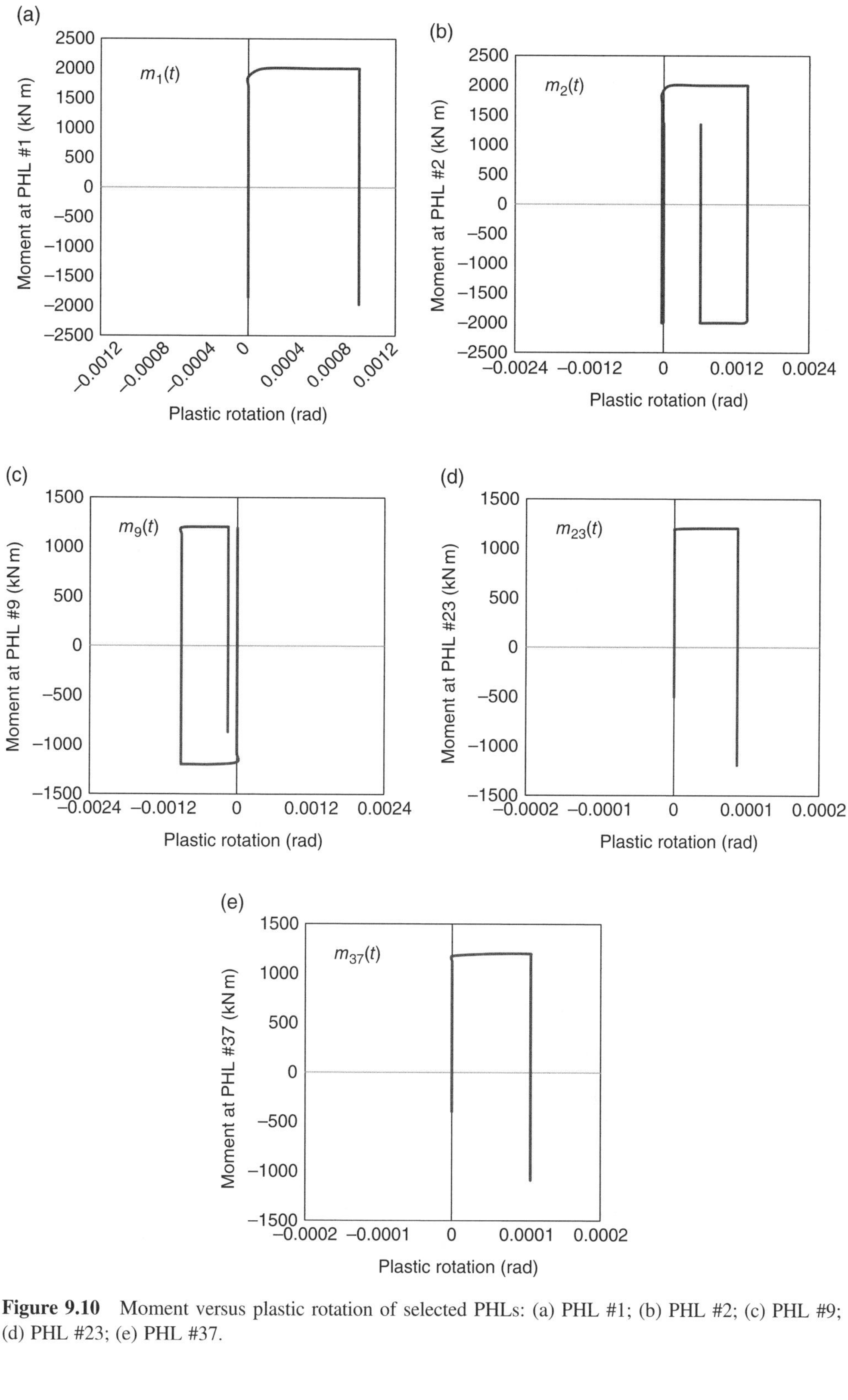

Figure 9.10 Moment versus plastic rotation of selected PHLs: (a) PHL #1; (b) PHL #2; (c) PHL #9; (d) PHL #23; (e) PHL #37.

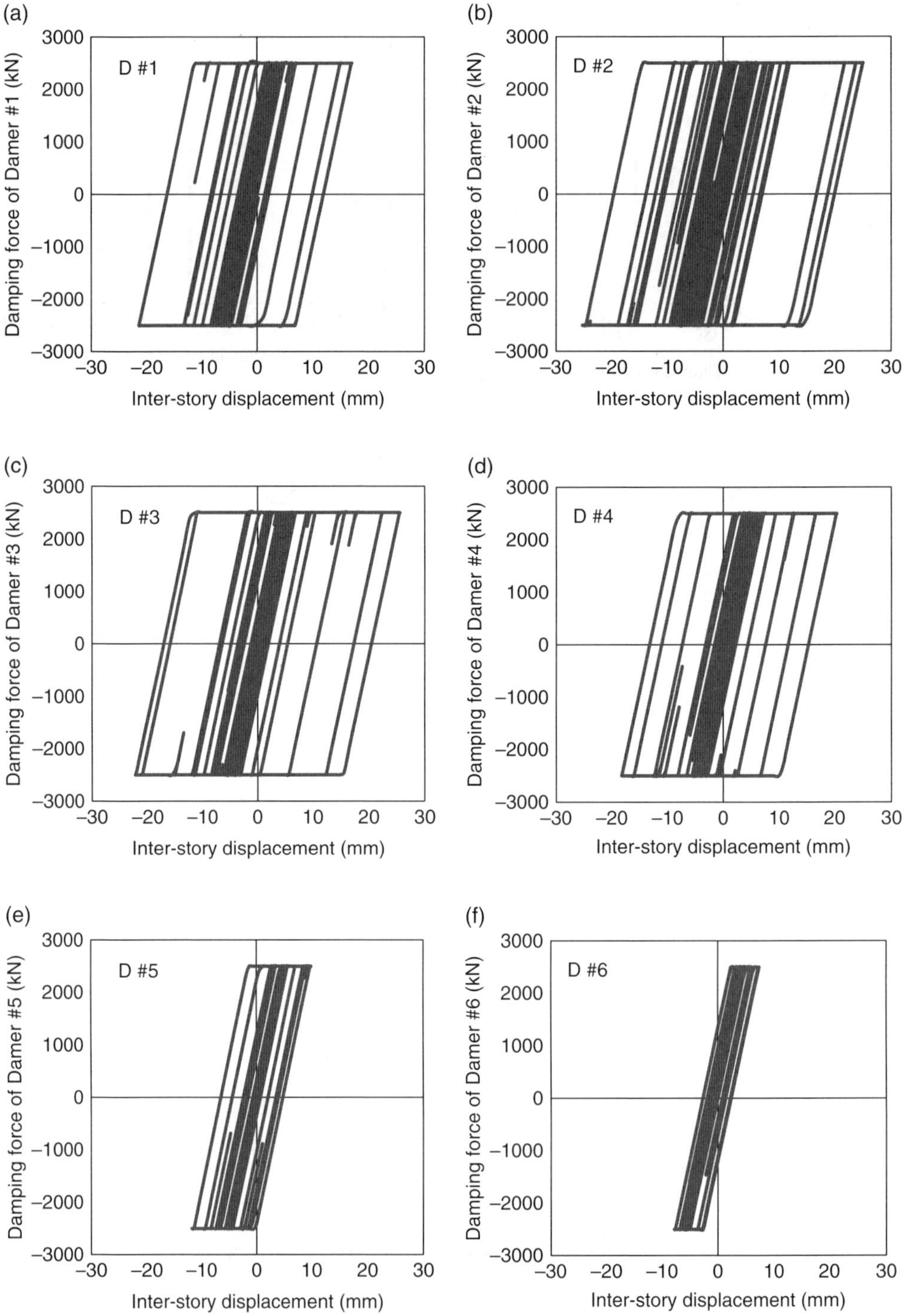

Figure 9.11 Hysteretic loops of seven yielded dampers: (a) Damper #1; (b) Damper #2; (c) Damper #3; (d) Damper #4; (e) Damper #5; (f) Damper #6;

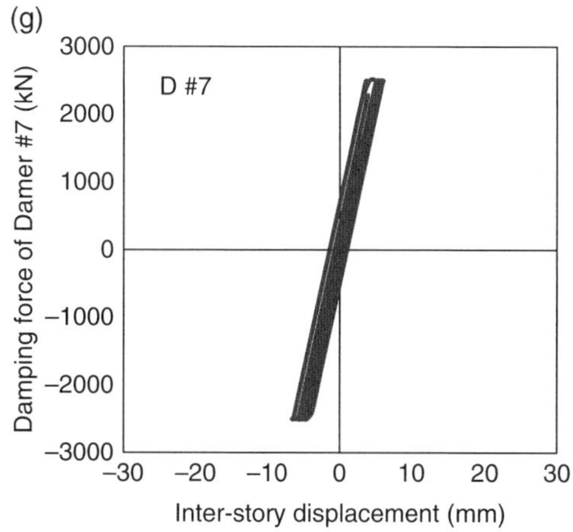

Figure 9.11 (*Continued*) (g) Damper #7.

a market-based energy-control approach. In the field of economics, the free market can set up an effective distribution of resources by the means of price strategy. The complicated market economy can be idealized by the behavior of market institutions such as consumer (buyers), firms (sellers), trade unions and governments. Mastering the basic concept as follows is very important to understand the MBC theory. The amount of commodity, which is scheduled to sell by sellers in some time at the market price is called 'Supply', which is affected by the market price and produced power. The aim of the supply is to make the profit maximal. The amount of commodity, which is scheduled by buyers in the same period at market price is called "Demand". Making the utility maximal is the buyers' motivation and goal during trade in the marketplace. The price is paid for goods or the workers' salary. It makes the free market optimally 'controlled'. Because the price is also consistent with the controlled force from actuator, once the demand in the market or environmental load increases, the price and wealth consumption become larger. Change regularity of the price and consumption are consistent with the ones of demand in the market and environmental loads in structural control. Therefore, the precision and effectiveness of control strategy are judged by comparing the change of price and wealth consumption to demand or environmental load. The competitive mechanisms of a free market can be extended for the application to the structural control field. In the control marketplace, the roles of market buyers and sellers are assumed by system actuators and power sources, respectively. The foundation of applying the MBC strategy to the civil engineering is that the decentralized complex system with the market model uses the merchandise's supply–demand law to distribute the control energy, where the energy system and controlled system are replaced by the seller and consumer. The behavior of a buyer is defined by the individual utility function, J_D, which measures the amount of utility derived by the buyer from purchasing the market commodity. The utility is a function of the price per unit commodity, p, the amount of purchased commodity, Q_D, and the response measures of the dynamic system, $Y(t)$. Similarly, a seller is governed by the individual profit function, J_s, that measures the amount of profit derived by the seller from selling the commodity. The profit is modeled as a function of the price per unit commodity, p, and sold commodity Q_S. The goal of the seller is to maximize their profit that is constrained by the maximum amount of commodity they possess Q_{max}. Meanwhile, the maximization of the buyer' utility is constrained by the total purchase cost, $p{\cdot}Q_D$ to be less than their fictitious wealth W_{max}:

$$\begin{array}{llll} \max\ J_{S1}(Q_{S1},p) & Q_{S1} \le Q_{max1} & \max\ J_{D1}(Q_{D1},p,Y_1(t)) & p{\cdot}Q_{D1} \le W_{max1} \\ \max\ J_{S2}(Q_{S2},p) & Q_{S2} \le Q_{max2} & \max\ J_{D2}(Q_{D2},p,Y_2(t)) & p{\cdot}Q_{D2} \le W_{max2} \\ \vdots & & \vdots & \end{array} \tag{9.9}$$

The system is constrained by the conservation of total energy, where the m systematic buyers purchase all the power produced by the system's n suppliers. According to the price rule and supply–demand law, the equation can be expressed as follows

$$\sum_{j=1}^{n} Q_{Sj} = \sum_{i=1}^{m} Q_{Di} \tag{9.10}$$

At each point in time, the solution of Eqs. (9.9) and (9.10) is shown to be the Pareto Optimal Solution. That is to say, an equalization price is obtained according to the equations of supply and demand in each time step, and the distribution of scarce resources produce maximized benefits under the optimal price. In the fictitious market, there are no strict relations between the supply and demand. The commodity price, p, is a key factor with regard to the supply function, Q_S, and the demand function, Q_D. The supply function is usually chosen as a linear function as to the amount of supply rises with increasing price, and the demand function can be selected in several models for a high speed of calculation. Based on the forms of demand supply, the supply–demand mode can be classified into four types, such as Linear-supply and linear-demand model, Linear-supply and power-demand model (LPM), Advanced Linear-supply and power-demand model (ALPM), and Linear-supply and exponential-demand model (LEM) (Li 2011, 2013).

9.2.2 *Force Analogy Method in Market-Based Control*

The application of FAM in the MBC can be depicted as shown as in Figure 9.12.

In order to interpret the MBC clearly, the state space formulation method and the FAM theory are mentioned again.

Equation (9.8) is changed to the first-order linear differential equation as:

$$\dot{\mathbf{z}}(\mathbf{t}) = \mathbf{A}\mathbf{z}(\mathbf{t}) + \mathbf{B}\mathbf{F_d}(\mathbf{t}) + \mathbf{H}\ddot{x}_g(t) + \mathbf{F_p^c}(\mathbf{t})\mathbf{X}''(\mathbf{t}) \tag{9.11}$$

where $\mathbf{z}(\mathbf{t}) = \begin{Bmatrix} \mathbf{X}(\mathbf{t}) \\ \dot{\mathbf{X}}(\mathbf{t}) \end{Bmatrix}$, $\mathbf{A} = \begin{bmatrix} 0 & \mathbf{I} \\ -\mathbf{M}^{-1}\mathbf{K} & -\mathbf{M}^{-1}\mathbf{C} \end{bmatrix}$, $\mathbf{B} = \begin{bmatrix} 0 \\ \mathbf{M}^{-1} \end{bmatrix}$, $\mathbf{H} = \begin{bmatrix} 0 \\ -1 \end{bmatrix}$, $\mathbf{F_p^c} = \begin{bmatrix} 0 \\ \mathbf{M}^{-1}\mathbf{K} \end{bmatrix}$.

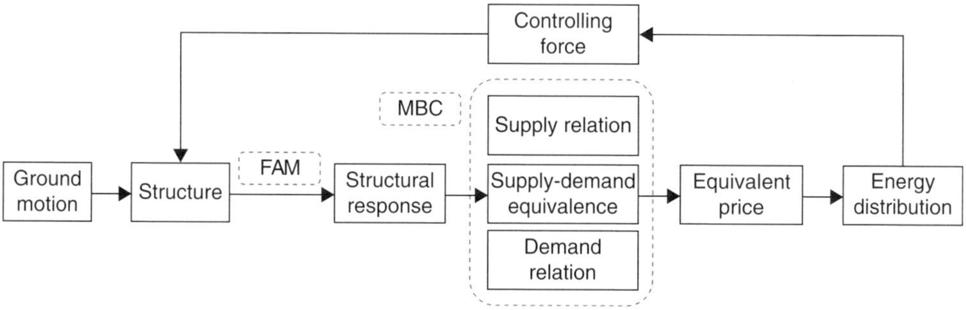

Figure 9.12 Schematic diagram of MBC nonlinear control strategy based on FAM.

The solution of Eq. (9.11) can be obtained as follows:

$$\mathbf{z}(\mathbf{t}) = e^{\mathbf{A}\Delta t}\mathbf{z}(t-\Delta t) + e^{\mathbf{A}t}\int_{t-\Delta t}^{t} e^{-\mathbf{A}s}\left[\mathbf{B}\mathbf{F}_{\mathbf{d}}(\mathbf{t}) + \mathbf{H}\ddot{\mathbf{x}}_{\mathbf{g}}(\mathbf{s}) + \mathbf{F}_{\mathbf{p}}^{\mathbf{c}}\mathbf{X}''(\mathbf{s})\right]ds \tag{9.12}$$

where Δt is the integration time step. Let $t_{k-1} = t - \Delta t$ and $t_k = t$. The discrete state solution of Eq. (9.12) becomes

$$\mathbf{z}(k+1) = e^{\mathbf{A}\Delta t}\mathbf{z}(\mathbf{k}) + e^{\mathbf{A}\Delta t}\mathbf{B}\Delta t\mathbf{F}_{\mathbf{d}}(t) + e^{\mathbf{A}\Delta t}\mathbf{H}\Delta t\ddot{\mathbf{x}}_{\mathbf{g}}(\mathbf{k}) + e^{\mathbf{A}\Delta t}\mathbf{F}_{\mathbf{p}}^{\mathbf{c}}\Delta t\mathbf{X}''(\mathbf{t}) \tag{9.13}$$

Assuming $\mathbf{A_d} = e^{\mathbf{A}\Delta t}$, $\mathbf{B_d} = e^{\mathbf{A}\Delta t}\mathbf{B}\Delta t$, $\mathbf{H_d} = e^{\mathbf{A}\Delta t}\mathbf{H}\Delta t$ and $\mathbf{F_{pd}} = e^{\mathbf{A}\Delta t}\mathbf{F}_{\mathbf{p}}^{\mathbf{c}}\Delta t$, Eq. (9.13) can be simplified as

$$\mathbf{z}(k+1) = \mathbf{A_d}\mathbf{z}(\mathbf{k}) + \mathbf{B_d}\mathbf{F_d}(\mathbf{k}) + \mathbf{H_d}\ddot{\mathbf{x}}_{\mathbf{g}}(\mathbf{k}) + \mathbf{F_{pd}}\mathbf{X}''(\mathbf{k}) \tag{9.14}$$

The solution for this control force vector can be given in terms of the state vector **z(k)** using the supply–demand law. Take the ALPM as an illustration here to investigate how to use the supply–demand model in the FAM.

For a system with n degrees of freedom, the linear supply function can be written as

$$Q_{S,j} = \eta_j \cdot p \tag{9.15}$$

The power demand function can be written as

$$Q_{D,i} = \frac{W_i\left|c_i x_{d,i} + d_i \dot{x}_{d,i}\right|}{p} \tag{9.16}$$

where η_j is the parameter that reflects the energy supply, $x_{d,i}$ and $\dot{x}_{d,i}$ are the story displacement and story velocity of the ith floor, $c_i \geq 0$ and $d_i \geq 0$ are the weighting coefficients, respectively.

Based on the optimal price solved from Eq. (9.10), the control force with the direction can be drawn as follows

$$U_i = -K_g \cdot \frac{W_i \cdot \left(c_i x_{d,i} + d_i \dot{x}_{d,i}\right)}{p} \tag{9.17}$$

where $K_g > 0$ is a gain coefficient related to the actuator.

The equation at the time step of t_i becomes

$$\sum_{j=1}^{n} \eta_j p(t_i) = \sum_{i=1}^{n} \frac{W_i(t_i) \cdot \left|c_i x_{d,i}(t_i) + d_i \dot{x}_{d,i}(t_i)\right|}{p(t_i)} \tag{9.18}$$

The solution of this equation is given by

$$p(t_i) = \sqrt{\sum_{i=1}^{n} \frac{W_i(t_i) \cdot \left|c_i x_{d,i}(t_i) + d_i \dot{x}_{d,i}(t_i)\right|}{\sum_{j=1}^{n} \eta_j}} \tag{9.19}$$

Substituting Eq. (9.19) back into Eq. (9.17), then the solution of the control force is obtained as

$$\mathbf{F_d}(t_i) = -K_g \cdot \mathbf{Q_D}(t_i) \tag{9.20}$$

In which $\mathbf{Q_D}(t_i) = [Q_{D,1}(t_i) \;\; Q_{D,2}(t_i) \;\; \cdots \;\; Q_{D,N}(t_i)]^{\mathrm{T}}$.

Example 9.2 Market-based control for a one-story moment-resistant frame

A one-story moment-resistant frame with magnetorheological damper controlled by the MBC method is considered (Li *et al.*, 2011b). Assume four PHLs exist in this frame, as shown in Figure 9.13. The floor masses are assumed to be 16,000 kg, and other structural parameters are shown in Table 9.2. A Magnetorheological Damper (MRD) is installed as an actuator into

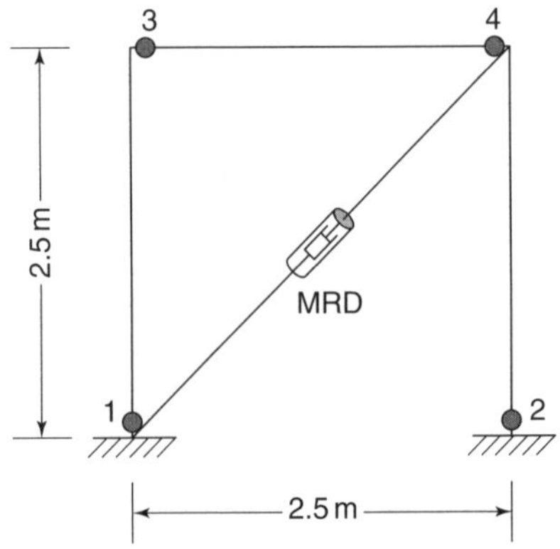

Figure 9.13 The model of structure with magnetorheological damper. Gang Li, Qifeng Liu, Hongnan Li. Inelastic structural control based on MBC and FAM, Mathematical Problems in Engineering, vol. 2011, Article ID 460731, 18 pages, 2011. doi:10.1155/2011/460731.

Table 9.2 Parameters of the structural component.

Structural component	Elastic modulus EI (N-m^2)	Yielding moment M (N-m)
columns	2.87×10^6	3.107×10^5
beams	2.87×10^6	3.107×10^5

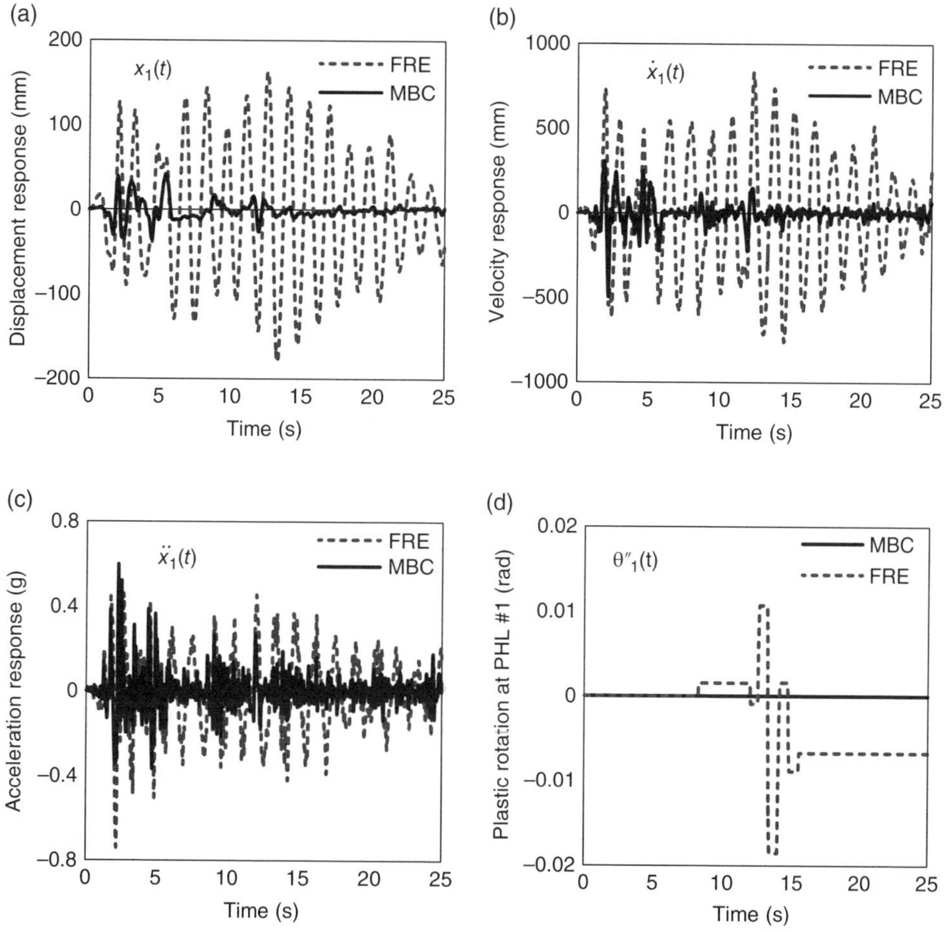

Figure 9.14 Responses of the structure with and without controlling: (a) displacement; (b) velocity; (c) acceleration; (d) plastic rotation at PHL #1. Gang Li, Qifeng Liu, Hongnan Li. Inelastic structural control based on MBC and FAM, Mathematical Problems in Engineering, vol. 2011, Article ID 460731, 18 pages, 2011. doi:10.1155/2011/460731.

the frame. The structure is excited by ground motions, El Centro earthquake (NS, May 18,1940, the acceleration peak of which is adjusted to 0.5 g).

To understand the performance of the inelastic control method combing the FAM and MBC, Figure 9.14 shows the displacement, velocity, acceleration and plastic rotation of the structure with and without MRD (named as FRE and MBC). As shown in these figures, the structure without MRD has entered into plasticity and the structure with MRD remains in elasticity because of the MBC providing an effective control.

Example 9.3 Market-based control for a Six-story moment-resistant structure
Consider a six-story structure (Li *et al.* 2011b), which masses are assumed to be 10,800 kg on every floor, and other structural parameters are shown in Table 9.3. The PHLs and joint numbers of structural elements are illustrated in Figure 9.15. The structure is excited by a ground motion,

Table 9.3 the main parameters of the structure.

Structural component	Elastic modulus EI (N-m^2)	Yielding moment M (N-m)
1st to 2nd rows columns	7.680×10^8	0.953×10^6
3rd to 4th rows columns	6.800×10^8	0.844×10^6
5th to 6th rows columns	4.800×10^8	0.596×10^6
1st to 3rd rows beams	2.604×10^9	3.388×10^6
4th rows beams	1.808×10^9	2.363×10^6
5th rows beams	1.560×10^9	2.070×10^6
6th rows beams	0.654×10^9	1.313×10^6

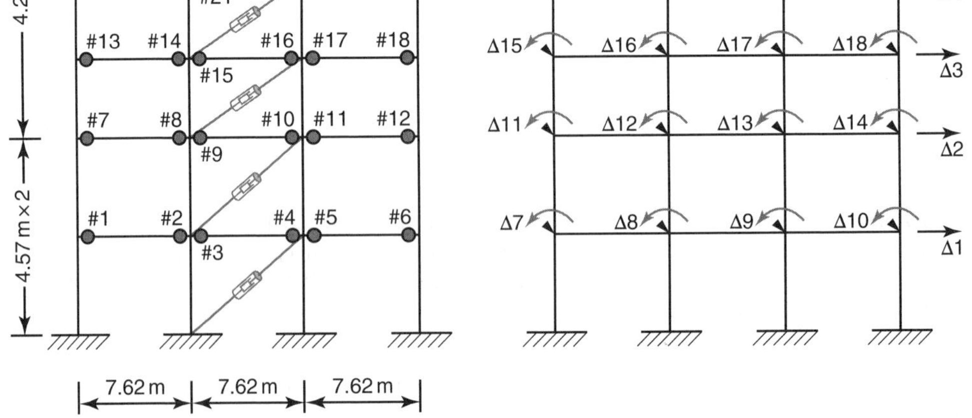

Figure 9.15 A six-story structure with dampers model. Gang Li, Qifeng Liu, Hongnan Li. Inelastic structural control based on MBC and FAM, Mathematical Problems in Engineering, vol. 2011, Article ID 460731, 18 pages, 2011. doi:10.1155/2011/460731.

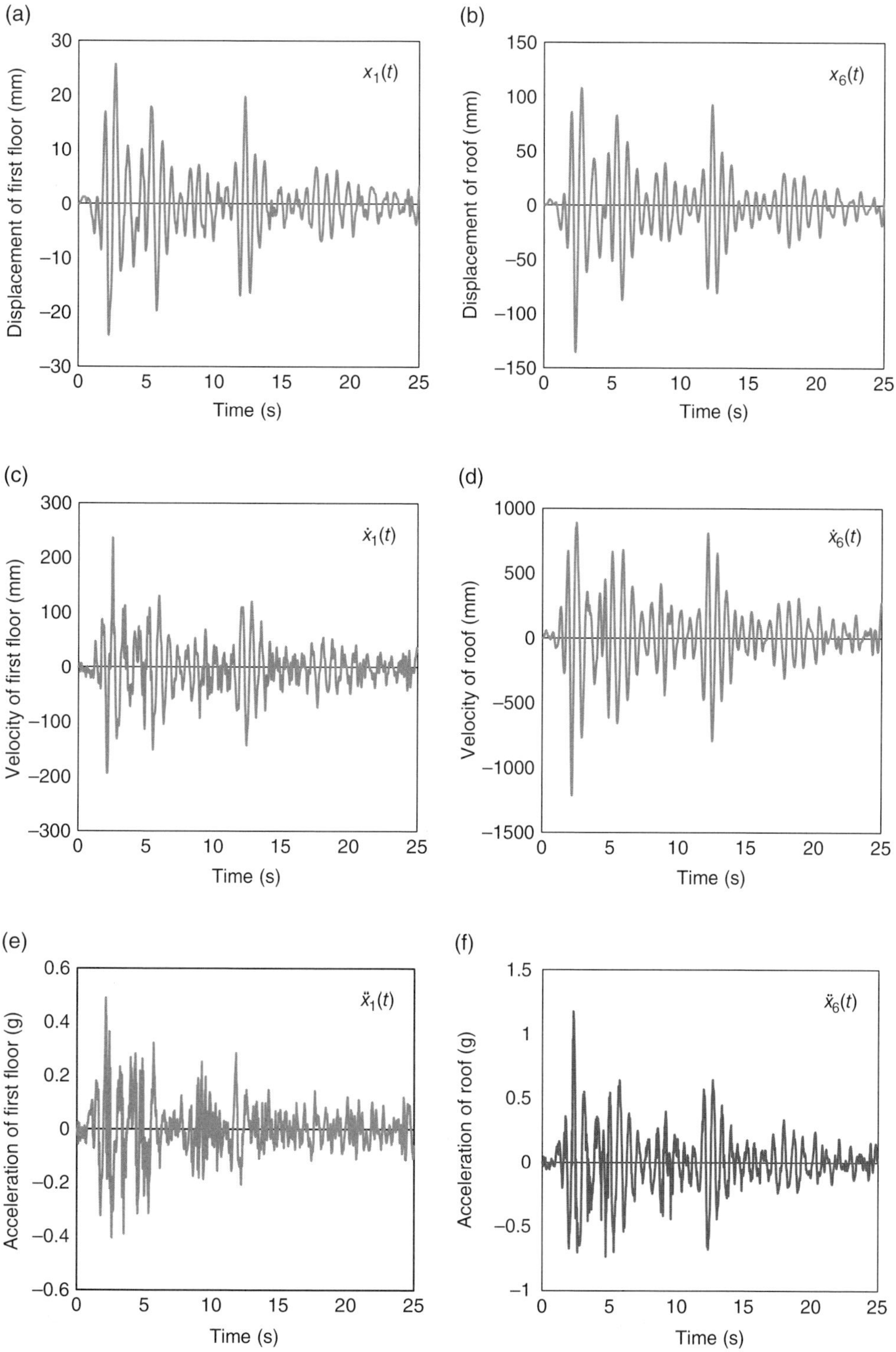

Figure 9.16 Response of the structure: (a) first floor displacement; (b) roof displacement; (c) first floor velocity; (d) roof velocity; (e) first floor acceleration; (f) roof acceleration. Gang Li, Qifeng Liu, Hongnan Li. Inelastic structural control based on MBC and FAM, Mathematical Problems in Engineering, vol. 2011, Article ID 460731, 18 pages, 2011. doi:10.1155/2011/460731.

El Centro (NS, May 18, 1940), the acceleration peak of which is adjusted to 0.5 g. The actuators are the same to that in Example 9.2.

Figure 9.16 shows the displacement, velocity, acceleration and plastic rotation of the selected floors responses. Figure 9.17 shows the rotations versus time history of the PH #1, #3, #7 and #9 under the earthquake motion.

Figure 9.18 shows the relationship diagrams of MRD damping force versus story displacement and story velocity with the MBC strategy. The conclusion can be obtained, that the direction of MRD damping force is opposite to the story velocity.

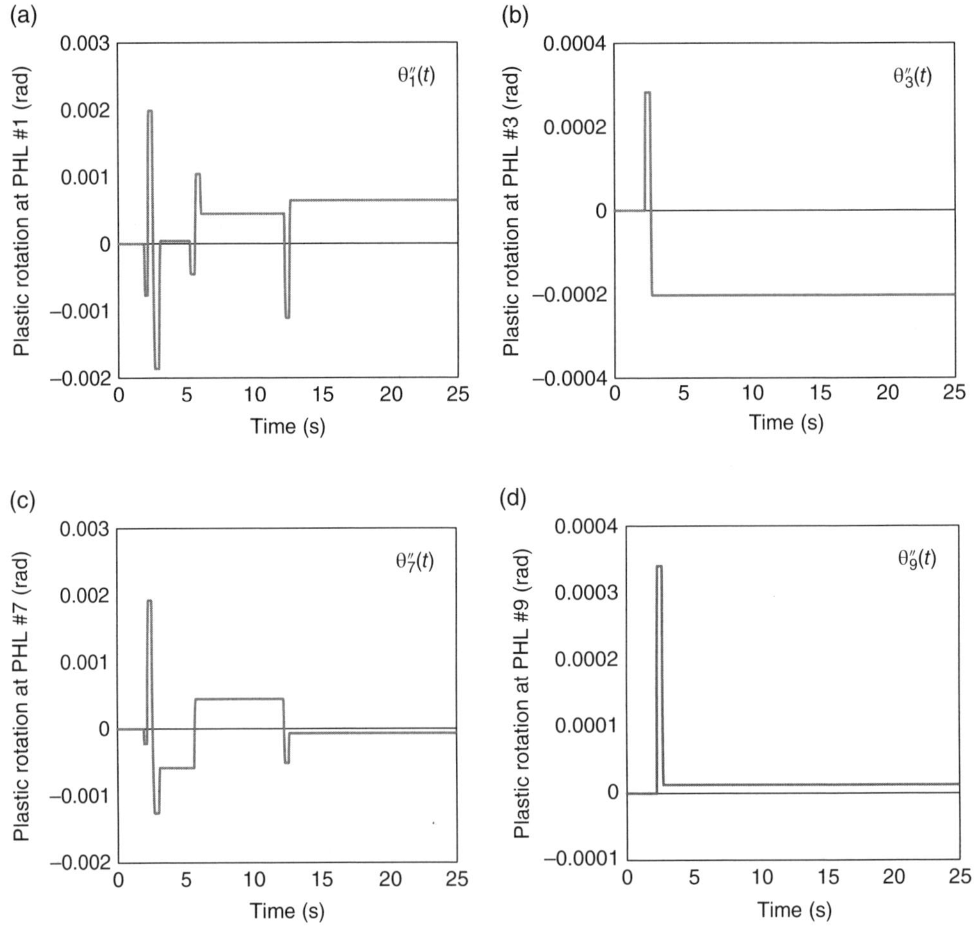

Figure 9.17 Plastic rotation of selected PHs: (a) #1; (b) #3; (c) #7; (d) #9. Gang Li, Qifeng Liu, Hongnan Li. Inelastic structural control based on MBC and FAM, Mathematical Problems in Engineering, vol. 2011, Article ID 460731, 18 pages, 2011. doi:10.1155/2011/460731.

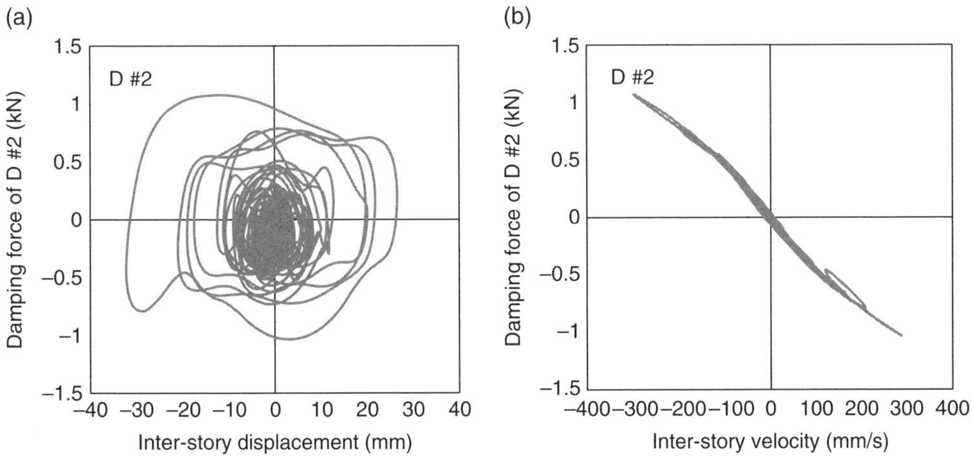

Figure 9.18 Damping force versus inter-story displacement and velocity relation of the MRD in second floor: (a) D#1; (b) D#2. Gang Li, Qifeng Liu, Hongnan Li. Inelastic structural control based on MBC and FAM, Mathematical Problems in Engineering, vol. 2011, Article ID 460731, 18 pages, 2011. doi:10.1155/2011/460731.

References

Clearwater S. (1996). *Market-based Control: A Paradigm for Distributed Resource Allocation.* World Scientific Publishing, Singapore.

Lynch J.P. and Law K.H. (2002). Market-Based Control of Linear Structural Systems. *Earthquake Engineering and Structural Dynamics*, 31(10): 1855–1877.

Li G. and Li H.N. (2011a). "Seismic response analysis of structure with energy dissipation devices using force analogy method". *Struct, Design Tall Spec, Build.*, 20(3): 291–313.

Gang Li, Qifeng Liu and Hongnan Li, (2011b). Inelastic structural control based on MBC and FAM. Mathematical Problems in Engineering, Article ID 460731, 18 pages,doi:10.1155/2011/460731.

Index

Theory of Nonlinear Structural Analysis: The Force Analogy Method for Earthquake Engineering, First Edition. Gang Li and Kevin K.F. Wong.